Numerical methods for fluid dynamics II

Based on the proceedings of a conference on Numerical Methods for Fluid Dynamics held in Reading in April 1985

Edited by

K. W. MORTON
University of Oxford

and

M. J. BAINES
University of Reading

CLARENDON PRESS · OXFORD · 1986

Oxford University Press, Walton Street, Oxford OX2 6DP
Oxford New York Toronto
Delhi Bombay Calcutta Madras Karachi
Kuala Lumpur Singapore Hong Kong Tokyo
Nairobi Dar es Salaam Cape Town
Melbourne Auckland
and associated companies in
Beirut Berlin Ibadan Nicosia

Oxford is a trademark of Oxford University Press

Published in the United States
by Oxford University Press, New York

British Library Cataloguing in Publication Data
Numerical methods for fluid dynamics II: based on a conference on Numerical Methods for Fluid Dynamics held in Reading in April 1985. – (The Institute of Mathematics and its Applications conference series. New series; no. 7)
1. Fluid dynamics 2. Numerical calculations
I. Morton, K. W. II. Baines, M. J.
III. Series
532'.05'015194 QA911
ISBN 0 19 853610 0

Printed in Great Britain by St Edmundsbury Press, Bury St Edmunds, Suffolk

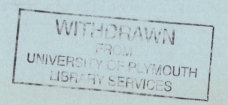

The Institute of Mathematics
and its Applications
Conference Series

The Institute of Mathematics
and its Applications
Conference Series

Previous volumes in this series were published by Academic Press to whom all enquiries should be addressed. Forthcoming volumes will be published by Oxford University Press throughout the world.

NEW SERIES

1. *Supercomputers and parallel computation* Edited by D. J. Paddon
2. *The mathematical basis of finite element methods* Edited by David F. Griffiths
3. *Multigrid methods for integral and differential equations* Edited by D. J. Paddon and H. Holstein
4. *Turbulence and diffusion in stable environments* Edited by J. C. R. Hunt
5. *Wave propagation and scattering* Edited by B. J. Uscinski
6. *The mathematics of surfaces* Edited by J. A. Gregory
7. *Numerical methods for fluid dynamics II* Edited by K. W. Morton and M. J. Baines

PREFACE

Since 1978 three successful conferences in computational fluid dynamics have been organised at Reading by the Institute of Mathematics and its Applications. Of the last two, that in 1981 was specifically on aeronautical fluid dynamics while that in 1982 was aimed at exploring the common features, developments and difficulties of numerical methods for the whole range of fluid flow applications. At that meeting it was proposed that the next conference should be expanded to combine the coverage of these two.

The Conference on which the present book is based was the result. It was organised by the ICFD (Institute for Computational Fluid Dynamics, sponsored by the SERC at the Universities of Reading and Oxford) in association with the IMA.

The aim of the conference was to bring together mathematicians and engineers in the field of computational aerodynamics and computational fluid dynamics to review recent advances in mathematical and computational techniques for modelling fluid flows. The subject area is very large with very many active researchers in industry, government laboratories and universities working on a wide variety of methods and applications. Thus to give coherence to the meeting three main themes were selected:-

Alternative Data Representations - by which is meant the various forms of discretisation that may be used to model a fluid - such as particle, spectral or vortex models, finite difference and finite element viewpoints and alternative choices of dependent variables.

Adaptive Modelling - that is modelling which in some way responds to the features of the flow as part of its solution procedure.

Solution of Systems of Linear and Non-linear equations - particularly those arising from discretised models of fluid flow.

Particular thanks are due to the organising committee, Dr. N.E. Hoskin (AWRE), Dr. J. Rae (Harwell) and Professor P.L. Roe (Cranfield) for suggesting themes, speakers, etc. Thanks are also due to the speakers for taking up the themes of the Conference so enthusiastically. The Conference was most competently organised by Anne-Marie Crawley, the Secretary of the ICFD.

Financial support for the Conference included grants from the Royal Society and the United States Air Force.

The organisation of the Proceedings is in two sections. The first section consists of Invited Papers, in the order in which they were presented at the Conference. The second section contains Contributed Papers, organised in the same way. In order to obtain rapid publication of these Proceedings, within a year of the date of the Conference, close cooperation has been necessary between the authors, the IMA, who retyped all the papers, and the publishers, Oxford University Press. We should like to express our thanks to all the individuals concerned.

K.W. Morton
M.J. Baines

CONTENTS

CONTENTS

ACKNOWLEDGEMENTS

The Institute thanks the authors of the papers, the editors, Dr. M.J. Baines (University of Reading) and Professor K.W. Morton (University of Oxford) and also Miss P. Irving, Miss K. Jenkins and Mrs. J. Robinson for typing the papers.

CONTRIBUTORS

S. ABARBANEL; *Tel-Aviv University, Tel-Aviv, Israel and ICASE, NASA Langley, VA 23665, USA.*

C.M. ALBONE; *Aerodynamics Division, RAE, Farnborough, GU14 6TD.*

W. ARTER; *Culham Laboratory, Euratom/UKAEA Fusion Association, Abingdon, OX14 3DB.*

M.J. BAINES; *Institute for Computational Fluid Dynamics, Department of Mathematics, University of Reading, Whiteknights, P.O. Box 220, Reading, RG6 2AX.*

S.P. BALLARD; *Meteorological Office, London Road, Bracknell, RG3 6JN.*

J.P. BENQUÉ; *Electricite de France, Laboratoire National d'Hydraulique, 78400 Chatou, France.*

V. BILLEY; *AMD/BA, 78 Quai Carnot, 92214 St. Cloud, France.*

P. BONTOUX; *IMFM, Université d'Aix-Marseille II, Marseille, France.*

A.D. BURNS; *Computer Sciences and Systems Division, AERE, Harwell, OX11 ORA.*

J. CAHOUET; *Electricité de France, Laboratoire National d'Hydraulique, 78400 Chatou, France.*

M.A. CHRISTIE; *BP Research Centre, Chertsey Road, Sunbury-on-Thames, TW16 7LN.*

R.A. CLARK; *Computational Physics Group X-7, MS B257, Los Alamos National Laboratory, Los Alamos, New Mexico 89745, USA.*

K.A. CLIFFE; *Theoretical Physics Division, AERE, Harwell, OX11 ORA.*

M.W. COLLINS; *Thermofluids Engineering Research Centre, The City University, London, EC1.*

M.J.P. CULLEN; *Meteorological Office, London Road, Bracknell, RG3 6JN.*

A. DERVIEUX; *INRIA, Route de Lucioles, Sophia Antipolis, 06560 Valbonne, France.*

P.M. DE ZEEUW; *Centre for Mathematics and Computer Science, P.O. Box 4079, 1009 AB Amsterdam, Netherlands.*

J.W. DOLD; *School of Mathematics, Bristol University, University Walk, Bristol, BS8 1TW*

J.W. EASTWOOD; *Culham Laboratory, Euratom/UKAEA Fusion Association, Abingdon, OX14 3DB.*

F. EL DABAGHI; *INRIA, Domaine de Voluceau, Rocquecourt, BP 105, 78153 Le Chesnay, France.*

C.L. FARMER; *Oil Recovery Projects Division, AEE Winfrith, Dorchester, DT2 8DH.*

B. FORNBERG; *Exxon Research and Engineering Company, Annandale, New Jersey 08801, USA.*

C.R. FORSEY; *Aircraft Research Association Ltd., Manton Lane, Bedford, MK41 7PF.*

L. FUCHS; *Department of Gasdynamics, Royal Institute of Technology, S-10044, Stockholm, Sweden.*

D. GOTTLIEB; *Tel-Aviv University, Tel-Aviv, Israel and ICASE, NASA Langley, VA 23665, USA.*

J. GOUSSEBAILE; *Electricité de France, Laboratoire National d'Hydraulique, 78400 Chatou, France.*

J.M.R. GRAHAM; *Department of Aeronautics, Imperial College, London, SW7 2BY.*

M.G. HALL; *Aerodynamics Division, RAE, Farnborough, GU14 6TD.*

A. HAUGUEL; *Electricité de France, Laboratoire National d'Hydraulique, 78400 Chatou, France.*

A. JAMESON; *Princeton University, New Jersey, USA.*

A.D. JEPSON; *Department of Computer Science, University of Toronto, Toronto, MSS 1A4, Canada.*

I.P. JONES; *Computer Sciences and Systems Division, AERE, Harwell, OX11 ORA.*

J.R. KIGHTLEY; *Computer Sciences and Systems Division, AERE, Harwell, OX11 ORA.*

L. KONG; *Institute for Numerical Methods in Engineering, University College, Swansea, SA2 8PP.*

R. LOHNER; *Institute for Numerical Methods in Engineering, University College, Swansea, SA2 8PP.*

K. MORGAN; *Institute for Numerical Methods in Engineering, University College, Swansea, SA2 8PP.*

S.P. NEWMAN; *Rolls-Royce Ltd., P.O. Box 31, Derby, DE2 8BJ.*

W.F. NOH; *Lawrence Livermore National Laboratory, Livermore, California 94550, USA.*

R.A. NORMAN; *Oil Recovery Projects Division, AEE Winfrith, Dorchester, DT2 8DH.*

B. PALMERIO; *Université de Nice, IMSP, Parc Valrose, 06034, Nice, France.*

M. PANDOLFI; *Dipartimento di Ingegneria Aeronautica e Spaziale, Politecnico di Torino, Corso Duca degli Abruzzi, Torino, Italy.*

A.K. PARROTT; *BP Exploration, Britannic House, Moor Lane, London, EC27 9BU.*

J. PERAIRE; *Institute for Numerical Methods in Engineering, University College, Swansea, SA2 8PP.*

D.H. PEREGRINE; *School of Mathematics, Bristol University, University Walk, Bristol, BS8 1TW.*

J. PERIAUX; *AMD/BA, 78 Quai Carnot, 92214 St. Cloud, France.*

R. PEYRET; *Départment de Mathématiques, Université de Nice, Parc Balrose, 06034 Nice, France.*

O. PIRONNEAU; *Université Paris XIII, 93430 Villetaneuse, France and INRIA.*

G. POIRIER; *AMD/BA, 78 Quai Carnot, 92214 St. Cloud, France.*

P.A. RAVIART; *Analyse Numerique, Université Pierre et Marie Curie, Paris, France.*

A. RIZZI; *FFA, The Aeronautical Research Institute of Sweden, S-161 11 Bromma, Sweden.*

P.L. ROE; *College of Aeronautics, Cranfield Institute of Technology, Cranfield, Bedford, MK43 OAL.*

J.S. ROLLETT; *Oxford University Computing Laboratory, 8-11 Keble Road, Oxford, OX1 3QD.*

B.W. SCOTNEY; *Department of Mathematics, University of Ulster, Coleraine, County Londonderry, Northern Ireland, BT52 1SA.*

J.A. SHAW; *Aircraft Research Association Ltd., Manton Lane, Bedford, MK41 7PF.*

S. SIVALOGANATHAN; *Oxford University Computing Laboratory, 8-11 Keble Road, Oxford, OX1 3QD.*

P. SONNEVELD; *Department of Mathematics and Informatics, Delft University of Technology, P.O. Box 356, 2600 AJ Delft, Netherlands.*

A. SPENCE; *School of Mathematics, University of Bath, Claverton Down, Bath, BA2 7AY.*

P. STOW; *Rolls-Royce Ltd., P.O. Box 31, Derby, DE2 8BT.*

E. SULI; *Oxford University Computing Laboratory, 8-11 Keble Road, Oxford, OX1 3QD.*

E. TADMOR; *Tel-Aviv University, Tel-Aviv, Israel and ICASE, NASA Langley, VA 23665, USA.*

C.P. THOMPSON; *Computer Sciences and Systems Division, AERE, Harwell, OX11 ORA.*

H. TREASE; *Computational Physics Group X-7, Los Alamos National Laboratory, Los Alamos, New Mexico 89745, USA.*

H.M. TSAI; *Turbulence Unit, Queen Mary College, University of London, London, E2.*

J.M. VANEL; *Département de Mathématiques, Université de Nice, Parc Balrose, 06034 Nice, France.*

B. VAN LEER; *Delft University of Technology, Department of Mathematics and Computer Sciences, P.O. Box 356, 2600 AJ Delft, Netherlands.*

P.R. VOKE; *Turbulence Unit, Queen Mary College, University of London, London, E2.*

N.P. WEATHERILL; *Aircraft Research Association Ltd., Manton Lane, Bedford, MK41 7PF.*

M.F. WEBSTER; *Institute for Computational Fluid Dynamics, Department of Mathematics, University of Reading, Whiteknights, P.O. Box 220, Reading, RG6 2AX.*

P. WESSELING; *Delft University of Technology, Department of Mathematics and Computer Sciences, P.O. Box 356, 2600 AJ Delft, Netherlands.*

N.S. WILKES; *Environmental Sciences Division, AERE, Harwell, OX11 ORA.*

P.R. WOODWARD; *Lawrence Livermore National Laboratory, CP Division L-297, P.O. Box 808, Livermore, California, USA.*

L. ZANNETTI; *Dipartimento di Ingegneria Aeronautica e Spaziale, Politecnico di Torino, Corso Duca degli Abruzzi, Torino, Italy.*

O.C. ZIENKIEWICZ; *Institute for Numerical Methods in Engineering, University College, Swansea, SA2 8PP.*

LOCALLY ADAPTIVE MOVING FINITE ELEMENTS

M.J. Baines
(Department of Mathematics, University of Reading)*

1. INTRODUCTION

Computational Fluid Dynamics covers a wide range of modelling activity. It includes complex codes based on traditional methods for difficult and pressing engineering problems and, at the other end of the spectrum, the development of numerical methodology stimulated by the desire for flexible and efficient new algorithms. It is with the latter area that this paper is concerned, in particular with the development of algorithms which adapt in some way to the flow field. By considering moving grids we overcome some of the limitations peculiar to fixed grids while raising other issues. Here we describe and develop the Moving Finite Element (MFE) technique in relation to hyperbolic conservation laws.

We begin in one dimension by summarising the MFE method for piecewise linear approximations as originally presented in (Miller, 1981) but without his use of penalty functions: see (Wathen and Baines, 1985). We then describe an alternative formulation, based on element basis functions, which is equivalent to the Miller formulation but involves only local projections as in (Baines, 1985). From this formulation we obtain ordinary differential equations in time for the evolution of each element segment in terms of its velocity and angular speed. The determination of the nodal velocities from the motion of the segments (also local) is then described, and the resulting method is used to

---

*The work reported here forms part of the research programme of the Oxford/Reading Institute for Computational Fluid Dynamics.

derive the decomposition of the standard MFE matrix.

We then describe singularities of the method. In the rather rare event of the nodes becoming collinear (parallelism) one of the local matrices becomes singular, and a special non-local procedure is needed which is however simple to implement. The only other singularity of the method occurs when nodes overtake one another. This arises if the time step (in the numerical integration of the semi-discrete problem) is so large as to destroy the accuracy. However it may also arise (correctly) in hyperbolic problems when a shock forms. In that case the appropriate jump conditions may be used to continue the solution.

Time-stepping strategies are then discussed in relation to accuracy and practical results. So far the forward Euler explicit method has been found to be sufficient, as in (Wathen, 1984 and Johnson, 1984). Results are given for two non-linear examples, the inviscid Burgers' equation (for which the MFE method gives the exact solution) and the Buckley-Leverett equation. Both programs run in BASIC on a home computer.

Generalising to systems of equations we describe a method which uses separate grids for each conserved variable as given in (Baines and Wathen, 1985). Results are given for a shock tube problem. Again a small microcomputer is sufficient to obtain a reasonable approximate solution to this problem.

The method has considerable potential in two dimensions, see (Wathen, 1984). However care is required in that the time derivative v_t of the piecewise linear continuous function v in two dimensions now belongs only to a subspace of the space S_ϕ spanned by the element basis functions. Hence a constrained projection is considered which gives the standard Miller method. The approach again demonstrates that the MFE matrix has a decomposition similar to that in one dimension. A local segment velocity can also be derived which assists in determining the rules for shock formulation. Results are given for a non-linear problem in two dimensions, a generalisation of the Buckley-Leverett equation.

2. STANDARD MFE APPROACH

We first describe the standard Moving Finite Element method as presented in (Miller, 1981). We begin by approximating u in the scalar equation

$$u_t + f(u)_x = 0 \qquad (2.1)$$

by the piecewise linear function v given by

$$v = \sum_j a_j \alpha_j \qquad (2.2)$$

where the a_j are nodal coefficients and the α_j are linear basis functions as shown in Figure 1(a).

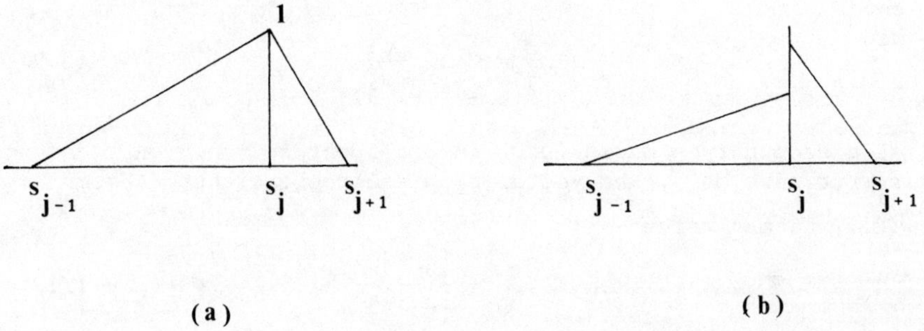

Fig. 1. Basis functions (a) α_j and (b) β_j

For fixed finite elements a_j depends on time t and α_j depends on the space co-ordinate x. In the MFE method α_j depends also on the time t through the nodal co-ordinates s_i $(i = 0,1,\ldots,N+1)$. In order to study the solution of the conservation law (2.1) we differentiate v with respect to time. Since t appears twice in the expression for v we have

$$v_t = \sum_j (\dot{a}_j \alpha_j + a_j \dot{\alpha}_j) \qquad (2.3)$$

where the dot denotes differentiation with respect to time. Now, by the chain rule,

$$\dot{\alpha}_j = \sum_i \dot{s}_i \frac{\partial \alpha_j}{\partial s_i} \qquad (2.4)$$

so that (2.3) becomes

$$v_t = \sum_j \dot{a}_j \alpha_j + \sum_j a_j \sum_i \dot{s}_i \frac{\partial \alpha_j}{\partial s_i} \tag{2.5}$$

By interchanging the order of summation we obtain

$$v_t = \sum_j (\dot{a}_j \alpha_j + \dot{s}_j \beta_j) , \tag{2.6}$$

where

$$\beta_j = \sum_i a_i \frac{\partial \alpha_i}{\partial s_j} \tag{2.7}$$

is a second type basis function dependent not only on t and on s_i but also on $\underline{a}$, the vector of nodal coefficients. From (2.2) we may write

$$\beta_j = \frac{\partial v}{\partial s_j} \tag{2.8}$$

which in the case of piecewise linear basis functions can be written (after some manipulation) as

$$\beta_j = -m\alpha_j , \tag{2.9}$$

where m is the local slope of v: (see also (Lynch, 1981)).

Thus the second type basis function β_j has components which are multiples of α_j and have the same support as α_j. (The result is true for linear basis functions in any number of dimensions (see (Baines, 1985)). Diagrams illustrating the α_j and β_j in one dimension are shown in Figure 1.

Hence in one dimension

$$v_t = \sum_j (\dot{a}_j \alpha_j + \dot{s}_j \beta_j) \tag{2.10}$$

belongs to the space of piecewise linear discontinuous functions $S_{\alpha\beta}$, say.

If $f(v)_x$ also belongs to $S_{\alpha\beta}$ we immediately obtain from (2.1) ordinary differential equations for a_j and s_j in time.

For example if

$$f(u) = \frac{1}{2} u^2 \tag{2.11}$$

the ODE's are

$$\dot{a}_j = 0 \qquad \dot{s}_j = a_j , \tag{2.12}$$

which are readily integrated to give a_j and s_j. In this particular case the trajectories of the nodes are the characteristics of the partial differential equation as in (Wathen, 1984). This is an exceptional case however.

More generally we may project $f(v)_x$ into the space $S_{\alpha\beta}$ by minimising the L_2 norm of the residual

$$||v_t + f(v)_x||_2 \tag{2.13}$$

over the variables $\dot{a}_j$ and $\dot{s}_j$. This leads to the double set of Galerkin equations

$$\left.\begin{aligned} &<\alpha_j, v_t + f(v)_x> = 0 \\ &<\beta_j, v_t + f(v)_x> = 0 \end{aligned}\right\} \quad \forall_j \tag{2.14}$$

Substituting for v from (2.10) we obtain the matrix system of ODE's

$$A\dot{\underline{y}} = \underline{g} \tag{2.15}$$

where

$$\dot{\underline{y}}^T = [\ldots, \dot{a}_j, \dot{s}_j, \ldots] \tag{2.16}$$

$$A = \{A_{ij}\}, \qquad A_{ij} = \begin{bmatrix} <\alpha_i, \alpha_j> & <\alpha_i, \beta_j> \\ <\beta_i, \alpha_j> & <\beta_i, \beta_j> \end{bmatrix} \tag{2.17}$$

and $$\underline{g} = \{g_i\} \qquad g_i = <{}^{\alpha_i}_{\beta_i}, -f(v)_x> . \tag{2.18}$$

The matrix A is symmetric block 2x2 tri-diagonal, positive semi-definite, depending on $\underline{y}$. Equation (2.15) gives the MFE equations as derived in (Miller, 1981) without the use of penalty functions.

It has been shown in (Morton, 1982) that the MFE equations (2.15) carry the best least squares fit to the exact solution in the case of a scalar conservation law.

3. LOCAL APPROACH

We now adopt instead a local elementwise approach. The function v_t in the space of piecewise linear discontinuous functions may be re-parameterised in the form

$$v_t = \sum_j (\dot{a}_j \alpha_j + \dot{s}_j \beta_j) = \sum_k (w_{k1}\phi_{k1} + w_{k2}\phi_{k2}) \qquad (3.1)$$

using element basis functions ϕ_{ki} as shown in Figure 2. The w_{ki} are the coefficients of the ϕ_{ki} in the expansion which can be related to the $\dot{a}_j, \dot{s}_j$ (see below). Denote by S_ϕ the space spanned by the basis functions ϕ_{ki}: this is the same space as $S_{\alpha\beta}$ in the one-dimensional case.

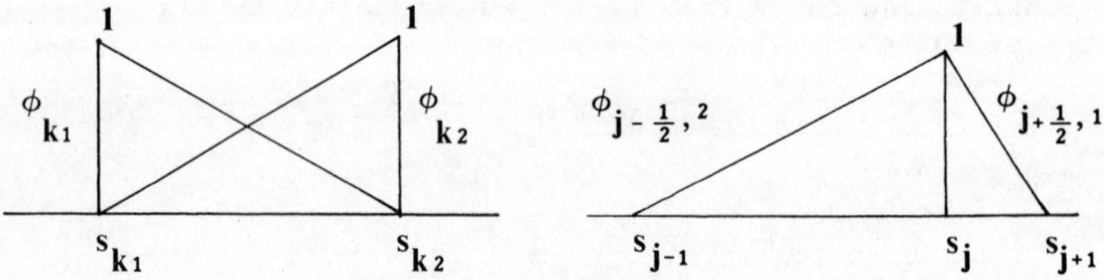

Fig. 2. Basis functions ϕ_{k1} and ϕ_{k2}
(a) on an element, (b) around a node.

Again, if $f(v)_x$ belongs to S_ϕ ($\equiv S_{\alpha\beta}$) we obtain w_{k1}, w_{k2} at once and consequently $\dot{a}_j$ and $\dot{s}_j$. More generally we may again project $f(v)_x$ into S_ϕ. This can be done locally within each element by minimising the local element L_2 norm

$$||v_t + f(v)_x||_2 \tag{3.2}$$

over w_{k1}, w_{k2}. We thus obtain alternative Galerkin equations

$$\begin{aligned} &\langle \phi_{k1}, v_t + f(v)_x \rangle = 0 \\ &\langle \phi_{k2}, v_t + f(v)_x \rangle = 0 \end{aligned} \tag{3.3}$$

for each element which can be written in the form

$$C_k \underset{\sim}{w}_k = \underset{\sim}{b}_k , \tag{3.4}$$

a 2x2 system. In (3.4)

$$\underset{\sim}{w}_k^T = [w_{k1}, w_{k2}] \tag{3.5}$$

$$C_k = \begin{bmatrix} \langle \phi_{k1}, \phi_{k1} \rangle & \langle \phi_{k1}, \phi_{k2} \rangle \\ \langle \phi_{k2}, \phi_{k1} \rangle & \langle \phi_{k2}, \phi_{k2} \rangle \end{bmatrix} \tag{3.6}$$

and

$$\underset{\sim}{b}_k = \langle \begin{matrix} \phi_{k1} \\ \phi_{k2} \end{matrix} , -f(v)_x \rangle . \tag{3.7}$$

To relate the w_k's to $\dot{a}_j$ and $\dot{s}_j$ we use the fact that

$$\alpha_j = \phi_{j-\frac{1}{2},2} + \phi_{j+\frac{1}{2},1} \qquad \beta_j = -m_{j-\frac{1}{2}}\phi_{j-\frac{1}{2},2} - m_{j+\frac{1}{2}}\phi_{j+\frac{1}{2},1} , \tag{3.8}$$

where elements $j-\frac{1}{2}$, $j+\frac{1}{2}$ are adjacent to node j, as in Figure 2. Then using (3.1) we have the correspondence

$$\left.\begin{aligned} \dot{a}_j - m_{j-\frac{1}{2}}\dot{s}_j &= w_{j-\frac{1}{2},2} \\ \dot{a}_j - m_{j+\frac{1}{2}}\dot{s}_j &= w_{j+\frac{1}{2},1} \end{aligned}\right\} \forall j . \tag{3.9}$$

This can be written in the form

$$M_j \, \dot{\underline{y}}_j = \underline{w}_j \, , \tag{3.10}$$

another 2x2 system, where

$$\dot{\underline{y}}_j^T = [\dot{a}_j, \dot{s}_j] \tag{3.11}$$

$$M_j = \begin{bmatrix} 1 & -m_{j-\frac{1}{2}} \\ 1 & -m_{j+\frac{1}{2}} \end{bmatrix} \tag{3.12}$$

$$\underline{w}_j^T = [w_{j-\frac{1}{2},2}, \; w_{j+\frac{1}{2},1}] \, . \tag{3.13}$$

Since both the Miller method and the local elementwise method minimise the same residual in the same space the MFE equations derived from the two methods must be identical. It follows that, by writing

$$C = \begin{bmatrix} \ddots & & 0 \\ & C_k & \\ 0 & & \ddots \end{bmatrix} , \quad M = \begin{bmatrix} \ddots & & 0 \\ & M_j & \\ 0 & & \ddots \end{bmatrix} , \tag{3.14}$$

we have from (2.15), (2.16), (3.11), (3.10), (3.4), (3.7), (3.8) and (2.18) the decomposition of the global MFE matrix A

$$A = M^T C M \quad , \tag{3.15}$$

where both C and M are diagonal block 2x2 matrices (apart from end effects).

A particular property shared by the methods is conservation. Since both the α_j's and the ϕ_{ki}'s are a partition of the unit function, summation of the first of (2.14) or (3.3) gives

$$\int_{s_0}^{s_{N+1}} [v_t + f(v)_x]dx = 0 , \qquad (3.16)$$

from which we may deduce that

$$\frac{d}{dt}\left[\int_{s_0}^{s_{N+1}} v\,dx\right] = -\left[f(v)\right]_{s_0}^{s_{N+1}} \qquad (3.17)$$

and consequently that $\int_{s_0}^{s_{N+1}} v\,dx$ is constant in time apart from boundary effects.

Boundary conditions may be imposed locally on the element adjacent to boundaries as in (Baines, 1985). In the case of a Neumann condition at the end $j = 0$ we have

$$\dot{s}_0 = 0 \qquad (3.18)$$

which, in conjunction with the second of (3.9), gives

$$\dot{a}_0 = w_{\frac{1}{2},1} \qquad (3.19)$$

and hence the motion of the boundary element.

If the boundary condition at the end $j = 0$ is Dirichlet then, because we cannot impose both

$$\dot{s}_0 = 0, \quad \dot{a}_0 = 0 \qquad (3.20)$$

simultaneously and preserve the projection, a special constrained projection has to be carried out in the end element. The result is that

$$w_{\frac{1}{2},1} = 0, \; w_{\frac{1}{2},2} = 3\, b_{\frac{1}{2},2}/(s_1 - s_0) , \qquad (3.21)$$

where $b_{\frac{1}{2},2}$ is the second of (3.7) for the end element. This is consistent with (3.20).

4. LOCAL ELEMENT MOTION

We now show that the motion of a local segment of the approximating piecewise linear function may be obtained entirely from the local projection step (3.4).

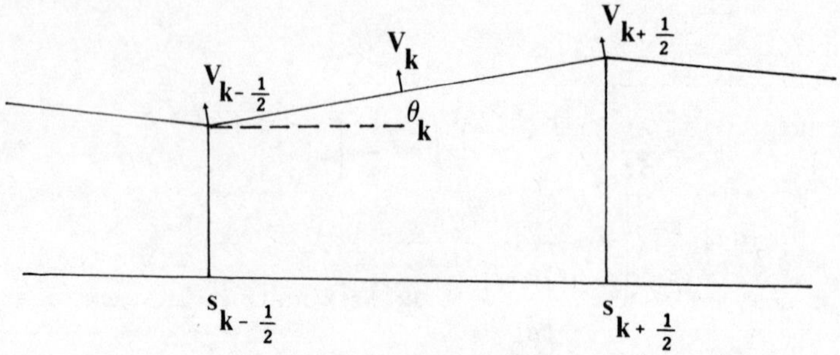

Fig. 3. Local segment angle and velocities

Let V_k be the velocity of the mid-point of the segment in the direction perpendicular to the segment and let θ_k be the angle between the segment and the x-axis, as shown in Figure 3. Then the equations (3.9) can be written in the form

$$\left.\begin{aligned} \dot{a}_{k-\frac{1}{2}} \cos\theta_k - \dot{s}_{k-\frac{1}{2}} \sin\theta_k &= w_{k1} \cos\theta_k \\ \dot{a}_{k+\frac{1}{2}} \cos\theta_k - \dot{s}_{k+\frac{1}{2}} \sin\theta_k &= w_{k2} \cos\theta_k \end{aligned}\right\} \forall k \qquad (4.1)$$

where nodes $k-\frac{1}{2}$, $k+\frac{1}{2}$ are the ends of element k, as in Figure 3. The left hand sides of (4.1) are the velocities $V_{k-\frac{1}{2}}$, $V_{k+\frac{1}{2}}$ (due to the motion of the single element k only) of the ends of the element k in Figure 3 at right angles to the element. (The full velocity of a node will be a combination of two such element end velocities from adjacent segments).

Let
$$m_k = \frac{a_{k+\frac{1}{2}} - a_{k-\frac{1}{2}}}{s_{k+\frac{1}{2}} - s_{k-\frac{1}{2}}} \, . \qquad (4.2)$$

Then these end velocities may be written

$$\begin{bmatrix} V_{k-\frac{1}{2}} \\ V_{k-\frac{1}{2}} \end{bmatrix} = \begin{bmatrix} w_{k1} \\ w_{k2} \end{bmatrix} \cos\theta_k = \frac{1}{\sqrt{(1+m_k^2)}} C_k^{-1} \begin{bmatrix} b_{k1} \\ b_{k2} \end{bmatrix} = \frac{-1}{\sqrt{(1+m_k^2)}} C_k^{-1} \begin{bmatrix} \langle \phi_{k1}, f(v)_x \rangle \\ \langle \phi_{k2}, f(v)_x \rangle \end{bmatrix} \tag{4.3}$$

using (3.4) and (3.7). Since $[1,1]^T$ is a eigenvector of the symmetric matrix C_k with eigenvalue $\frac{1}{2}(s_{k+\frac{1}{2}} - s_{k-\frac{1}{2}}) = \frac{1}{2}\Delta s_k$, say, we obtain

$$[1,1] \begin{bmatrix} V_{k-\frac{1}{2}} \\ V_{k+\frac{1}{2}} \end{bmatrix} = \frac{-1}{\sqrt{(1+m_k^2)}} \frac{2}{\Delta s_k} [1,1] \begin{bmatrix} \langle \phi_{k1}, f(v)_x \rangle \\ \langle \phi_{k2}, f(v)_x \rangle \end{bmatrix}$$

$$= \frac{-1}{\sqrt{(1+m_k^2)}} \frac{2}{\Delta s_k} \langle 1, f(v)_x \rangle = \frac{-2}{\Delta s_k \sqrt{(1+m_k^2)}} \int_{s_{k-\frac{1}{2}}}^{s_{k+\frac{1}{2}}} f(v)_x dx \tag{4.4}$$

Hence

$$V_k = \tfrac{1}{2}(V_{k-\frac{1}{2}} + V_{k+\frac{1}{2}}) = \frac{-1}{\sqrt{1+m_k^2)}} \frac{1}{\Delta s_k} \{(f(v))_{k+\frac{1}{2}} - (f(v))_{k-\frac{1}{2}}\} \tag{4.5}$$

which gives the normal velocity of the mid-point of the segment in Figure 3. Putting $\Delta f_k = \{(f(v))_{k+\frac{1}{2}} - (f(v))_{k-\frac{1}{2}})\}$ this may be written in the alternative forms

$$V_k = -\frac{\Delta f_k}{\Delta s_k} \cos\theta_k = -\frac{\Delta f_k}{PQ} = -\frac{\Delta f_k}{\Delta v_k} \sin\theta_k \tag{4.6}$$

where $\Delta v_k = \Delta a_k = a_{k+\frac{1}{2}} - a_{k-\frac{1}{2}}$ and PQ are as shown in Figure 3. Thus we have the important result that the speed of the mid-point of the segment in Figure 3 in the direction normal to the segment is consistent with the local average speed $\Delta f_k/\Delta v_k$ in the element.

Subtracting pairs of the equations (3.9) we obtain another important result, namely,

$$\dot{a}_{k+\frac{1}{2}} - \dot{a}_{k-\frac{1}{2}} - m_k (\dot{s}_{k+\frac{1}{2}} - \dot{s}_{k-\frac{1}{2}}) = w_{k2} - w_{k1} \tag{4.7}$$

or, using (4.2),

$$\frac{dm_k}{dt} = \frac{1}{\Delta s_k} \; [-1 \; 1]\underline{w}_k \tag{4.8}$$

Since [-1 1] is also an eigenvector of C_k with eigenvalue $\frac{1}{6}\Delta s_k$ we obtain

$$\frac{dm_k}{dt} = -\frac{6}{[\Delta s_k]^2} \; [-1 \; 1] \begin{bmatrix} \langle\phi_{k1}, f(v)_x\rangle \\ \langle\phi_{k2}, f(v)_x\rangle \end{bmatrix} = -\frac{6}{[\Delta s_k]^2} \int_{s_{k-\frac{1}{2}}}^{s_{k+\frac{1}{2}}} (\phi_{k2} - \phi_{k1}) f(v)_x dx \tag{4.9}$$

This leads to the alternative forms

$$\frac{dm_k}{dt} = \frac{12}{[\Delta s_k]^2} (\hat{f} - \bar{f}) = -f_{xx}(\eta_k) = -m_k^2 \; f''(\eta) \tag{4.10}$$

where

$$\hat{f} = \frac{1}{\Delta s_k} \int_{s_{k-\frac{1}{2}}}^{s_{k+\frac{1}{2}}} f(v)dx, \quad \bar{f} = \tfrac{1}{2}((f(v))_{k-\frac{1}{2}} + (f(v))_{k+\frac{1}{2}})$$

$$\text{and } \eta_k \in (s_{k-\frac{1}{2}}, s_{k+\frac{1}{2}}). \tag{4.11}$$

Hence we have the result that the rate of change of the slope of the solution in an element is equal to the second space derivative of the flux function. In other words the solution segment rotates in response to the local convexity of f. Another form of this result is

$$\frac{d\theta_k}{dt} = \frac{-m_k^2}{1 + m_k^2} \; f''(\eta_k). \tag{4.12}$$

As the segments move the intersections (nodes) also move, giving the nodal velocities. Note that the segments have lengths that vary with time. The movement of a node is thus the locus of the intersection of adjacent elements (see Figure 4).

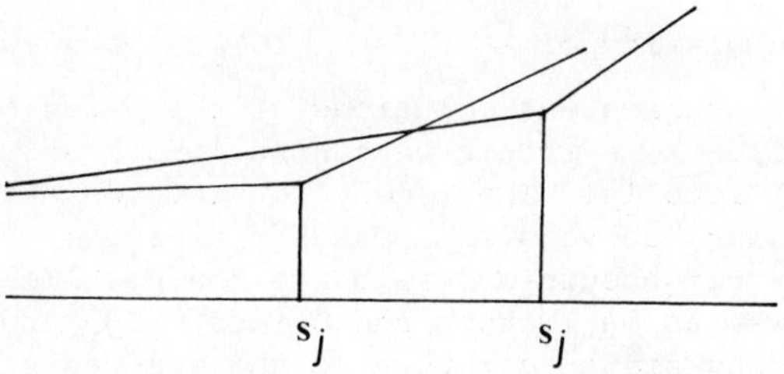

Fig. 4. Nodal movement as locus of segment intersection

The results (4.6) and (4.10) show clearly the element behaviour in terms of the flux function.

5. SUMMARY AND PARALLELISM

We give now a summary of the local elementwise method which is complete provided that none of the matrices involved are singular. We then go on to discuss the treatment of such singularities.

Summary

(1) Evaluate $\underline{b}_k = \langle \phi_k, -f(v)_x \rangle$

(2) Solve $C_k \underline{w}_k = \underline{b}_k$ for $\underline{w}_k$ $\qquad \forall$ elements k

(3) Pair off the w's nodewise

(4) Solve $M_j \, \dot{\underline{y}}_j = \underline{w}_k$ for $\dot{\underline{y}}_j = [\dot{a}_j, \dot{s}_j]^T$ $\forall$ nodes j.

Note that the method involves the inversion of 2x2 matrices only: little storage is required and the algorithm can be run on a small computer.

The method may break down if any of the matrices C_k or M_j is singular. Singularity of M_j corresponds to collinearity of nodes, usually described by the term parallelism. This occurs when

$$m_{k-\frac{1}{2}} = m_{k+\frac{1}{2}} \tag{5.1}$$

in (3.12). Singularity of C_k arises only if $\Delta s_k = 0$ (see (3.6)).

Consider singularity of M_j first. In this case we find that we obtain inconsistent solutions of the pairs of equations (3.10). As a result we can no longer solve equation (3.10) locally. The remedy is to temporarily fix any parallel nodes, solve over a patch consisting of these nodes and their neighbours and relocate the parallel nodes in some averaged way as in (Wathen and Baines, 1985 and Baines, 1985). It is convenient to return to the α,β basis by combining equations (3.4) in staggered pairs. Then in the event of parallelism at a single node we retain the combination of these equations corresponding to the basis function α_j and replace the second by

$$\dot{s}_j = 0. \tag{5.2}$$

This gives

$$\frac{1}{6}(\Delta_{j-\frac{1}{2}}s)w_{j-\frac{1}{2},1} + \frac{1}{3}(\Delta_{j-\frac{1}{2}}s)w_{j-\frac{1}{2},2} + \frac{1}{3}(\Delta_{j+\frac{1}{2}}s)w_{j+\frac{1}{2},1} + \frac{1}{6}(\Delta_{j+\frac{1}{2}}s)\;w_{j+\frac{1}{2},2} = b_{j-\frac{1}{2},2} + b_{j+\frac{1}{2},1} \tag{5.3}$$

Since now $\dot{s}_j = 0$ we have $\dot{a}_j = w_{j-\frac{1}{2},2} = w_{j+\frac{1}{2},1}$ from (3.9) in the reduced problem which yields

$$\dot{a}_j = \frac{b_{j-\frac{1}{2},2} + b_{j+\frac{1}{2},1} - \frac{1}{6}\{(\Delta_{j-\frac{1}{2}}s)w_{j-\frac{1}{2},1} + (\Delta_{j+\frac{1}{2}}s)w_{j+\frac{1}{2},2}\}}{\frac{1}{3}(\Delta_{j-\frac{1}{2}}s + \Delta_{j+\frac{1}{2}}s)} \tag{5.4}$$

$$\dot{s}_j = 0$$

as the solution for the modified system.

The null space of the singular matrix M_j is spanned by the vector $[m,1]^T$ (where $m_{j-\frac{1}{2}} = m_{j+\frac{1}{2}} = m$) and an appropriate multiple of this vector may be added to satisfy an externally imposed averaged velocity or position.

If several nodes become parallel simultaneously a number of equations of the type (5.3) will occur and it may be necessary to solve a tri-diagonal system if the nodes are adjacent to one another.

Parallelism is a rather rare occurrence and is usually associated with the curvature of the solution changing sign during the evolution, or perhaps with the evolution to a steady state.

We go now to consider singularity of C_k. Since this type of singularity is associated with node overtaking as a result of time stepping we discuss it in the next section in association with time integration.

6. TIME STEPPING AND NODE OVERTAKING

The MFE method is semi-discrete and gives rise to ordinary differential equations in time which require integration to obtain the full solution.

We have already seen that in the case of the inviscid Burgers' equation the nodes move along characteristics, while for the general scalar hyperbolic law the least squares best fit to the exact solution is carried asymptotically for small time steps. Approximate time stepping will degrade the latter property if the time step is too large, while the former property will be lost for more general flux functions.

It has been found that using the Euler explicit forward difference method is sufficient in the examples tried so far by Wathen (1984) and Johnson (1984). In no case do implicit methods give any advantage. We therefore use

$$\begin{pmatrix} a_j^{n+1} \\ s_j^{n+1} \end{pmatrix} = \underline{y}_j^{n+1} = \underline{y}_j^n + \Delta t\, M_j^{-1}\, \underline{w}_j = \begin{pmatrix} a_j^n \\ s_j^n \end{pmatrix} + \Delta t\, M_j^{-1}\, \underline{w}_j \tag{6.1}$$

(see (3.10)).

As far as the choice of Δt is concerned we still require an accuracy criterion. Algorithms for accuracy are not well developed but in view of the simplicity of the method we can afford to be generous in taking a trial and error approach. A possible algorithm compares the result of one MFE step with that of two half steps and continues halving the step until the difference between the two results is acceptable, as in (Baines, 1985).

A major difficulty with time stepping is that the nodes may overtake one another if Δt is not small enough.

In problems whose solution is expected to be smooth this gives an implicit condition on the time step. For hyperbolic problems which admit shocks however we expect discontinuities to form and we can take advantage of node overtaking to model shocks in an effective way. Moreover the condition on Δt is now explicit in that it is the shortest time taken for one node to catch up with the next as in (Wathen and Baines, 1985).

As the separation of nodes goes to zero the element segment becomes vertical (parallel to the u-axis) and from (4.6) with $\theta_k \to \pi/2$ the normal velocity of the mid-point of the segment tends smoothly to the shock speed, at least in the semi-discrete case. The shock speed may then be "frozen", ie. imposed on both nodes of the shocked element: this acts as an internal boundary condition and the solution on the adjacent elements to left and right may proceed separately. The procedure is also feasible when nodes run into shocks or when shocks overtake shocks, in which cases a node can be deleted.

From equation (4.12) we see that the angular speed of the segment is non-zero when the shock forms ($m_k \to \infty$) so that there is a change of state at this instant. The manner in which the shock forms in the MFE method is consistent with the (Oleinik, 1956) entropy inequality in the semi-discrete case. Moreover this is also true for expansions.

Accuracy in space is of course determined by the number of nodes used to represent the solution and this is decided when carrying out the projection of the initial data into the piecewise linear space. Less obviously it has been found that it is crucial how these nodes are distributed in space, equidistribution of $(u'')^{\frac{1}{2}}$ being an effective choice: see (Herbst, 1982).

We show numerical results for two scalar problems, the inviscid Burgers' equation and the Buckley-Leverett equation. These are shown in Figure 5. The first gives the exact solution for a convex flux function while the second is an approximate solution in the case of a non-convex flux function.

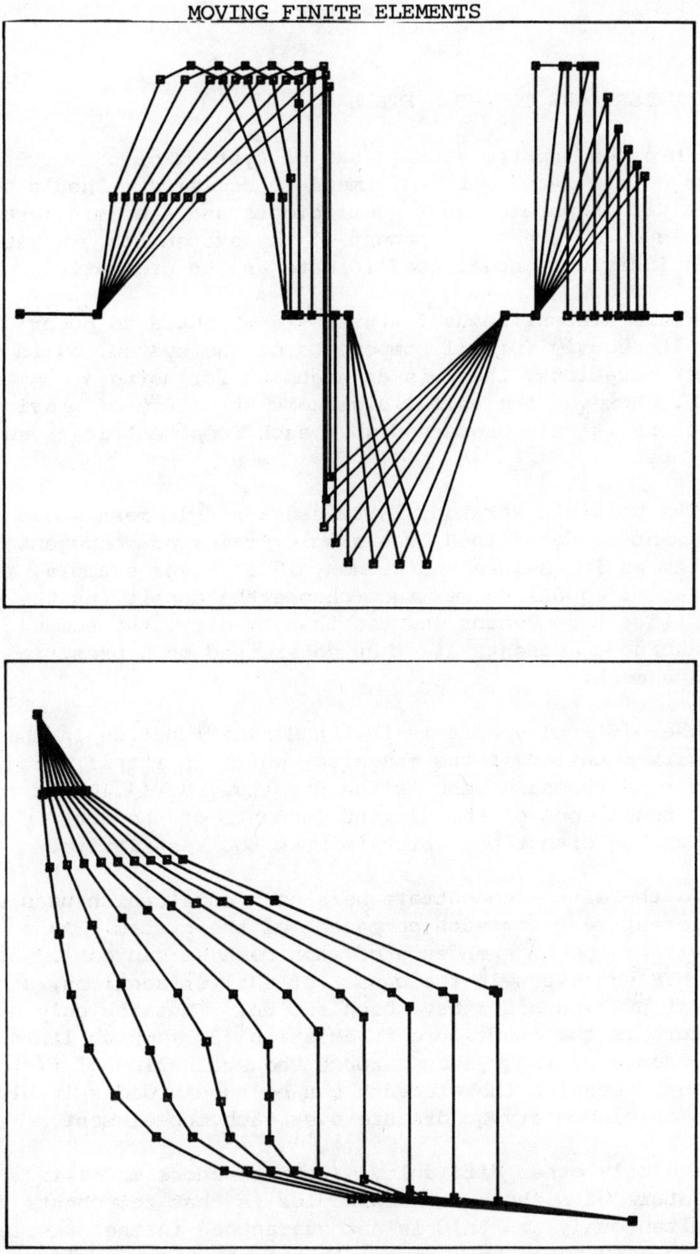

(a)

(b)

Fig. 5. MFE solutions for the scalar non-linear equation

$$u_t + f_x = 0 \ .$$

(a) $f(u) = \frac{1}{2}u^2$ (Burgers')

(b) $f(u) = \dfrac{u^2}{u^2 + \frac{1}{2}(1-u)^2}$ (Buckley-Leverett)

with initial data as in (Wathen and Baines, 1985).

7. EXTENSIONS TO 1-D SYSTEMS

In extending the above ideas of systems of conservation laws we are faced with an immediate decision. Should we work with separate nodal coefficients and a common mesh or should we give each component of the system its own mesh with individual nodal coefficients and co-ordinates?

Where discontinuous features are expected to occur simultaneously for all components of the system, as in the Euler equations, there is an argument for using a common mesh. However the useful algebraic structure of previous sections is only preserved when each component is given its own mesh.

One possible strategy is to use a single mesh whose movement is determined by a single preferred component of the system as in (Baines and Wathen, 1985). For example, in the Euler equations we might choose the density as the most significant component and use that to drive the nodes. The remaining components are then determined on a prescribed moving mesh.

The difficulty here is that the flux function in the density equation is the momentum, which is itself piecewise linear on the same mesh as the density. By (4.10) it follows that the slopes of the element segments of the density do not change with time, which is much too restrictive.

We therefore concentrate here on a model which uses a different mesh for each component of the system. As a result of the independence of each mesh we can easily solve the MFE equations in the manner of earlier sections, once the right hand sides have been set up. Thus the only new feature is the quadrature in equation (3.7) which links the components of the system through the evaluation of $\underline{f}(\underline{v})$. In one dimension the elements can be subdivided suitably with an elementary quadrature over each sub-element.

The only other difficulty arises in shock modelling. A feature of a shock in gasdynamics is that components shock simultaneously and this is not guaranteed in the numerical method. An additional device is therefore needed in general to ensure that when a shock occurs it is simultaneous in the appropriate components.

We give numerical results for the well known shock tube problem used as a basis for comparisons by (Sod, 1978). This is shown in Figure 6.

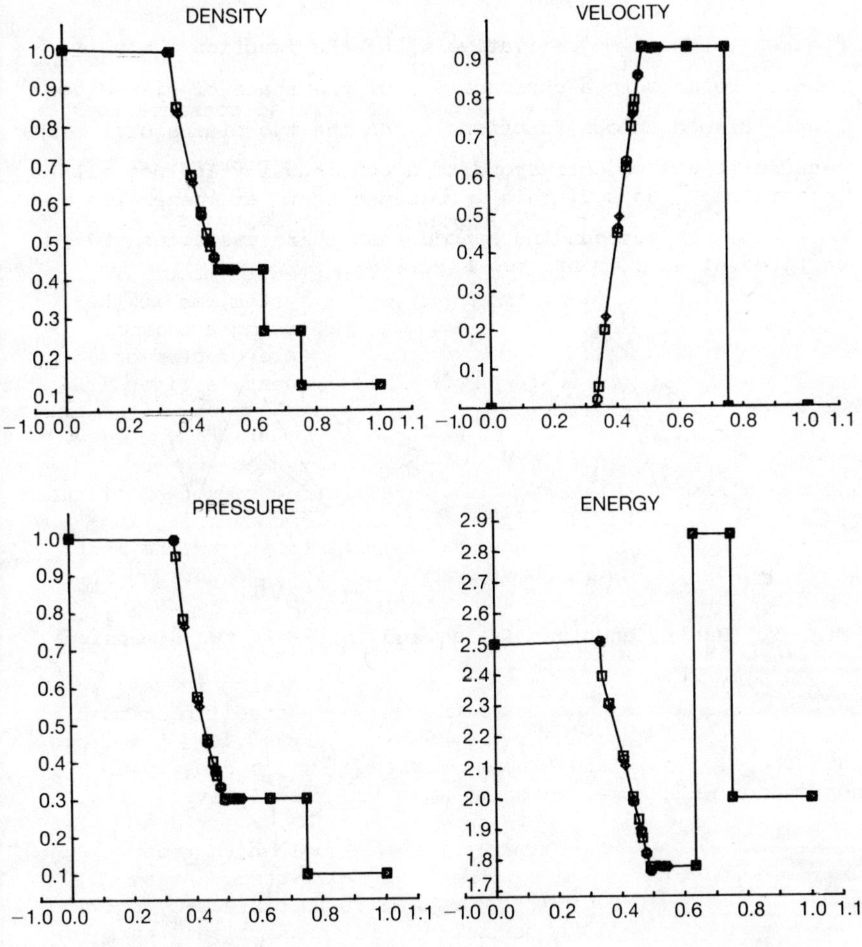

Fig. 6. MFE solutions for the Sod shock tube problem at $t = 0.144$ from initial data at $t = 0.1$ as in (Baines and Wathen, 1985).

8. THE METHOD IN HIGHER DIMENSIONS

One of the most promising aspects of the MFE method is its straightforward generalisation to higher dimensions. We consider here the two dimensional scalar conservation law

$$u_t + f_x + g_y = 0 . \qquad (8.1)$$

We again approximate the function u by a piecewise linear function v given by equation (2.2), where the basis functions are now two dimensional "pyramid" functions as shown in

Figure 7. The time derivative v_t of the function v now however belongs to a subspace $S_{\alpha\beta}$ of the space of piecewise linear discontinuous functions S_ϕ on the two dimensional mesh, because it has to correspond to a continuous v and not all members of S_ϕ do so. This is because there are generally more elements surrounding a node than there are nodes at vertices of an element (see Figure 8).

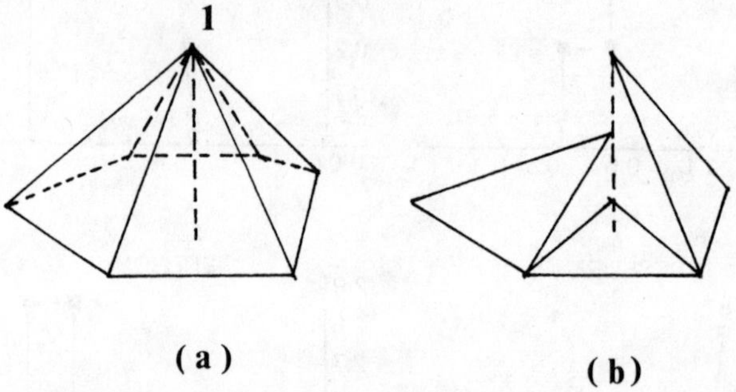

Fig. 7. Basis functions (a) α_j and (b) β_j in two dimensions.

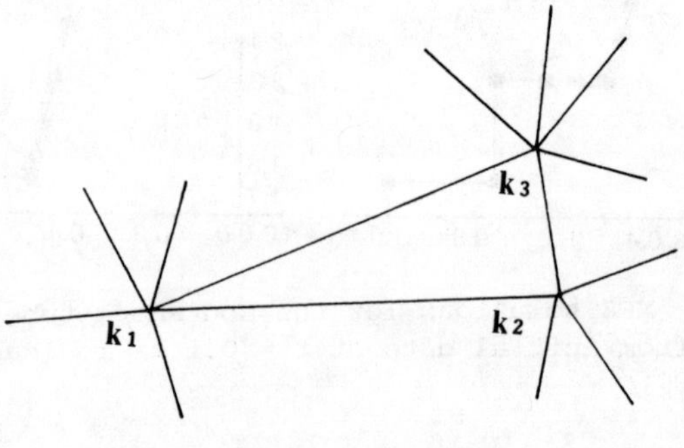

Fig. 8. Nodes/elements in two dimensions.

From a local elementwise point of view we can readily calculate the $\underline{b}_k$ in the two dimensional generalisation of equation (3.4), which is now a 3x3 system. But in order to obtain the nodal velocities, which are evaluated from the union of equations (3.10), we require $\underline{w}$, the union of the $\underline{w}_k$'s, to lie in the range space of M (see (3.14)), which is now rectangular.

It is therefore necessary to constrain the projection which leads to the vector $\underline{w}$. The result of this constrained projection is that $\underline{w}$ satisfies

$$M^T C\underline{w} = M^T\underline{b} \tag{8.2}$$

which, since
$$M^T\underline{b} = \underline{g} \tag{8.3}$$

leads to the familiar form

$$M^T CM\dot{\underline{y}} = \underline{g} , \tag{8.4}$$

ie. it is equivalent to the standard method (see for example (Wathen and Baines, 1985)).

Hence the decomposition

$$A = M^T CM \tag{8.5}$$

holds in higher dimensions, but although the matrix C is square 2x2 block diagonal (using elementwise numbering) the matrix M is rectangular. For example, in two dimensions using nodewise numbering M takes the form N where

$$N = \text{diag}\,\{N_j\} \quad N_j = \begin{bmatrix} 1 & -m_{j1} & -n_{j1} \\ 1 & -m_{j2} & -n_{j2} \\ 1 & -m_{jI} & -n_{jI} \end{bmatrix} \tag{8.6}$$

where m_{ji}, n_{ji} are the slopes v_x, v_y of the function v in the x and y directions within the element.

Using a permutation matrix Q to map between the elementwise numbering and the nodewise numbering we obtain the consistent decomposition

$$A = N^T Q^T CQN , \tag{8.7}$$

where $M = QN$.

It has been shown by Wathen (1984) using this decomposition that if D is the matrix consisting of diagonal blocks of the MFE matrix A then the eigenvalues of the matrix $D^{-1}A$ lie in the real interval

$$[\tfrac{1}{2}, 1 + \frac{d}{2}] , \qquad (8.8)$$

where d is the number of dimensions. There is therefore every reason for using a conjugate gradient method with D^{-1} as preconditioner to invert the MFE matrix A.

It is interesting to note that the MFE equations (8.4) also arise from first carrying out an unconstrained projection for $\underline{w}$ into the space S_ϕ and then doing a least squares projection of the equation

$$N\dot{\underline{y}} = \underline{w} \qquad (8.9)$$

weighted by the matrix $C^{\frac{1}{2}}Q$ as in (Baines, 1985).

If $S_\phi = S_{\alpha\beta}$ then by a calculation similar to that in Section 4 we again find as in (Baines, 1985) that V_k, the velocity of the centroid of the triangular segment of the solution in the direction perpendicular to the segment is given by equation (4.6), where now

$$\Delta f_k = \text{the outward flux of } (f,g) \text{ across the element boundary,} \qquad (8.10)$$

$$\Delta v_k = \text{the vertical profile of the element } k \qquad (8.11)$$

and the angle θ_k is given by

$$\tan \theta_k = |\underline{\nabla} v| . \qquad (8.12)$$

There is also a corresponding result on the segment rotation similar to equation (4.10) but apart from noting that the convexity of the flux function is again involved we do not set down the details.

9. SINGULARITIES IN HIGHER DIMENSIONS

Because of the decomposition (8.5) the singularities of the method in higher dimensions correspond to those in one dimension, namely, they are singularity of C when the element size goes to zero and rank deficiency of M when nodes become coplanar: see (Wathan and Baines, 1985).

In the latter case of parallelism the remedy is the same as in one dimension, namely to remove the (linearly dependent) equations causing the parallelism and to solve a reduced system, adding a suitable multiple of the null space at the end. Some care is required with the larger blocks occurring in higher dimensions; it is safest to transform the local system to upper triangular form, which avoids problems of ill-conditioning which can arise if an arbitrary equation is omitted.

Singularity of C arises when the size of an element goes to zero as a result of a node running into the opposite side of a triangle. Even though we do not have a proved best fit property in more than one dimension we may still conjecture that this occurrence corresponds to the formation of a shock. Technical difficulties arise in determining how the shocked triangle should move subsequently but we have the following lead in the case $S_\phi = S_{\alpha\beta}$.

We recall the result stated above on the normal velocity V_k of the centroid of the triangle segment perpendicular to itself. It can be seen that this velocity tends to the local shock speed as the segment becomes vertical in the sense that the triangle sweeps out "mass" at the correct rate. It appears that we should therefore impose this shock speed on all points of the vertical line through the centroid of the triangle in the subsequent motion with the shock in place. This still allows possible rotation of the triangle about the line through the centroid which must be determined from the solution of the rest of the system. We do not go into the details here.

We show results for one problem, a generalisation of the Buckley-Leverett equation in one dimension. The governing equation is

$$u_t + \underline{\nabla} \cdot \left[\frac{u^2}{u^2 + \frac{1}{2}(1-u)^2} \right] = 0 \tag{8.13}$$

and Figure 9 shows both the initial data and the solution at a later time. It is worth noting that even though the triangles of the mesh become highly distorted there is no ill-conditioning of the MFE matrix.

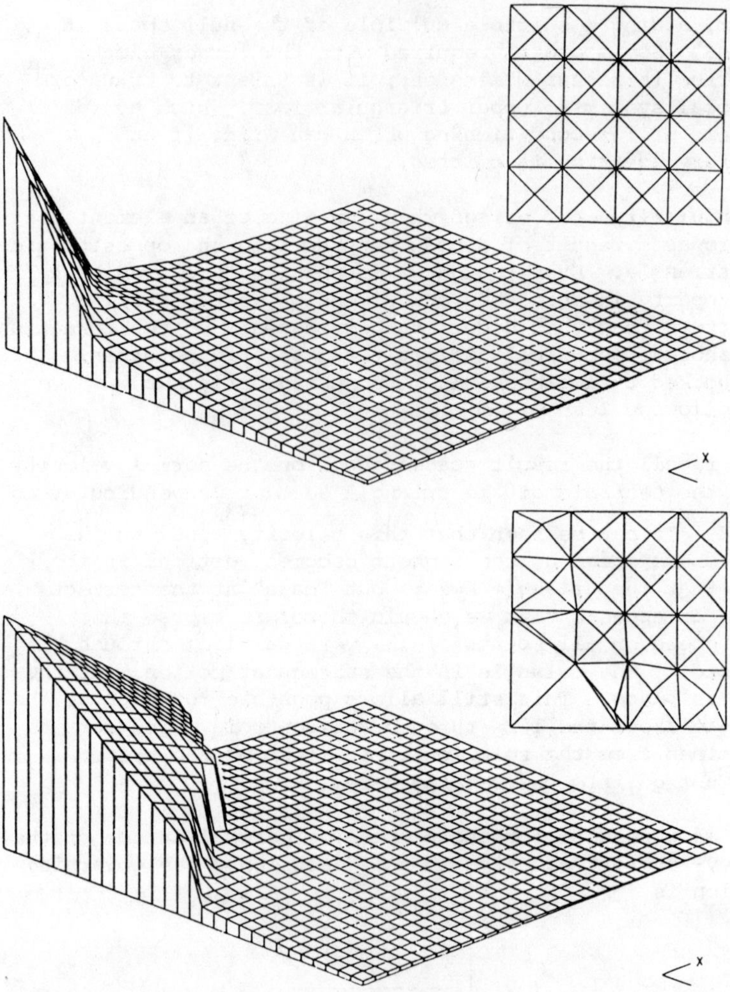

Fig. 9. MFE solution for the two dimensional equation (8.13): initial data and solution (isoplots) as in (Wathen, 1984).

10. CONCLUSION

In this paper we have shown that the MFE method in one dimension is a local method which gives the motion of individual element segments in terms of elementary properties of the flux function. As a result non-linear problems involving scalar conservation laws can be solved simply on a small computer. Shocks are particularly well resolved by the method, the entropy rules being respected.

In higher dimensions the local nature of the method is constrained in such a way that an additional least squares fit is required to map the local element motion onto the nodewise velocities. The technical problem of dealing with shocks is assisted by the consistency of the element segment velocity with a local wave speed.

11. ACKNOWLEDGEMENTS

I should like to thank Professor K.W. Morton and Dr. A.J. Wathen for many illuminating discussions.

12. REFERENCES

BAINES, M.J. (1985) "On Approximate Solutions of Time-Dependent Partial Differential Equations by the Moving Finite Element Method", Numerical Analysis Report 1/85, Department of Mathematics, University of Reading.

BAINES, M.J. and WATHEN, A.J. (1985) "Moving Finite Element Modelling of Compressible Flow", Numerical Analysis Report 4/85, Department of Mathematics, University of Reading.

HERBST, B.M. (1982) "Moving Finite Element Methods for the Solution of Evolution Equations", Ph.D Thesis, University of the Orange Free State, South Africa.

JOHNSON, I.W. (1984) "The Moving Finite Element Method for the Viscous Burgers' Equation", Numerical Analysis Report 3/84, Department of Mathematics, University of Reading.

LYNCH, D.R. (1982) "Unified Approach to Simulation on Deforming Elements with Application to Phase Change Problems", *J. Comput. Phys.* **47**, 387-411.

MILLER, K. (1981) "Moving Finite Elements, Part I" (with R.N. Miller), *SIAM J. Numer. Anal.* **18**, 1019-1032: "Moving Finite Elements, Part II", *SIAM J. Numer. Anal.* **18**, 1033-1057.

MORTON, K.W. (1982) Private Communication. See also (Wathen, 1984).

OLEINIK, O.A. (1957) "Discontinuous Solutions of Nonlinear Differential Equations", *Amer. Math. Soc. Trans. Ser. 2, No. 26*, 95-172.

SOD, G.A. "A Survey of Several Finite Difference Methods for Systems of Nonlinear Hyperbolic Conservation Laws", *J. Comput. Phys.* 27 , 1-31.

WATHEN, A.J. (1984) "Moving Finite Elements and Oil Reservoir Modelling", Ph.D. Thesis, University of Reading.

WATHEN, A.J. and BAINES, M.J. (1985) "On the Structure of the Moving Finite Element Equations", *IMAJNA* 5, 161-182.

FINITE ELEMENT METHODS FOR COMPRESSIBLE FLOW

R. Löhner, K. Morgan, J. Peraire, O.C. Zienkiewicz and L. Kong
(Institute for Numerical Methods in Engineering, University College, Swansea)

1. INTRODUCTION

In this paper we review the on-going research work at Swansea into the application of finite element based methods to the solution of problems involving compressible fluid flow. From the outset, our main area of activity has been the development of algorithms for modelling supersonic flows, with the ultimate aim of producing a computer code for predicting aerodynamic heating rates over bodies of complex geometrical shape at high Mach number.

The finite element method can be used to discretise accurately any arbitrary-shaped region but the consequence is, in general, that the resulting finite element mesh will be completely unstructured. An example of such a mesh is given in fig. 1, which shows a typical discretisation of a 2D region using 3-noded triangular elements, and it may be observed how the mesh is lacking in structure e.g. the number of elements surrounding a node varies with position in the mesh.

The use of the finite element method in this manner should ease the problems frequently associated with discretising geometrically complex domains, as the process of mesh generation is decoupled from the solution algorithm, and will enable local refinement of the mesh to be made in a straightforward fashion. However, as a consequence, we must restrict consideration to solution algorithms which can be implemented on such unstructured meshes in two and three dimensions. This means that optimal implicit schemes, such as those employing factorisation methods (e.g. MacCormack (1981) or Beam and Warming (1978)), cannot be used and neither can methods relying on line or surface relaxation (e.g. Jameson (1983)). In addition, methods which are based upon the extension of strictly

1D concepts into two and three dimensions cannot be considered (e.g. Colella (1984)).

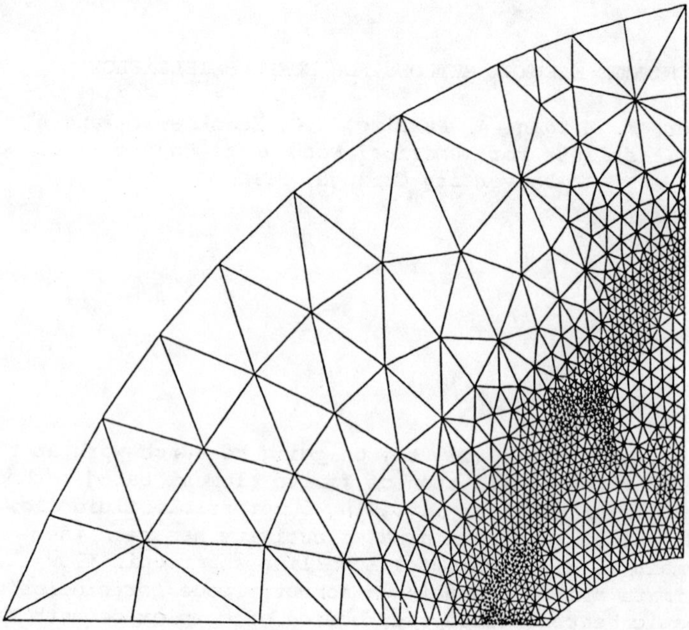

Fig. 1 A typical unstructured finite element mesh

The approach chosen is therefore the use of a simple explicit time-stepping scheme with the capability of producing time-accurate solutions for transient flows and with the hope of coupling the scheme with a multigrid procedure (such as that employed by Ni (1981)), to obtain rapid solutions to steady-state problems. A computer code based upon such a scheme will be capable of a high degree of vectorisation and vector/scalar speed-up ratios of around 20/1 will be obtainable on a modern supercomputer such as the CYBER 205. However, it must be accepted that a program using the finite element method on an unstructured grid will be capable of achieving only 0.1-0.25 of the maximum Mflop rate for such a machine, because of the penalty associated with the large amount of gather and scatter operations which are required. For this reason, when the solution of large realistic problems is attempted using this approach, it is essential to try and ensure that the best accuracy is obtained for a given number of grid points. Our hope is that this goal can be attained by the incorporation of adaptive refinement procedures.

2. INVISCID FLOWS

2.1 *The Solution Algorithm*

We begin by considering problems governed by the two-dimensional compressible Euler equations in the form

$$\frac{\partial \underline{U}}{\partial t} + \frac{\partial \underline{F}_j}{\partial x_j} = \underline{O} \qquad (2.1.1)$$

where the summation convention has been employed and

$$\underline{U} = \begin{bmatrix} \rho \\ \rho u_1 \\ \rho u_2 \\ \rho \epsilon \end{bmatrix} \qquad \underline{F}_j = \begin{bmatrix} \rho u_j \\ \rho u_1 u_j + p\delta_{j1} \\ \rho u_2 u_j + p\delta_{j2} \\ u_j(\rho\epsilon + p) \end{bmatrix} \qquad (2.1.2)$$

Here ρ, p and ϵ denote the density, pressure and specific total energy of the fluid respectively and u_j is the component of the fluid velocity in the direction x_j of a Cartesian coordinate system. The equation set is completed by the addition of an equation of state

$$p = (\gamma - 1)\rho\,[\epsilon - \frac{1}{2} u_j u_j] \qquad (2.1.3)$$

which is valid for a perfect gas, where γ is the ratio of the specific heats.

Our initial approach (Löhner et al (1984a)) to the solution of this equation set was an implementation of the second order Euler-Taylor-Galerkin method introduced by Donéa(1984). The resulting algorithm takes the form

$$\int_\Omega (\underline{U}^{m+1} - \underline{U}^m) N_i \, d\Omega = -\Delta t \int_\Omega \frac{\partial \underline{F}_j^m}{\partial x_j} N_i \, d\Omega - \frac{\Delta t^2}{2} \int_\Omega \underline{A}_k^m \frac{\partial \underline{F}_j^m}{\partial x_j} \frac{\partial N_i}{\partial x_k} \, d\Omega$$
$$+ \frac{\Delta t^2}{2} \int_\Gamma n_k \underline{A}_k^m \frac{\partial \underline{F}_j^m}{\partial x_j} N_i \, d\Gamma \qquad (2.1.4)$$

where Ω is the solution domain, Γ is the boundary curve of Ω, $\underline{n} = (n_1, n_2)$ is the unit outward normal to Γ, and

$$\underline{A}_k = \frac{d\underline{F}_k}{d\underline{U}} \qquad (2.1.5)$$

is the jacobian matrix. In this expression the superscript m denotes an evaluation at time $t = t_m$ and the timestep $\Delta t = t_{m+1} - t_m$. The solution vector $\underline{U}$ and the flux vectors $\underline{F}_j$

are nodally interpolated using 3-noded linear triangular elements, with nodal shape functions N_i, while a piecewise constant representation is adopted for the jacobian matrices, with element shape functions P_e. The integrations in equation (2.1.4) may then be performed exactly and lead to an assembled equation system of the form

$$\underline{M}\ \delta\underline{U} = \underline{f}^m \tag{2.1.6}$$

where $\underline{M}$ is a consistent mass matrix and $\delta\underline{U}$ is the vector of changes in the nodal values of $\underline{U}$ over the timestep. To maintain the explicit character of the scheme, this equation system is solved using the iteration process

$$\underline{M}_L[\delta\underline{U}^{(r)} - \delta\underline{U}^{(r-1)}] = \underline{f}^m - \underline{M}\ \delta\underline{U}^{(r-1)} \tag{2.1.7}$$

starting with $\delta\underline{U}^{(o)} = \underline{O}$, where $\underline{M}_L$ is the lumped (diagonal) mass matrix. This process converges in three iterations and is employed when computing true transients or steady solutions to problems involving flow at high Mach number. If only steady state solutions at low supersonic Mach number are of interest, then one pass through the iteration procedure is used and this is equivalent to solving equation (2.1.6) directly with $\underline{M}$ replaced by $\underline{M}_L$. In this case, in 1D the scheme reduces to the Lax-Wendroff method.

Finite difference workers have consistently avoided the use of one-step schemes of this type because of the CPU penalty associated with the evaluation and multiplication of the jacobian matrices in equation (2.1.4). We can proceed in a similar fashion here and develop a two-step scheme which aims to remove these troublesome terms from the algorithm of equation (2.1.4). In the first step, we use the Taylor expansion

$$\underline{U}^{m+\frac{1}{2}} = \underline{U}^m - \frac{\Delta t}{2}\frac{\partial \underline{F}_j^m}{\partial x_j} \tag{2.1.8}$$

and approximate $\underline{U}^{m+\frac{1}{2}}$ in a piecewise constant fashion, with $\underline{U}^m$ and $\underline{F}_j^m$ interpolated as before. The elementwise constant values of $\underline{U}^{m+\frac{1}{2}}$ can be obtained from the weighted residual statement

$$\int_\Omega \underline{U}^{m+\frac{1}{2}} P_e\, d\,\Omega = \int_\Omega \underline{U}^m P_e\, d\,\Omega - \frac{\Delta t}{2}\int_\Omega \frac{\partial \underline{F}_j^m}{\partial x_j} P_e\, d\Omega \tag{2.1.9}$$

The solution of this equation system completes the first step of the procedure.

In the second step we note that, correct to first order,

$$\underline{F}_k^{m+\frac{1}{2}} = \underline{F}_k^m + \frac{\Delta t}{2} \frac{\partial \underline{F}_k^m}{\partial t} = \underline{F}_k^m - \frac{\Delta t}{2} \underline{A}_k^m \frac{\partial \underline{F}_j^m}{\partial x_j} \qquad (2.1.10)$$

and hence

$$\underline{A}_k^m \frac{\partial \underline{F}_j^m}{\partial x_j} = -2 \frac{(\underline{F}_k^{m+\frac{1}{2}} - \underline{F}_k^m)}{\Delta t} \qquad (2.1.11)$$

A weighted residual form of this expression can now be used to obtain an alternative representation for the troublesome terms appearing in equation (2.1.4), which is then replaced by

$$\int_\Omega (\underline{U}^{m+1} - \underline{U}^m) N_i \, d\Omega =$$

$$\Delta t \int_\Omega \underline{F}_j^{m+\frac{1}{2}} \frac{\partial N_i}{\partial x_j} d\Omega - \Delta t \int_\Gamma \underline{F}_n^m N_i \, d\Gamma - \Delta t \int_\Gamma (\underline{F}_n^{m+\frac{1}{2}} - \bar{\underline{F}}_n^m) Ni \, d\Gamma \qquad (2.1.12)$$

where the overbar denotes an average element value. In this expression the normal flux $\underline{F}_n$ is defined by

$$\underline{F}_n = n_j \underline{F}_j \qquad (2.1.13)$$

and the boundary conditions are applied by ensuring a correct interpretation of the boundary integrals: see (Peraire et al., 1985). Again an equation system of the form of equation (2.1.6) results and this can be solved by the iterative process of equation (2.1.7). In 1D, this two step scheme reduces to the method due to Burstein (1967).

When the equation system (2.1.1) is linear, the one and two step schemes can be shown to be exactly equivalent. For the Euler equations, the two schemes appear to be numerically equivalent. The two step method is therefore to be preferred in practice, due to its speed and beneficial vectorisable properties.

2.2 Artificial Viscosity

When the two step scheme is used for the analysis of problems involving strong shocks, it must be stabilised by the incorporation of an artificial viscosity. This is accomplished by replacing the computed values $\underline{U}^{m+1}$ by smoothed values of $\underline{U}_S^{m+1}$ calculated according to

$$\underline{U}_s^{m+1} - \underline{U}^{m+1} = \Delta t \frac{\partial}{\partial \ell} \left(k^{\ell\ell} \frac{\partial \underline{U}^{m+1}}{\partial \ell} \right) \qquad (2.2.1)$$

where ℓ denotes the direction of the local maximum rate of change of the absolute value of the velocity i.e.

$$\underline{\ell} = \text{grad}\, |\underline{u}| / \, |\text{grad}|\underline{u}|| \qquad (2.2.2)$$

and the diffusion coefficient, $k^{\ell\ell}$, is calculated as

$$k^{\ell\ell} = Ch_e^2 \left| \frac{\partial}{\partial \ell} (\underline{u}.\underline{\ell}) \right| \qquad (2.2.3)$$

In this expression, C is a constant whose value must be specified and h_e is a representative element length. This form for the artificial viscosity is rotationally symmetric and reduces to the well-known model of Lapidus (1967) for 1D problems. In addition, this model should also be useful when viscous problems are modelled, since it will produce only a small smearing of boundary layers, where $\underline{u}.\underline{\ell} \approx 0$. An investigation of the behaviour of this model and a comparison with familiar models is given in (Löhner et. al., 1985a).

2.3 Stability

The explicit character of the solution scheme implies that it will be subject to a Courant-type stability criterion. In practice, the timestep Δt is chosen so that

$$\Delta t \leq \frac{s\, \alpha\, h_e}{[\, |\underline{u}| + c]} \qquad (2.3.1)$$

where c is the local sound speed, $\alpha = 1$ when one pass through the iteration scheme of equation (2.1.7) is employed, $\alpha = 1/\sqrt{3}$ otherwise and s is a safety factor.

2.4 Domain Splitting

For purposes of computational efficiency and accuracy, a domain splitting procedure has been proposed (for full details see (Löhner et. al., 1984b) which groups together elements which allow approximately the same time step. The solution is then advanced with different timesteps on different portions of the mesh in a time accurate fashion. This technique is employed in all the numerical examples reported here.

2.5 Adaptive Refinement for Steady Problems

The two step solution algorithm described in the previous sections has been coupled with an adaptive refinement procedure which is designed to improve the quality of the solutions calculated for the steady state Euler equations. The basic idea is that the solution is advanced towards steady state on an original mesh and then, at prescribed times, the mesh is automatically refined in an adaptive fashion and the computation continued. In (Löhner et. al., 1985b) the authors investigated the use of mesh refinement via mesh movement. However, because of its additional flexibility, we are currently favouring the use of local mesh enrichment and it is this algorithm which will be described here.

We should like to ensure that if E is the error in our calculation then

$$\max_{e} \| E \|^{e} \longrightarrow \text{minimum} \tag{2.5.1}$$

where $\|.\|^{e}$ denotes some suitable measure of the error over element e and the maximum is sought over all elements in the mesh. Following Oden (1983) we attempt to achieve the satisfaction of this condition by requiring that

$$\|E\|^{e} = \text{constant} \tag{2.5.2}$$

on each element. There are several possible methods for obtaining a suitable measure of the error and these have been discussed in (Löhner et. al., 1985c). However, our experience indicates that the local use of classic theoretical error estimates works satisfactorily and is very economical in the finite element context. For the Euler equations, the density ρ is identified as the key-variable and the measure of the error in the solution is taken to be

$$\|E\|^{e} = |\rho - \hat{\rho}|^{e} \tag{2.5.3}$$

where $|.|^{e}$ denotes the L_2 - norm over element e, and $\hat{\rho}$ is the approximate solution. From interpolation theory (see e.g. (Strang and Fix, 1973)), we can then write

$$\|E\|^{e} \leqslant C\, h_e^{k}\, |\rho|_k^{e} \tag{2.5.4}$$

where $|.|_k^{e}$ denotes the kth seminorm over element e, h_e is a representative element length and C is a constant. Equation (2.5.2) is now replaced by the requirement that

$$h_e^{k}\, |\rho|_k^{e} = \text{constant} \tag{2.5.5}$$

or, in practice , since the exact solution for the density ρ is unknown,

$$h_e^k \, |\hat{\rho}|_k^e = \text{constant} \tag{2.5.6}$$

A mesh enrichment algorithm can be devised based upon the satisfaction of this requirement. At a prescribed stage in the calculation we determine

$$\bar{\psi} = \max_e \, h_e^k \, |\hat{\rho}|_k^e \tag{2.5.7}$$

and then refine all elements e for which

$$h_e^k \, |\hat{\rho}|_k^e > a \, \bar{\psi} \tag{2.5.8}$$

where a is a constant whose value must be specified. The refinement of an element is accomplished by adding a node to the mid-point of each side and so sub-dividing the element into four. There will be some problems associated with 'hanging nodes' but these can be removed by surrounding the region to be refined with a layer of transition elements and sub-dividing some of these elements into two, as required (Löhner et. al., 1984c). After refinement, the mesh can be smoothed locally and any badly deformed elements which remain are removed (Löhner et al., 1985b). Alternatively, a mesh reconnection algorithm can be used to prevent the appearance of unacceptably large angles in the mesh and the need for smoothing can then be avoided.

The value that should be used for the constant a in equation (2.5.8) has been the subject of some experimentation. We are currently choosing the value by using a method similar to that proposed in a finite difference context by Dannenhoffer and Baron (1985). In fig. 2 we plot a typical curve showing the variation in the ratio of the number of elements refined to the total number of elements in the mesh for different choices of a. For an economical refinement process, the value of a used is chosen to be in the vicinity of the point a*, which is the point where the line, which is parallel to the line joining the two ends of the curve, is tangent to the curve.

The value of k remains to be specified and either the use of k=1 or k=2 are possibilities here. With k=1, the requirement becomes

$$h_e \left[\int_{\Omega_e} \frac{\partial \hat{\rho}}{\partial x_i} \frac{\partial \hat{\rho}}{\partial x_i} \, d\Omega \right]^{\frac{1}{2}} = \text{constant} \tag{2.5.9}$$

or, since the gradient of $\hat{\rho}$ is constant over each element,

$$\sum_i h_e^2 \left| \frac{\partial \hat{\rho}}{\partial x_i} \right|_e = \text{constant} \tag{2.5.10}$$

With k=2, the requirement is that

$$h_e^2 \left[\int_{\Omega_e} \frac{\partial^2 \hat{\rho}}{\partial x_i \partial x_j} \frac{\partial^2 \hat{\rho}}{\partial x_i \partial x_j} d\Omega \right]^{\frac{1}{2}} = \text{constant} \qquad (2.5.11)$$

where the second derivatives are interpolated in terms of nodal values which are obtained by a variational recovery process (Löhner et al., 1985c). The behaviour of both requirements has been investigated and our current preference is for equation (2.5.11), because of the quality of meshes which are produced.

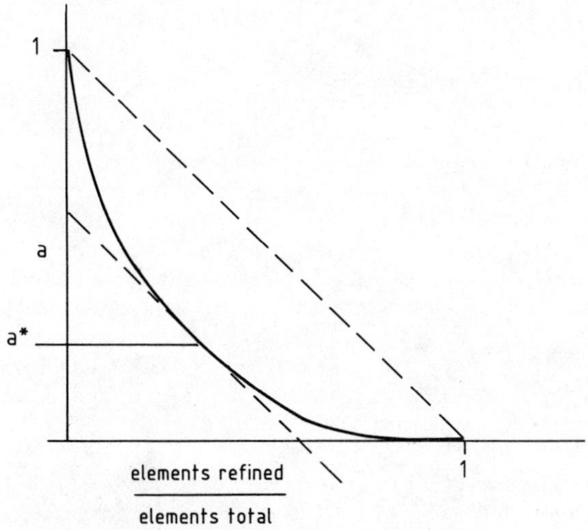

Fig. 2 Determination of the value of a* for the refinement procedure

2.6 Numerical Results

To illustrate the performance of the adaptive mesh refinement procedures which have been introduced above, we consider the solution of two problems involving steady high speed inviscid flows.

The first problem is that of Mach 3 flow past a wedge of angle 20°. The initial mesh, and corresponding density contours after 100 steps, is shown in fig. 3a. The adaptive refinement procedure with local mesh smoothing produces the refined mesh of fig. 3b and the density contours produced after a further 100 steps on this mesh are also shown. This mesh is now refined and smoothed and the resulting mesh, together with the final solution produced, is shown in fig. 3c. A nice feature of the mesh enrichment procedure is that the initial calculation can be made on a relatively coarse mesh - all we need ensure is that the mesh is sufficiently fine to

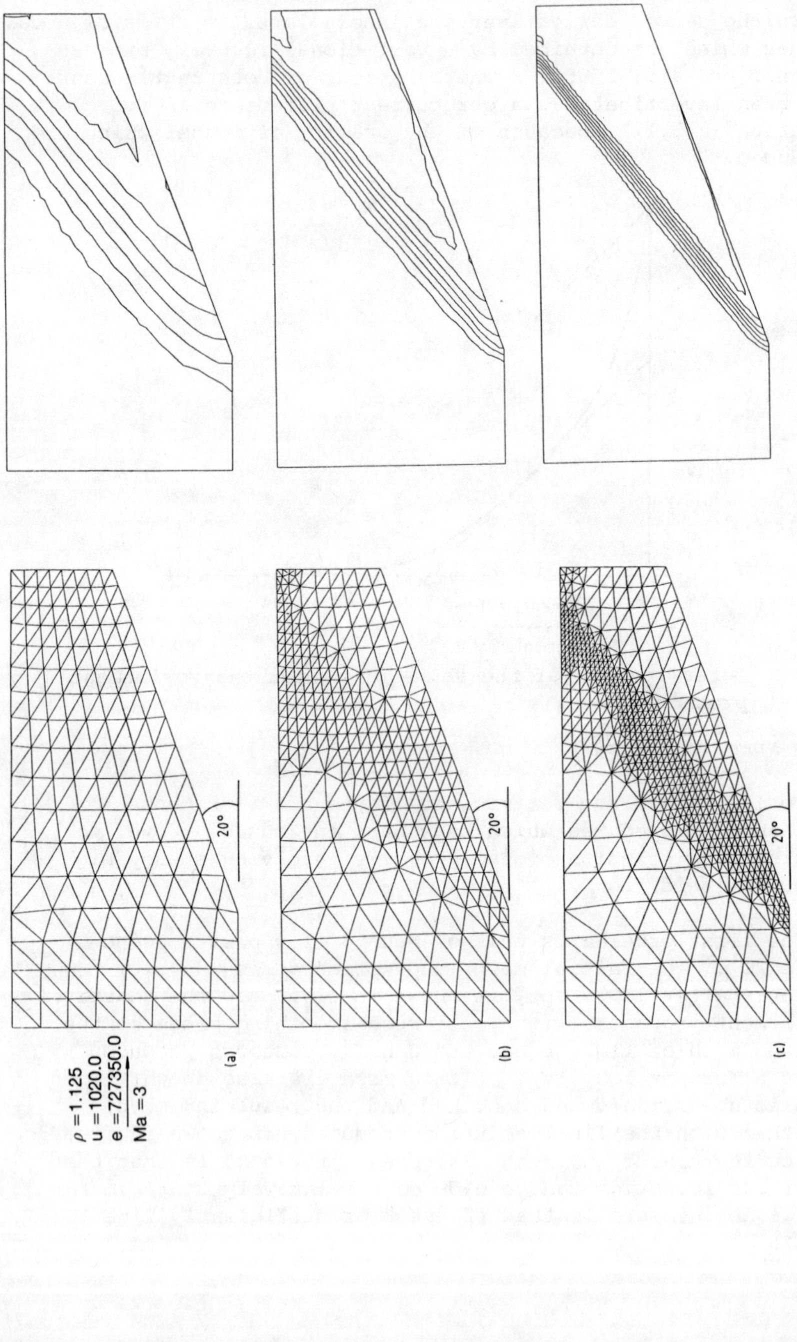

Fig. 3 Steady inviscid Mach 3 flow past a wedge. Sequence of meshes and corresponding density contours.

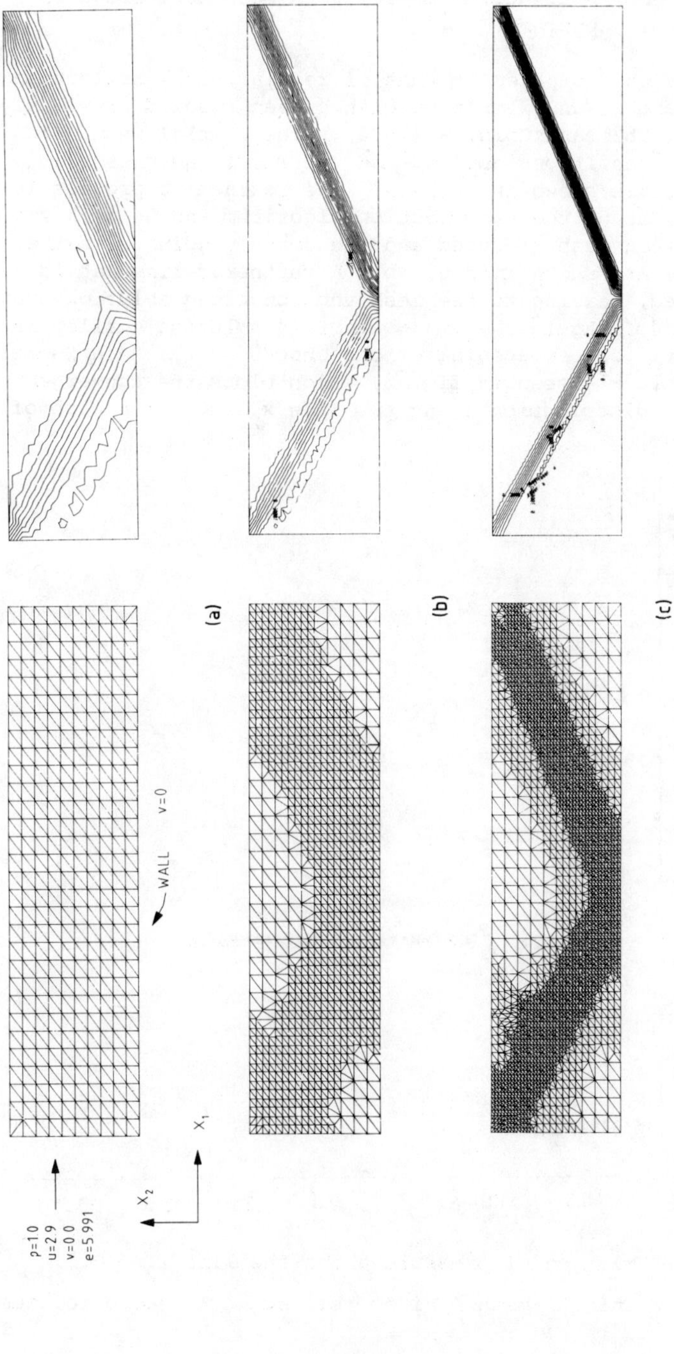

Fig. 4 Regular shock reflection at a wall. Sequence of meshes and corresponding pressure contours.

pick up the main features of the flow. The significant improvement in solution quality with mesh enrichment is immediately apparent.

The second problem is that of regular shock reflection at a wall and is an example which has been studied previously by Harten (1982) and Colella (1984). The initial mesh, the boundary conditions applied and the resulting pressure contours are shown in fig. 4a. The refinement process is now entered, using the reconnection algorithm and no mesh smoothing. The refined mesh produced and the corresponding pressure contours are shown in fig. 4b. A further refinement is performed, leading to the mesh and converged solution shown in fig. 4c. Again the improvement in solution quality is apparent. The steepening of the shocks as the mesh is refined can be clearly seen in fig. 5, which plots the converged pressure distribution along the line $x_2 = 0.5$ for each of the meshes.

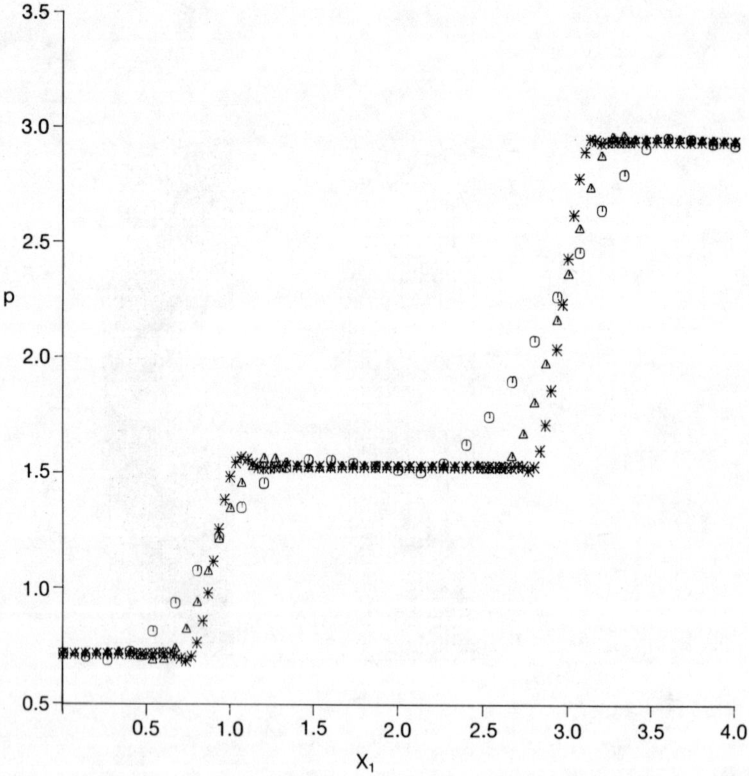

Fig. 5 Variation of pressure along the line $x_2 = 0.5$.

⊙ Initial mesh △ First refinement * Second refinement

3. VISCOUS FLOW

3.1 The Solution Algorithm

The compressible Navier-Stokes equations can be written in the form

$$\frac{\partial \underline{U}}{\partial t} + \frac{\partial \underline{F}_j}{\partial x_j} = \frac{\partial \underline{G}_j}{\partial x_j} \qquad (3.1.1)$$

where $\underline{U}$ and $\underline{F}_j$ are defined previously in equation (2.1.2) and

$$\underline{G}_j = \begin{bmatrix} 0 \\ \sigma_{1j} \\ \sigma_{2j} \\ u_1\sigma_{1j} + u_2\sigma_{2j} + k\,\frac{\partial T}{\partial x_j} \end{bmatrix} \qquad (3.1.2)$$

Here k denotes the thermal conductivity of the fluid, T is the temperature defined by

$$T = c_v\,[\in - \frac{1}{2}\,u_i u_i] \qquad (3.1.3)$$

c_v is the specific heat at constant volume and the components of the viscous stress tensor σ_{ij} are given by

$$\sigma_{ij} = \mu\,[\frac{\partial u_i}{\partial x_j} + \frac{\partial u_j}{\partial x_i}] + \lambda\,\frac{\partial u_k}{\partial x_k}\,\delta_{ij} \qquad (3.1.4)$$

For present purposes it is assumed that the Stokes hypothesis is valid so that the viscosity coefficients λ and μ are related by

$$\lambda = -\,\frac{2}{3}\,\mu \qquad (3.1.5)$$

The solution approach adopted here is to view equation (3.1.1) in an operator-split fashion, treating the viscous terms in a simple explicit manner and the advective terms as before. The result is that equation (2.1.12) is replaced by

$$\int_\Omega (\underline{U}^{m+1} - \underline{U}^m) N_i\, d\Omega = \Delta t \int_\Omega \underline{F}_j^{m+\frac{1}{2}} \frac{\partial N_i}{\partial x_j}\, d\Omega - \Delta t \int_\Gamma \underline{F}_n^m N_i\, d\Gamma$$
$$- \Delta t \int_\Gamma (\underline{F}_n^{m+\frac{1}{2}} - \underline{\bar{F}}_n^m) N_i\, d\Gamma - \Delta t \int_\Omega \underline{G}_j^m \frac{\partial N_i}{\partial x_j}\, d\Omega + \Delta t \int_\Gamma \underline{G}_n^m N_i\, d\Gamma \qquad (3.1.6)$$

and the solution $\underline{U}^{m+1}$ may again be determined by a two-step approach similar to that outlined above.

3.2 Stability

The timestep limit used comes from a direct generalisation of the result obtained by applying these ideas to the solution of the linear one dimensional convection-diffusion equation. With this equation written in the form

$$\frac{\partial \phi}{\partial t} + a \frac{\partial \phi}{\partial x} = b \frac{\partial^2 \phi}{\partial x^2} \tag{3.2.1}$$

the scheme will be stable, with a lumped mass matrix, provided that

$$\Delta t \leqslant \frac{h_e^2}{[2b + ah_e]} \tag{3.2.2}$$

The generalisation to the Navier-Stokes equations replaces a by $|\underline{u}| + c$ and b by max $(\mu/\rho,\ k/\rho c_p)$, where c_p is the specific heat at constant pressure. For the computations to be reported here, the Prandtl number, Pr, has been taken to be constant and equal to 0.72 which implies that

$$(\mu/\rho) \Big/ ({}^{k}/\rho c_p) = 0.72 \tag{3.2.3}$$

In practice, therefore, a stability requirement of the form

$$\Delta t \leqslant \frac{s\ h_e^2\ \rho c_p}{[2k + (|\underline{u}| + c) h_e \rho c_p]} \tag{3.2.4}$$

is employed where s is a safety factor.

3.3 Numerical Results

The first problem considered is the steady state solution of supersonic viscous flow over a flat plate. The flow conditions correspond indentically to one of the cases considered by Carter (1972), using a finite difference scheme. The free stream Mach number is 3 and the Reynolds number based on the plate length is 1000. The temperature of the plate is assumed constant. The Sutherland viscosity law (see White, 1974) is used and the initial conditions are chosen to be appropriate to the case of a flat plate impulsively inserted into the free stream. Free stream boundary conditions were enforced along the inflow boundary, whereas extrapolation type boundary conditions (see (Peraire et. al., 1985)) were applied to the outflow and upper boundaries. A symmetry type boundary condition is applied to the portion of the boundary ahead of the leading edge of the plate. The mesh used, with 1051 elements and 581 nodes, is displayed in fig. 6a and the general features of the solution can be appreciated by examining the density contours shown in fig. 6b. Comparison of the computed wall pressure distribution with the results of Carter (1972) is given in fig. 7. Fig. 8 shows profiles of velocity, temperature and density at the outlet section. The agreement between the two solutions is good, with a small discrepancy being apparent in the profiles of u_2 and ρ.

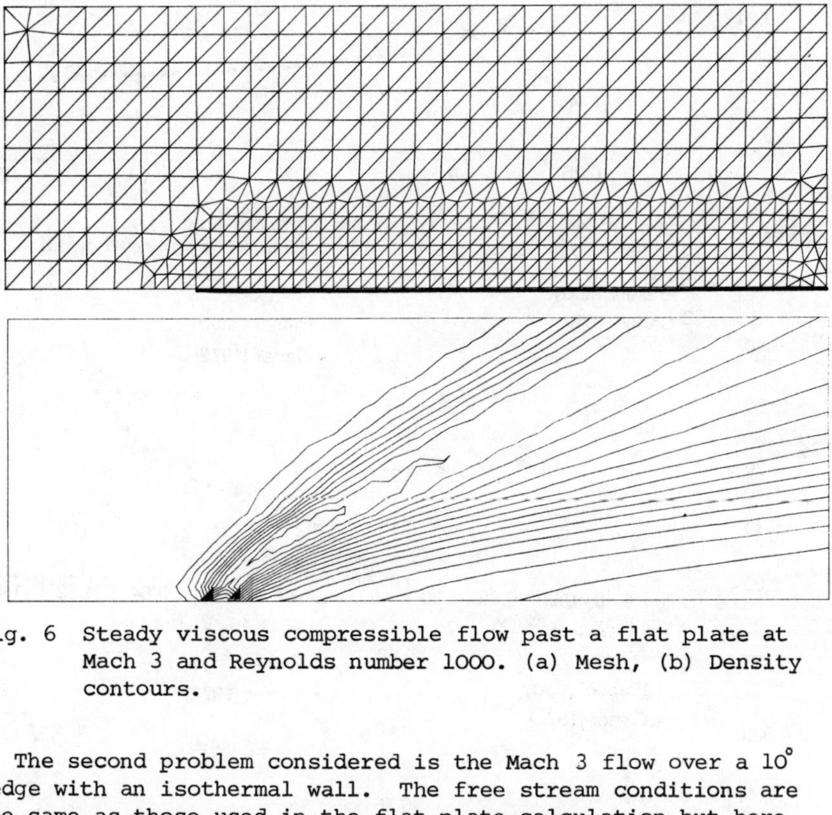

Fig. 6 Steady viscous compressible flow past a flat plate at Mach 3 and Reynolds number 1000. (a) Mesh, (b) Density contours.

The second problem considered is the Mach 3 flow over a 10° wedge with an isothermal wall. The free stream conditions are the same as those used in the flat plate calculation but here the computational domain has been enlarged resulting in a Reynolds number of 1.68×10^4 relative to the distance between the leading edge and the compression corner. The identical problem has also been studied by Carter (1972) and Hung and MacCormack (1976). The problem specification and the mesh used are shown in fig. 9a. The computed velocity vectors and contours for density, pressure and temperature are displayed in fig. 9b-9e.

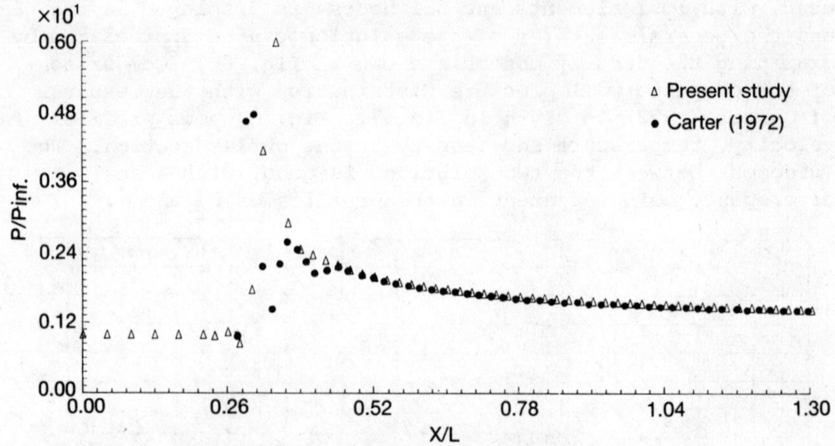

Fig. 7 Computed wall pressure distribution for viscous compressible flow past a flat plate, compared with the results of Carter (1972)

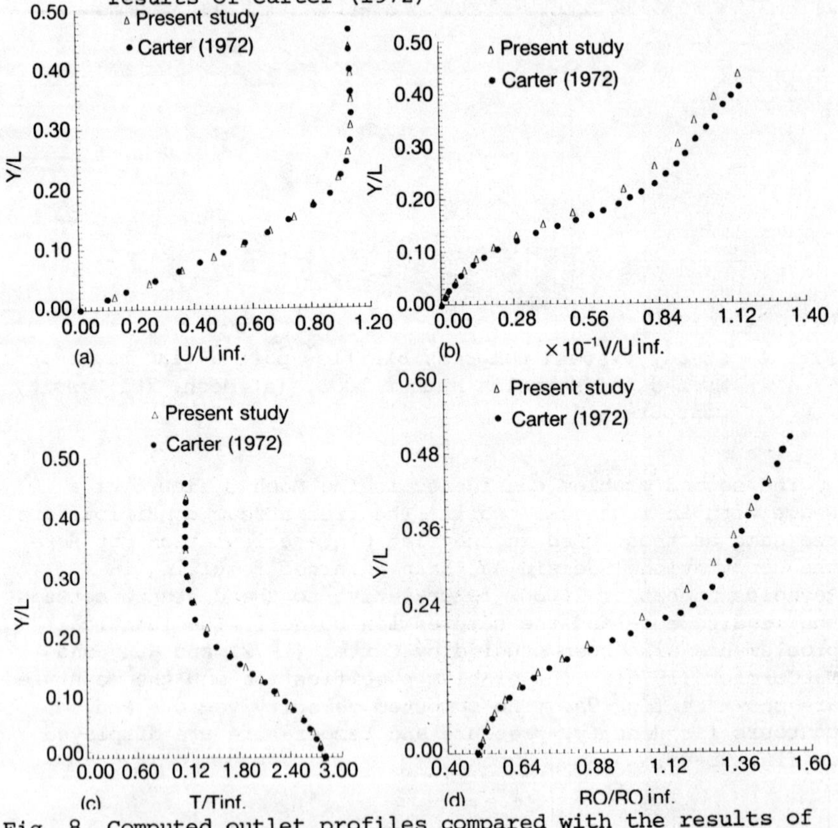

Fig. 8 Computed outlet profiles compared with the results of Carter (1972) for viscous compressible flow past a flat plate (a) velocity parallel to the wall, (b) velocity perpendicular to the wall, (c) temperature, (d) density.

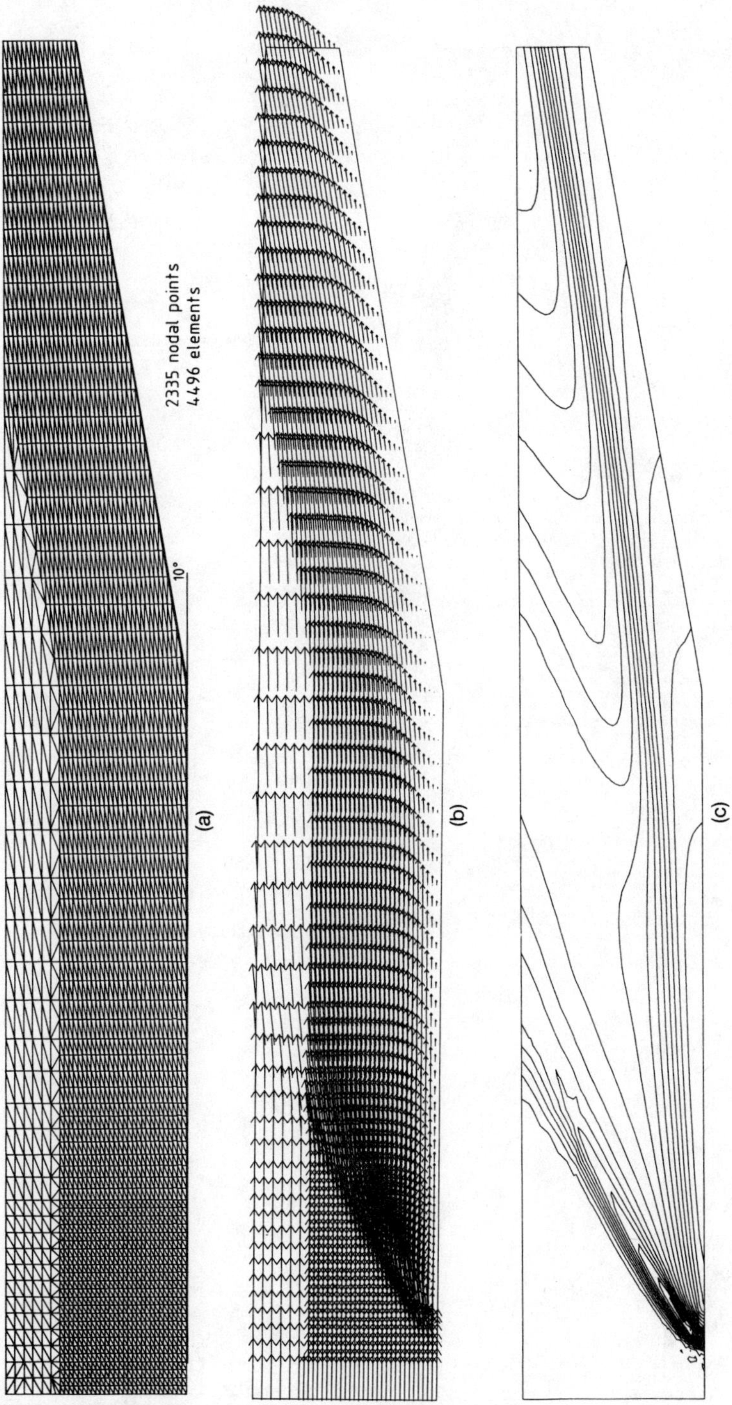
2335 nodal points
4496 elements
10°
(a)
(b)
(c)

Figs. 9a-9c

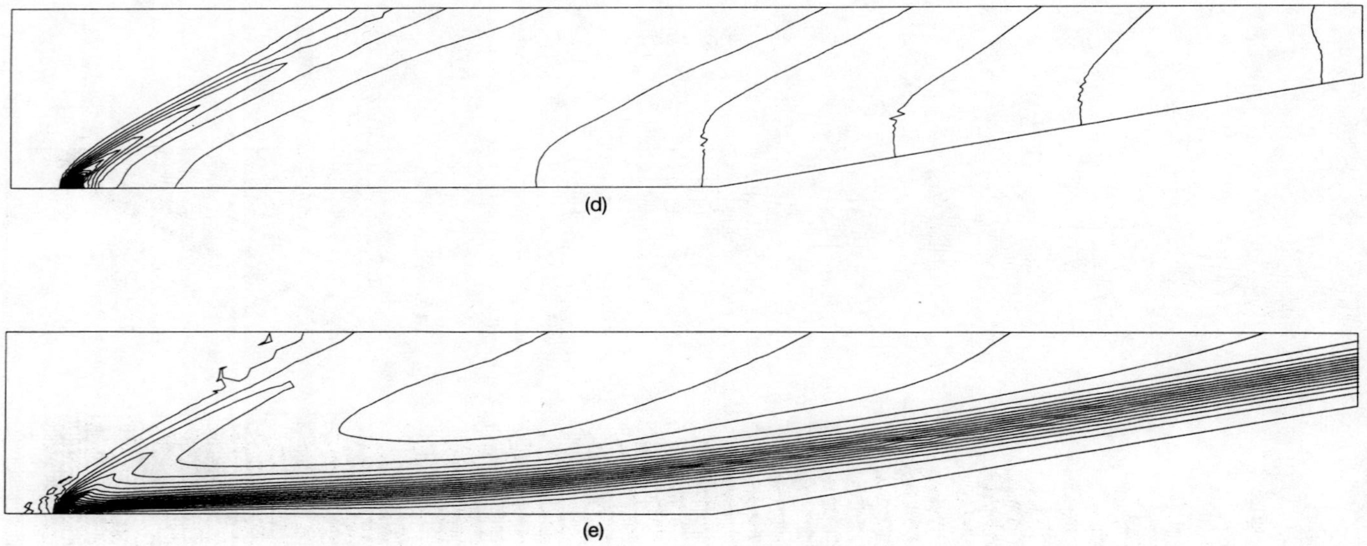

Fig. 9d-9e

Steady viscous compressible flow past a compression corner at Mach 3 and Reynolds number 1.68×10^4.
(a) Mesh, (b) Velocity vectors, (c) Density contours, (d) Pressure contours, (e) Temperature contours.

3.4 Adaptive Refinement for Steady Problems

The refinement procedure outlined previously does not appear to be ideal for viscous flow problems, as it tends to continuously refine shocks, whereas in certain problems we may require a better definition of other flow-field features, such as boundary layers. In this section, therefore, we describe some preliminary attempts at producing refinement indicators which are capable of identifying shocks and boundary layers and so enable these features to be separately refined if so desired.

Fig. 10a shows an idealised situation in the vicinity of a shock and it can be seen that the direction of the vector $\underline{\ell}$ defined as in equation (2.2.2), will be approximately normal to the shock and hence the quantity $\partial u_\ell/\partial \ell$ will be significant near the shock, where $u_\ell = \underline{u}.\underline{\ell}$. This suggests that we replace the requirement of equation (2.5.10) by

$$h_e^2 \left| \frac{\partial u_\ell}{\partial \ell} \right|_e = \text{constant} \qquad (3.4.1)$$

if we want to give preference to shocks in the refinement process.

For boundary layers, the situation is somewhat different and fig. 10b shows an idealised flow in the vicinity of a wall. Now the direction $\underline{\ell}$ is almost normal to the wall, while it is the quantity $\partial u_t/\partial \ell$ which will be significant, where $u_t = \underline{u}.\underline{t}$ and $\underline{t}$ is a unit vector such that $\underline{\ell}.\underline{t} = 0$. A refinement requirement which gives preference to boundary layers might then be

$$h_e^2 \left| \frac{\partial u_t}{\partial \ell} \right|_e = \text{constant} \qquad (3.4.2)$$

To illustrate the performance of these refinement indicators we return to the problem of steady compressible flow past a flat plate considered previously. The solution shown in fig. 6b and obtained on the mesh of fig. 6a is subjected to the requirement of equation (3.4.1). The resulting mesh is shown in fig. 11a, which clearly gives a refinement in the vicinity of the shock. When the same solution is subjected to the requirement of equation (3.4.2) the mesh which results is shown in fig. 11b, in which a portion of the boundary layer has been refined. Employing both criteria simultaneously produces the mesh shown in fig. 11c and the corresponding converged solution for the density on this mesh is displayed in fig. 11d.

Clearly, much remains to be done in this area and improved indicators will no doubt result, but these initial investigations have shown much promise.

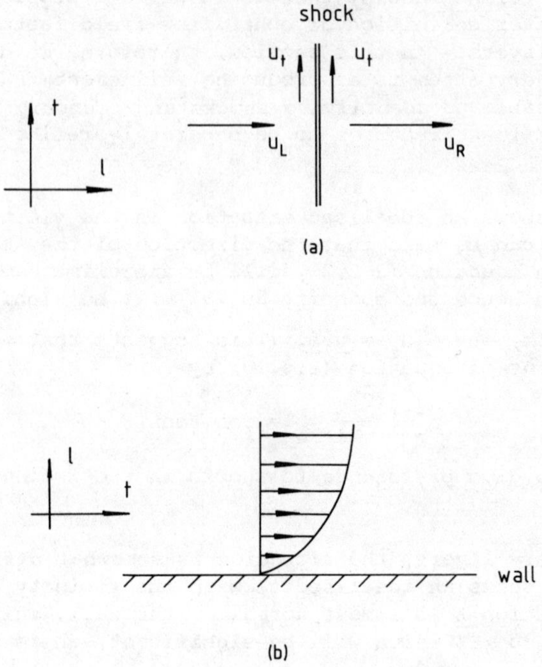

Figure 10 Direction of ℓ and t in the vicinity of (a) a shock and (b) a boundary layer.

Fig. 11a-11d

Steady viscous flow past a flat plate, (a) Refined mesh produced by shock criterion, (b) Refined mesh produced by boundary layer criterion, (c) Refined mesh produced by combining these two, (d) Corresponding density contours.

(figure facing)

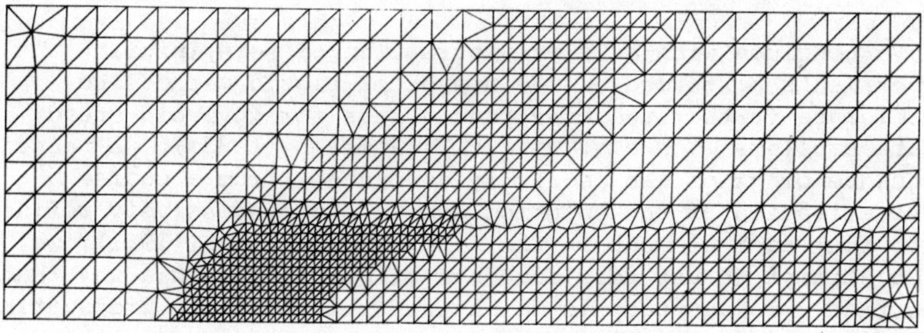

(a)

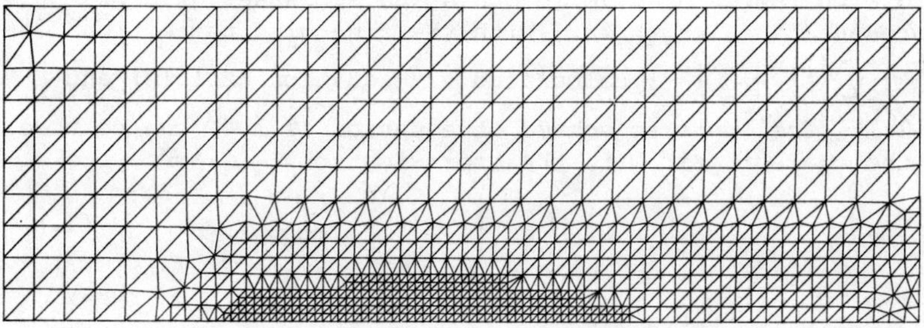

(b)

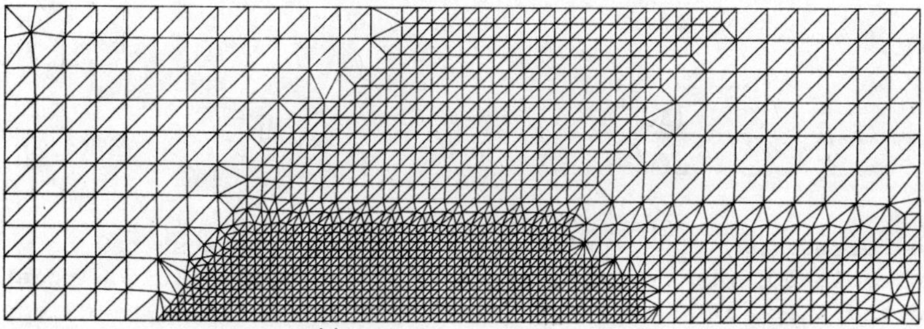

(c)

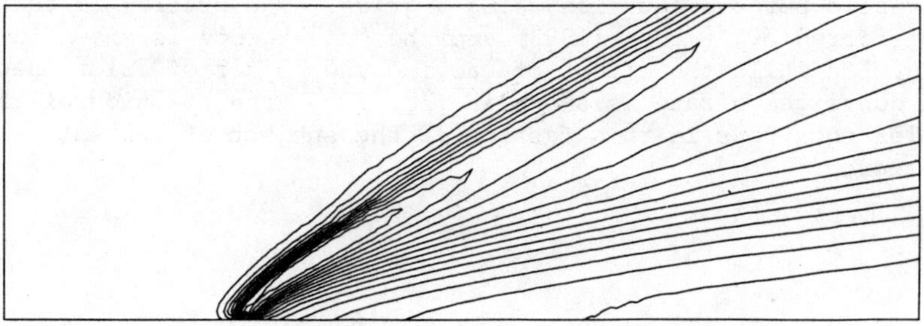

(d)

4. MULTIGRID METHODS

We conclude with a brief description of some initial experiments performed on the use of multigrid methods to accelerate further the convergence rate for inviscid steady state problems (see e.g. Ni(1981), Jameson (1983)). A nested sequence of grids can be easily generated, from a coarse discretisation of the solution domain, by successively refining into four each element in the mesh. Following Ni(1981), we then attempt to use this sequence of grids in such a way that the fine grid accuracy is maintained while rapidly propagating the fine grid corrections through the mesh via the coarser grids. For illustration, consider the use of two grids, a fine grid with representative length h and a coarser grid of representative length 2h. With the fine grid correction $\Delta\underline{U}^h$ $(=\underline{U}_s^{m+1}|^h - \underline{U}^m|^h)$ calculated via equation (2.1.12) and (2.2.1), the one step scheme of equation (2.1.4) can be used on the coarser grid with the approximation

$$- \frac{\partial F_j^m}{\partial x_j}\Bigg|^h \approx \frac{\Delta\underline{U}^h}{\Delta t^h} \tag{4.1}$$

The result is that

$$\int_\Omega \Delta\underline{U}^{2h} N_i^{2h} d\Omega = \frac{\Delta t^{2h}}{\Delta t^h} \int_\Omega \Delta\underline{U}^h N_i^{2h} d\Omega - \frac{(\Delta t^{2h})^2}{2\Delta t^h} \int_\Omega \underline{A}_k^m\Big|^{2h} \Delta\underline{U}^h \frac{\partial N_i^{2h}}{\partial x_k} d\Omega$$

$$+ \frac{(\Delta t^{2h})^2}{2\Delta t^h} \int_\Gamma n_k \underline{A}_k^m\Big|^{2h} \Delta\underline{U}^h N_i^{2h} d\Gamma \tag{4.2}$$

The correction, $\Delta\underline{U}^{2h}$, calculated on the coarse mesh can be interpolated onto the fine mesh, the fine mesh solution updated and the steps repeated until convergence results. The extension to the case of several grids is straightforward. As an example of this scheme in operation, we consider the solution of the inviscid Burgers equation using 5 grids. The problem is that considered by Jameson (1983) and the convergence is shown in fig. 12a while the effect of varying the number of grids used on convergence rate is shown in fig. 12b. The solution of the Euler equations in this fashion is the subject of current research.

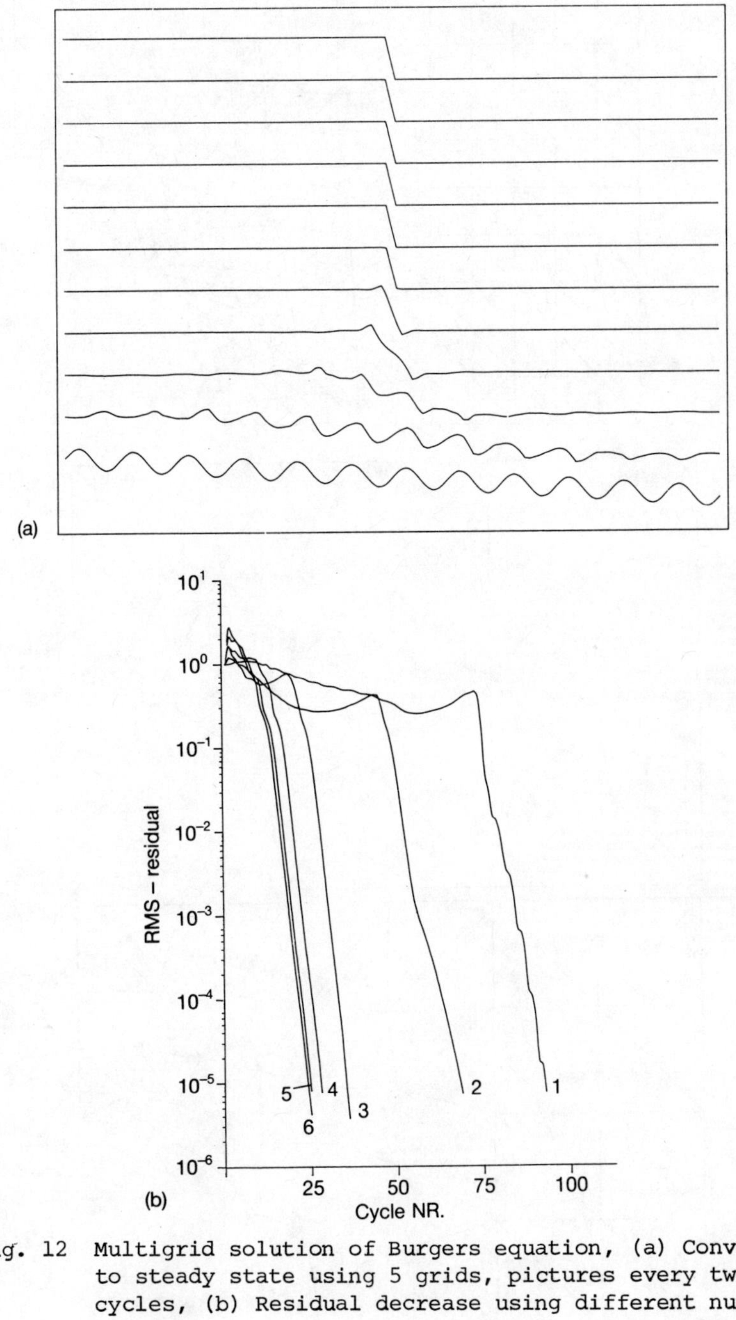

Fig. 12 Multigrid solution of Burgers equation, (a) Convergence to steady state using 5 grids, pictures every two cycles, (b) Residual decrease using different numbers of grids.

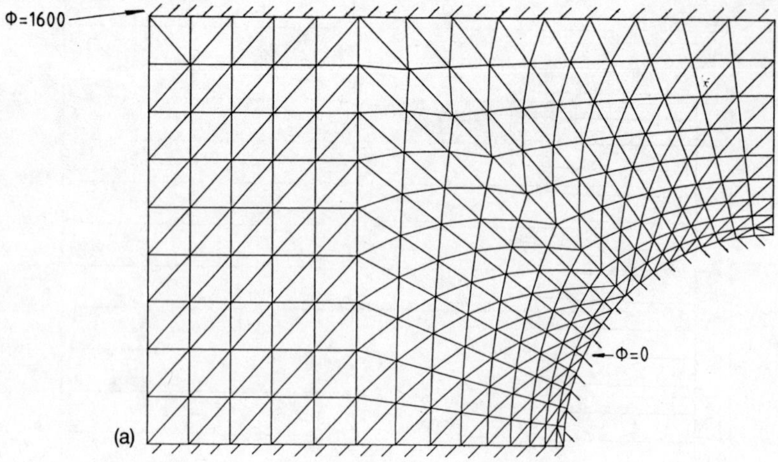

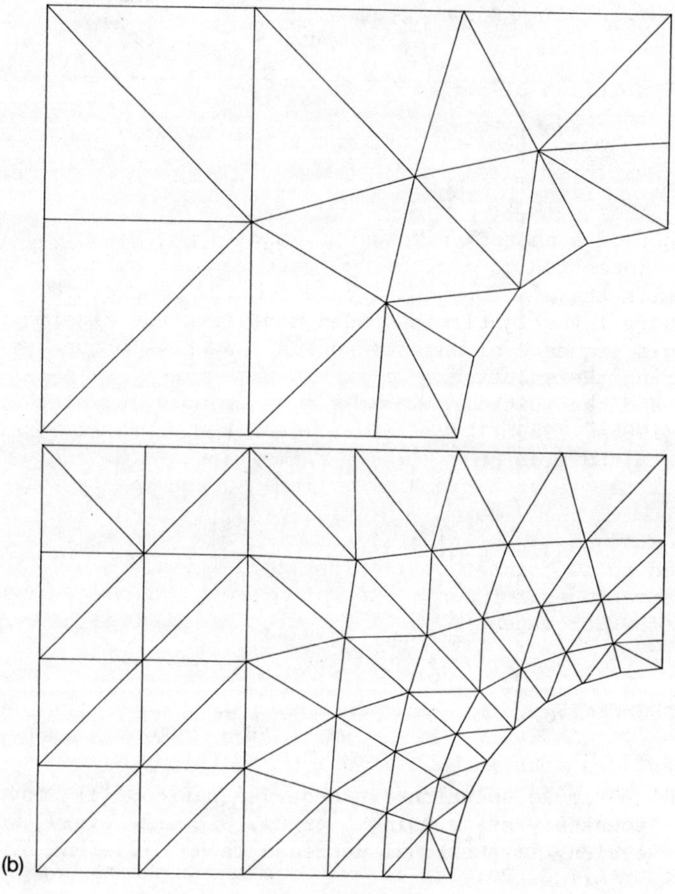

Fig. 13a-13b

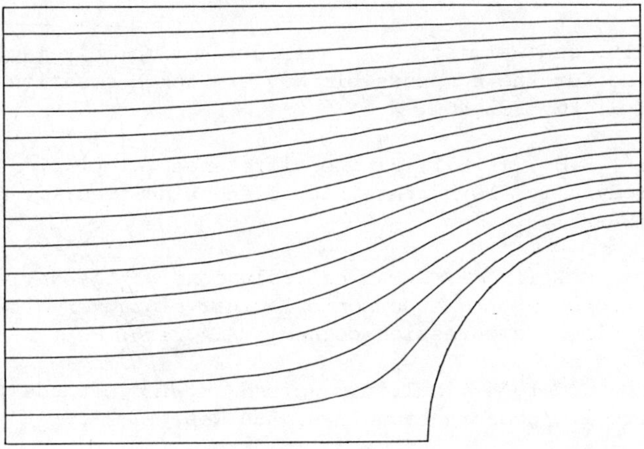

Fig. 13c

Fig. 13 Solution of potential flow past a cylinder using an unnested multigrid procedure, (a) Finest mesh and problem boundary conditions, (b) Coarser meshes employed, (c) Converged solution.

The use of a nested sequence of grids appears to conflict with our intention to accurately represent the solution domain with the finest, unstructured, grid. Therefore, a preliminary investigation has been made into the feasibility of using a sequence of unnested grids. We have begun by considering the solution of elliptic problems in this fashion and the initial results are encouraging (Löhner and Morgan (1985a, 1985b)). Here we demonstrate the procedure applied to the problem of determining the stream function for potential flow past a circular cylinder. The region analysed together with the sequence of unnested grids used is shown in fig. 13a and fig. 13b. The solution shown in fig. 13c was converged to 10^{-6} in 30 work units. This approach clearly offers advantages for finte element based procedures and will receive further attention.

ACKNOWLEDGEMENTS

The authors would like to thank the Aerothermal Loads Branch of the NASA Langley Research Centre for partial support of this research under Grant No. NAGW-478, and especially A.R. Wieting and K.S. Bey for their continued interest and encouragement. J. Peraire ackowledges the support given by the Ministerio de Educacion y Ciencia of Spain.

REFERENCES

Beam, R.M. and Warming, R.F., (1978), An implicit factored scheme for the compressible Navier-Stokes equations, *AIAA J.* **16**, 393-401.

Burstein, S.Z., (1967), Finite difference calculations for hydrodynamic flows containing discontinuities, *J. Comp. Phys.* **2**, 198-222.

Carter, J.E., (1972), Numerical solutions of the Navier-Stokes equations for the supersonic laminar flow over a two-dimensional compression corner, *NASA Tech. Rep.* R-385.

Colella, P., (1984), Multidimensional upwind methods for hyperbolic conservation laws, *LBL Rep.* 17023.

Dannenhoffer, J.F. and Baron, J.R., (1985), Grid adaptation for the 2-D Euler equations, *AIAA*-85-0484.

Donéa, J., (1984), A Taylor-Galerkin method for convective transport problems, *Int. J. Num. Meth. Engng.* **20**, 101-119.

Harten, A., (1982), On second order accurate Godunov-type schemes. *NASA Ames Rep.* NCA2-ors25-201.

Hung, C.M. and MacCormack, R.W., (1976) Numerical solution of supersonic and hypersonic laminar compression corner flows, *AIAA J.* **4**, 475-481.

Jameson, A., (1983), Transonic flow calculations, Princeton University MAE Rep. 1651.

Lapidus, A., (1967), A detached shock calculation by second-order finite differences, *J. Comp. Phys.* **2**, 154-177.

Löhner, R. and Morgan, K., (1984a), Unstructured multigrid methods: first experiences, In Proc. Conference on Numerical Methods in Thermal Problems, Pineridge Press, Swansea.

Löhner, R. and Morgan, K., (1984b), An unstructured multigrid method for elliptic problems , University College of Swansea Rep. C/R/515/85.

Löhner, R., Morgan, K. and Zienkiewicz, O.C., (1984a), The solution of non-linear hyperbolic equation systems by the finite element method, *Int. J. Num. Meth. Fluids* **4**, 1043-1063.

Löhner, R., Morgan, K. and Zienkiewicz, O.C., (1984b), The use of domain splitting with an explicit hyperbolic solver, *Comp. Meth. Appl. Mech. Engng.* **45**, 313-329.

Löhner, R., Morgan, K. and Peraire, J., (1985a), A simple extension to multidimensional problems of the artificial viscosity model due to Lapidus, *Comm. Appl. Num. Meth.* **1**, 141-147.

Löhner, R., Morgan, K. and Zienkiewicz, O.C., (1985b), Adaptive grid refinement for the compressible Euler equations, In Accuracy Estimates and Adaptivity for Finite Elements, Wiley, Chichester (to appear).

Löhner, R., Morgan, K. and Zienkiewicz, O.C., (1985c), An adaptive finite element method for high speed compressible flow. In Lecture Notes in Physics, 218, pp. 388-392 Springer-Verlag, Berlin.

Löhner, R., Morgan, K. and Zienkiewicz, O.C., (1985d), An adaptive finite element procedure for compressible high speed flows. *Comp. Meth. Appl. Mech. Engng.* (to appear).

MacCormack, R.W., (1981), A numerical method for solving the equations of compressible viscous flow. AIAA-81-0110.

Ni, R.H., (1981). A multiple grid scheme for solving the Euler equations, Proc. AIAA 5th Computational Fluid Dynamics Conference, pp. 257-264.

Oden, J.T., (1983). Notes on grid optimisation and adaptive methods for finite element methods. TICOM Report, University of Texas.

Peraire, J., Löhner, R., Morgan, K. and Zienkiewicz, O.C., (1985), A finite element method for high speed inviscid and viscous flow. *Int. J. Num. Meth. Engng.* (to appear).

Strang, G. and Fix, G.J., (1973), An analysis of the finite element method. Prentice-Hall, Englewood Cliffs.

White, F.M., (1974), Viscous fluid flow, McGraw-Hill, New York.

A BASIS FOR UPWIND DIFFERENCING OF THE TWO-DIMENSIONAL UNSTEADY EULER EQUATIONS

P.L. Roe
(College of Aeronautics, Cranfield Institute of Technology)

1. INTRODUCTION

One of the unifying themes suggested by the organisers of this conference is the study of algorithms which adapt themselves to suit the problem being solved. Such an adaptation is the hallmark of upwind-differencing schemes. The theory of such schemes has reached a high level of sophistication in the context of hyperbolic systems of partial differential equations in one space dimension. That is, in seeking numerical solutions to

$$\underline{w}_t + A(\underline{w})\ \underline{w}_x = 0 \tag{1.1}$$

where $\underline{w}$ is a vector of unknowns, and $A(\underline{w})$ is a matrix with a complete set of real eigenvectors. It is well known that each eigenvector corresponds to a pattern of disturbances associated with the passage of waves of a particular family. The various upwind schemes consist of "pattern-recognition" mechanisms which identify these waves in the data, followed by "evolution" mechanisms which use these patterns to update the solution. The design principle which they all invoke is that information should be propagated numerically in the same direction as it would propagate physically. Thus, left- and right-moving waves have to be distinguished and treated differently. It is in this way that the algorithm adapts to the data.

The topic addressed in this paper is the nature of the analogous process in computing two-dimensional flow (and, although we do not treat the details, higher dimensions also). Since wave propagation is then possible in infinitely many directions, we need somehow to select a finite set of directions for special treatment in the algorithm. Loosely expressed, out of the infinite set of possible patterns, we seek the most significant. This difficulty is present even if the equations being studied are linear, so this will be the only case studied here. Our results relate chiefly to the Euler equations which describe inviscid compressible gasdynamics, although a similar analysis for other systems of equations would be possible.

2. ONE-DIMENSIONAL ANALYSIS

To point out the contrast with the two-dimensional case, we must first review the well-known analysis for one-dimensional flow. Since this paper is limited to a linearised analysis, we study

$$\underline{w}_t + A\,\underline{w}_x = 0 \tag{2.1}$$

where A is a constant matrix. In principle, $\underline{w}$ can be chosen to comprise any three independent flow quantities, but certain choices both simplify the algebra, and illuminate the results. We choose $\underline{w} = (p,u,\rho)^T$, where p = pressure, u = velocity, and ρ = density. Then the matrix A is

$$A = \begin{vmatrix} u & \rho a^2 & 0 \\ 1/\rho & u & 0 \\ 0 & p & u \end{vmatrix} \tag{2.2}$$

where a = sound speed = $(\gamma p/\rho)^{\frac{1}{2}}$, γ = ratio of specific heats. The constant quantities which appear in A represent some average state of which the flow is a small perturbation. A simple wave solution of (2.1) is defined to be of the form

$$\underline{w}(x,t) = \underline{w}(x - \lambda t) \tag{2.3}$$

and substituting this into (2.1) gives

$$(A - \lambda I)\,d\underline{w} = 0 \tag{2.4}$$

showing that $d\underline{w}$ is an eigenvector of A, and λ the corresponding eigenvalue. Clearly there are three sorts of simple wave solution, corresponding to the three eigenvectors of (2.2), which are

$$\underline{r}_1 = \begin{vmatrix} \rho a^2 \\ -a \\ \rho \end{vmatrix}, \quad \underline{r}_2 = \begin{vmatrix} 0 \\ 0 \\ \rho \end{vmatrix}, \quad \underline{r}_3 = \begin{vmatrix} \rho a^2 \\ a \\ \rho \end{vmatrix} \tag{2.5}$$

with eigenvalues

$$\lambda_1 = u - a \quad \lambda_2 = a \quad \lambda_3 = u + a \tag{2.6}$$

The complete solution to the initial-value problem (given $\underline{w}(x,0)$, find $\underline{w}(x,t)$ for $t > 0$) is easily written in terms of these basic solutions. The data is uniquely representable as

$$\underline{w}(x,0) = \sum_{k=1}^{k=3} w^{(k)}(x)\, \underline{r}^{(k)} \tag{2.7}$$

where the $w^{(k)}$ are scalar functions of x.

The solution is then found as

$$\underline{w}(x,t) = \sum_{k=1}^{k=3} w^{(k)}(x - \lambda^{(k)} t)\, \underline{r}^{(k)} \tag{2.8}$$

Such a straightforward process cannot be applied to nonlinear equations,but by using techniques based on a local linearisation, successful numerical algorithms (Roe 1981, Harten 1983), employing the above ideas, can be created. These algorithms differ very little between the linear and non-linear cases, so that although in this paper we confine our attention to the linear case, we may be hopeful about making the nonlinear extensions.

The first stage of the algorithm is the "pattern-recognition" stage. From the data $w_i^n = w(i\Delta x, n\Delta t)$, given over some range of i at a particular time level n, we compute the set of 'fluctuations'

$$\underline{\Phi}_{i+\frac{1}{2}} = \underline{w}_{i+1}^n - \underline{w}_i^n \tag{2.9}$$

and find their projections onto the eigenvectors of A.

$$\underline{\Phi}_{i+\frac{1}{2}} = \sum_k \alpha_{i+\frac{1}{2}}^{(k)}\, \underline{r}^{(k)} = \sum \underline{\Phi}_{i+\frac{1}{2}}^{(k)} \tag{2.10}$$

In (2.1.10) the $\{\alpha\}$ are local wavestrengths.

They are easily evaluated as

$$\left.\begin{aligned}\alpha^{(1)}_{i+\frac{1}{2}} &= \frac{1}{2a^2}[\Delta p - \rho a\Delta u]\\ \alpha^{(2)}_{i+\frac{1}{2}} &= \frac{1}{a^2}[a^2\Delta\rho - \Delta p]\\ \alpha^{(3)}_{i+\frac{1}{2}} &= \frac{1}{2a^2}[\Delta p + \rho a\Delta u]\end{aligned}\right\} \quad (2.11)$$

where $\Delta(\cdot) = (\cdot)_{i+1} - (\cdot)_i$.

We can think of these local wave solutions as a set of model flows which are equivalent to the data. Knowing how these model flows would evolve over a short period of time guides our choice of the evolution algorithm. There are at least two possible choices. In one dimension they are conceptually different but lead to identical results. In two dimensions, as we shall see, they are much more distinct.

In one dimension the simplest approach is the 'fluctuation-signal' concept (Roe, 1981, Roe and Baines, 1981, 1983). Each component of the fluctuation is thought of as associated with a signal equal to

$$\frac{\Delta t}{\Delta x}\lambda^{(k)}\ \underline{\Phi}^{(k)}_{i+\frac{1}{2}} = \nu^{(k)}\ \underline{\Phi}^{(k)}_{i+\frac{1}{2}}$$

where $\nu^{(k)}$ is the Courant number of the kth wave. Then if $\nu^{(k)}$ is positive, the signal is subtracted from $\underline{w}_{i+1}$, and if $\nu^{(k)}$ is negative, the signal is subtracted from w_i. When all signals have been despatched, the solution has been updated to time level (n + 1).

An alternative approach is to define a set of interface states $\underline{w}^{n+\frac{1}{2}}_{i+\frac{1}{2}}$, and then to update the solution by means of

$$\underline{w}^{n+1}_i - \underline{w}^n_i = -\frac{A\Delta t}{\Delta x}\left[\underline{w}^{n+\frac{1}{2}}_{i+\frac{1}{2}} - \underline{w}^{n+\frac{1}{2}}_{i-\frac{1}{2}}\right] . \quad (2.12)$$

Note that (2.12) is most usually applied to conservation laws, for which $A\,d\underline{w} = d\underline{F}$, where $\underline{F}$ is a flux vector, so that (2.12) is written

$$\underline{w}_i^{n+1} - \underline{w}_i^{n} = -\frac{\Delta t}{\Delta x}\left[\underline{F}_{i+\frac{1}{2}}^{n+\frac{1}{2}} - \underline{F}_{i-\frac{1}{2}}^{n+\frac{1}{2}}\right] . \tag{2.13}$$

In the present instance, we are working chiefly with non-conservative variables, so that no $\underline{F}$ exists, but (2.12) can still be used. To use the model flow to find $\underline{w}_{i+\frac{1}{2}}^{n}$, it is usually assumed (e.g. Harten 1983) that the states $\underline{w}_i^n$, $\underline{w}_{i+1}^n$ are constant either side of $x = (i+\frac{1}{2})\Delta x$ at $t = n\Delta t$, so that the flow coincides locally with the solution to a Riemann problem. On the interface $x = (i+\frac{1}{2})\Delta x$, for $t > n\Delta t$, the state is a constant one which differs from $\underline{w}_i$ by virtue of the left-running waves, and from $\underline{w}_{i+1}$ by virtue of the right-running waves. We can easily show in fact that

$$\underline{w}_{i+\frac{1}{2}}^{n+\frac{1}{2}} = \tfrac{1}{2}(\underline{w}_i^n + \underline{w}_{i+1}^n) - \tfrac{1}{2}\sum_k (\mathrm{sgn}\lambda^{(k)})\ \alpha_{i+\frac{1}{2}}^{(k)}\ \underline{r}^{(k)} . \tag{2.14}$$

With this definition of the interface states, it will be found that the algorithm (2.12) gives exactly the same result as the 'fluctuation-signal' algorithm.

Starting with these basic ideas, the one-dimensional algorithms can be refined considerably. When the problem is nonlinear, the $\underline{r}^{(k)}$ and $\lambda^{(k)}$ are no longer constant, but must be evaluated within each interval (Roe, 1981). The schemes can be modified to achieve higher-order accuracy without sacrificing monotone behaviour at discontinuities (Harten, 1983, Roe and Pike, 1984). Versions exist which can be guaranteed to produce the physically-correct (i.e. entropy-satisfying) alternative, whenever ambiguous solutions arise (Osher, 1984, Tadmor, 1985).

In the present paper, however, we wish to avoid these complexities, concentrating instead on a more basic issue which arises when we attempt to apply the upwind concept to flow in two (or more) space dimensions. This is the problem of ensuring that the numerical processes really do imitate accurately the richer variety of flow phenomena to be found in two dimensions. Difficulties associated with the pattern-recognition stage are discussed in Section 3, and with the evolution stage in section 4.

3. TWO-DIMENSIONAL ANALYSIS

3.1 Simple wave flows

To analyse the two-dimensional problem, we retain the same variables, but add a second velocity component, so that $\underline{w} = (p,u,v,\rho)^T$ and the equations of motion become

$$\underline{w}_t + A\,\underline{w}_x + B\,\underline{w}_y = 0 \tag{3.1}$$

where

$$A = \begin{vmatrix} u & \rho a^2 & 0 & 0 \\ 1/\rho & u & 0 & 0 \\ 0 & 0 & u & 0 \\ 0 & \rho & 0 & u \end{vmatrix}, \qquad B = \begin{vmatrix} v & 0 & \rho a^2 & 0 \\ 0 & v & 0 & 0 \\ 1/\rho & 0 & v & 0 \\ 0 & 0 & \rho & v \end{vmatrix} \tag{3.2}$$

Simple wave solutions of (3.1) are of the form

$$\underline{w}(x,y,t) = \underline{w}(x\cos\theta + y\sin\theta - \lambda t) \tag{3.3}$$

which implies that

$$(A\cos\theta + B\sin\theta - \lambda I)\,d\underline{w} = 0\ . \tag{3.4}$$

Hence $d\underline{w}$ is an eigenvector of $(A\cos\theta + B\sin\theta)$ and λ is its eigenvalue. (For more detail on simple wave solutions in the non-linear case, the reader may consult the recent monograph by Majda, 1984).

The eigenvectors of $(A\cos\theta + B\sin\theta)$ are of three kinds. Acoustic disturbances are represented by

$$\underline{r}_a(\theta) = \begin{vmatrix} \rho a^2 \\ a\cos\theta \\ a\sin\theta \\ \rho \end{vmatrix} \tag{3.5}$$

with eigenvalue

$$\lambda_a(\theta) = u\cos\theta + v\sin\theta + a\ . \tag{3.6}$$

'Entropy waves', across which the pressure and velocity do not change, are represented by

$$\underline{r}_e = \begin{vmatrix} 0 \\ 0 \\ 0 \\ \rho \end{vmatrix} \tag{3.7}$$

with eigenvalue

$$\lambda_e = u \cos\theta + v \sin\theta \ . \tag{3.8}$$

Finally, 'shear waves' across which the tangential component of velocity changes, and which have no analogue in one-dimensional flow, are represented by

$$\underline{r}_s = \begin{vmatrix} 0 \\ -a \sin\theta \\ a \cos\theta \\ 0 \end{vmatrix} \tag{3.9}$$

with the same eigenvalue (3.8), i.e. $\lambda_s = \lambda_e$.

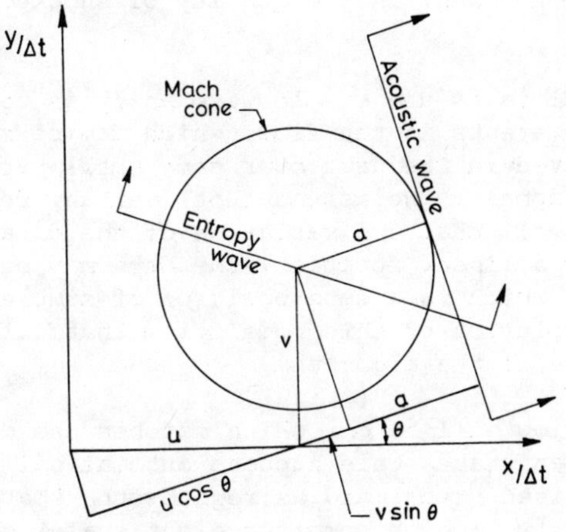

Fig. 1 The geometry of plane unsteady waves in two-dimensional flow.

The geometrical interpretation of the eigenvalue is shown in Fig. 1. Disturbances which pass through x = 0, y = 0, at t = 0 with constant velocity can be displayed in coordinates (x/t, y/t). On such a diagram the fluid velocity is the point (u,v) and the Mach circle has centre (u,v) and radius a. An arbitrary tangent to that circle represents an acoustic wave moving with speed $\lambda_a(\theta)$. An arbitrary line through the centre may represent either an entropy wave or a shear wave, moving with speed λ_e.

Clearly, the search for simple wave solutions has much in common with the search for characteristic surfaces (Courant and Friedrichs, 1954). Indeed, a plane in (x,y,t) which can be a surface of constant state in a simple wave is also a plane within which a characteristic equation can be written. There is, however, a crucial distinction. Characteristic analysis is concerned only with the equations of motion, and tells us only that certain paths may be used to transmit information. It does not tell us which paths are actually being used, and to obtain that information we need both the differential equation and the data to which it is applied. Since the number of possible paths is infinite, corresponding to the arbitrary tangents and diameters in Fig. 1, some restriction is needed if the theory is to be of assistance in numerical work. Davis (1984) sought a single dominant direction which would indicate the presence of a shockwave, and obtained an encouraging improvement in the quality of shock-dominated calculations.

Our proposal is to use simple wave analysis to provide a local model of events in the flow, which may be matched to patterns observed in the data over some small part of the computational domain. We assume that over any region sufficiently small that the variation of the data is accurately represented by a linear function, the data may be replaced by a 'model flow' which is a superposition of simple waves. It will be the evolution of this model flow that guides the evolution stage of the algorithm.

The model flow will be one which matches the data both as regards its mean state (this happens automatically in the present linearised study) and as regards the spatial derivatives $\underline{w}_x$, $\underline{w}_y$ (which comprise eight scalar quantities). The model flow must therefore contain eight free parameters, but it is not obvious what set of parameters should be chosen. The following properties seem natural requirements:

(i) If the data are sampled from a flow which is one-dimensional in some direction other than the coordinate directions, the model flow shall be the one-dimensional flow in that direction. This requirement recognizes that two-dimensional flows contain one-dimensional features (such as shocks) and asks that our description of them shall be, in a sense, coordinate-free.

(ii) Whatever data are provided, the process of matching the model to them must be fast and direct. This remark is made because it invalidates many previous, unpublished efforts by the author where the parameters of the model turned on solving a cubic equation, or worse. Even if the use of such models improved the quality of the results, they would be too expensive for use in practical codes.

(iii) The parameters of the model must be real-valued for arbitrary data. This obvious requirement has also been a source of past failures. For example, a natural model might be composed of two acoustic waves, one entropy wave, and one shear wave, which is indeed how any steady supersonic flow will appear. Matching the parameters of this model to arbitrary data (as is done in Section 3.3) leads to a set of eight equations whose solution depends on the roots of a quadratic equation. Unfortunately it is easy to construct data for which real roots do not exist.

3.2 Operator splitting

The simplest possible model flow is one which is merely the sum of two one-dimensional flows in the coordinate directions. Formally, we set $\theta = 0$ or $\theta = \frac{1}{2}\pi$ in (3.5) to (3.9) to obtain eigenvectors

$$\underline{r}_a^{(x\pm)} = \begin{vmatrix} \rho a^2 \\ \pm a \\ 0 \\ \rho \end{vmatrix}, \quad \underline{r}_e^{(x)} = \begin{vmatrix} 0 \\ 0 \\ 0 \\ \rho \end{vmatrix}, \quad \underline{r}_s^{(x)} = \begin{vmatrix} 0 \\ 0 \\ a \\ 0 \end{vmatrix}. \qquad (3.10)$$

If we set $\theta = \pm \pi/2$, we obtain

$$\underline{r}_a^{(y\pm)} = \begin{vmatrix} \rho a^2 \\ 0 \\ \pm a \\ \rho \end{vmatrix}, \quad \underline{r}_e^{(y)} = \begin{vmatrix} 0 \\ 0 \\ 0 \\ \rho \end{vmatrix}, \quad \underline{r}_s^{(y)} = \begin{vmatrix} 0 \\ a \\ 0 \\ 0 \end{vmatrix}. \qquad (3.11)$$

These are merely the one-dimensional wave patterns, with the additional feature of shear waves. For example, a discontinuity in v, i.e., $(\underline{r}_s^{(x)})$ appears as a vertical line propagating in the x-direction with velocity u. One way to generate a model flow is to assume that the x-gradients are accounted for by (3.10) and the y-gradients by (3.11). If we have data at the four points ABCD as shown, we may approximate

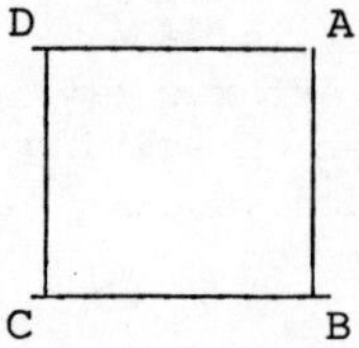

$$\underline{w}_x = \frac{1}{2\Delta x}(\underline{w}_A + \underline{w}_B - \underline{w}_C - \underline{w}_D) \; , \qquad (3.12)$$

$$\underline{w}_y = \frac{1}{2\Delta y}(\underline{w}_A - \underline{w}_B - \underline{w}_C + \underline{w}_D) \; , \qquad (3.13)$$

and set

$$\underline{w}_x = \alpha_1\underline{r}_a^{(x-)} + \alpha_2\underline{r}_e^{(x)} + \alpha_3\underline{r}_s^{(x)} + \alpha_4\underline{r}_a^{(x+)} \; , \qquad (3.14)$$

$$\underline{w}_y = \beta_1\underline{r}_a^{(y-)} + \beta_2\underline{r}_e^{(y)} + \beta_3\underline{r}_s^{(y)} + \beta_4\underline{r}_a^{(y+)} \; . \qquad (3.15)$$

In effect, this is what is done in many existing codes which apply upwind methods to multidimensional problems (Sells (1980), Lytton (1984), Woodward and Colella (1983)), but it does not meet our first requirement . Suppose that the data ABCD is taken from a region across which an acoustic wave is propagating at an angle θ. Both $\underline{w}_x$ and $\underline{w}_y$ will be multiples of $\underline{r}_a(\theta)$ (see eqn (3.5)).

The calculation (3.14) will result in

$$\underline{w}_x = k\begin{vmatrix} \rho a^2 \\ a\cos\theta \\ a\sin\theta \\ \rho \end{vmatrix} = \tfrac{1}{2}k(1-\cos\theta)\begin{vmatrix} \rho a^2 \\ -a \\ 0 \\ \rho \end{vmatrix} + k\sin\theta\begin{vmatrix} 0 \\ 0 \\ a \\ 0 \end{vmatrix}$$

$$+ \tfrac{1}{2}k(1+\cos\theta)\begin{vmatrix} \rho a^2 \\ a \\ 0 \\ \rho \end{vmatrix} \qquad (3.16)$$

and will indicate non-zero strengths for three waves whose velocities are $u - a$, u, $u + a$, instead of a single wave with velocity $(u \cos\theta + v \sin\theta + a)$. This will frequently result in spurious signals being sent in inappropriate directions.

We conclude this section with the remark that the above objection to operator-splitting would be totally lacking in force if the matrices A, B in (3.1) happened to commute. It would then follow (Strang, 1980, p. 193) that A and B shared the same set of eigenvectors, so that the pattern-recognition problem would then be the same in any direction, since $A \cos\theta + B \sin\theta$ will also have the same eigenvectors. In practice, A and B seldom do commute. In the present case of the Euler equations, the sets (3.10), (3.11) have only one common element (unless $a = 0$). It is worth bearing this point in mind, since theoretical studies of operator-splitting (e.g. Crandall and Majda, 1980) have considered mostly scalar problems. Even in that case, non-linearity introduces difficulties, but the additional complications due to lack of commutativity cannot be revealed.

3.3 Alternative models

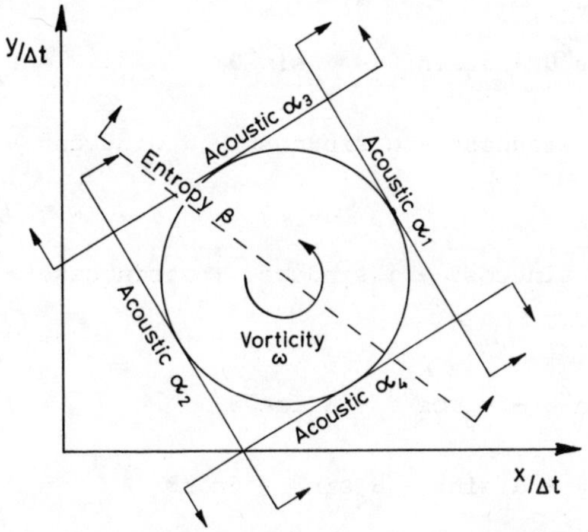

Fig. 2 The waves used to generate the discrete flow model.

To avoid the undesirable features of operator splitting, it seems necessary to leave at least some of the wave directions unspecified, as part of the eight free parameters of the model. The following model, taken from (Roe, 1985) has useful properties. The acoustic part of the fluctuations in the data is represented by four plane waves (3.5) travelling orthogonally to each other (see Fig. 2). The four amplitudes of these waves ($\alpha_1,\alpha_2,\alpha_3,\alpha_4$) and the one independent direction (θ) contribute five parameters to the model. An entropy wave (3.7) of unknown amplitude β and the direction ϕ contributes two more, and the one remaining parameter can be taken as an unknown vorticity, ω. Note that on the small scale where all variations appear linear, uniform vorticity cannot be distinguished from the effect of two shear waves, of equal strength, intersecting at right angles, but otherwise of arbitrary orientation. This observation allows the effects of vorticity on the evolution of the flow to be handled numerically in the same way as the effects of acoustic and entropy waves.

If the flow gradients due to each assumed component of the model are added together and equated to the gradients observed in the data, the following system of eight equations is obtained:

$$\alpha_1\cos\theta + \alpha_2\cos\theta - \alpha_3\sin\theta - \alpha_4\sin\theta = P_x\,, \tag{3.17a}$$

$$\alpha_1\sin\theta + \alpha_2\sin\theta + \alpha_3\cos\theta + \alpha_4\cos\theta = P_y\,, \tag{3.17b}$$

$$\alpha_1\cos^2\theta - \alpha_2\cos^2\theta + \alpha_3\sin^2\theta - \alpha_4\sin^2\theta = U_x\,, \tag{3.17c}$$

$$\alpha_1\sin\theta\cos\theta - \alpha_2\sin\theta\cos\theta - \alpha_3\sin\theta\cos\theta + \alpha_4\sin\theta\cos\theta - \tfrac{1}{2}\omega/a = U_y\,, \tag{3.17d}$$

$$\alpha_1\sin\theta\cos\theta - \alpha_2\sin\theta\cos\theta - \alpha_3\sin\theta\cos\theta + \alpha_4\sin\theta\cos\theta + \tfrac{1}{2}\omega/a = V_x\,, \tag{3.17e}$$

$$\alpha_1\sin^2\theta - \alpha_2\sin^2\theta + \alpha_3\cos^2\theta - \alpha_4\cos^2\theta = V_y\,, \tag{3.17f}$$

$$\alpha_1\cos\theta + \alpha_2\cos\theta - \alpha_3\sin\theta - \alpha_4\sin\theta + \beta\cos\phi = R_x\,, \tag{3.17g}$$

$$\alpha_1\sin\theta + \alpha_2\sin\theta + \alpha_3\cos\theta + \alpha_4\cos\theta + \beta\sin\phi = R_y\,. \tag{3.17h}$$

Here, the gradients are nondimensionalised as follows

$$P_x = \frac{1}{\rho a^2}\frac{\partial p}{\partial x}\,,\quad U_x = \frac{1}{a}\frac{\partial u}{\partial x}\,,\quad R_x = \frac{1}{\rho}\frac{\partial \rho}{\partial x}\,,\ \text{etc.}$$

Note that these equations are linear in the six unknowns $(\alpha_1,\alpha_2,\alpha_3,\alpha_4,\beta,\omega)$ but non-linear in the two unknowns (θ,ϕ). Such non-linearity is the inevitable consequence of admitting unknown directions into the model. In principle, it means that a set of real-valued parameters matching the data may not exist, or may fail to be unique. Actually, equations (3.17) do have an extremely simple explicit solution which gives real, unique, and physically significant, parameters for any data. It seems that the assumption of orthogonality of the acoustic waves is crucial in producing equations for which a tidy solution is possible.

We have immediately

$$\omega = a(V_x - U_y) \tag{3.18}$$

and

$$\beta\cos\phi = R_x - P_x , \tag{3.19a}$$

$$\beta\sin\phi = R_y - P_y , \tag{3.19b}$$

whence β,ϕ. Thus the terms which model the vorticity and entropy effects emerge simply and clearly. The next step is to solve for θ. If (3.17f) is subtracted from (3.17c), and (3.17d) is added to (3.17e), we find,

$$\tan 2\theta = \frac{V_x + U_y}{U_x - V_y} . \tag{3.20}$$

This equation has an encouraging kinematic interpretation. Consider the fluid particles which at some instant form a small circle of radius ε around some arbitrary point. Relative to that point, their coordinates are

$$\{\varepsilon\cos\psi, \quad \varepsilon\sin\psi\} .$$

After a small time Δt, they have moved to new coordinates

$$\{\varepsilon\cos\psi + (u + \varepsilon\cos\psi u_x + \varepsilon\sin\psi u_y)\Delta t,$$
$$\varepsilon\sin\psi + (v + \varepsilon\cos\psi v_x + \varepsilon\sin\psi v_y)\Delta t\}.$$

Relative to the point $(u\Delta t, v\Delta t)$, their new configuration is a curve whose polar equation is:

$$r^2(\psi) = \varepsilon^2[((1 + u_x\Delta t)\cos\psi + u_y\Delta t\sin\psi)^2 +$$

$$((1 + v_y\Delta t)\sin\psi + v_x\Delta t\cos\psi)^2] . \quad (3.21)$$

To first order in Δt, this is an ellipse:-

$$r^2(\psi) = \varepsilon^2[1 + 2(u_x\cos^2\psi + (u_y + v_x)\sin\psi\cos\psi + v_y\sin^2\psi)\Delta t] . \quad (3.22)$$

The principal axes of this ellipse are found by setting $dr/d\psi = 0$, which yields eqn (3.20). Therefore the acoustic waves produced by the model are aligned with the directions of maximum and minimum strain rates in the fluid. It is also relevant to compute what those extremal strain rates are. Inserting (3.20) into (3.22) gives

$$r^2_{max/min} = \varepsilon^2[1 + (u_x - v_y \pm aR)\Delta t] \quad (3.23)$$

where

$$a^2R^2 = (v_x + u_y)^2 + (u_x - v_y)^2 . \quad (3.24)$$

These expressions appear when we compute the amplitudes of the acoustic waves. Rewrite (3.17c,f) as:

$$(\alpha_1 - \alpha_2)\cos^2\theta + (\alpha_3 - \alpha_4)\sin^2\theta = U_x , \quad (3.25a)$$

$$(\alpha_1 - \alpha_2)\sin^2\theta + (\alpha_3 - \alpha_4)\cos^2\theta = V_y , \quad (3.25b)$$

and solve for $(\alpha_1 - \alpha_2)$, $(\alpha_3 - \alpha_4)$. Thus

$$\alpha_1 - \alpha_2 = \frac{U_x\cos^2\theta - V_y\sin^2\theta}{\cos^2\theta - \sin^2\theta}$$

$$= \frac{\frac{1}{2}U_x(1 + \cos 2\theta) - \frac{1}{2}V_y(1 - \cos 2\theta)}{\cos 2\theta}$$

$$= \tfrac{1}{2}[U_x + V_y + R] . \quad (3.26a)$$

Similarly,

$$\alpha_3 - \alpha_4 = \tfrac{1}{2}[U_x + V_y - R] . \quad (3.26b)$$

Compare these results with (3.23). To complete the analysis we find expressions for $\alpha_1 + \alpha_2$, $\alpha_3 + \alpha_4$, using (3.17a,b):

$$\alpha_1 + \alpha_2 = P_x \cos\theta + P_y \sin\theta \ , \qquad (3.27a)$$

$$\alpha_3 + \alpha_4 = P_y \cos\theta - P_x \sin\theta \ . \qquad (3.27b)$$

The R.H.S. of these equations are simply the pressure gradients along the principal strain axes. All the parameters of the model flow therefore relate very simply to quantities of established significance in fluid dynamics.

It is apparent from the construction of the model that if the flow were locally one-dimensional, comprising two acoustic waves and an entropy wave moving in the same direction, then that flow would be exactly reproduced. This is a big improvement over the operator-splitting model. However, a shear wave (3.9) is not correctly recognised. In the general case, with the wave inclined at an angle δ, the observed disturbance would consist of velocity gradients

$$U_x = -k \sin\delta\cos\delta \ ,$$

$$U_y = -k \sin^2\delta \ ,$$

$$V_x = k \cos^2\delta \ ,$$

$$V_y = k \sin\delta\cos\delta \ ,$$

and inserting these into (3.18), (3.20), (3.24) and (3.26) would yield

$$\omega = R = k \ ,$$

$$\theta = \delta - \pi/4 \ ,$$

$$\alpha_1 = \alpha_3 = -\alpha_2 = -\alpha_4 = k/4 \ .$$

The disturbance would therefore be seen as vorticity, together with four acoustic waves of equal strength, inclined at 45° to the direction of shear. It would be a little fanciful to associate this representation with the sound which is radiated from shear layers as a consequence of their instability. Nevertheless, the model just described has a far better chance of describing the flow realistically than has the operator-split model.

4. TWO-DIMENSIONAL EVOLUTION

4.1 Preliminary remarks

Just as the pattern-recognition half of an upwind algorithm requires new properties when deployed in more than one dimension, so the evolution part also raises new and unexpected problems. These can be illustrated in a setting even simpler than the linearised Euler equations. We consider the equation of uniform linear advection

$$u_t + au_x + bu_y = 0 \tag{4.1}$$

where u is a scalar unknown, and a, b are constants. The one-dimensional version (b = 0) of (4.1) has been an unending source of inspiration for methods which can be applied to more complex one-dimensional problems, but (4.1) itself has been less fruitful. In this section we explore some possible reasons.

Note that we must restrict ourselves to solving (4.1) by methods which hold some promise of being extended to more general problems. For example, the operator-splitting technique is an excellent way of solving (4.1) (whose 'matrices' a, b clearly commute), but we reject it because of the objections in Section 3. We also restrict ourselves to seeking only the simplest first-order scheme, analogous to that described at the end of Section 2. The supposition is that once a satisfactory first-order scheme is available, higher-order extensions can then be made.

4.2 Strategies for linear advection

Equation (2.9) is a measure of the lack of equilibrium within a one-dimensional interval [i, i + 1]. An analogous method within a two-dimensional 'interval' [i, i + 1] x [j, j + 1] (square cell ABCD) would be

$$Q_{i+\frac{1}{2},j+\frac{1}{2}} = \frac{A}{2\Delta x}(\underline{w}_A + \underline{w}_B - \underline{w}_C - \underline{w}_D) + \frac{B}{2\Delta y}(\underline{w}_A - \underline{w}_B - \underline{w}_C + \underline{w}_D) \,. \tag{4.2}$$

In a time Δt this requires the solution to change (somewhere) by $Q_{i+\frac{1}{2},j+\frac{1}{2}}\Delta t$, and this change, for the linear advection equation (4.1) would be

$$\delta u = Q_{i+\frac{1}{2},j+\frac{1}{2}}\Delta t = \tfrac{1}{2}(\nu_1 + \nu_2)(u_A - u_C) + \tfrac{1}{2}(\nu_1 - \nu_2)(u_B - u_D) \tag{4.3}$$

where

$$\nu_1 = A\Delta t/\Delta x \ , \quad \nu_2 = B\Delta t/\Delta y \ .$$

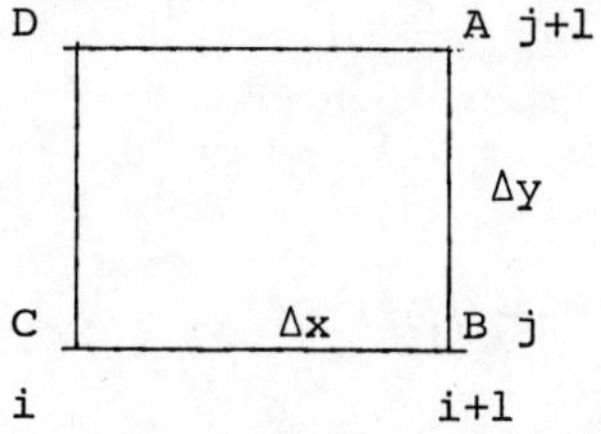

Now let this change be divided into four parts with weights $\alpha,\beta,\gamma,\delta$ where $\alpha + \beta + \gamma + \delta = 1$, so that $\alpha\delta u$ would be subtracted from u_A, $\beta\delta u$ from u_B, etc. This seems to be a natural analogue of the upwind scheme described for one dimension in Section 2. We may imagine, for example, choosing $(\alpha,\beta,\gamma,\delta)$ according to the signs of (ν_1,ν_2) so as to create an upwind scheme. Now a property which enables a scheme to represent discontinuous solutions without creating spurious oscillations is monotonicity, defined as follows. Let $u^{n+1}_{i,j}$ be computed as some function of various u values at time level n, i.e.

$$u^{n+1}_{i,j} = H(h_1, h_2, h_3 \ . \ . \ . \ h_N) \tag{4.4}$$

where each argument of the function H is $u^n_{p,q}$ for some (p,q).

Then the scheme is monotone if, for all k,

$$\frac{\partial H}{\partial h_k} \geqslant 0 \ . \tag{4.5}$$

<u>Theorem</u> There is no set of constant weights $(\alpha,\beta,\gamma,\delta)$ which yields a monotone algorithm, consistent with the differential equation.

<u>Proof</u> In the general case with $\alpha,\beta,\gamma,\delta$ all non-zero, $u^{n+1}_{i,j}$ will depend on nine items of data. With notation as in the diagram, taking

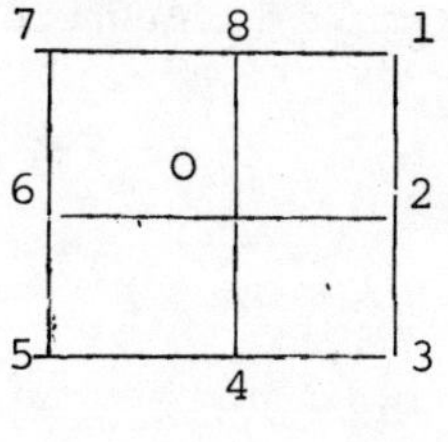

$u_{i,j} = u_o$, we have

$$u^{n+1}_o = \sum_{k=0}^{k=8} c_k u^n_k \tag{4.6}$$

where the c_k are easily found to be

$$
\left.
\begin{aligned}
c_o &= 1 - \tfrac{1}{2}\alpha(\nu_1 + \nu_2) + \tfrac{1}{2}\beta(\nu_2 - \nu_1) + \tfrac{1}{2}\gamma(\nu_1 + \nu_2) + \tfrac{1}{2}\delta(\nu_1 - \nu_2) \\
c_1 &= -\tfrac{1}{2}\gamma(\nu_1 + \nu_2) \\
c_2 &= \tfrac{1}{2}\gamma(\nu_2 - \nu_1) - \tfrac{1}{2}\delta(\nu_1 + \nu_2) \\
c_3 &= \tfrac{1}{2}\delta(\nu_2 - \nu_1) \\
c_4 &= \tfrac{1}{2}\alpha(\nu_2 - \nu_1) + \tfrac{1}{2}\delta(\nu_1 + \nu_2) \\
c_5 &= \tfrac{1}{2}\alpha(\nu_1 + \nu_2) \\
c_6 &= \tfrac{1}{2}\alpha(\nu_1 - \nu_2) + \tfrac{1}{2}\beta(\nu_1 + \nu_2) \\
c_7 &= \tfrac{1}{2}\beta(\nu_1 - \nu_2) \\
c_8 &= -\tfrac{1}{2}\beta(\nu_1 + \nu_2) + \tfrac{1}{2}\gamma(\nu_1 - \nu_2).
\end{aligned}
\right\} \qquad (4.7)
$$

To achieve monotonicity, we seek $(\alpha,\beta,\gamma,\delta)$ such that for given (ν_1,ν_2) none of the c_k is negative. Now we observe that

$$c_2 + c_4 + c_6 + c_8 = 0 \;,$$

and if no c_k are negative, this implies

$$c_2 = c_4 = c_6 = c_8 = 0 \;,$$

whence

$$\frac{\alpha}{\nu_1 + \nu_2} = \frac{\beta}{\nu_2 - \nu_1} = \frac{-\gamma}{\nu_1 + \nu_2} = \frac{\delta}{\nu_1 - \nu_2} \;. \qquad (4.8)$$

But this would give

$$\alpha + \beta + \gamma + \delta = 0 \;,$$

whereas for consistency with the differential equation we require

$$\alpha + \beta + \gamma + \delta = 1 .$$

Having failed to generalise to two dimensions the first of the one-dimensional upwind schemes mentioned in Section 2, we now attempt to generalise the 'interface flux' approach. There seem to be two natural ways of doing this, and the first of these is illustrated on the left. At the centre of each cell, define average quantities

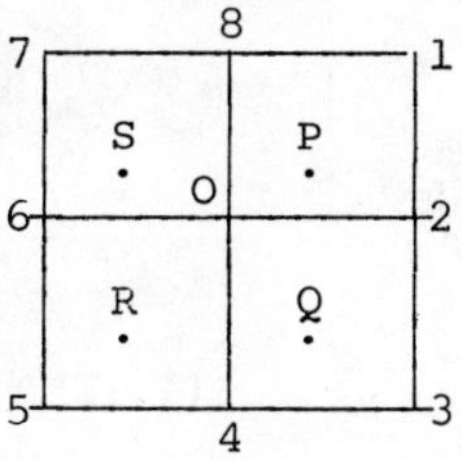

$$\begin{aligned} u_P &= \alpha u_O + \beta u_8 + \gamma u_1 + \delta u_2 , \\ u_Q &= \alpha u_4 + \beta u_O + \gamma u_2 + \delta u_3 , \\ u_R &= \alpha u_5 + \beta u_6 + \gamma u_O + \delta u_4 , \\ u_S &= \alpha u_6 + \beta u_7 + \gamma u_8 + \delta u_O , \end{aligned} \tag{4.9}$$

where again $\alpha, \beta, \gamma, \delta$ are arbitrary weights summing to unity. Now update u_O according to

$$\begin{aligned} u_O^{u+1} - u_O^u &= -\frac{\nu_1}{2}(u_P + u_Q - u_R - u_S) - \frac{\nu_2}{2}(u_P + u_S - u_R - u_Q) \\ &= -\frac{\nu_1 + \nu_2}{2}(u_P - u_R) - \frac{\nu_1 - \nu_2}{2}(u_Q - u_S) . \end{aligned} \tag{4.10}$$

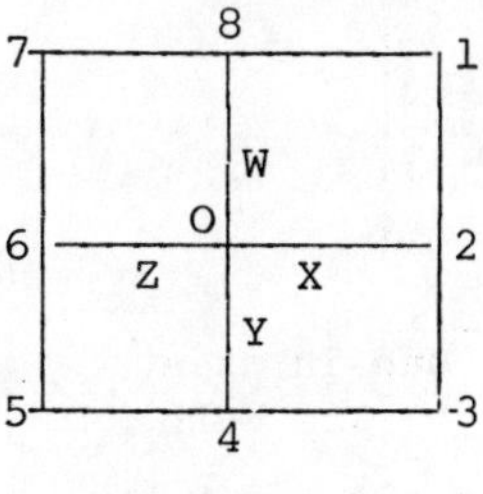

Unfortunately this is not an independent approach. Substitution of (4.8) into (4.9) merely reproduces (4.4). Therefore, we have to attempt a third class of scheme, illustrated in the last diagram. Here we begin by assigning suitable mean values to the mid-points of the cell sides, W,X,Y,Z, thus:

$$\left.\begin{aligned} u_X &= \alpha u_1 + \beta u_2 + \gamma u_3 + \delta u_4 + \varepsilon u_O + \phi u_8 ,\\ u_Z &= \alpha u_8 + \beta u_O + \gamma u_4 + \delta u_5 + \varepsilon u_6 + \phi u_7 ,\\ u_W &= a u_7 + b u_8 + c u_1 + d u_2 + e u_O + f u_6 ,\\ u_Y &= a u_6 + b u_O + c u_2 + d u_3 + e u_4 + f u_5 , \end{aligned}\right\} \quad (4.11)$$

where $\alpha + \beta + \gamma + \delta + \varepsilon + \phi = a + b + c + d + e + f = 1$. (4.12)

Then u_O is updated by the formula

$$u_O^{n+1} - u_O^n = -\nu_1(u_X - u_Z) - \nu_2(u_W - u_Y) . \quad (4.13)$$

This approach will again lead to a nine-point scheme of the form (4.4) but

$$\left.\begin{aligned} c_O &= 1 - (\beta + \varepsilon)\nu_1 - (b + e)\nu_2 ,\\ c_1 &= -\alpha\nu_1 - c\nu_2 ,\\ c_2 &= -\beta\nu_1 + (c - d)\nu_2 ,\\ c_3 &= -\gamma\nu_1 + d\nu_2 ,\\ c_4 &= (\gamma - \delta)\nu_1 + e\nu_2 ,\\ c_5 &= \delta\nu_1 + f\nu_2 ,\\ c_6 &= \varepsilon\nu_1 + (a - f)\nu_2 ,\\ c_7 &= \phi\nu_1 - a\nu_2 ,\\ c_8 &= (\alpha - \phi)\nu_1 - b\nu_2 . \end{aligned}\right\} \quad (4.14)$$

This approach does yield monotone schemes, and in great variety. For example, if it is given that ν_1, ν_2 are both positive, then we can easily ensure $c_k \geqslant 0$ for all odd k by taking (α,γ,a,c) all negative, and (δ,ϕ,d,f) all positive. Positivity of c_0 is assured for ν_1,ν_2 sufficiently small, and of the remaining even coefficients by choosing the last four parameters to satisfy

$$\beta\nu_1 \lessgtr (c - d)\nu_2 \lessgtr 0 \quad ,$$

$$\varepsilon\nu_1 \gtrless (f - a)\nu_2 \gtrless 0 \quad ,$$

$$b\nu_2 \lessgtr (\alpha - \phi)\nu_1 \lessgtr 0 \quad ,$$

$$e\nu_2 \gtrless (\delta - \gamma)\nu_1 \gtrless 0 \quad .$$

The simplest example of a monotone scheme in this class would be the straightforward upwind scheme defined by $\varepsilon = e = 1$, with all other parameters zero, but it would obviously be possible to find many others.

Evidently we may create similar schemes based on non-quadrilateral meshes. The plane could be divided into arbitrary polygons, and values of u could be assigned to their vertices. In this situation we can devise algorithms corresponding to each of the above types.

Type A Evaluate a fluctuation over each polygon and send signals to each vertex of that polygon.

Type B Evaluate an average state at the centre of each polygon, and update each vertex by computing the fluctuation over the centres of the polygons which meet there.

Type C Evaluate an average state at the centre of each edge, and update each vertex by computing the fluctuation over the centres of the edges which meet there.

Although we have no proof, we conjecture that there are never any monotone schemes of Types A or B. We have not found any for triangular or hexagonal meshes. However, we can exhibit a monotone algorithm of Type C for a fairly arbitrary mesh, (see Fig. 3).

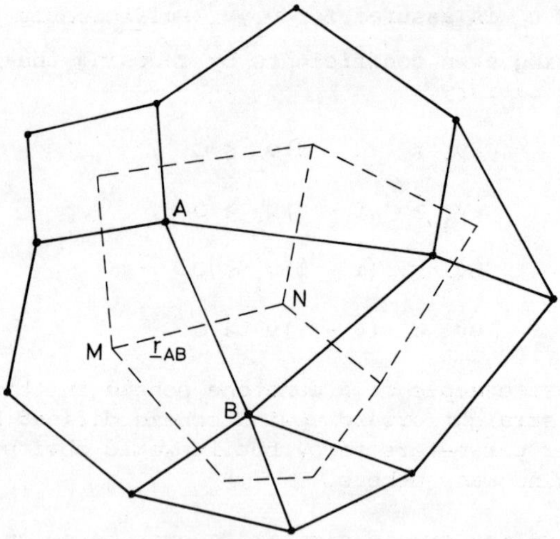

Fig. 3 Part of a general tessellation, and its dual.

Inside each polygon construct an interior point. Join these up to form a dual mesh so that each vertex of the original mesh lies inside a cell of the dual mesh. (If this cannot be done, the mesh was probably not sensible anyway.) Assume u = constant inside each of the dual cells. Then the flux across the boundary between cells A and B is

$$Q_{AB} = (\underline{r}_{AB}.\underline{a})u_A \quad \text{or} \quad (\underline{r}_{AB}.\underline{a})u_B \tag{4.15}$$

where $\underline{r}_{AB}$ is the vector representing the interface MN between A and B, $\underline{a}$ is the advection velocity (a,b) and we select u_A if $\underline{a}$ is directed out of A, otherwise u_B. Now let ΣQ_{AX} be the sum of all such terms over the edges of the dual cell surrounding A, whose area is S_A. Then

$$\Sigma Q_{AX} = S_A \frac{\partial u_A}{\partial t} \tag{4.16}$$

defines a semi-discrete evolution scheme for advancing the solution in time. If we make a sufficiently small finite time-step in the simplest way

$$u_A^{n+1} - u_A^n = \frac{\Delta t}{S_A} \Sigma Q_{AX} \tag{4.17}$$

it is clear that the new value of u_A will be some average, with positive weights summing to unity, of the old values of u at A and certain neighbouring cells. The scheme is therefore monotone, and hence stable. Now that we have it, the assumption of piecewise constant states may be forgotten. It was merely a device to derive the scheme.

To effect a connection with conservation laws, we write

$$Q_{AB} = (\underline{r}_{AB}.\underline{a})u_{AB}$$

where u_{AB} is one of u_A, u_B, and express u_{AB} in a form that can be generalised to yield an analogue of (2.16). We write

$$u_{AB} = \tfrac{1}{2}(u_A + u_B) - \tfrac{1}{2}\mathrm{sgn}(\underline{r}_{AB}.\underline{a})(u_A - u_B) \tag{4.18}$$

and note that

$$u_A - u_B \simeq \underline{S}_{AB}.\mathrm{grad}\ u \tag{4.19}$$

where $\underline{S}_{AB}$ is the vector representing AB. Thus

$$u_{AB} = \tfrac{1}{2}(u_A + u_B) - \tfrac{1}{2}\mathrm{sgn}(\underline{r}_{AB}.\underline{a})(\underline{S}_{AB}.\mathrm{grad}\ u) \tag{4.20}$$

which closely resembles (2.14), and would reduce to it for a one-dimensional case.

In general, we will not know grad u except in so far as we can deduce it from the values given at the vertices. Unless the variation of u over each polygon is exactly linear, the use of an estimated gradient will leave (4.19, 4.20) only approximately satisfied. In consequence, the scheme based on (4.20) is not monotone on general grids. However, it is monotone on triangular grids, because u can then be assumed linear over each cell without inconsistency.

4.3 Application to systems of equations

The ingredients needed to construct a genuinely two-dimensional monotone upwind scheme for the linearised Euler equations can now be assembled, see (Fig. 4). Construct a triangular mesh with flow states $\underline{w}$ given at its vertices. Within each triangle compute the gradients $\underline{w}_x$, $\underline{w}_y$. Using the results from Section 3.3 find the local flow model corresponding to these gradients. Now find the dual mesh as follows. Within each triangle join the mid points of the sides to the centroid of the vertices, creating three equal sub-areas. This particular interior point is chosen because we can imagine the same integrated quantity of $\underline{w}$ either linearly distributed or constant in each sub-area. We then associate with each edge of the mesh the state

$$\underline{w}_{AB} = \tfrac{1}{2}(\underline{w}_A + \underline{w}_B) - \tfrac{1}{4}\Sigma\,\mathrm{sgn}(\underline{r}_{AB}.\underline{a}^{(k)})(\underline{S}_{AB}.\underline{e}^{(k)})\underline{r}^{(k)} \tag{4.21}$$

where $\underline{a}^{(k)}$ is, as before, the vector describing propagation speed, and $\underline{e}^{(k)}$ is the vector $(\alpha^{(k)}\cos\theta^{(k)}, \alpha^{(k)}\sin\theta^{(k)})$ in which α = wave strength, θ = wave orientation, as in Section 3. Finally $\underline{r}^{(k)}$ is the eigenvector giving the distribution of the wave's effects over $\underline{w}$, so that the product $\underline{e}^{(k)}\,\underline{r}^{(k)}$ represents the flow gradients due to the kth wave. The summation is over all waves present in either of the cells adjacent to AB. The factor ¼ appears in (4.21) rather than ½ in (4.20) because in regions of uniform gradient there will be contributions from both cells. In general the boundary of the dual cell will change direction as it crosses AB; this does not affect the legitimacy of (4.20), since the fluxes across all curves with the same endpoints are equal.

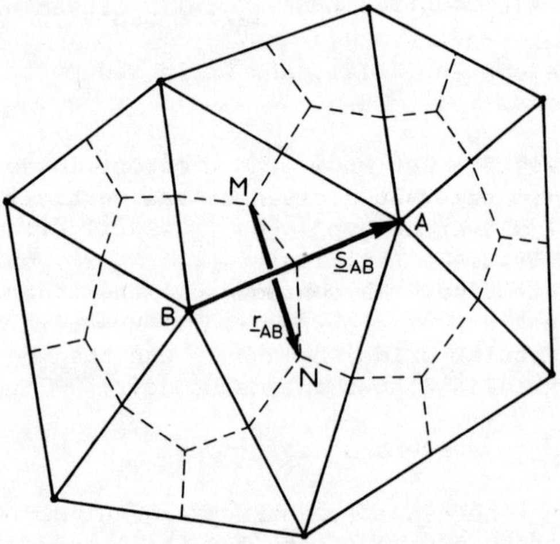

Fig. 4 Part of a triangular tessellation, and its dual

To apply these ideas to a set of conservation laws,

$$\underline{w}_t + \underline{F}_x + \underline{G}_y = 0 \tag{4.22}$$

the formula (4.21) can be modified as follows. We now need to associate with each edge AB a value of $\underline{F}$ and a value of $\underline{G}$. The simplest way to do this is just to take

$$\underline{F}_{AB} = \underline{F}(\underline{w}_{AB}) \ , \quad \underline{G}_{AB} = \underline{G}(\underline{w}_{AB}) \ . \tag{4.23}$$

Then the total flux through XY is

$$\underline{Q}_{AB} = \underline{F}_{AB}(y_M - y_N) - \underline{G}_{AB}(x_M - x_N) \tag{4.24}$$

and the formula to update $\underline{w}_A$ is

$$\underline{w}_A^{uM} - \underline{w}_A^{n} = - \frac{\Delta t}{S_A} \Sigma \underline{Q}_{AX} \ . \tag{4.25}$$

This formula defines a scheme which is conservative in the traditional sense, monotone in the linear scalar case, and strictly upwind in any region of simple wave flow.

5. CONCLUDING REMARKS

The purpose of this paper has been to provide a critique of the way that upwind- differencing is usually applied to two-dimensional problems. It has been shown that the operator-splitting method can produce misleading results. An alternative proposal is made to base the algorithm on a local flow model which attempts to recognize flow patterns associated with two-dimensional simple waves. These patterns are strongly linked to familiar analysis concerning the local kinematics of the flow.

Once the nature of the local flow is established, we wish to exploit it in an evolution algorithm, creating to begin with a monotone first-order scheme, as a foundation for future. work. Through exploring the simplest linear advection equation, it transpires that the possibilities are quite limited. The author feels that the formula presented in Section 4.2 may well represent the most natural way to create a two-dimensional upwind algorithm.

REFERENCES

Courant, R. and Friedrichs, K.O., (1948) Supersonic Flow and Shock Waves, Interscience, New York. Republished as Applied Mathematical Sciences 21 (1976) Springer-Verlag, New York.

Crandall, M. and Majda, A., (1980) The method of fractional steps for conservation laws. *Numerische Math.* **34**, 285-314.

Davis, S.F., (1984). A rotationally biased upwind scheme for the Euler equations. *J. Comput. Phys.* **56** 65-92.

Harten, A., (1983) High resolution schemes for hyperbolic conservation laws. *J. Comput. Phys.* **49**, 357-372.

Lytton, C.C., (1984) Solution of the Euler equations for transonic flow past a lifting aerofoil - the Bernoulli formulation. Royal Aircraft Establishement TR 84080.

Majda, A., (1984) Compressible Fluid Flow and Systems of Conservation Laws in Several Space Variables. Applied Mathematical Sciences 53, Springer-Verlag, New York.

Osher, S., (1984) Riemann solvers, the entropy condition, and difference approximations. *SIAM J. Numer. Anal.* **18**, 217-235.

Roe, P.L., (1981) Fluctuations and signals, a framework for numerical evolution problems. In Numerical Methods for Fluid Dynamics (K.W. Morton and M.J. Baines, eds) Academic Press, London.

Roe, P.L. (1985) Discrete models for the numerical analysis of time-dependent multidimensional gas dynamics. ICASE Report 85-18, to appear in *J. Comput. Phys.*

Roe, P.L. and Baines, M.J., (1982) Algorithms for advection and shock problems. In Proc. 4th GAMM-Conference on Numerical Methods in Fluid Dynamics, (H. Viviand, ed) Vieweg, Braunschweig.

Roe, P.L. and Baines, M.J., (1984) Asymptotic behaviour of some non-linear schemes for linear advection. In Proc 5th GAMM-Conference on Numerical Methods in Fluid Dynamics (M. Pandolfi, R. Piva, eds.) Vieweg, Braunschweig.

Roe, P.L. and Pike, J., (1984) Efficient construction and utilisation of approximate Riemann solutions. In Computing Methods in Applied Sciences and Engineering VI (R. Glowinski and J-L Lions, Eds.) North-Holland, Amsterdam.

Sells, C.C.L., (1980) Solution of the Euler equations for transonic gas flow past a lifting aerofoil. Royal Aircraft Establishment TR 80065.

Strang, G., (1980) Linear Algebra and its Applications. Academic Press, New York.

Tadmor, E., (1984) Numerical viscosity and the entropy condition for conservative difference schemes. *Math. Comp.* **43** 369-381.

Woodward, P. and Collella, P., (1983) The numerical simulation of two-dimensional flow with strong shocks. *J. Comput. Phys.* **54** 115-173.

METHODS FOR THE PREDICTION OF "HYPERBOLIC FLOWS" BASED ON PHYSICAL ARGUMENTS

M. Pandolfi
(Dipartimento di Ingegneria Aeronautica e Spaziale, Politecnico di Torino, Italy)

1. INTRODUCTION

The "hyperbolic flows" we are considering here are the unsteady and steady supersonic flows of a compressible inviscid gas.

Investigations on unsteady flows are required in order to provide a physical description of actual transients as well to give steady transonic flow configurations achieved through "time-dependent" or "pseudo time-dependent" techniques. The "hyperbolic" coordinate is represented by the time and the solution is obtained by marching along this coordinate.

In supersonic regions the steady flow problem also has a "hyperbolic" character. This fact enables us to obtain a solution of the flow field by marching along a space coordinate, and the description of the flow evolution will then be provided in this direction. This space coordinate is then the "hyperbolic" one.

Even if trivial, it should be remembered that 2D unsteady flow is a phenomenon in which we consider two dimensions in space and look for the evolution in time, whereas 3D steady supersonic flow is still a 2D "hyperbolic" problem in the sense that we look for a description of the flow field over a two-dimensional plane and aim to give the evolution in space by proceeding in the third space dimension. So, as regards the "hyperbolic" nature of the problem, 2D unsteady and 3D steady supersonic flows are equivalent.

In the following we will emphasise unsteady flows and say only a few words on steady supersonic flows just at the end. However almost everything that will be said for the former flows can be easily translated into the domain of the latter.

"Hyperbolic" flows are characterized by the propagation of waves, as is well known. The numerical procedures developed on the basis of the method of characteristics aim to follow and describe explicitly these waves during their propagation. While these procedures are very accurate and consistent with the physics in the case of 1D "hyperbolic" problems (that is 1D unsteady and 2D steady supersonic flows), they become less appealing in multidimensional flows for topological reasons.

In the last decades numerical procedures have been developed which are labelled as "finite-difference" (or finite-volume, or finite element) approaches to the approximation of the original equations on the basis of a discrete number of points where the flow properties are represented. Most of these procedures have more or less neglected the conceptual effort made in the past to describe the propagation of waves and attention has been focused mainly on the development of numerical aspects. By doing this, the physics has quite often been ignored.

Somewhat different approaches were proposed many years ago (Courant et. al. (1952), Gordon (1968) and Godunov (1959)). They attempted to include the concept of wave propagation into finite-difference approximations. More recently this last kind of investigation has again received attention and many such approaches are proposed today for the prediction of unsteady flows. They are generally known as "upwind" methods.

Their common feature is the way they are conceived. First the Euler equations are analysed emphasizing the propagation of waves. The important role of the domains of dependence is recognised together with the nature of the convection of signals through the flow field. Furthermore, important suggestions are made for avoiding the introduction of artificial conditions at the boundaries. Moreover it follows that some of the variables appearing in the original equations are not the most significant and that new variables express better the mechanism of the convection of signals. We will call this preliminary step of investigation the "formulation".

Once the formulation has been worked out we can proceed into the numerics by approximating derivatives with finite differences. But, whatever the numerical algorithm might be (first, second or higher order of accuracy, explicit or implicit procedures, fast or accelerating solvers, multigrid methods and so on), it should preserve and retain the features emphasized in the previous formulation, namely the character of the convection of signals: in other words the "upwind" idea should be incorporated into the numerical approximation.

In the following we will consider two approaches belonging to this class of methodologies. We have been working on both for some years and found them rewarding.

The first is generally called the "lambda" scheme. Here we will call it rather the "lambda" formulation, because its most important feature is the preliminary work done on the original equations (formulation). After that any more or less efficient numerical scheme may be applied.

The second approach may be called the "flux-difference splitting" (hereafter referred as F.D.S.) formulation: here the work done in the formulation is mainly related to the task of solving Riemann problems (R.P.) in order to recognize the propagation of signals. Even here different numerical schemes may be introduced as regard the numerics.

At first glance the difference between these two approaches seems to be considerable. Nevertheless they are strongly related to each other. For example, in the case of the linearized version of the Euler equations it turns out that the two coincide exactly from the conceptual and formal points of view.

In fact the main difference between the two approaches lies in the original equations on which the formulation is based. The first (lambda) is founded on the differential equations written in the so called non-conservative form and in terms of the primitive variables (the gas velocity and two thermodynamic properties, for example the speed of sound and the entropy). The numerical solutions are then confined to the field of smooth solutions. The second (F.D.S.) is based on the conservation form of the equations and provides a numerical approximation of the weak solution of the laws of conservation expressed by the Euler equations. For this reason the two approaches differ with respect to the strategies for predicting gasdynamical discontinuities such as shock waves.

In the following the presentation will be organized thus:

§ 2. The "lambda" formulation in the 1D problem.
§ 3. The "F.D.S." formulation in the 1D problem.
§ 4. The multidimensional problem.
§ 5. A suggestion for a hybrid formulation.
§ 6. The extension to supersonic steady flows.

2. THE "LAMBDA" FORMULATION IN THE 1D PROBLEM

The Euler equations written in non-conservative form are:

$$a_t + ua_x + \frac{\gamma-1}{2} au_x = 0$$

$$u_t + uu_x + \frac{2}{\gamma-1} aa_x - \frac{a^2}{\gamma(\gamma-1)} S_x = 0 \qquad (2.1)$$

$$S_t + uS_x = 0$$

where the primative variables (speed of sound, gas velocity and entropy) are denoted by a, u, and S.

By rearranging equations (2.1), two new variables may be defined, which replace the primitives ones a and u:

$$dR_1 = d\left(\frac{2}{\gamma-1} a-u\right) - \frac{a}{\gamma(\gamma-1)} dS$$

$$\qquad (2.2)$$

$$dR_3 = d\left(\frac{2}{\gamma-1} a + u\right) - \frac{a}{\gamma(\gamma-1)} dS$$

Therefore equations (2.1) become:

$$R_{1t} + (u-a)R_{1x} = 0$$

$$R_{3t} + (u+a)R_{3x} = 0 \qquad (2.3)$$

$$S_t + uS_x = 0$$

These equations now reveal the wavelike nature of the problem. In fact they describe the convection of signals.

The first two of the equations (2.3) refer to the propagation of "acoustic" waves. The corresponding signals are defined in equation (2.2) and coincide with the Riemann invariants in the case of homoentropic flow ($S_x = 0$). For the less restrictive assumption of isentropic flow ($S_t + uS_x = 0$), it is no longer possible to define the classical Riemann invariants and the "acoustic" signals can only be defined in a differential form (equation (2.2). Of course the entropy gradients can only originate at the incoming flow boundary and not from the generation of entropy through shocks, equations

(2.1) being written in non-conservative form. The third of equations (2.3) represents the convection of the entropy wave.

As the result of the definition in equations (2.2), the set of equations (2.3) can be finally written as

$$a_t = -\frac{1}{2}\frac{\gamma-1}{2}\left[(u-a)R_{1x} + (u+a)R_{3x}\right] + \frac{a}{2\gamma}S_t$$

$$u_t = -\frac{1}{2}\left[-(u-a)R_{1x} + (u+a)R_{3x}\right] \qquad (2.4)$$

$$S_t = -u\,S_x$$

On the basis of equations (2.4), we can now describe the evolution in time of the flow after the evaluation of the space derivatives R_{1x}, R_{3x} and S_x. These have to be regarded as one-sided derivatives, the side being determined by the sign of the relative velocity of propagation of the signals (respectively u-a, u+a and u).

All the above considerations represent what we have called previously the "lambda" formulation. After an initial tentative suggestion by Pandolfi and Zannetti (1978), the basics of this formulation have been well laid down by Moretti (1979).

We can now proceed to the numerics. The space derivatives will be approximated by finite differences along the space coordinate x with one-sided differences, in order to retain in the algorithm the physical description of the propagating signals. Algorithms of any required order of accuracy can be built up. They may be explicit or implicit. In the present case we use the simple first order "upwind" scheme, an algorithm equivalent to the basic version of the method of the characteristics.

It will be shown by Zannetti (1985), in the case of homoentropic flow, that equations (2.4) satisfy also the jump relationships of space derivatives across characteristics. Therefore the "lambda" formulation agrees with the double physical interpretation of the characteristics as lines where the signals propagate and where jump relationships of derivatives are prescribed.

A further important feature of the "lambda" formulation emerges when we deal with the boundaries. It can be seen that whenever a space difference is required behind a boundary (which therefore is not available), a physical boundary condition will take its place. So no artificial or additional boundary conditions will be introduced. The "post-correction" technique, proposed by De Neef and Moretti (1980), should be considered as the basic contribution to this important point, even in connection with different formulations or procedures.

As can be argued by looking at equations (2.4) the number of operations involved in a numerical algorithm based on the "lambda" formulation is quite small and this contributes greatly to confining the required computational time to very low levels. Furthermore the accuracy of the numerical results, for a given numerical scheme and computational grid, turns out to be very satisfactory, mainly because we have selected as dependent variables the most significant ones. Therefore the computational grid can be kept coarse and the computational time further reduced.

However, the solution of equations (2.4) being confined to smooth flows, we cannot deal correctly with discontinuities, such as shock waves, other than working out some explicit shock treatment, such as shock fitting or tracking techniques. This is needed to generate the entropy jump, not provided by the original equations.

3. THE "F.D.S" FORMULATION IN THE 1D PROBLEM

The Euler equations are now written in the conservation form related to the laws of conservation

$$W_t + F_x = 0 \tag{3.1}$$

where:

$$W = \begin{vmatrix} \rho \\ \rho u \\ e \end{vmatrix} \quad \text{and} \quad F = \begin{vmatrix} \rho u \\ -p + \rho u^2 \\ u(p+e) \end{vmatrix}$$

Here ρ, p, and e denote respectively the density, the pressure and the total internal energy per unit volume.

Let us consider the flow configuration at the time t. The flow properties being given at a discrete number of points, we approximate the distribution of any variable along x with the value at the point N kept constant on the two half intervals in front and behind it. The picture of this representation at the time t, over the interval bounded by the points N and N+1, is shown in Fig. 1. The discontinuity located at the middle of the interval will then evolve in time, on the basis of the initial conditions prescribed in the regions a and b (corresponding to the points N and N+1). Three waves (1,2,3) will be generated and the new regions c and d will appear. Waves (1,3) correspond to the "acoustic" waves and wave (2) is the entropy wave.

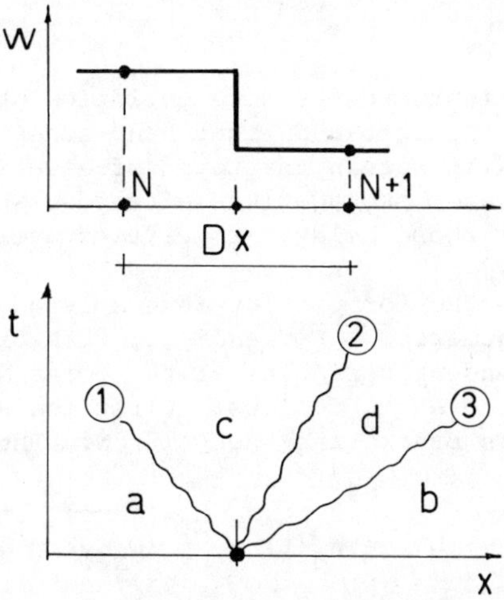

Fig. 1 Solution of the Riemann Problem

The description of this evolution is provided by the solution of a Riemann problem (R.P.), which can be obtained exactly following classical procedures. So the pattern of the waves (1,2,3) and the flow in the regions c and d can be evaluated. At this stage we assume that the unsteadiness caused by the development of the discontinuity will affect the evolution in time of the flow at the computational points, as follows. Those signals which are related to the waves propagating rightwards will contribute to the evolution in time at points located on the right hand side of the interval, and the corresponding will be done for the signals related to the waves travelling to the left. The signals are determined by the difference of the flow in the regions a, b, c, and d.

The formulation will be then defined as it follows. Any difference of the flux over an interval Dx:

$$D_N F = F_{N+1} - F_N$$

will be divided into three terms:

$$D_N F = (F_b - F_d) + (F_d - F_c) + (F_c - F_a) \tag{3.2}$$

Each term is a signal and corresponds to one of the waves (1,2,3). Then we split the difference of the flux $D_N F$ in two parts:

$$D_N F = \overleftarrow{D_N F} + \overrightarrow{D_N F}$$

which will contribute to the time evolution at points located respectively on the left and right hand sides of the interval. The first part will retain the terms of equations (3.2) which are related to waves propagating leftwards, whereas the second will account for those related to waves travelling rightwards.

At this stage the formulation is completed and we may proceed to the numerics. For example, with the first order scheme, the dependent variables at the point N and the time t+Dt are evaluated as it follows. First the R.P. is solved in the two intervals bracketing the point N. Then we have:

$$W_N(t + Dt) = W_N(t) - \frac{Dt}{Dx} (\overrightarrow{D_{N-1} F} + \overleftarrow{D_N F}) \qquad (3.3)$$

More accurate schemes can be built up, always based on the "upwind" idea for the split terms $\overleftarrow{DF}$ and $\overrightarrow{DF}$.

The exact solution of the R.P. (Fig. 1) can be generally obtained only at the price of tedious and long computations, because of the iterations required when one or both of the waves (1,3) are shocks. It will then be convenient to look for approximate solutions of the R.P. in order to save computational time. Two different procedures have been proposed. According to the suggestion made by Roe (1981) it is convenient to replace the original non-linear R.P. by a linear one with certain appealing features (the solution will be exact if the wave pattern exhibits a single shock wave). The solution is then obtained quickly. On the other hand following the procedure indicated by Osher and Solomon (1982), we assume that the "acoustic" waves (1,3) are isentropic (even if in fact one or both can actually be shocks). Under this assumption the solution is worked out easily in a classical way. A third approximated solution (in fact closely related to the one proposed by Osher and Solomon (1982) and physically similar to the one suggested by Roe (1981)) was given later by Pandolfi (1984 a), who presented also some remarks on the previous two. In a few words it is conceived as follows.

As suggested by Osher and Solomon (1982) we assume as isentropic the "acoustic" waves (1,3). The flow properties being defined in the regions a and b (speed of sound, gas velocity and entropy) we look for the same variables in the new regions c and d. It follows that the six equations given below provide the solution in a simple fashion.

Conservation of the Riemann invariant R_3 and of the entropy across the wave 1:

$$\frac{2}{\gamma-1} a_c + u_c = \frac{2}{\gamma-1} a_a + u_a \ ; \quad S_c = S_a$$

Conditions on the contact surface (that is across the wave 2):

$$u_c = u_d \ ; \quad p_c = -p_d, \text{ i.e. } a_c = a_d \exp \left\{ \frac{S_c - S_d}{2\gamma} \right\}$$

Conservation of the Riemann invariant R_1 and of the entropy across the wave 3:

$$\frac{2}{\gamma-1} a_d - u_d = \frac{2}{\gamma-1} a_b - u_b \ ; \quad S_d = S_b$$

If the waves (1,3) are expansion or compression isentropic fans, it may occur that one characteristic imbedded in the fan is vertical, denoting a sonic transition. In this case a further splitting is done on this wave, in order to determine which part of it is propagating to the left and which to the right (see Pandolfi (1984 a)).

We have performed several numerical experiments by operating the splitting according the the exact solution and the three above mentioned approximate solutions of the R.P. and have found the difference in the numerical results quite negligible. Therefore, as also anticipated by Roe (1981), we can avoid the exact solution and apply any approximate procedure without penalty in the accuracy of the results.

The numerics based on the "F.D.S." formulation provide good capability of capturing shocks, which come out very sharp and neat. For steady configurations shock waves are always described in two or three intervals, even with the ordinary first order scheme. Second order schemes give better sharpness to the shocks, but some flux-limiters or monotonicity criteria must be included to avoid spurious oscillations in the neighbourhood of shock waves.

However, on the basis of our experience, the overall level of accuracy is not remarkable, at least compared with that achievable by the "lambda" formulation. This result could be ascribed to the fact that in the "F.D.S." formulation the convection of the signals is hidden in the split parts of the difference of the flux and not treated explicitly.

Furthermore there are several operations involved in the preliminary work of carrying out the splitting on each interval and the computing time is penalized.

4. THE MULTIDIMENSIONAL PROBLEM

We have been so far how to deal with the 1D problem, where the analysis is quite clear and without any ambiguity. We now try to go from the 1D case to the multidimensional cases.

The diagrams in Fig. 2 show how the picture changes in the transition to multidimensional problems. We aim to describe the evolution at the point P from time t to t+Dt. In the 1D problem (Fig. 2a) the variation of the flow is determined by the merging at P(t+Dt) of the three characteristics (carrying the two "acoustic" signals and the entropy) leaving three definite locations at the time level t. In the 2D case (Fig. 2b), the entropy is still carried on a well defined ray (the path of the particle reaching P(t+Dt)), but there are an infinity of "acoustic" signals merging at P(t+Dt), which propagate on the straight lines which define the Mach cone. All these signals leave the plane (x,y) at the time t from a circle with radius equal to the speed of sound a. Going now to the 3D case (Fig. 3c) the picture cannot anymore be represented on the paper and we shall imagine the signals leaving the space (x,y,z) at the time t from a sphere (with radius a) and merging at the point P(t+Dt) along rays propagating in a fourth dimension (the time).

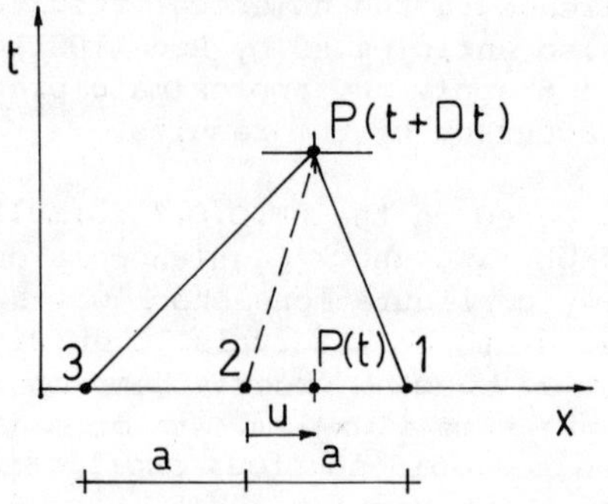

Fig. 2a Evolution in time for the 1D unsteady flow.

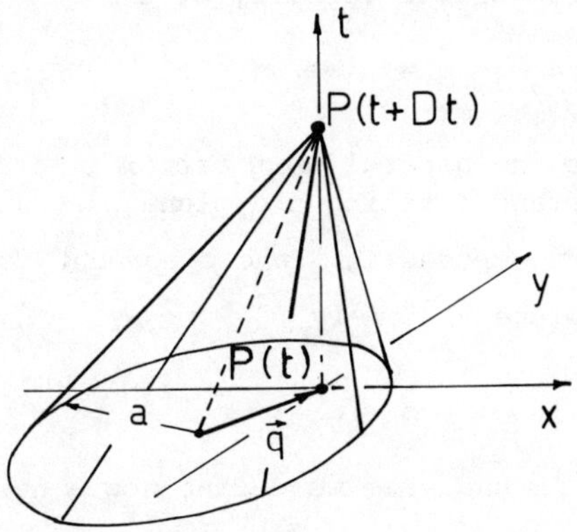

Fig. 2b Evolution in time for the 2D unsteady flow.

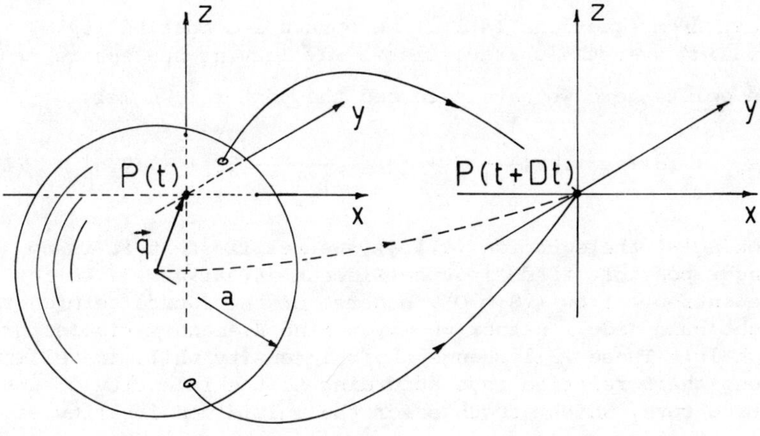

Fig. 2c Evolution in time for the 3D unsteady flow.

4.1 The "Lambda" Formulation in Multidimensional Problems

Let us consider the non-conservative form of the Euler equations in a cartesian frame of reference. They are:

$$\left.\begin{aligned} & a_t + \vec{q}.\nabla a + \frac{\gamma-1}{2}\, a\, \nabla.\vec{q} = 0 \\ & \vec{q}_t + (\vec{q}.\nabla)\vec{q} + \frac{2}{\gamma-1}\, a\nabla a - \frac{a^2}{\gamma(\gamma-1)}\, \nabla S = 0 \\ & S_t + \vec{q}.\nabla S = 0 \end{aligned}\right\} \qquad (4.1.1)$$

where now $\vec{q}$ represents the velocity of the gas in the form

$$\vec{q} = u\vec{i} + v\vec{j} + w\vec{k}$$

Let us now define the general unit vector $\vec{\xi}$ normal to the sphere of Fig. 2c and defining the plane μ normal to it. The velocity may be decomposed into the components q_ξ along $\vec{\xi}$ and $\vec{q}_\mu$ lying on the plane μ, namely

$$\vec{q} = q_\xi \vec{\xi} + \vec{q}_\mu$$

According to the 1D analysis we define now a new variable, related to the vector $\vec{\xi}$ by

$$dR_\xi = d\left(\frac{2}{\gamma-1}\, a + q_\xi\right) - \frac{a}{\gamma(\gamma-1)}\, dS$$

Then, from equations (4.1.1) we obtain a compatibility equation over the characteristic ray leaving the sphere from the point where we have selected the vector $\vec{\xi}$, namely

$$R_{\xi t} + (\vec{q} + a\vec{\xi}).\left[\nabla\left(\frac{2}{\gamma-1}\, a + q_\xi\right) - \frac{a}{\gamma(\gamma-1)}\nabla S\right] = -\, a\nabla.\vec{q}_\mu \qquad (4.1.2)$$

Looking at the equation (4.1.2), we realize that it is no longer possible to define the Riemann invariants, even for the homoentropic flow ($\nabla S = 0$), because of the source term on the right hand side. Rather we may define Riemann variables or signals. These will change their intensity while travelling along their relative rays according to the intensity of the source term, which accounts for the multidimensionality of the flow. It is also clear that the choice of the vector $\vec{\xi}$ will provide a compatibility equation related to the projection of the momentum equation in the direction of $\vec{\xi}$.

By following the line taken by Zannetti and Colasurdo (1981) we select for the vector $\vec{\xi}$ six values, corresponding to $\pm\vec{i}$, $\pm\vec{j}$, and $\pm\vec{k}$. This choice takes into account the equilibrium between the pressure gradient and the inertial forces in the directions of the three coordinates. We then have

$$a_t = -\frac{1}{2}\frac{\gamma-1}{2}\left[(u-a)R_{-ix} + (u+a)R_{ix} + (v-a)R_{-jy} + (v+a)R_{jy} + (w-a)R_{-kz} + (w+a)R_{kz}\right] + \frac{a}{2\gamma}S_t$$

$$u_t = -\frac{1}{2}\left[-(u-a)R_{-ix} - vR_{-iy} - wR_{-iz} + (u+a)R_{ix} + vR_{iy} + wR_{iz}\right] \tag{4.1.3}$$

$$v_t = -\frac{1}{2}\left[-uR_{-jx} - (v-a)R_{-jy} - wR_{-jz} + uR_{jx} + (v+a)R_{jy} + wR_{jz}\right]$$

$$w_t = -\frac{1}{2}\left[-uR_{-kx} - vR_{-ky} - (w-a)R_{-kz} + uR_{kx} + vR_{ky} + (w+a)R_{kz}\right]$$

$$S_t = -\left[u\,S_x + v\,S_y + w\,S_z\right]$$

The set of equations (4.1.3) provides the description of the evolution in time of the flow on the basis of space derivatives. As in the 1D case, they have to be regarded as one-sided derivatives, the side being determined by the sign of their coefficient.

A very large number of numerical experiments have been carried out in the past with numerical schemes of second order accuracy for 2D and 3D flows. In particular we refer to Pandolfi and Colasurdo (1980) as far as 3D flows are concerned and point out the capability of predicting correctly non-trivial secondary flows, where the role of the rotationality is predominant. The ability of the "lambda" formulation in multidimensional flows to agree with the physical interpretation of the characteristic surfaces will be discussed by Zannetti (1985).

4.2 The "F.D.S." Formulation in Multidimensional Problems

The Euler equations are now considered in conservative form for the 3D flows described in a cartesian frame of reference, i.e.

$$W_t + F_x + G_y + H_z = 0,$$

where

$$W = \left|\begin{matrix} \rho \\ \rho u \\ \rho v \\ \rho w \\ e \end{matrix}\right| ; \quad F = \left|\begin{matrix} \rho u \\ p+\rho u^2 \\ \rho uv \\ \rho uw \\ u(p+e) \end{matrix}\right| ; \quad G = \left|\begin{matrix} \rho v \\ \rho uv \\ p+\rho v^2 \\ \rho vw \\ v(p+e) \end{matrix}\right| ; \quad H = \left|\begin{matrix} \rho w \\ \rho uw \\ \rho vw \\ p+\rho w^2 \\ w(-p+e) \end{matrix}\right|$$

Following what has been done for the 1D case, we shall now proceed to the splitting of the difference of F,G and H over the intervals Dx, Dy and Dz. This can be done by solving Riemann problems on each interval.

For a Dx interval the R.P. will refer to a discontinuity of all the variables (a,u,v,w,S) and will be solved in a 1D fashion, by taking the velocity components normal to the x-axis (v,w) unchanged through the "acoustic" waves (u-a and u+a), just as is done for the entropy. The approximated solution worked out for the 1D case can be easily transfered here (see Pandolfi (1984 b)). On the basis of the solution of this particular R.P. we can proceed to the splitting of the difference of the flux P over the Dx intervals.

The same procedure will be followed for splitting the differences of the flux G and H, respectively over the intervals Dy and Dz.

It can be easily recognized that such a procedure makes the "F.D.S." formulation coincide with the "lambda" formulation, if the linearized version of the Euler equations is used.

Quite often some transformation of coordinates is needed, which will generate a non-orthogonal grid in the physical region. In this case R.P.s have to be solved over a given interval if we desire to reproduce in the "F.D.S." formulation the features of the "lambda" formulation. In general it can be seen that each test on the sign of a given component of the velocity of a signal propagating on a characteristic ray, in order to determine the "upwind" side for the relative derivative, will require, as counterpart in the "F.D.S.", the solution of a corresponding R.P. approach. The reader may refer for this point to Pandolfi (1984 b).

5. A SUGGESTION FOR A HYBRID FORMULATION

From what has been said on the "lambda" and the "F.D.S." formulations it may be concluded that the first presents many advantages, such as precise description of the physical phenomenon, a natural way of imposing the boundary conditions, accuracy of the numerical results, simplicity in the coding and low computational times. Unfortunately it does not give correct capturing of discontinuities. On the other hand the "F.D.S." formulation seems to be particularly suitable for providing good capturing of discontinuities. Furthermore the two formulations belong to the same family of approaches and coincide for linear problems.

From our past experience in both approaches it turns out to be natural to think about a hybrid formulation, capable of picking up the positive features of the previous two.

The hybrid formulation is conceived as follows. At the beginning of each integration step we set at all the computational points an index IFORM=0. Then a sweeping is done over the whole computational region to locate in which intervals a sonic transition from the supersonic to the subsonic regime occurs. At the four points bracketing these intervals we redefine the index IFORM=1.

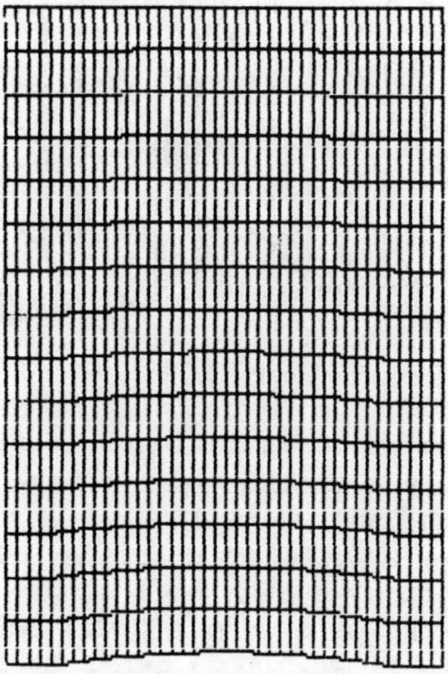

Fig. 3 Computational grid (15 x 40 intervals).

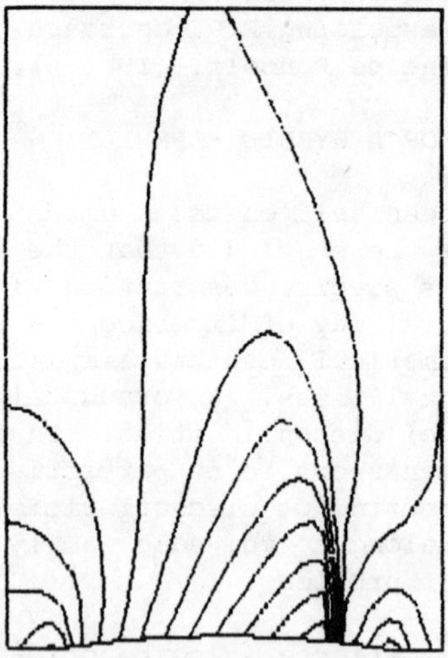

Fig. 4 Mach number contours.

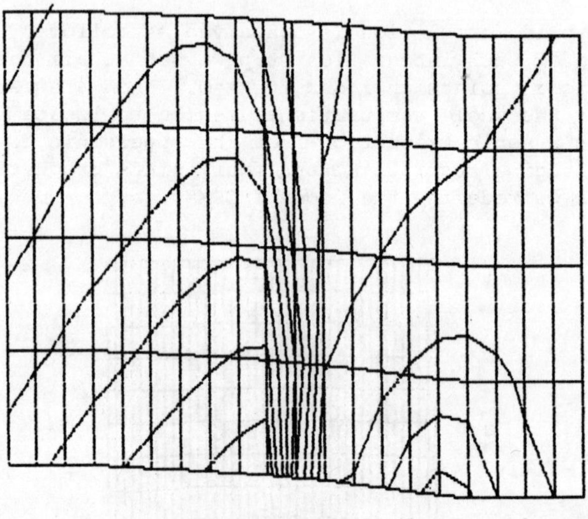

Fig. 5 Detail of the shock at the wall (Mach number contours).

Then we proceed with the computation. At those points flagged with IFORM=0 we apply a second order scheme based on the "lambda" formulation, and where IFORM=1 we adopt a numerical scheme based on the "F.D.S". formulation. In computing the results shown hereafter we have used simple first order scheme for the "F.D.S". Perhaps a second order scheme could be an advantage in certain flows. This seems to be the case for the transonic flow over airfoils, according to some preliminary results we have achieved so far.

We show here some numerical results obtained on the basis of the hybrid formulation. They refer to the test case proposed by Foerster (1978), namely the internal flow through a parallel channel with a circular bump on the lower wall. The grid we used is shown in Fig. 3. It has 15x40 intervals, equally spaced in the physical space. It is a very primitive grid and it may be modified easily by adding simple stretching both in x (to move the inlet and exit boundaries farther from the bump) and in y (to cluster more points near the bump).

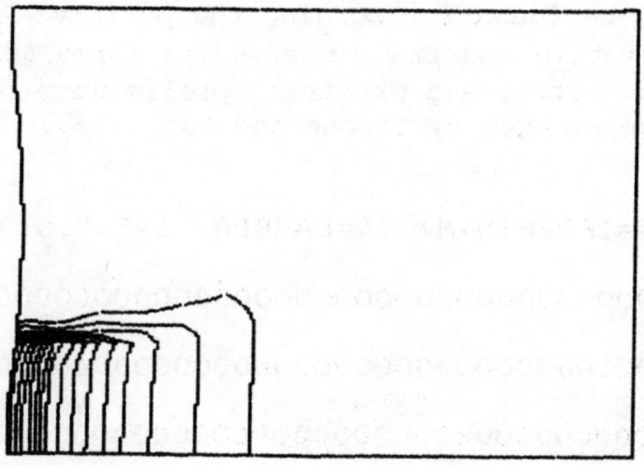

Fig. 6 Entropy contours.

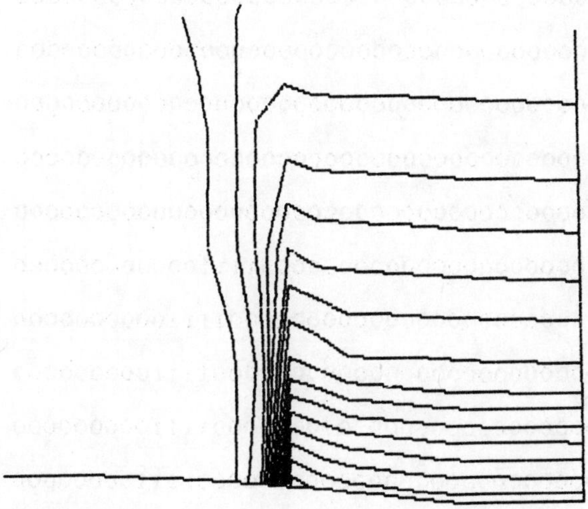

Fig. 7 Detail of the shock at the wall (Entropy contours).

The distribution of the Mach number is shown in Fig. 4. A detailed picture is given in Fig. 5, to show the structure of the captured shock. The isolevel lines of the entropy are given in Fig. 6 and the corresponding detailed picture on Fig. 7 shows the description of the rotational flow behind the shock. Finally Table 1 gives the value of the index IFORM, at the computational points. Only 24 points are computed with the time consuming "F.D.S." formulation, whereas the others are predicted with the efficient "lambda" formulation.

Other ways can be found for merging the positive features of the non-conservative and the conservative formulations. A very interesting suggestion and promising preliminary results have been recently presented by Dacone and Yagi (1985).

```
======IFORM(N,M) (O=LAMBDA  1=F.D.S.)===

0000000000000000000000000000000000000000

0000000000000000000000000000000000000000

0000000000000000000000000000000000000000

0000000000000000000000000000000000000000

0000000000000000000000000000000000000000

0000000000000000000000000000000000000000

0000000000000000000000000000000000000000

0000000000000000000000000000000000000000

0000000000000000000000000000000000000000

0000000000000000000000000000000000000000

0000000000000000000000001111000000000000

0000000000000000000000000011110000000000

0000000000000000000000000001111000000000

0000000000000000000000000001111000000000

0000000000000000000000000000111100000000

0000000000000000000000000000111100000000
```

Table 1

6. THE EXTENSION TO THE STEADY SUPERSONIC FLOWS

As mentioned in the introduction, almost everything discussed so far for the unsteady flow can be easily transfered to the supersonic steady flow.

The extension of the "lambda" formulation to supersonic problems was proposed many years ago by Moretti (1979) and after that several codes were developed for practical purposes with very good results (for instance see Moretti (1982) and Marconi (1984)).

In the field of the "F.D.S." formulation the practical extension to supersonic flows also appears quite easy. The only additional problem comes from the fact that in the approximate solution of the R.P. the simple form of the classical Riemann invariants is replaced by the Prandtl-Meyer relationship. However a linearization of this formula makes the solution of the R.P. as easy as for the unsteady flow, without any penalty in the final numerical results.

A "F.D.S." formulation and relative second order schemes are presented in a detailed form with numerical results by Pandolfi (1985) in the simple case of 2D steady supersonic flows. Work is under way for extending the procedure to 3D flows.

7. CONCLUSIONS

"Hyperbolic flows" can be investigated by numerical techniques aiming to describe the physical feature of the propagation of waves. Two steps are required in defining these methodologies. First the formulation, in which the original equations are rearranged in order to emphasize the role of the propagation of signals on characteristic rays. Then a numerical scheme is applied which should retain in the numerical approximation the "upwind" nature recognized in the previous formulation.

Two approaches have been considered here. The "lambda" and the "flux-difference splitting" formulations. The first is very appealing from all points of view except for the capability of correct shock capturing. Therefore some shock fitting procedure is required when this discontinuity appears. The second provides the numerical approximation of the weak solution of the laws of conservation and a very sharp and neat description of shock waves. However it does not perform, as far as accuracy and computational time are concerned, at the best level.

A "hybrid" formulation is proposed to pick up the positive features of the two previous approaches. It requires the "F.D.S." procedure just where a "strong" shock is detected, leaving to the "lambda" procedure the responsibility in the rest of the flow field. The preliminary results achieved in a classical test case for internal flows are quite promising and the approach looks efficient as regard accuracy and computational time. In particular no spurious sources of entropy are generated, as often happens when working only with the equations written in the conservative form.

8. ACKNOWLEDGEMENTS

The theory and results reported in the present paper come from a research activity supported partly by the "Ministero della Pubblica Istruzione" and partly by the "Fiat Aviazone".

REFERENCES

Courant, R., Isaacson, E. and Rees, M., (1952) "On the solution of nonlinear hyperbolic differential equations by finite differences," *Comm. Pure Appl. Math.*, **5**, 243.

Dadone, A. and Magi, V., (1985) "A Quasi-conservative Lambda Formulation," AIAA Paper N. 85-0088.

De Neef, T. and Moretti, G., (1980) "A Shock fitting for everybody," *Computers and Fluids*, **8**.

Foerster, K., (1978) "Boundary Algorithms for multidimensional inviscid hyperbolic flows," Notes on Numerical Fluid Dynamics, Vol. 1, Vieweg.

Godunov, S.K., (1959) "A finite-difference method for the numerical computation of discontinuous solutions of the equations of fluid dynamics". *Mat. Sbornik*, 47, 3.

Gordon, P., (1968) "The diagonal form of quasi-linear hyperbolic systems as a basis for difference equations" General Electric Final Report, NOL Contract N. 60921-7164.

Marconi, F., (1984) "Supersonic conical separation due to shock vorticity," *AIAA Journal*, **22**, 8.

Moretti, G., (1979) "The lambda scheme," *Computer and Fluids*, 7.

Moretti, G., (1982) "Calculation of three-dimensional inviscid supersonic steady flows." NASA C.R. 3573.

Osher, S. and Solomon, F., (1982) "Upwind difference schemes for hyperbolic system of conservation laws," *Maths. of Comp.*, **38**.

Pandolfi, M. and Zannetti, L., (1978) "Some tests on finite difference algorithms for computing boundaries in hyperbolic flows," Notes on Numerical Fluid Dynamics, Vol. 1 Vieweg.

Pandolfi, M. and Colasurdo, G., (1980) "Three-dimensional inviscid compressible rotational flows. Numerical results and comparison with analytical results." Flow in Primary, Non-rotating Passages in Turbomachines, ASME.

Pandolfi, M., (1984 a) "A contribution to the numerical prediction of unsteady flows," *AIAA Journal*, **22**, 5.

Pandolfi, M., (1984 b) "On the flux-difference splitting method in multidimensional unsteady flows," AIAA Paper 84-0166.

Pandolfi, M., (1985) "Computation of steady supersonic flows by a flux-difference splitting method," to appear in *Computers and Fluids*.

Roe, P.L., (1981) "Approximate Riemann solvers, parameters vectors, and differences schemes," *Journal of Computational Physics*, **43**, 2.

Zannetti, L. and Colasurdo, G., (1981) "Unsteady compressible flows: a computational method consistent with the physical phenomena," *AIAA Journal*, **19**, 7.

Zannetti, L., (1985) "On the features of the lambda scheme applied to 1D problems. Special considerations when moving to 2D," to appear.

SPLITTING UP TECHNIQUES FOR COMPUTATIONS OF INDUSTRIAL FLOWS

J.P. Benqué, J. Cahouet, J. Goussebaïle and A. Hauguel,
(Laboratoire National d'Hydraulique, E.D.F., Chatou - France)

1. INTRODUCTION

The purpose of the numerical development at the << Laboratoire National d'Hydraulique >> during the last ten years is the realistic simulation of industrial flows (Benqué et al., 1983, Grégoire et al., 1984, 1985); in practice this purpose implies the treatment of complex equations with several variables discretised over a large number of nodes. The trend is now to develop finite element codes given that the use of finite elements allows a natural treatment of complex geometry with easy local refinements. Moreover the use of the splitting up techniques seems the best way in order to offset the increase in computational requirements and also to secure the most effective solvers: actually nearly all numerical models developed in our laboratory during the decade are based on the simple idea that an algorithm is efficient when it has a single objective. Progress has been achieved over this decade regarding the precision on linking up solutions. Simple fractional steps were employed previously and practically transparent formulations are in use today. As regards industrial computations, it is convenient to draw most benefit from the regularity of the grid (as in finite differences) while preserving an accurate description of the boundaries (as in finite elements); an attractive solution then appears in the use of geometrical substructures with specific solutions. The use of splitting up techniques in the finite elements context, firstly developed for the Navier-Stokes equations, has been recently applied for more complex models such as $k - \varepsilon$ model for turbulent flow. This article is a review of the basic principles and their most recent applications concerning such various fields as domain decomposition or turbulence modelling. In section 4, these developments are illustrated by some industrial computations in 2-D or 3-D.

2. SPLITTING-UP OF OPERATORS - RONAT ALGORITHM (Benqué and Ronat, 1981)

The method of fractional steps was used very early because of the initial choice made in our laboratory: transient equations for proper processing of non-linearity; implicit methods (or to be more precise, no time-step limitation required).

The principle is quite simple; for instance to solve the classical transport diffusion equation:

$$\frac{\partial f}{\partial t} + \nabla .(\underset{\sim}{u}\, f) - K\, \Delta\, f = 0. \tag{2.1}$$

The basic idea is to make advective terms explicit by using an upwind scheme as physical as possible, and to build an accurate scheme of high order which guarantees unconditional stability with low numerical diffusion. This is obtained by upwinding along the physical characteristics defined by:

$$\begin{cases} \dfrac{dx}{dt} = u, \; t \in [t^n, t^{n+1}] \\ x^{n+1} = M, \text{ each node of the natural grid.} \end{cases} \tag{2.2}$$

The computation of these curves is a simple problem reversed in time, solved by applying a Runge Kutta method of order 2. In the finite-element approach, the idea of upwinding led RONAT (Benqué and Ronat, 1981) to introduce the use of test functions, transported along the characteristic curves.

$$\begin{cases} \dfrac{\partial \psi}{\partial t} + \underset{\sim}{u}.\nabla\psi = 0, \; t \in [t^n, t^{n+1}] \\ \psi(t = t^{n+1}, x) = \psi^{n+1}(x) = \phi(x) \quad \text{(classical Galerkin shape function)} \end{cases} \tag{2.3}$$

$\{\psi\}$ then defines two grids, the natural one at time t^{n+1} and the upwind grid at time t^n, constructed at the foot of the characteristic curves. It can be observed that this weak formulation leads to the projection of the velocity field on the space associated with the upwind grid.

At each time-step, the treatment of advective terms leads to computing a right hand term and then solving the problem as shown below in the standard equation (2.5).

After integrating the weak formulation of equation (2.1) in space and over time between t^n and t^{n+1}, and after some mathematical handling, taking equation (2.3) into account, we have:

$$\int_\Omega f^{n+1}\psi^{n+1}\,d\Omega - \int_\Omega f^n\psi^n\,d\Omega + \int_{t^n}^{t^{n+1}} dt\int_\Omega K\nabla f\nabla\psi\,d\Omega = \text{Boundary term.} \tag{2.4}$$

With a Crank-Nicolson discretisation of the diffusion term, we have:

$$\int_\Omega f^{n+1}\psi^{n+1}\,d\Omega + \frac{\Delta t}{2}\int_\Omega K\nabla f^{n+1}\nabla\psi^{n+1}\,d\Omega = \int_\Omega f^n\psi^n\,d\Omega + \frac{\Delta t}{2}\int_\Omega K\nabla f^n\nabla\psi^n\,d\Omega + \text{Boundary term.} \tag{2.5}$$

The equation of diffusion and transport then takes the form of a simple diffusion equation. The whole advective effect has been done away with by the use on the right-hand side of the upwind test functions ψ^n. This new formulation splits up the operators but is of second order over time. It is possible to devise an exact integration rule for a one-dimensional computation, and the scheme then shows outstanding properties: negligible damping and phase shift for a wave length which is greater than five mesh steps. Numerical integration is of easier use for two-dimensional computations, but introduces a degree of damping. Nevertheless, performance remains very good. Many results on classical test cases, such as the rotating cone, have been already presented (Benqué et al., 1981, 1984, 1985).

3. APPLICATION TO NAVIER-STOKES EQUATIONS

For Navier-Stokes equations, the treatment of the advective term leads to the computation of a right hand side term, as mentioned earlier, followed by the solving of a Stokes-type problem:

$$\begin{cases} \frac{1}{\Delta t} \underset{\sim}{u}^{n+1} - \nu \Delta \underset{\sim}{u}^{n+1} + \nabla p^{n+1} = \underset{\sim}{S}^{n} / \Omega & (3.1a) \\ \nabla . (\underset{\sim}{u}^{n+1}) = 0 / \Omega & (3.1b) \\ \underset{\sim}{u}^{n+1} = \underset{\sim}{u}_d / \Gamma & (3.1c) \end{cases}$$

Algorithms allowing, despite (3.1), an efficient uncoupled computation of pressure and of each velocity component were sought following the general principles mentioned above.

The CHORIN-TEMAM projection algorithm, which implies an approximate boundary condition for pressure, is one of them. It is based on a natural decomposition of $L^2(\Omega)$ (Chorin, 1968 , Temam, 1971)

$$L^2(\Omega) = H \oplus H^{\perp}$$

with

$$H = \{\underset{\sim}{u} \in L^2(\Omega) \mid \nabla . (\underset{\sim}{u}) = 0, \ \gamma_n(\underset{\sim}{u}) = 0\}$$

where γ_n is a trace operator defined on:

$$E(\Omega) = \{\underset{\sim}{u} \mid \underset{\sim}{u} \in L^2(\Omega), \ \nabla . (\underset{\sim}{u}) \in L^2(\Omega)\}$$

by: $$\int_\Gamma \gamma_n(\underset{\sim}{u}) w \, d\Gamma = \int_\Omega \underset{\sim}{u} . \nabla w d\Omega + \int_\Omega \nabla . (\underset{\sim}{u}) w \, d\Omega$$

Hence $\gamma_n(\underset{\sim}{u})$ is tantamount to $\underset{\sim}{u}.\underset{\sim}{n}$ for smooth $\underset{\sim}{u}$.

$$H^{\perp} = \{\underset{\sim}{u} \in L^2 (\Omega) \mid \underset{\sim}{u} = \nabla p, \ p \in H_1(\Omega) \}$$

To use this decomposition, we transform in (3.1) variable $\underset{\sim}{u}$ in $\underset{\sim}{v}$:

$$\underset{\sim}{u}^{n+1} = \underset{\sim}{v}^{n+1} + \nabla p_o$$

with ∇p_o, a lifting of $\underset{\sim}{u}_d.\underset{\sim}{n}$ on to Ω, the solution of:

$$\begin{cases} \Delta p_o = 0/\Omega \\ \frac{\partial p_o}{\partial n} = \underset{\sim}{u}_d.\underset{\sim}{n}/\Gamma \end{cases} \qquad (3.2)$$

Problem (3.1) becomes: Find $(v^{n+1}, \nabla p^{n+1}) \in H \times H^{\perp}$ such that:

$$\begin{cases} \frac{1}{\Delta t} \underset{\sim}{v}^{n+1} - \nu\Delta \underset{\sim}{v}^{n+1} + \nabla p^{n+1} = \underset{\sim}{S}^{n} - \frac{1}{\Delta t} p_o - \nu\Delta\nabla p_o/\Omega \\ \underset{\sim}{v}^{n+1} = \underset{\sim}{u}_d - \nabla p_o/\Gamma \end{cases} \tag{3.3}$$

The CHORIN-TEMAM algorithm is then the following:

For a given estimate of pressure p_m^{n+1}:

. Firstly compute $\underset{\sim}{v}^{n+1}$ solution of (3.3)

. Secondly consider the quantity:

$$\underset{\sim}{f}_m^{n+1} = \frac{1}{\Delta t} \underset{\sim}{v}_m^{n+1} + \nabla p_m^{n+1} \tag{3.4}$$

At convergence the right-hand side will be the decomposition of $\underset{\sim}{f}^{n+1}$ on H and $H^{\perp}$. Thus, project f_m^{n+1} on $H^{\perp}$ to obtain a better estimate of the pressure, and solve

$$\int_\Omega \nabla p_m^{n+1} \nabla\psi \, d\Omega = \int_\Omega \underset{\sim}{f}_m^{n+1} . \nabla\psi \, d\Omega = \int \frac{1}{\Delta t} \underset{\sim}{v}_m^{n+1} \nabla\psi d\Omega + \int_\Omega \nabla p_m^{n+1} \nabla\psi \, d\Omega. \tag{3.5}$$

$\nabla\psi$ is a genuine function of $H^{\perp}$, when ψ is a shape function of $H_1(\Omega)$, the space where we seek the pressure.

. Thirdly iterate eventually on pressure.

An other approach is the UZAWA algorithm. Actually (3.1) is a saddle point problem on $(\underset{\sim}{u}^{n+1}, p^{n+1})$. For a given pressure p_m^{n+1} and the associated velocity field $\underset{\sim}{u}_m^{n+1}$ (solution of (3.1a) - (3.1c)), the quantity $\nabla(u_m^{n+1})$ is the gradient of the functional to be minimised in order to reach the saddle point. UZAWA algorithm is then a gradient algorithm to solve (3.1); actually we develop a conjugate gradient version. In order to accelerate the convergence we have searched an efficient preconditioner, a natural idea is then to use the previous projection operator:

$$\begin{cases} \int_\Omega \nabla s_{m+1} \nabla\psi \, d\Omega = \int_\Omega \nabla (\underset{\sim}{u}_m^{n+1}) \psi \, d\Omega \\ d_{m+1} = s_{m+1} + \beta_m d_m \\ p_{m+1} = p_m + \rho_m d_{m+1} \end{cases} \tag{3.6}$$

UZAWA algorithm with preconditioning improves the efficiency of CHORIN-TEMAM algorithm by use of optimal direction and optimal steepness parameter (Labadie and Lasbleiz, 1983).

4. DOMAIN DECOMPOSITION

Jointly to this study of efficient Navier-Stokes solvers for 3-D problems, have been developed new algorithms of domain decomposition. They allow one to divide a computational domain into substructures and obtain independent problems on each subdomain by the use of a simple symmetrical boundary operator.

The main aim of the geometric splitting is to pull out the most available regular part (i.e. topologically equivalent to a rectangle) from the initial domain.

Although industrial cases require locally accurate discretisation these regular substructures can represent from 50% to 90% of the computational domain, the lower bound corresponding to confined flows (like the study of thermal-hydraulics of fast breeder reactors, cf (Esposito and Hauguel, 1983) and the upper bound to environmental problems (like modelling of the flume of a cooling tower, cf. (Benqué et al., 1983), (Grégoire and Goussebaile, 1984).

On these regular parts, we will be able to use specific tools like grid discretisation with IJK indexation, multigrids methods, ... which allow a great reduction in computational cost even in a finite-element discretisation concept.

Further, because of their high degree of parallelism, these methods are well suited for solution by multiprocessor computers and let one solve problems involving several mathematical models according to the region under consideration (a typical example is the matching of a viscous flow and inviscid flow for aerodynamic problems).

Given the concept of operator splitting mentioned above, the main difficulty lies in matching the solution of the Stokes problem. Since characteristic curves have been used explicitly, advective terms are discarded on the right hand side; near an inside junction boundary of a subdomain where there is an inflow the advective term is as easy to compute as in any other part of the subdomain; indeed, the time-space transformation applied to the contribution of the boundary term allows its computation to be effected over the neighbouring subdomain using the characteristic curves (see Cahcuet (1985) for further detail).

As regards the Stokes problem, adequate boundary conditions should be determined along the junction boundaries to uncouple the solution operation in each subdomain. The analysis of the continuous equations shows that the global solution obtained on the whole domain is the unique solution which satisfies the Stokes problem on each subdomain and ensures the continuity of velocity and stresses along the junction: from this main result due to MARTINAUD (1984) follows clearly two formulations; seek the Dirichlet conditions relating to velocity which give equality of the stresses (resulting from Stokes equations) on each side of the junction (primal method) or vice versa (dual method). These two approaches can be formulated in terms of linear boundary operators which are symmetric and V-elliptic.

For the sake of simplicity, here we only consider the primal appproach applied on a domain Ω split in two parts Ω_i $i = 1,2$ without overlapping. As shown in Fig. 1, Σ and Γ_i denote respectively the junction boundary and the physical boundary of the subdomain Ω_1.

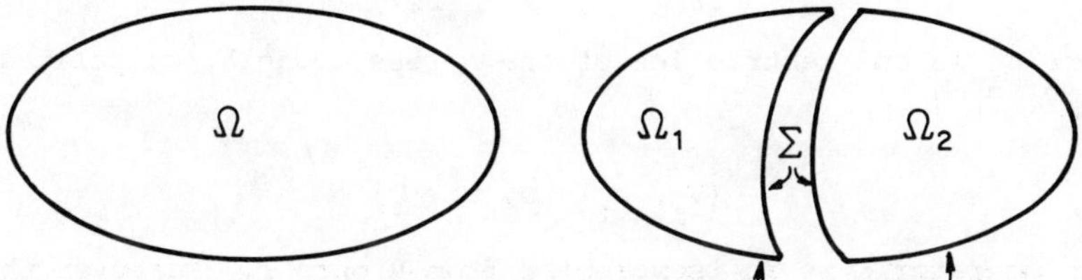

Fig. 1 Splitting of domain Ω.

In order to simplifiy the required expressions we introduce the following notation:

$$(\underset{\sim}{V}_i, p_i) \{\underset{\sim}{u}_{\Gamma_i}, \underset{\sim}{u}_\Sigma, \underset{\sim}{S}_i\}$$

which are the velocity and pressure solutions of Stokes problems on Ω_i (cf. equation above) corresponding to $\underset{\sim}{u}_\Sigma$ and $\underset{\sim}{u}_\Gamma$ boundary conditions on $\partial\Omega_i$ and second member $\underset{\sim}{S}_i$ on Ω_i;

and two function spaces in conjunction with the restrictions of the velocity $\underset{\sim}{V}_i$ and the stresses $\underset{\sim}{\sigma}_i$ along Σ:

$$V = \{\underset{\sim}{v} \in H^{1/2}_{oo} , \int_\Sigma \underset{\sim}{v}.\underset{\sim}{n} \, d\Gamma = 0\}$$

where $H^{1/2}_{oo}$ denotes the interpolate between the classical Sobolev spaces $H^1_o(\Sigma)^N$ and $L^2(\Sigma)^N$, (these conditions taking into account the decreasing of velocity near $\Gamma_i \wedge \Sigma$),

$$T = H^{-1/2} (\Sigma)^N / \mathcal{R}$$

where $\mathcal{R}$ denotes the equivalence relation:

$$\forall (\underset{\sim}{\sigma}\ \underset{\sim}{\tau}) \in H^{-1/2}(\Sigma)^{2N}, \ \underset{\sim}{\tau}\mathcal{R}\underset{\sim}{\sigma} \Leftrightarrow \underset{\sim}{\tau} - \underset{\sim}{\sigma} = C\,\underset{\sim}{n}, \ C \in \mathbb{R}.$$

We are able now to introduce the primal matching operator K mapping V onto T as

$$K: \ V \rightarrow T$$

$$\underset{\sim}{u}_\Sigma \rightarrow \underset{\sim}{K}(u_\Sigma) = \underset{\sim}{\sigma}_2 - \underset{\sim}{\sigma}_1,$$

where σ_i is the restriction of the stress along Σ, calculated from

$$(\underset{\sim}{V}_i, p_i) \{\underset{\sim}{0}, \underset{\sim}{u}_\Sigma, \underset{\sim}{0}\}$$

Then operator K is an isomorphism from V onto T. Moreover the associate bilinear form k(.,.) defined by:

$$k: \ V \times V \rightarrow \mathbb{R}$$

$$(\underset{\sim}{u}_\Sigma, \underset{\sim}{v}_\Sigma) \rightarrow k(\underset{\sim}{u}_\Sigma, \underset{\sim}{v}_\Sigma) = \langle K\, \underset{\sim}{u}_\Sigma, \underset{\sim}{v}_\Sigma \rangle$$

(where <.,.> denotes the duality pairing between V and T) is continuous symmetric and V elliptic (see Martinaud, (1984) for proof). Let us now introduce the stress $\underset{\sim}{\sigma}_1$ obtained from ($\underset{\sim}{V}_i$ and P_i) $\{\underset{\sim}{u}_{\Gamma i}, \underset{\sim}{0}, \underset{\sim}{S}_i\}$: it is obvious considering the linearity of the Stokes equations that the sought boundary condition is the unique solution of:

$$K(\underset{\sim}{u}_\Sigma) = -(\underset{\sim}{\sigma}_2 - \underset{\sim}{\sigma}_1)$$

because of the continuity of the stresses along Σ.

Thus the numerical resolution of time dependent Navier-Stokes equations requires at each time step four stages:

A - Calculation of S_i, the right hand side resulting from advection;

B - Calculation of the stresses jump along the junction by solution of a Stokes problem on each subdomain;

C - Inversion of the discretized matching operator (which is a symmetric positive definite matrix whose rank is equal to the number of velocity degrees of freedom on Σ) to obtain $\underset{\sim}{u}_\Sigma$

D - Calculation of the final velocity and pressure fields on each subdomain with the previously computed boundary condition.

Note: Steps A, B and D which represent more than 95% of CPU cost, can be carried out in parallel on multi-processor computers; only step C requires a task synchronization.

As regard to the good properties of the matching matrix K, it is advisable to use an iterative solver like conjugate gradient with scaling, which avoids assembling, factorization and storage of a full matrix.

Many numerical tests of domain decomposition applied to the Navier-Stokes and shallow water equations (cf. Cahcuet, (1985)) have proved the strength and efficiency of these methods. The first example in figure 2 compares to a reference solution (global computation of the << driven cavity >> flow) the velocity and pressure fields obtained with a 3 subdomain decomposition using a Taylor-Hood finite element discretization. Then in figure 3 is considered the Stokes flow around a cylinder, a configuration which allows an important regular subdomain (75% of the whole domain).

These tests and many others show that domain matching remains costly if the regularity of the subdomains is not exploited about five time more in CPU cost for half memory used. This ratio decreases when the size of the mesh grows (due to the growth of the time consumed for a linear system solution) and these methods become fully competitive for industrial cases, if the use of regularity can provide a reduction by a factor ten of CPU cost on regular subdomains.

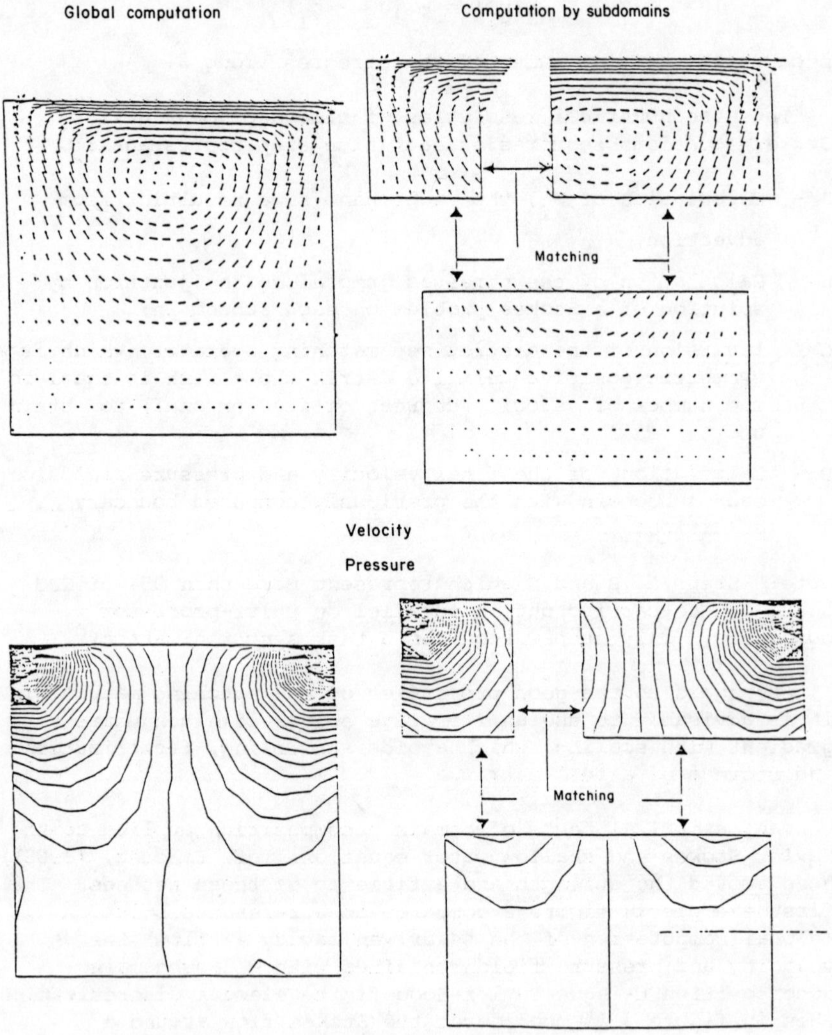

Fig. 2 Solving the Stokes problem with domain decomposition: the wall driven cavity test case

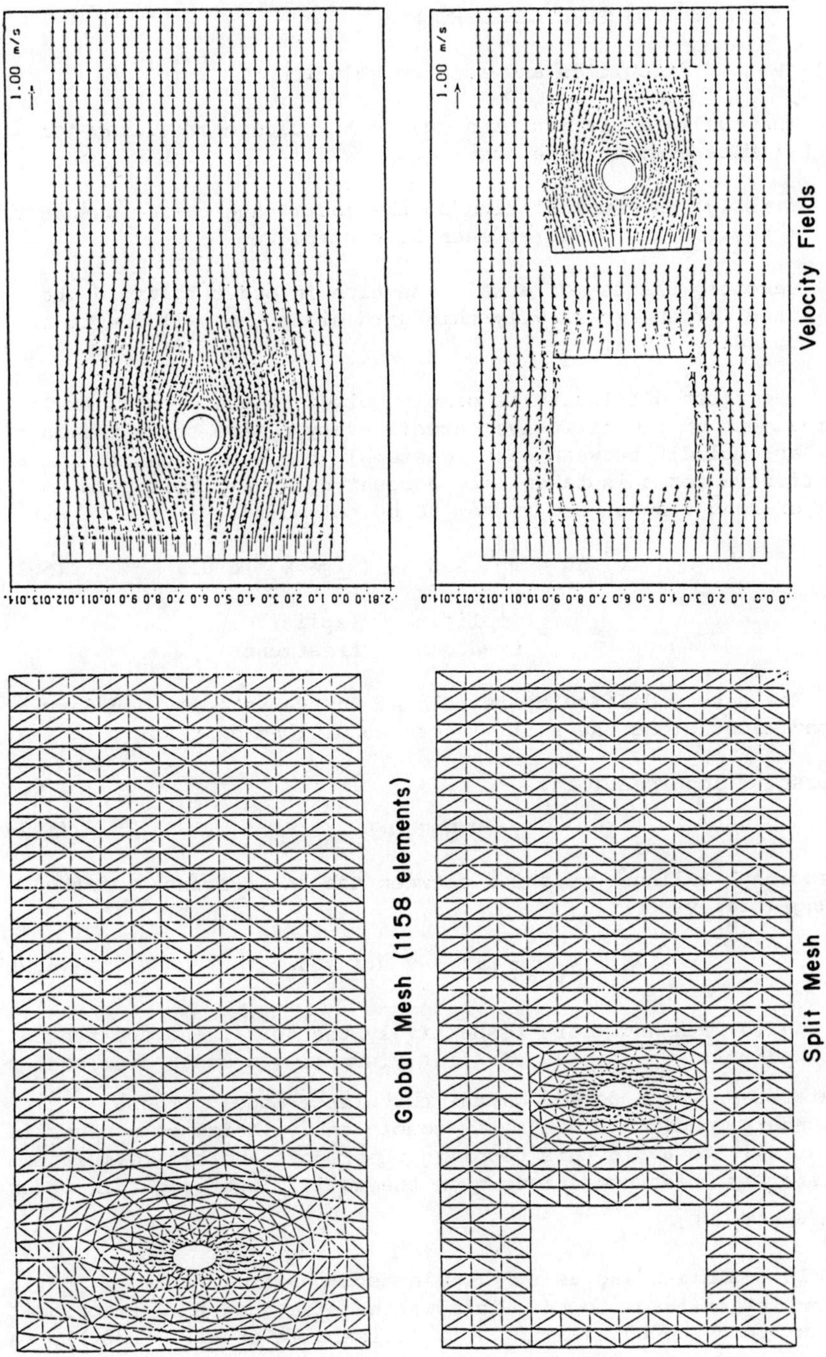

Fig. 3 Example of domain decomposition for a Stokes problem: flow past a cylinder.

5. TREATMENT OF TURBULENCE MODELS

a) Main difficulties and adopted solutions

The two main difficulties due to turbulence modelling for a 3 D finite element code are:

- firstly the recalculation of the matrix due to the change of the viscosity field at each time step,
- secondly the use of a wall function in order to avoid the need for an excessively thin grid along the wall due to the boundary layer.

The first difficulty is simply solved by use of semi explicit treatment of the diffusive terms: every p time steps the matrix is assembled; between each reassembly the evolution of diffusive terms is taken into account by an explicit term on the right-hand side of the system to be solved:

$$\nu_t^n \Delta \underset{\sim}{u} = \underbrace{\nu_t^{n_o}}_{\text{implicit treatment}} \Delta \underset{\sim}{u} + \underbrace{(\nu_t^n - \nu_t^{n_o})}_{\text{explicit treatment}} \Delta \underset{\sim}{u}. \tag{5.1}$$

The second difficulty is related to the velocity boundary conditions along the wall. These conditions express:

Firstly, impermeability:

$$\underset{\sim}{u}.\underset{\sim}{n} = 0. \tag{5.2a}$$

Secondly, a linear relation between the shear stress and the tangential velocity:

$$\nu_t \frac{\partial}{\partial n} \underset{\sim}{u}.\underset{\sim}{\tau} + \alpha\, \underset{\sim}{u}.\underset{\sim}{\tau} = \beta. \tag{5.2b}$$

In (5.2) $\underset{\sim}{n}$ and $\underset{\sim}{\tau}$ are respectively the normal and the tangent to the wall, ν_t is the turbulent viscosity, α and β are local parameters which depend on the turbulence model. These boundary conditions naturally couple the velocity components in general; to avoid this coupling we propose another formulation where Dirichlet conditions $\underset{\sim}{q}$ on the wall Γ_w are sought in order to verify (5.2).

In a similar way as for domain decomposition, this formulation leads to the minimization of a quadratic functional on $\underset{\sim}{q}$ along the boundary Γ_w (cf. Goussebaile et al., (1984)). A conjugate gradient algorithm, of the following form, is then justified:

For a given estimate of $\underset{\sim}{q}$, $\underset{\sim}{q}_m$, along the boundary Γ_w:

- Firstly compute by UZAWA algorithm the associated velocity and pressure fields solutions of (3.1a) and (3.1b),
- Secondly compute the boundary error in the profile along Γ_w using (5.2) and deduce a new estimate $\underset{\sim}{q}_{m+1}$ of $\underset{\sim}{q}$
- Thirdly iterate until convergence on (5.2b).

In this form the algorithm is still expensive so we have empirically developed an alternative which consists in going only once through the UZAWA algorithm for pressure at each iteration on q. This alternative form is then following:

$$\underset{\sim}{u}(\underset{\sim}{q}_m, p_m) \underset{(5.2b)}{\rightarrow \text{boundary error} \rightarrow} \underset{\sim}{u}(\underset{\sim}{q}_{m+1}, p_m)$$

$$\underset{(3.1b)}{\rightarrow \text{divergence error} \rightarrow} \underset{\sim}{u}(\underset{\sim}{q}_{m+1}, p_{m+1})$$

The algorithm stays very close to the original UZAWA algorithm and saves much CPU time, its efficiency seems to be due to the relative independence between pressure and tangential component of the normal stress.

Before introducing any refined eddy viscosity model, our purpose was to study the behaviour of this algorithm. In the test case, presented in figure 5, a simple mixing length turbulence model and boundary conditions such as (5.2) on the walls were used. The computation of the eddy viscosity through more complex models ($k - \varepsilon$) does not introduce any new difficulty compared to the first approach. Test cases concern inflow through tubes cluster (fig. 5) and two cases have been treated: smooth wall (pure slip conditions ($\alpha = 0$, $\beta = 0$ along tubes in (5.2) and rough wall ($\alpha = 20$, $\beta = 0$ along tubes in (5.2)). The influence of α is very important for the structure of flow; a rough wall implies development of large vortices between tubes; the influence of turbulence modelling is very limited due to the small Reynolds number (10^2) chosen in this test. We notice in figure 4 the quality of convergence of this algorithm concerning divergence error and velocity boundary error whatever the value of α.

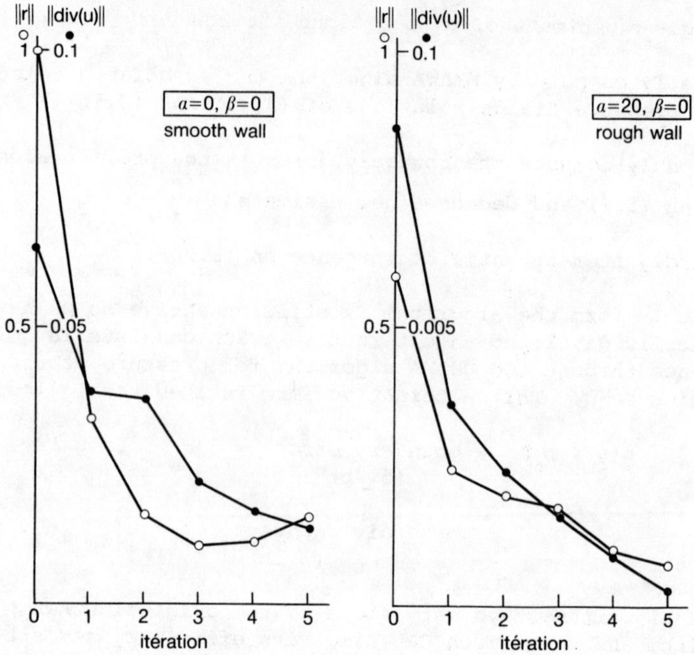

Fig. 4 Convergence of the preconditioned UZAWA algorithm associated with the treatment of the wall function

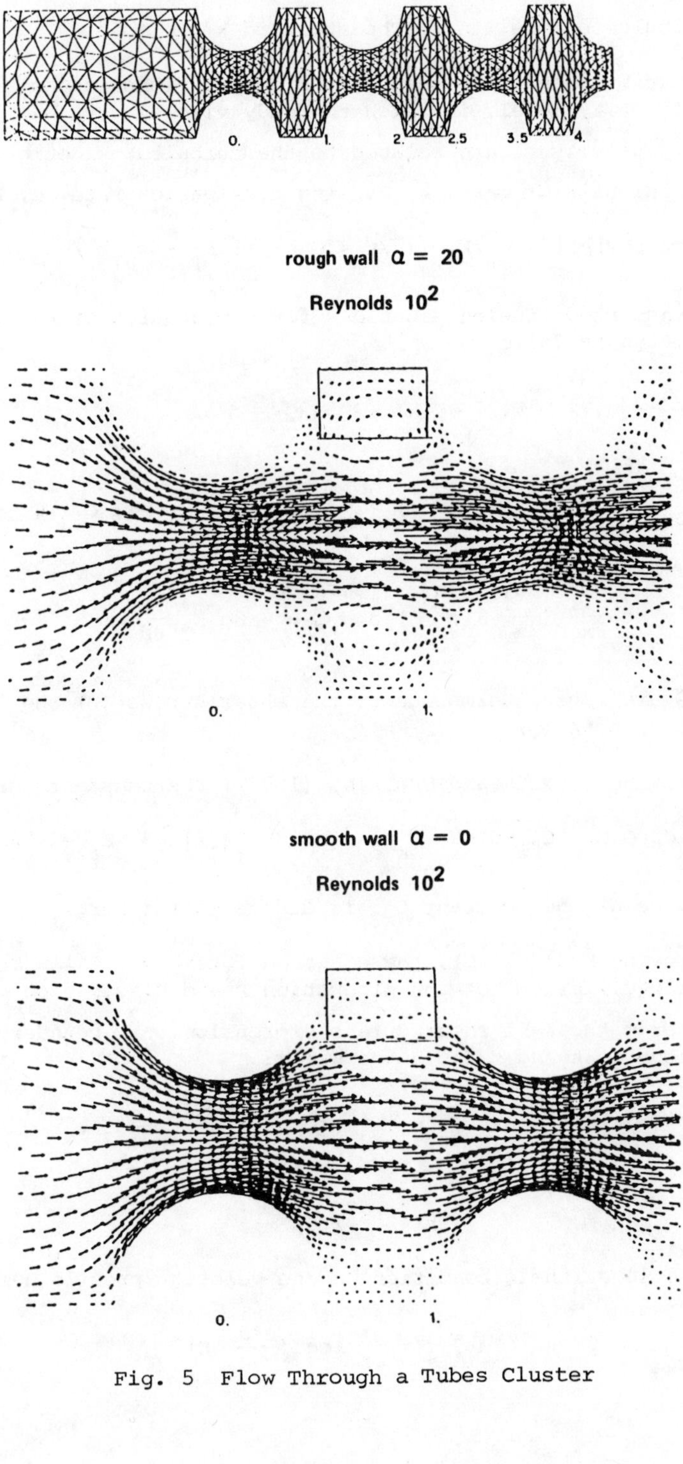

Fig. 5 Flow Through a Tubes Cluster

b) Turbulence modelling. The standard k - ε model

The turbulence model used in the computations is the k - ε eddy viscosity model, in which the eddy viscosity ν_t and the eddy diffusivity K_T are related to the turbulent kinetic energy per unit of mass, k, and its dissipation rate, ε, by:

$$\nu_t = C\mu[V][L] = C_\mu\sqrt{k}.\ \sqrt{k}^3/\varepsilon = C_\mu k^2/\varepsilon;\ K_T = c_{\mu T}k^2/\varepsilon \qquad (5.3)$$

The transport-diffusion equations for computation of k and ε are written as follows:

$$\begin{cases} \dfrac{\partial k}{\partial t} + \underset{\sim}{u}.\nabla k = \nabla.(\dfrac{\nu t}{\sigma_k}\nabla k) + \mathbb{P} + \mathbb{G} - \varepsilon & (5.4a) \\ \dfrac{\partial \varepsilon}{\partial t} + \underset{\sim}{u}.\nabla\varepsilon = \nabla.(\dfrac{\nu t}{\sigma\varepsilon}\nabla\varepsilon) + \dfrac{\varepsilon}{k}(C_{\varepsilon 1}\mathbb{P} + C_{\varepsilon 1}(1-C_{\varepsilon 3})\mathbb{G} - C_{\varepsilon 2}\varepsilon) & (5.4b) \end{cases}$$

with $\mathbb{P} = \nu t\ \dfrac{\partial u_i}{\partial x_j}(\dfrac{\partial u_i}{\partial x_j} + \dfrac{\partial u_j}{\partial x_i})$ and $\mathbb{G} = -\beta g\ K_T\dfrac{\partial T}{\partial z}$

$\mathbb{P}$ and $\mathbb{G}$ are source terms due to the shear stress ($\mathbb{P}$) and thermal effects ($\mathbb{G}$).

Following LAUNDER and SPALDING (1974), the constants are:

$$C\mu=0.09 \quad C\mu_T=0.09 \quad \sigma_k=1 \ \sigma_\varepsilon=1.3 \quad C\varepsilon_1=1.44 \quad C_{\varepsilon 2}=1.92$$

The choice of the constant $C_{\varepsilon 3}$ is discussed further.

Following Rodi (1980), the equation for k near walls reduces a local equilibrium between production $\mathbb{P}$ and dissipation ε; we deduce from the wall shear stress expression u^{*2}, boundary conditions on k and ε along the walls:

$$\varepsilon = \mathbb{P} = u^{*2}\frac{\partial}{\partial n}\underset{\sim}{u}.\underset{\sim}{\tau} = u^{*4}/\nu t = u^{*4}.\ \frac{\varepsilon}{C_\mu k^2}.$$

Hence

$$k = u^{*2}/\sqrt{C\mu} \qquad (5.5)$$

From the logarithmic condition on the velocity profile near the walls:

$$\underset{\sim}{u}\ .\ \underset{\sim}{\tau} = \frac{u^*}{K}\log(\frac{u^*d}{\nu}E) \qquad (5.6)$$

with d: distance to the wall; K: Karman constant, E: roughness parameter we deduce << boundary condition >> on ε << near >> walls:

$$\varepsilon = u^{*3}/(K.d) \tag{5.7}$$

Boundary conditions on velocity along the walls are the following:

$$\begin{cases} \tilde{u} \cdot \tilde{n} = 0 \\ \nu \dfrac{\partial \tilde{u}}{\partial \tilde{n}} \cdot \tilde{\tau} = - u^* \, |u^*| \end{cases} \tag{5.8}$$

The variables k and ε are treated quadratically. Actually a standard explicit treatment of the source terms in the k and ε equations being unstable, these terms are upstreamed by integration along the characteristics during the advective step.

Finally in the case of k - ε turbulence modelling the algorithm becomes the following:

At each time step t^n:

Firstly deduce the eddy viscosity ν_t and the eddy diffusivity α_t by projecting k^2/ε; during this step ν_t and α_t are treated up to now as quadratic unknowns at the left hand side and k and ε are computed explicitly at the right hand side.

Secondly compute u* along the walls by solving iteratively (5.6) at each wall node using the explicit values of the velocity.

Thirdly compute the advection of all the variables; during this step the source terms of the k and ε equations are integrated along the characteristics.

Fourth compute the diffusion of all the variables other than velocity; at this point the temperature, the kinetic energy and the dissipation at time t^n are known.

Fifth solve the Stokes problem; this is accomplished by the previous algorithm which allows the decoupling of each velocity component despite the wall conditions.

A first illustration of this algorithm is given by the test case of the 2 D 'BOYLE and GOLAY' CELL; this test case was submitted at the last AIRH working group on refined flow modelling stated in Cadarache - France, January, 1985. The set of boundary conditions given on figure 6 were requested.

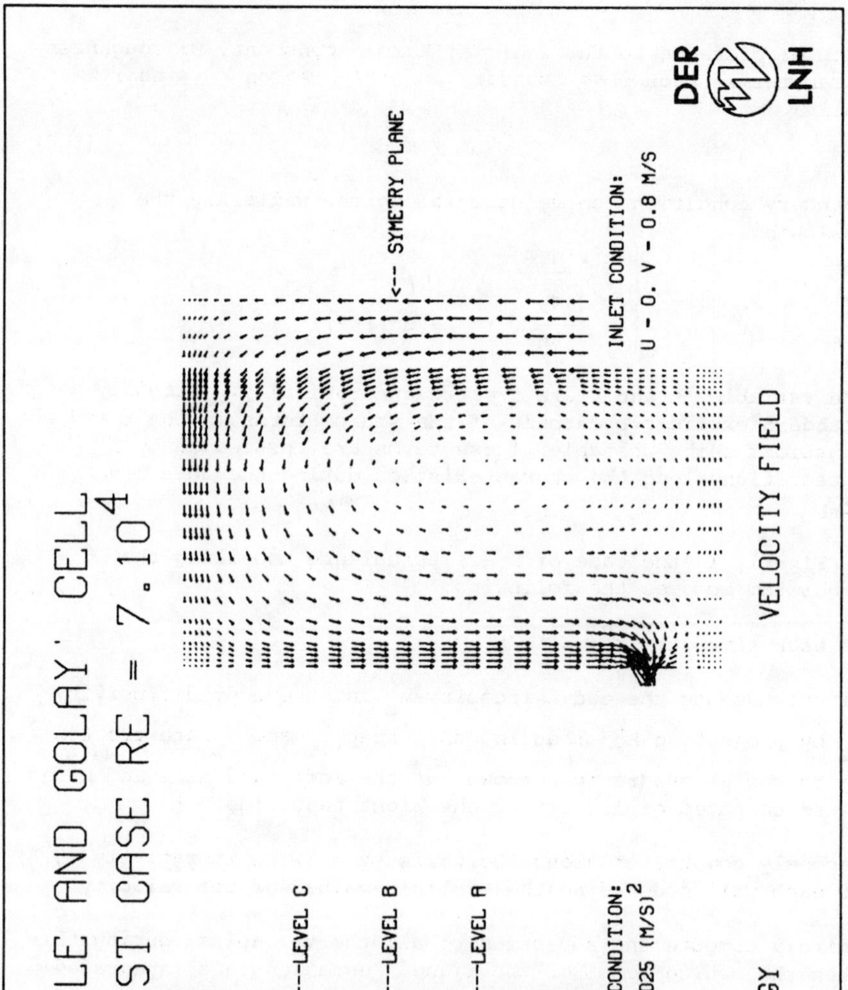

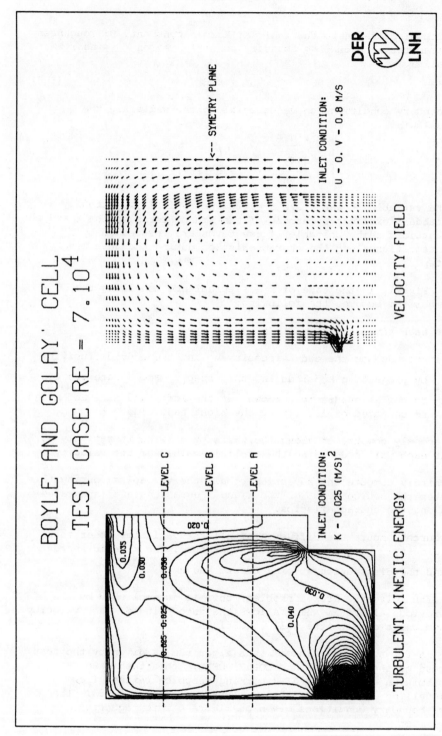

Fig. 6 Data for the test of Fig. 7.

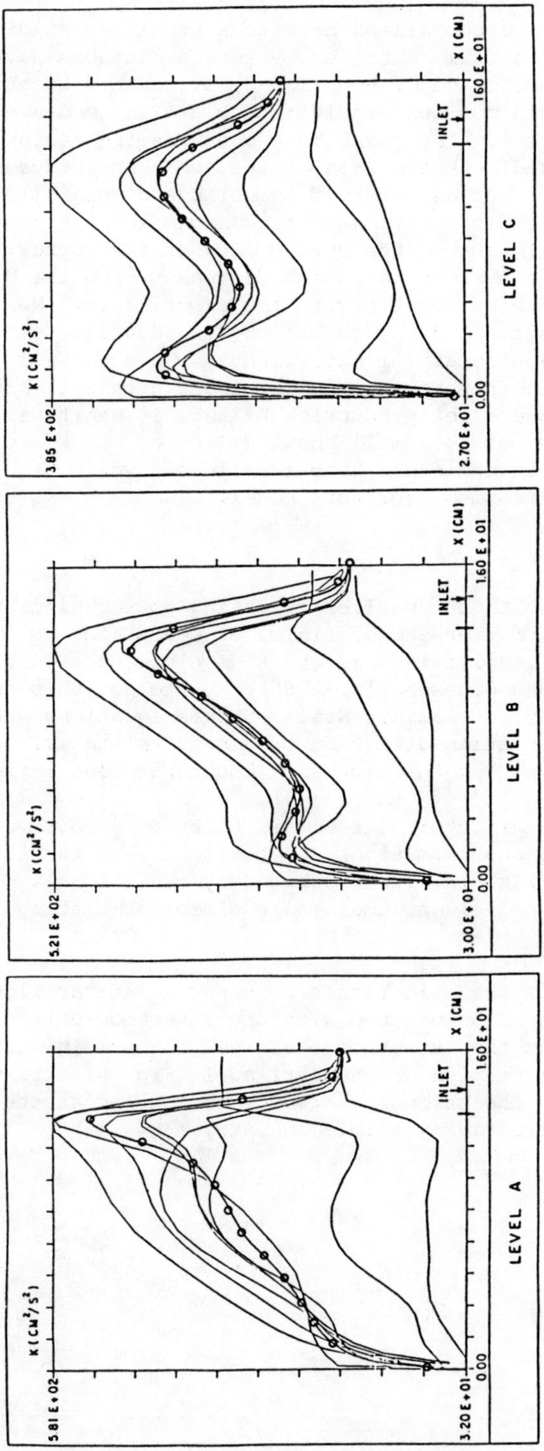

Fig. 7 Boyle and Golay' Cell: Turbulent Kinetic Energy

In figure 7, a comparison between k profiles, at different levels of the cell, obtained by several authors have been done by M. GRANDOTTO and C. BOUFFINIER, organizers of the Meeting. We have added our own results; they are in good accordance with the global set. The peaks in k occur in the regions where the velocity gradients are steepest, since k is produced by mean flow shear. But as observed by BOYLE and GOLAY (1983), the calculated k distributions differ significantly from the experimental data on the upflow side of the vortex; the magnitude is overestimated and decreases with the elevation of the jet which is contrary to the experiments. Near the outer wall we observe a decay of k along the downflow streamline in the same fashion as for experiments. This decay is due to the fact that in this region the dissipation rate of k has risen to the point where net production balance is negative. This test case points out some well known defect of the k - ε model in jet flow; on the other hand it allows a validation of our numerical treatment for this model.

6. INDUSTRIAL COMPUTATIONS

According to the earlier splitting up techniques, a three dimensional Navier-Stokes finite elements code N 3 S (Navier-Stokes, 3 D, Splitting) is under development at E.D.F. (Gregoire and Goussebaile, 1985). Up to now this code allowed the treatment of laminar Navier-Stokes equations associated with the advection-diffusion of a passive scalar. Recently a simple mixing length turbulence model has been implemented.

In the near future our aim is to treat an active scalar (Boussinesq approximation) and besides k - ε turbulence model associated with logarithmic boundary conditions. N 3 S will then be an efficient numerical tool for simulating industrial flows.

Up to now the main result concerns a laminar flow behind a cooling tower, associated with the advection-diffusion of a pollutant at the outlet of the tower. The main characteristics of this computation are summarized in Fig. 8; its main interest is the strongly three dimensional character of the flow in a relatively simple geometry.

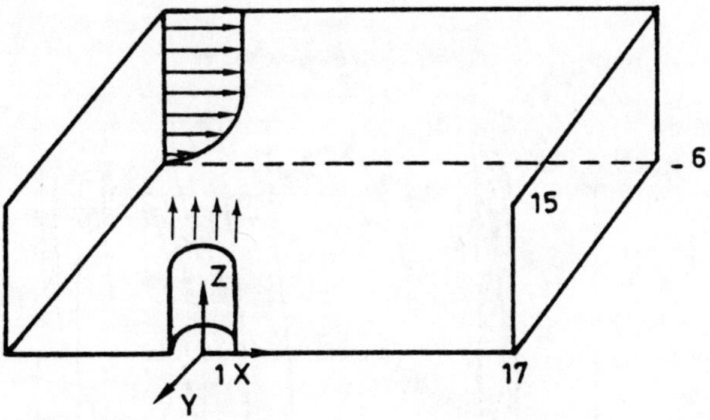

Fig. 8 Jet in a Crossing Wind

Mesh: 2844 tetrahedrons, 4653 nodes

Boundary conditions:

. crossing wind velocity at the inlet of the domain : 36ν/D

. velocity at the outlet of the cooling tower : 64ν/D

. purely slip conditions at the top of the domain

. no slip conditions at the bottom of the domain

. free outlet of the domain

Computer: Cray 1

CPU time for 1 time step:

Advection: 2.0 s Stokes problem: 11.3 s

Core requirements for matrix

Uzawa solver-packing mode: 0.4 million words.

From a physical point of view despite a rough mesh the computation reproduces quite well the shearing of the jet, due to the crossing wind. One can notice on Fig. 9 the ascent of the central part of the jet, while at the same time, the peripherical part is deflected by the wind; the computation reproduces the typical formation of rolls due to the shearing and to the deflection. Furthermore one can notice the separation of the flow and the recirculation behind the tower (figures 9, 10). The evolution of the pollutant essentially produced by the advection shows on figure 11 the evolution in << horseshoe >> of the jet. All these reults concerning a complex three dimensional flow are encouraging for the near future capabilities of N 3 S.

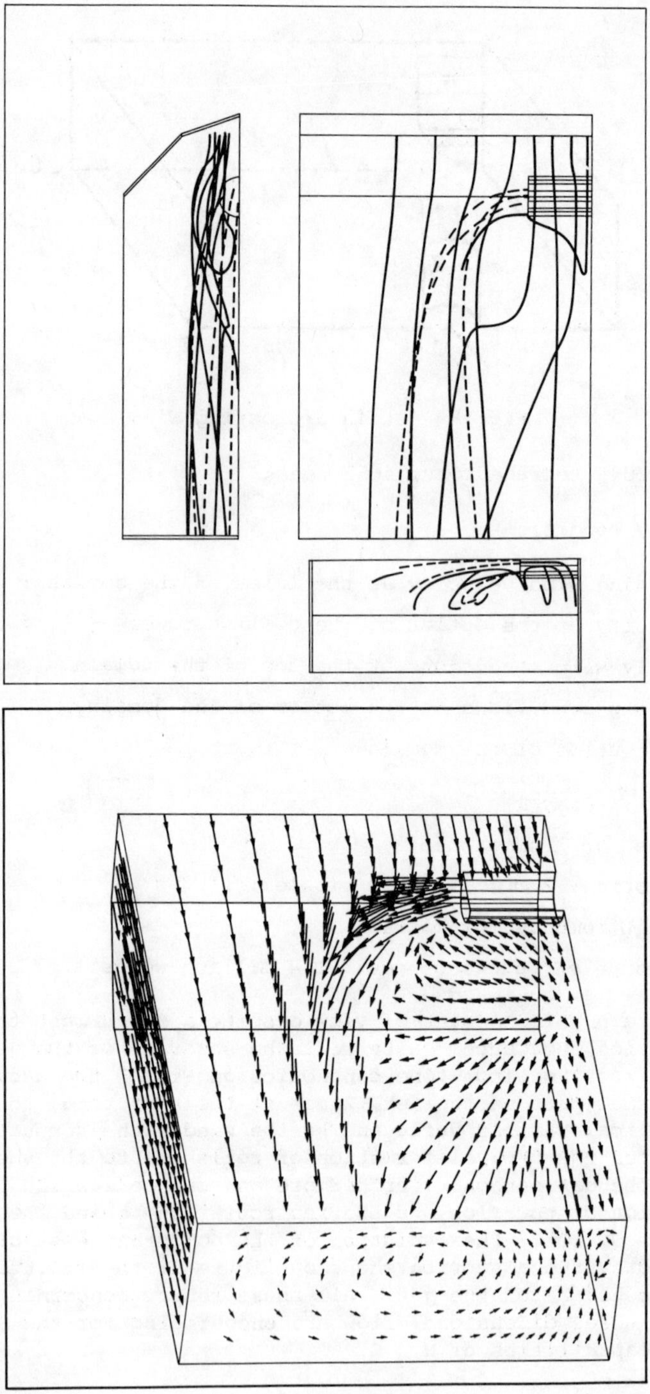

Fig. 9 and 10.

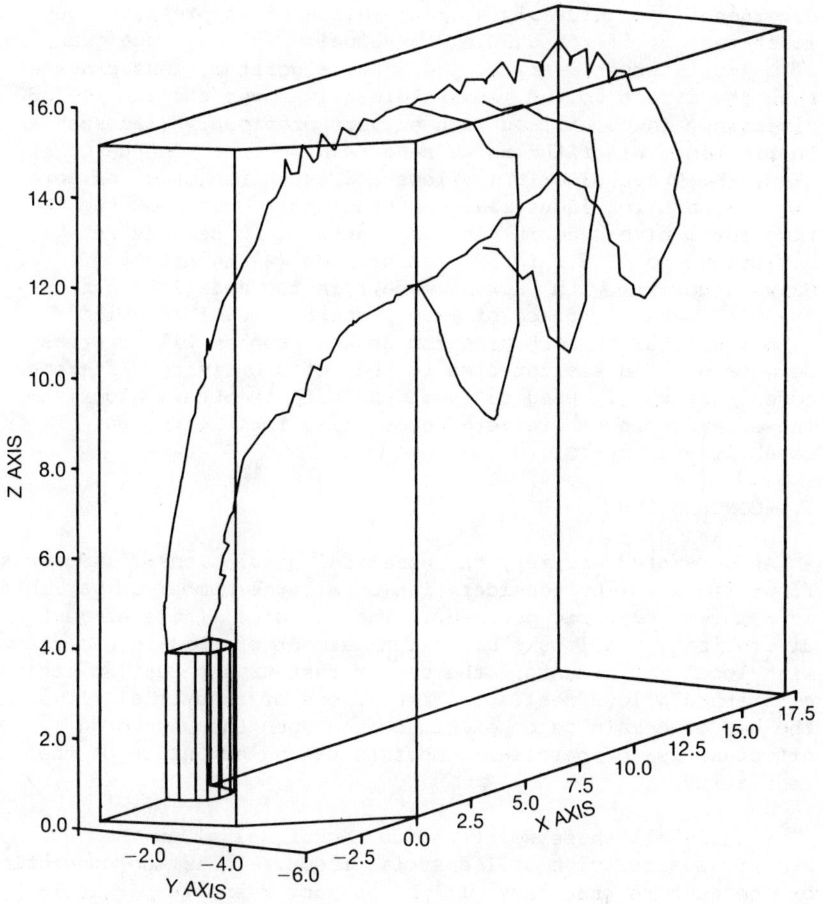

Fig. 11 Cooling Tower in a crossing wind: Advection - Diffusion of a pollutant

From the numerical point of view, the core requirements on Cray 1 is 0.4 million words for Uzawa algorithm using packing mode; the inner solver for linear systems is then a conjugate gradient preconditioned by an incomplete Cholesky factorization. This value is to be compared with the 17 million words needed by the global solver previously used on the same example: this solver used the Cholesky factorization of the global Stokes problem discretization and the resolution was made by blocks. Despite achieving a better numerical precision, all other things being equal, by use of a direct solver, the Uzawa algorithm with packing mode will be preferred: the loss of accuracy results from the divergence, for which great precision in thermohydraulic problems is not required; the error divergence has naturally a great influence on pressure, but a moderate one, if reasonable, on velocity which is the quantity of major interest. Anyway the Uzawa algorithm, less precise than the direct solver, nevertheless improves the quality of divergence in comparison with solvers previously used such as Chorin-Temam algorithm as we have seen before. On the other hand, the Uzawa algorithm allows a drastic reduction of core requirements for industrial computations. Up to now the CPU time for 1 time step in the computation is 2 seconds for advection and 11.3 s for Stokes problem (4 iterations of Uzawa algorithm with packing mode); in the near future a Cray XMP will take the place of Cray 1 and by use of a cabled gather-scatter the CPU time for Stokes problem will decrease down to 6 s: a similar time to that of 3 D finite difference codes intensively used today at LNH with the Uzawa algorithm. These performances are very encouraging for the future capability of N 3 S.

7. CONCLUSIONS

As we stated earlier, the numerical simulation of industrial flows requires the consideration of a large number of variables in complex geometric patterns. The use of a finite element discretization allows a natural treatment of complex geometry with local refinements; the use of fast transparent splitting algorithms allows realistic simulations of industrial flows; the use of domain decomposition could open the way to an efficient use of parallel computers taking advantage of the regularity.

Finally all these splitting up techniques allow the numerical simulation of industrial flows. It is the conviction of the authors that they will be in many respects particularly useful for fully 3 D complex flow simulations of high quality.

REFERENCES

Benqué, J.P., Daubert, O., Goussebaile, J. and Hauguel, A., (1984) Splitting up Techniques for computations of industrial flows - Rapport EDF HE/41/84.09.

Benqué, J.P., Grégoire, J.P., Hauguel, A. and Maxant, M., (1983) Application des méthodes de décomposition aux calculs numériques en hydraulique industrielle - Sixth International Symposium on Computing Methods in Applied Sciences and Engineering - Versailles, France.

Benqué, J.P., Labadie, G. and Ronat, J., (1985) A new finite element method for the Navier Stokes equations coupled with a temperature equation - To be published in <<International Journal for Numerical Methods in Fluids>>.

Benqué, J.P. and Ronat, J., (1981) Quelques difficultés des modèles numériques en hydraulique - Fifth International Symposium on Computing Methods in Applied Sciences and Engineering - Versailles, France.

Boyle, D.R. and Golay, M.W., (1983) Measurement of a recirculating, two dimensional, turbulent flow and comparison to turbulence model predictions - MIT - Journal of Fluid Engineering. Vol. 105, pages 439-454.

Cahouet, J., (1985) Méthodes de sous-structuration de domaine pour les équations de Navier-Stokes - Rapport EDF HE/41/85. 03.

Chorin, A.J., (1968) Numerical solution of incompressible flow problems - Numerical Annals, 2, pp.64-71.

Esposito, P. and Hauguel, A., (1983) Calcul des transitoires thermohydrauliques dans la cuve chaude des réacteurs à neutrons rapides - Twentieth IAHR Meeting - Moscou, - Rapport EDF HE/41/83.05.

Goussebaile, J., Hecht, F., Labadie, G. and Reinhart, L., (1984) Finite element solution of the shallow water equations by a quasi-direct decomposition procedure - <<International Journal for Numerical Methods in Fluids>> - Vol, 4, p. 1117-1136.

Grégoire, J.P., Benqué, J.P., Lasbleiz, P. and Goussebaile, J., (1984) 3 D Industrial Flows Calculations by Finite Element Method - Ninth International Conference on Numerical Methods in Fluid dynamics - Saclay, France.

Grégoire, J.P. and Goussebaile, J., (1984) Avancement du projet N3S de Mécanique des fluides - Rapport EDF HE/41/84.02 and HI/47/00.07.

Grégoire, J.P. and Goussebaile, J., (1985) Avancement du projet N3S de Mécanique des fluides - Rapport EDF HE/41/85.04 and HI/50/29.07.

Labadie, G. and Lasbleiz, P., (1983) Quelques méthodes de résolution du problème de Stokes en éléments finis - Rapport EDF HE/41/83.01 and HI/44/72.07.

Launder, B.E. and Spalding, D.B., (1974) The numerical computation of turbulent flows - <<Computing Methods in applied mathematics and engineering>>, Vol. 3. North Holland.

Martinaud, J.P., (1984) Approximation éléments finis d'équations aux dérivées partielles par une méthode de décomposition de domaines - Application aux équations de Stokes et Saint-Venant - Thèse de docteur de 3ème cycle - Université Paris 6 - Rapport EDF HE/41/84-06.

Rodi, W., (1980) Turbulence models and their application in Hydraulics - State of the art paper - AIRH.

Temam, R., (1977) Navier Stokes equations - Studies in mathematics and its applications - North Holland.

SPECTRAL METHODS FOR DISCONTINUOUS PROBLEMS

S. Abarbanel, D. Gottlieb and E. Tadmor
(Tel-Aviv University, Tel-Aviv, Israel
and
Institute for Computer Applications in Science and Engineering, USA)

1. INTRODUCTION

We show that spectral methods yield high-order accuracy even when applied to problems with discontinuities, though not in the sense of pointwise accuracy. Two different procedures are presented which recover pointwise accurate approximations from the spectral calculations.

Consider the evolution partial differential equation $u_t = Lu$, on a finite interval, where L is a hyperbolic operator. The solution u has a projection $P_N u$ on a finite subspace (which may for example consist of the first N modes in a Galerkin method, or N collocating points in the interval), and a numerical approximation u_N generated by some spectral method. For linear operators it is known from the Lax equivalence theorem that if the scheme is consistent and stable, then u_N approximates $P_N u$ in some appropriate norm. If u is smooth, then the theorem implies that u_N approximates the solution u in the same sense.

In practice, one looks at the point values of u_N at the grid points and takes it as an approximation to the values of the true solution u at these points. We shall call this approach the realization of the computed solution via its grid-points value. The aims of the paper are: 1) demonstrate that when u is a complicated function, this realization will not produce acceptable results; 2) to suggest different ways for the realization of the solution in such cases.

The following examples give a very clear illustration of the misleading results that may be obtained by pointwise realization.

Example 1

Consider the equation

$$\begin{aligned} u_t &= u_x \qquad 0 \leq x \leq 2\pi \\ u(x,0) &= u_0(x) \end{aligned} \tag{1.1}$$

where $u(x)$ and $u_0(x)$ are periodic functions and $u_0(x)$ is a discontinuous function. If we expand $u_0(x)$ in a Fourier series we get

$$u_0(x) = \sum_{k=-\infty}^{\infty} a_k \, e^{ikx} \tag{1.2a}$$

where

$$a_k = \frac{1}{2\pi} \int_0^{2\pi} u_0(x) e^{-ikx} . \tag{1.2b}$$

The solution $u(x)$ is thus given by

$$u(x) = \sum_{k=-\infty}^{\infty} a_k \, e^{ikt} \, e^{ikx} .$$

Suppose that (1.1) is solved numerically by the Fourier-Galerkin method, namely we seek a trigonometric polynomial of the form

$$u_N(x,t) = \sum_{k=-N}^{N} b_k(t) e^{ikx}$$

that satisfies

$$\begin{aligned} \left(\frac{\partial u_N}{\partial t} - \frac{\partial u_N}{\partial x} , e^{ikx} \right) &= 0 , \qquad -N \leq k \leq N \\ u_N(x,0) &= \sum_{k=-N}^{N} a_k \, e^{ikx} . \end{aligned} \tag{1.3}$$

From (1.3) it is clear that

$$\frac{db_k(t)}{dt} = ik\, b_k(t), \quad -N \leqslant k \leqslant N \tag{1.4}$$

and

$$b_k(0) = a_k$$

yielding the solution

$$b_k(t) = a_k\, e^{ikt}.$$

Therefore

$$u_N(x,t) = \sum_{k=-N}^{N} a_k\, e^{ikt}\, e^{ikx}. \tag{1.5}$$

Equation (1.5) implies that $u_N(x,t)$, obtained from the numerical solution (1.3), coincides with $P_N\, u(x,t)$, the Galerkin projection of u, thus yielding the best possible convergence of u_N to $P_N u$. However, since the Fourier series of u(x,t) converges very slowly, the point values $u_N(x_j,t)$ will not approximate well $u(x_j,t)$. In general, one would witness the Gibbs phenomenon of overshoot in the neighbourhood of the discontinuity and global oscillations all over the domain. In fact, even the initial approximation, $u_N(x,0)$, displays the same behaviour in relation to $u_0(x)$.

In the second example we show that the same phenomenon occurs even if the numerical initial point values do approximate the true initial point values to a high degree of accuracy.

Example 2

Consider the equation (1.1) where $u_0(x)$ is the saw-tooth function

$$u_0(x,\bar{x}) = \begin{cases} Ax & x < \bar{x} \\ A(2x - \pi) & x > \bar{x} \end{cases} \tag{1.6}$$

for some k, $0 < k < 2N-1$, $\bar{x} = \frac{\pi}{N}(k+\frac{1}{2})$.

In the pseudospectral Fourier method we seek a trigonometric polynomial $v_N(x,t)$

$$v_N(x,t) = \sum_{\ell=-N}^{N} b_\ell(t)e^{i\ell x} \tag{1.7}$$

such that

$$\frac{\partial v_N}{\partial t} = \frac{\partial v_N}{\partial x} \quad \text{at the points } x_j = \frac{\pi j}{N},\ j = 0,\ldots,2N-1 \tag{1.8a}$$

$$v_N(x,0) = u_0(x,\bar{x}) \tag{1.8b}$$

Since v_N is a polynomial of degree N, (1.8a) implies that

$$\frac{\partial v_N}{\partial t} = \frac{\partial v_N}{\partial x} \tag{1.9}$$

for all $x \in (0,2\pi)$. Moreover, from (1.8b) it is clear that $v_N(x,0)$ is the (unique) trigonometric polynomial of order N that interpolates $u_0(x)$ at the points x_j, $j = 0,\ldots,2N-1$; thus

$$v_N(x,0) = \sum_{\ell=-N}^{N} a_\ell(\bar{x})e^{i\ell x} = AF_N(x,\bar{x}) \tag{1.10}$$

where

$$a_\ell(\bar{x}) = \frac{1}{2Nc_\ell}\sum_{j=0}^{2N-1} u_0(x_j,\bar{x})e^{-i\ell x_j}. \tag{1.11}$$

Performing (1.11) we get

$$a_0(\bar{x}) = A\frac{\pi}{N}[k - N + .5] \tag{1.12}$$

$$a_\ell(\bar{x}) = A \frac{\pi}{2Nc_\ell} 2 \frac{1 - e^{\frac{-i\pi\ell}{N}(k+1)}}{1 - e^{\frac{-i\pi\ell}{N}}} + i\,\mathrm{ctn}\frac{\pi\ell}{N} - 1 , \quad \ell \neq 0 \tag{1.13}$$

where

$$c_{-N} = c_N = 2, \quad c_\ell = 1, \quad |\ell| \neq N.$$

The numerical solution $v_N(x,t)$ of (1.9), (1.10) is

$$v_N(x,t) = v_N(x + t,0) = AF_N(x + t,\bar{x}) \tag{1.14}$$

and upon manipulating (1.12), (1.13) one gets

$$v_N(x,t) = AF_N(x,\bar{x} - t) + At. \tag{1.15}$$

The trigonometric interpolant $F_N(x,\bar{x})$ collocates $u_0(x,\bar{x})$ at the grid points x_j. However, in between the grid points it oscillates. If we read the values of $v_N(x,t)$ at the grid points, then by (1.14)

$$v_N(x_j,t) = AF_N(x_j + t,\bar{x})$$

and unless $t = \frac{\pi m}{N}$ for some integer m, we will get a solution that looks oscillatory. Thus, even though the initial appromixation looks smooth at the grid points, when it evolves in time the oscillations will present themselves at the points x_j.

The conclusion one might draw from the above examples is that spectral methods (or any higher-order methods) are useless when applied to discontinuous functions. A different approach is to look at a different realization of the numerical solution rather than the pointwise one. We will argue that high-order accurate information is contained in the numerical solution and demonstrate how that information can be extracted in such a way that accurate pointwise approximation to the true solution can be obtained.

2. INFORMATION AND HOW TO EXTRACT IT

Consider the linear equation

$$\begin{aligned} u_t &= Lu \\ u(0) &= u_0 \end{aligned} \tag{2.1}$$

where L is a linear hyperbolic operator with variable coefficients and u_0 is a discontinuous function. For simplicity, we will restrict ourselves to a periodic one (space) dimensional problem, though the results are more general, (see Gottlieb and Tadmor (1985)). Let v be the solution of the auxiliary problem

$$\begin{aligned} v_t &= - L^* v \\ v(0) &= v_0, \end{aligned} \tag{2.2}$$

where v_0 is a C^∞ function. Because of the hyperbolicity of L, (2.2) is a well-posed problem. In Lemma 1 we quote the well-known Green's identity.

Lemma 1: Let u(t) and v(t) be the solutions of (2.1) and (2.2) at some level t, then

$$(u(t),\ v(t)) = (u_0, v_0). \tag{2.3}$$

Assume now that (2.1) and (2.2) are discretized by the Fourier-Galerkin method. That is, we seek u_N and v_N that are trigonometric polynomials of degree N such that for every k, $|k| \leqslant N$

$$\left(\frac{\partial u_N}{\partial t} - L\, u_N,\ e^{ikx}\right) = 0 \tag{2.4a}$$

$$\left(u_N(0) - u_0,\ e^{ikx}\right) = 0 \tag{2.4b}$$

$$\left(e^{ikx},\ \left(\frac{\partial v_N}{\partial t} + L^* v_N\right)\right) = 0 \tag{2.4c}$$

$$\left(e^{ikx},\ \left(v_N(0) - v_0\right)\right) = 0. \tag{2.4d}$$

We have also a Green's identity for u_N and v_N.

Lemma 2:

$$\left(u_N(t), v_N(t)\right) = \left(u_N(0), v_N(0)\right). \tag{2.5}$$

Proof: Since $v_N(t)$ and $u_N(t)$ are N^{th}-order trigonometric polynomials we use (2.4a) and (2.4c) to get

$$\left(\frac{\partial u_N}{\partial t} - Lu_N,\ v_N\right) = 0$$

$$\left(u_N,\ \frac{\partial v_N}{\partial t} + L^* v_N\right) = 0,$$

and therefore

$$\frac{\partial}{\partial t}\left(u_N, v_N\right) = \left(Lu_N, v_N\right) - \left(u_N, L^* v_N\right) = 0$$

which implies (2.5).

We will proceed by showing the relation of the RHS of (2.5) to that of (2.3).

Lemma 3:

$$\left(u_N(0), v_N(0)\right) = \left(u_0, v_0\right) + \varepsilon_1 \tag{2.6}$$

where

$$|\varepsilon_1| \leqslant K \frac{\|v_0\|_s}{N^s} \tag{2.7}$$

for every s.

Proof: From (2.4c) it is clear that

$$\left(u_N(0) - u_0, v_N(0)\right) = 0. \tag{2.8}$$

Also,

$$\left|\left(u_0, v_N(0) - v_0\right)\right| \leqslant K\|u_0\| \, \|v_N(0) - v_0\|$$

and since v_0 is a C^∞ function,

$$\|v_N(0) - v_0\| \leqslant K \frac{\|v_0\|_s}{N^s}, \text{ for every s.} \tag{2.9}$$

Now

$$\left(u_N(0), v_N(0)\right) = (u_0, v_0) + \left(u_N(0) - u_0, v_N(0)\right) + \left(u_0, v_N(0) - v_0\right)$$

and in view of (2.8) and (2.9)

$$\left(u_N(0), v_N(0)\right) = (u_0, v_0) + \varepsilon_1$$

where

$$|\varepsilon_1| \leqslant K \frac{\|v_0\|_s}{N^s}$$

and this proves the Lemma.

From Lemmas 1 - 3 we can conclude:

Theorem 1: Let u(t) and v(t) be the solutions of (2.1) and (2.2), respectively. Let $u_N(t)$ and $v_N(t)$ be the solutions of the Fourier-Galerkin approximations of (2.1) and (2.2). Then

$$\left|\left(u_N(t),v_N(t)\right) - \left(u(t),v(t)\right)\right| \leq K \frac{\|v_0\|_s}{N^s}, \text{ for every } s. \tag{2.10}$$

The proof is an immediate consequence of (2.3), (2.5) and (2.6).

Assume now that the Fourier-Galerkin method described in (2.4c) and (2.4d) is stable; then $v_N(t)$ approximates $v(t)$ within spectral accuracy, that is

$$\|v_N(t) - v(t)\| = \varepsilon_2 \leq K \frac{\|v\|_s}{N^{s-1}} .$$

We can, therefore, replace $v_N(t)$ in (2.10) and get

$$\left(u_N(t),v(t)\right) = \left(u(t),v(t)\right) + \varepsilon$$

where ε is spectrally small. We use now the fact that every C^∞ function $v(t)$ can be obtained from some v_0 in (2.2). This is, in fact, one of the definitions of hyperbolicity. We can, therefore, state:

Theorem 2: Let $u(t)$ be the (nonsmooth) solution of (2.1) and let $u_N(t)$ be the solution of the spectral Galerkin approximation to (2.1). Then for any C^∞ function $v(t)$

$$\left(u_N(t),v(t)\right) = \left(u(t),v(t)\right) + \varepsilon \tag{2.11}$$

where ε is spectrally small.

Thus, $u_N(t)$ approximates weakly $u(t)$ within spectral accuracy. It is in this sense that $u_N(t)$ contains highly accurate information about $u(t)$. We will show later how to use this information in order to obtain spectral accurate approximation to the grid-point values of $u(t)$.

We turn now to the pseudospectral Fourier case. Here we need some preprocessing of the initial data in order to prove the same result as in Theorem 2.

Theorem 3: Let $u_N(x,t)$ be a trigonometric polynomial of order N that satisfies

$$\frac{\partial u_N}{\partial t} = Lu_N \quad \text{at } x = x_j, \; x_j = \frac{\pi j}{N}, \; j = 0,\ldots 2N-1$$
$$\left(u_N(0) - u_0, e^{ikx}\right) = 0, \qquad |k| \leq N, \tag{2.12}$$

(ie., $u_N(x,t)$ is the solution of the pseudospectral Fourier scheme, but initially $u_N(x,0)$ is obtained by the Galerkin projection).

Then for every smooth function $u(x,t)$

$$\frac{\pi}{N}\sum_{j=0}^{2N-1} u_N(x_j,t)\; v(x_j,t) = \int_0^{2\pi} u(x,t)\; v(x,t)\, dx + \varepsilon \tag{2.13}$$

where ε is spectrally small, provided that the pseudospectral approximation is stable.

Proof: Let v_N be the solution of the pseudospectral Fourier approximation of (2.4a) and let $v_N(0)$ be the Galerkin projection of v_0, that is

$$\left(v_N(0) - v_0, e^{ikx}\right) = 0, \quad |k| \leq N. \tag{2.14}$$

From (2.12) and the analog equation for v_N, one gets

$$\frac{\pi}{N}\sum_{j=0}^{2N-1} u_N(x_j,t)\; v_N(x_j,t) = \frac{\pi}{N}\sum_{j=0}^{N} u_N(x_j,0)\, v_N(x_j,0). \tag{2.15}$$

From the exactness of the trapezoidal rule for polynomials of degree 2N, we conclude

$$\frac{\pi}{N}\sum_{j=0}^{2N-1} u_N(x_j,t)\; u_N(x_j,t) = \int_0^{2\pi} u_N(x,0)\; v_N(x,0)\, dx = \left(u_N(0), v_N(0)\right). \tag{2.16}$$

Note that the initial functions $v_N(x,0)$ and $u_N(x,0)$ are not the interpolants of u_0 and v_0 as in the usual pseudospectral methods but rather the Galerkin approximation to u_0 and v_0. We recall now Lemma 3 and equality (2.3) to establish (2.13). The proof is thus completed.

It is interesting to note the way in which the information is contained. The interpolant of u_0 looks smooth at the grid points, whereas the Galerkin approximation of u_0 looks oscillatory on the grid points. It means that in order to preserve the information one has to require initially oscillatory-looking solution. The information is preserved in the structure of the oscillations.

We will show now a way of using (2.11) and (2.13) in order to construct a better approximation to $u(x_j,t)$ than the one given by $u_N(x_j,t)$. (Here $u_N(x,t)$ is given by either the Galerkin method or the pseudospectral method).

From (2.13) and (2.11) it is clear that in order to get a good approximation to $u(y,t)$ at some point $y \in (0,2\pi)$, we need to find a function $v_y(x,t)$ such that

$$\int_0^{2\pi} u(x,t)\, v_y(x,t)\,dx = u(y,t) + \varepsilon_1,$$

where ε_1 is spectrally small. By (2.11) we will have

$$\int_0^{2\pi} u_N(x,t)\, v_y(x,t)\,dx = u(y,t) + \varepsilon + \varepsilon_1 \qquad (2.17)$$

for the Galerkin method and

$$\frac{\pi}{N} \sum_{j=0}^{2N-1} u_N(x_j,t)\, v_y(x_j,t) = u(y,t) + \varepsilon + \varepsilon_1 \qquad (2.18)$$

for the pseudospectral method.

For convenience we shift the interval $[0,2\pi]$ to $[-\pi,\pi]$. Let $\rho(x)$ be a C^∞-function vanishing outside the interval

$[-\pi,\pi]$ satisfying

$$\rho(0) = 1. \tag{2.19}$$

Let $D_p(x)$ be the Dirichlet kernel, namely

$$D_p(y) = \frac{1}{2\pi} \sum_{|k|\leq p} e^{ikx} = \frac{1}{2\pi} \frac{\sin(p+\frac{1}{2})y}{\sin(y/2)}. \tag{2.20}$$

We set now

$$v_y(x) = \psi^{\theta,p}(x) = \theta^{-1}\,\rho(\theta^{-1}y)D_p(\theta^{-1}y). \tag{2.21}$$

One can prove as in (Gottlieb et al., 1983) that

$$\int_{-\pi}^{\pi} u(x)\psi^{\theta,p}(y-x)\,dx = u(y) + \varepsilon_2 \tag{2.22}$$

where ε_2 is spectrally small.

Thus, it is possible to extract accurate pointwise values from $u_N(x)$.

3. NUMERICAL RESULTS

In this section we demonstrate the efficacy of the smoothing procedure outlined above. As a test function we have chosen the piecewise C^∞-function

$$f(x) = \begin{cases} \sin\frac{x}{2} & 0 \leq x \leq \pi \\ -\sin\frac{x}{2} & \pi \leq x \leq 2\pi. \end{cases} \tag{3.1}$$

Denote its spectral approximation by $\hat{f}_N(x)$, and let $\tilde{f}_N(x)$ be the pseudospectral approximation to $f(x)$. It is evident from the first column of Tables I and III that $\hat{f}_N(y_\nu)$ - the spectral approximation sampled at $y_\nu = \nu\pi/N$ - do not approximate $f(y_\nu)$ within spectral accuracy. In fact, the error committed by $\hat{f}_{128}(y_\nu)$ is only half of that committed by $f_{64}(y_\nu)$.

Regarding the pseudospectral approximation, $\tilde{f}_N(x)$, it, of course, collocates the exact values at the sampling grid points, $\tilde{f}_N(y_\nu) = f(y_\nu)$; yet, in between these gridpoints, $\tilde{f}_N\left(y_{\nu+\frac{1}{2}} = (\nu + \frac{1}{2})\pi/N\right)$ approximate $f(y_{\nu+\frac{1}{2}})$ within first-order accuracy only, as shown in the first column of Tables II and IV.

In order to construct our regularization kernel in (2.21), we define the cut-off function $\rho(\zeta) = \rho_\alpha(\zeta)$ to be

$$\rho_\alpha(\zeta) = \begin{cases} \exp \dfrac{\alpha\zeta^2}{\zeta^2 - 1} & |\zeta| < 1 \\ 0 & \text{otherwise} \end{cases} , \tag{3.2}$$

namely, $\rho_\alpha(\zeta)$ is a C^∞ - function whose support is the interval $|\zeta| < 1$. ψ to be used in (2.21) is of the form

$$\psi^{\theta,p}(y) = \frac{1}{2\pi\theta} \rho_\alpha (\theta^{-1} y) \frac{\sin (p + \frac{1}{2})y/\theta}{\sin y/2\theta}. \tag{3.3}$$

The post-processing procedure of the spectral approximation $\hat{f}_N$ involves convoluting $\hat{f}_N$ against $\psi^{\theta,p}$, namely

$$f(x) \sim \frac{1}{2\pi\theta} \int_0^{2\pi} \hat{f}_N(y) \rho\left(\frac{x-y}{\theta}\right) \frac{\sin (p + \frac{1}{2})y/\theta}{\sin (x-y)/2\theta} dy \tag{3.4}$$

where x is a fixed point of interest. (In practice, we use the trapezoidal rule to evaluate the right-hand-side of (3.4) taking a large number of quadrature points).

The parameter θ was chosen as

$$\theta = \pi \cdot |x - \pi|; \tag{3.5}$$

this guarantees that ψ is so localized that it does not interact with regions of discontinuity.

It should be noted, in this stage, that if θ was so chosen to be the same for each x, (and not as in (3.5), the formula

(3.4) admits a simpler form; that is, if

$$\psi^{\theta,p}(y) = \sum_{k=-\infty}^{\infty} \sigma_k e^{iky} \tag{3.6}$$

then

$$f(x) \sim \sum_{k=-N}^{N} \hat{f}(k)\sigma_k e^{ikx}. \tag{3.7}$$

This procedure can be carried out efficiently in the Fourier space.

Next, we turn to the post-processing for the pseudospectral approximation $\hat{f}_N(x)$ which is simpler than (3.4). In fact, in this case

$$f(x) \sim \frac{2\pi}{2N} \sum_{\nu=0}^{2N-1} \tilde{f}(y_\nu)\psi^{\theta,p}(x-y_\nu). \tag{3.8}$$

Note that carrying out the smoothing procedure defined in (3.8) does not involve any extra evaluation of $\tilde{f}(y)$ in points other than y_ν, in contrast to spectral smoothing procedures in (3.4). As before, the parameter θ was chosen according to (3.5). We have yet to determine the parameters p and α. The parameter p must be equal to N^β for $0 < \beta < 1$, in order to assure infinite accuracy. (In our computations, $\beta \approx .8$.) Finally, we feel that α is problem dependent and we chose $\alpha = 10$. We have not tuned the parameters to get optimal results; further tuning may improve the quality of our filtering procedure.

In Tables I, II, III, and IV we give the results of the smoothing procedure at several points in the domain. The pointwise values are now recovered with high accuracy. The first column in each table indicates the points in which the procedure was performed. We limited ourselves to four points in the interval $(0,\pi)$ because of the symmetry of the function f(x).

The second column gives either the spectral approximation $\hat{f}_N(x)$ or the pseudospectral approximation $\tilde{f}_N(x)$, N = 128 in Tables I and II and N = 64 in Tables III and IV. The third column gives the smoothed results, when filtered by (3.4) on (3.8), at the same points as in column I.

The results indicate the dramatic improvement obtained by the smoothing procedure. Moreover, note that the error committed by $\tilde{f}_{128}$ (or $\hat{f}_{128}$) is better than the one committed by $\tilde{f}_{64}$ (or $\hat{f}_{64}$) only by a factor of 2 whereas after the post-processing the error improves by a factor of 10^4.

Table I. Results of smoothing of the spectral approximation of f(x), N = 128

$x_\nu = \frac{\pi\nu}{8}$ ν equals	$\lvert f(x_\nu) - \hat{f}_N(x_\nu)\rvert$	$\lvert f - \hat{f}_N * \psi\rvert$ at $x - x_\nu$
2	3.2(-3)	5.8(-10)
3	5.2(-3)	7.9(-10)
4	7.8(-3)	6.3(-10)
5	1.1(-2)	1.1(-10)

Table II. Same as Table I for the pseudospectral approximation $\tilde{f}_N(x)$.

$x_{\nu+\frac{1}{2}} = \frac{\pi}{8}(\nu+\frac{1}{2})$ ν equals	$\lvert f(x_{\nu+\frac{1}{2}}) - \tilde{f}_N(x_{\nu+\frac{1}{2}})\rvert$	$\lvert f - \tilde{f}_N * \psi\rvert$ at $x = x_{\nu+\frac{1}{2}}$
2	5(-3)	7(-10)
3	8.1(-3)	7.9(-10)
4	1.2(-2)	6.4(-10)
5	1.8(-2)	1.2(-10)

Table III. Results of smoothing of the spectral approximation of f(x), N = 64

$x_\nu = \frac{\pi\nu}{8}$ ν equals	$\lvert f(x_\nu) - \hat{f}_N(x_\nu)\rvert$	$\lvert f - \hat{f}_N * \psi\rvert$ at $x = x_\nu$
2	6.4(-3)	4.8(-6)
3	1(-2)	5.9(-6)
4	1.5(-2)	7.7(-6)
5	2.3(-2)	8.9(-6)

Table IV. Same as Table III for the pseudospectral approximation, $\tilde{f}_N(x)$.

$x_{\nu+\frac{1}{2}} = \frac{\pi}{8}(\nu+\frac{1}{2})$ ν equals	$\lvert f(x_{\nu+\frac{1}{2}}) - \tilde{f}_N(x_{\nu+\frac{1}{2}})\rvert$	$\lvert f - \tilde{f}_N * \psi\rvert$ at $x = x_{\nu+\frac{1}{2}}$
2	1(-2)	4.1(-6)
3	1.6(-2)	6(-6)
4	2.4(-2)	7.8(-6)
5	3.6(-2)	8.9(-6)

4. A DIFFERENT METHOD FOR EXTRACTING INFORMATION

In this section we would like to present a different approach for extracting the information from an oscillatory solution. The idea is to subtract from the solution those oscillations that correspond to the saw-tooth function discussed in Example 2. This leads to the following procedure:

Let $u_N(x,t) = \sum_{\ell=-N}^{N} \hat{u}_\ell e^{i\ell x}$ be the solution of the pseudospectral approximation to a hyperbolic problem. We try to find an unknown smooth function and a (oscillatory) saw-tooth function $F_N(x-t,x_s)$ with an unknown jump $2\pi A$ at an unknown location x_s such that

$$H = \left[\sum_{j=0}^{2N-1} u_N(x_j,t) - AF_N(x_j,x_s) - c - \sum_{\substack{\ell=-p \\ \ell\neq 0}}^{P} b_\ell e^{i\ell k_j} \right]^2 \qquad (4.1)$$

is minimized. Note that we have 2p + 3 unknowns in (4.1): A, x_s, c and 2P values of b_ℓ ($\ell \neq 0$).

The conditions for local minima of H are found from the following 2p + 3 equations:

$$\frac{\partial H}{\partial A} = 0 \Rightarrow \sum_{j=0}^{2N-1} u_j F_j - AF_j^2 - cF_j - F_j \sum_{\substack{\ell=-p \\ \ell\neq 0}}^{p} b_\ell e^{i\ell x_j} = 0 \qquad (4.2)$$

where $F_j = F_N(x_j,x_s)$, $u_j = u_N(x_j,t)$. Also,

$$\frac{\partial H}{\partial c} = 0 \implies \sum_{j=0}^{2N-1} u_j - AF_j - c - \sum_{\substack{\ell=-p \\ \ell\neq 0}}^{p} b_\ell e^{i\ell x_j} = 0 \qquad (4.3)$$

$$\frac{\partial H}{\partial s} = 0 \implies \sum_{j=0}^{2N-1} F_j' u_j - AF_j' F_j - cF_j' - F_j' \sum_{\substack{\ell=-p \\ \ell\neq 0}}^{p} b_\ell e^{ix_j\ell} = 0 \qquad (4.4)$$

where $F_j' = \partial F_N x_j, x_s)/\partial s = \sum_{\ell=-N}^{N} \frac{\partial a_\ell(s)}{\partial s} \cdot e^{i\ell x_j}$; and

$$\frac{\partial H}{\partial b_m} = 0 \Longrightarrow b_m = \hat{u}_m - Aa_m, \quad |m| = 1,2,\ldots,p \tag{4.5}$$

where

$$\hat{u}_m = \frac{1}{2Nc_m} \sum_{j=0}^{2N-1} u_N(x_j) e^{-i\ell x_j}.$$

Substituting (4.5) into (4.2), (4.3) and (4.4) we get, respectively:

$$\hat{u}_0 - Aa_0 - c = 0 \tag{4.6}$$

$$\sum_{|\ell|>p} \left(c_\ell \, a_{-\ell} \, \hat{u}_\ell - A\right) \sum_{|\ell|>p} \left(c_\ell \, a_{-\ell} \, a_\ell\right) = 0 \tag{4.7}$$

$$\sum_{|\ell|>p} \left(c_\ell \, a'_{-\ell} \, \hat{u}_\ell - A\right) \sum_{|\ell|>p} \left(c_\ell \, a'_{-\ell} \, a_\ell\right) = 0 \tag{4.8}$$

where $a'(s) = \partial a_\ell(x)/\partial s$. Next, we combine (4.7) and (4.8) to get a single nonlinear equation for s:

$$\Sigma c_\ell \, a'_{-\ell} \, \hat{u}_\ell \, \Sigma c_\ell \, a_{-\ell} \, a_\ell - \Sigma c_\ell \, a_{-\ell} \, \hat{u}_\ell \, \Sigma c_\ell \, a'_{-\ell} \, a_\ell = 0 \tag{4.9}$$

where all sums run over $p < |\ell| \leq N$.

Equation (4.9) is solved iteratively for s. Having found s, one immediately obtains from Example 2 all $a_\ell(s)$'s. Then from (4.5) we have the b_m's, and A from (4.7). Finally, having A we find c from (4.6).

The minimum thus obtained may be a local one while we are seeking a global minimum. This means that in practice one searches for the global minimum.

We now give an example that illustrates the efficacy of the procedure. We solve the following problem:

$$\frac{\partial u_N}{\partial t} + \frac{\partial u_N}{\partial x} = 0, \qquad 0 < x < 2\pi,\ t > 0 \tag{4.10}$$

$$u_N(x,0) = \begin{cases} \sin\frac{x}{2} & 0 \leq x \leq \pi \\ -\sin\frac{x}{2} & \pi \leq x \leq 2\pi \end{cases} \tag{4.11}$$

$$u_N(0,t) = u_N(2\pi,t). \tag{4.12}$$

We ran the problem on several grids and exhibit here the numerical results for the case N = 8 (ie., 16 subintervals in the domain $(0,2\pi)$). The unadulterated results at $t = \pi/2N$ on the grid points are shown in Figure 1.

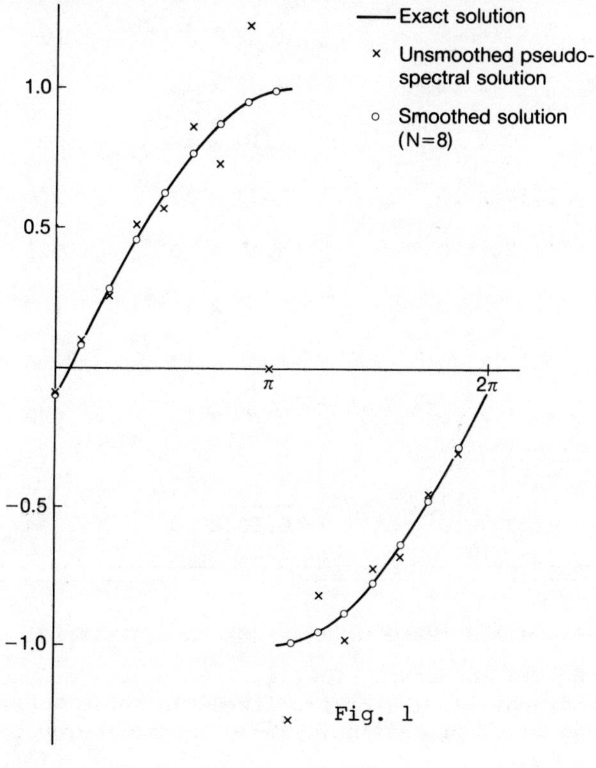

Fig. 1

Table V

j	exact Solution	error 1 = |exact-unsmoothed|	error 2 = |exact- smoothed|	error 1 / error 2
0	9.80×10^{-2}	5.86×10^{-5}	5.86×10^{-5}	1.00
1	9.80×10^{-2}	1.24×10^{-2}	5.86×10^{-5}	211
2	2.90×10^{-1}	2.57×10^{-2}	6.30×10^{-5}	408
3	4.71×10^{-1}	4.13×10^{-2}	7.33×10^{-5}	563
4	6.34×10^{-1}	6.15×10^{-2}	9.30×10^{-5}	661
5	7.73×10^{-1}	9.11×10^{-2}	1.31×10^{-4}	695
6	8.82×10^{-1}	1.43×10^{-1}	2.16×10^{-4}	662
7	9.57×10^{-1}	2.70×10^{-1}	4.42×10^{-4}	611
8	-9.95×10^{-1}	1.00×10^{0}	1.10×10^{-2}	91
9	-9.95×10^{-1}	2.68×10^{-1}	1.34×10^{-3}	200
10	-9.57×10^{-1}	1.42×10^{-1}	4.42×10^{-4}	321
11	-8.82×10^{-1}	9.07×10^{-2}	2.16×10^{-4}	420
12	-7.73×10^{-1}	6.12×10^{-2}	1.32×10^{-4}	464
13	-6.34×10^{-1}	4.11×10^{-2}	9.30×10^{-5}	442
14	-4.71×10^{-1}	2.55×10^{-2}	7.32×10^{-5}	348
15	-2.90×10^{-1}	1.22×10^{-2}	6.30×10^{-5}	194

We then post-processed these $u_N(x_j, \pi/2N)$ values according to the procedure described above. The filtered values are shown on the same graph, and the errors listed in Table V are computed before and after processing. The dramatic improvement is evident.

Next we demonstrate the procedure in the case of the Euler equation for gas dynamics. Because the physical problem involves inflow, outflow, and no-flow boundary conditions, periodicity could not be imposed and we use the Tchebyshev, rather than Fourier, pseudospectral method.

The physical problem is that of a wedge, inserted at a zero angle of attack, into a uniform supersonic flow of an ideal gas with $\gamma = 1.4$. An oblique shock develops in time and the flow reaches, after a while, a steady state. The time-dependent Euler equations in two-space dimensions were discretized by the pseudospectral Tchebyshev method in space with an 8x8 grid and a modified Euler scheme was used for the time discretization. Since we are interested in the steady state only, the accuracy for the time integration is of little importance. In order to be sure that a steady state is reached, the code was run until all physical quantities did not change to 11 significant figures over a span of 100 time steps. The values of the density in the steady state at the grid points together with the grid points themselves are given in Table VI.

Table VI

	ρ									Y
	1.862	1.851	1.869	1.871	1.837	1.865	1.892	1.885	1.878	1.
	1.862	1.870	1.867	1.820	1.870	1.954	1.899	1.803	1.759	.961
	1.862	1.854	1.852	1.904	1.877	1.770	1.782	1.864	1.900	.853
	1.862	1.871	1.876	1.812	1.838	1.969	1.975	1.884	1.841	.691
	1.862	1.848	1.842	1.935	1.899	1.703	1.710	1.890	1.984	.5
	1.862	1.883	1.894	1.729	1.832	2.429	2.994	3.255	3.316	.308
	1.862	1.808	1.810	2.387	3.133	3.375	3.224	3.054	3.002	.146
	1.862	2.115	2.868	3.288	3.176	2.965	3.006	3.136	3.187	.038
	1.862	3.083	3.046	2.975	3.087	3.108	3.024	3.013	3.016	0
x	0	.038	.146	.308	.5	.691	.853	.961	1.	

Note that the raw data in Table VI seem to indicate roughly the same y-shock location at $x_0 = 1$, $x_1 = .961$, and $x_2 = .853$, namely between the grid points $y_4 = .3086$ and $y_5 = .500$. This means that because of the coarse Tchebyshev grid the shock location cannot be resolved to better than 20% of the domain. In fact, the correct shock locations at those x-stations are $y = .434$ for x_0, $y = .417$ for x_1 and $y = .370$ for x_2.

In the present case it is not necessary to employ a saw-tooth piecewise smooth function, as was done in the previous section, because there is no need to preserve periodicity. Instead, we subtract from the oscillatory data an expansion of the Heaviside function, $S(y,y_s)$:

$$S(y,y_s) = \begin{cases} d_1 + d_2 & -1 \leq y \leq y_s \\ d_1 & y_s \leq y \leq 1 \end{cases} \qquad (4.13)$$

where d_1, the state ahead of the shock, and d_2, the magnitude of the discontinuity, are constant. The description here of $S(y,y_s)$, as if independent of x, has to do with the fact that the two-dimensional results of the pseudospectral algorithm were post-processed at fixed x stations. The expansion of $S(y,y_s)$ is given by

$$S_N(y,y_s) = \sum_{\ell=0}^{N} A_\ell(s) T_\ell(y) \qquad (4.14)$$

where $T_\ell(y)$ is the Tchebyshev polynomial of order ℓ, $T_\ell(y) = \cos\left[\ell \cos^{-1}(y)\right]$, and

$$A_0(s) = \left\{s + \frac{1}{2}\right\}/N$$

$$A_\ell(s) = \sin\left[\frac{\pi\ell}{N}\left(s + \frac{1}{2}\right)\right]/N \sin\frac{\pi\ell}{2N}; \qquad 1 \leq \ell \leq N-1$$

$$A_N(s) = \sin\left[\left(s + \frac{1}{2}\right)\right]/2N.$$

If s is an integer, then on the grid points, $y_j = \cos(\pi j/N)$,

$$S_N(y_j,y_s) = S(y_j,y_s). \tag{4.15}$$

The L_2-norm which we wish to minimize is now, at any given x-station:

$$H = \sum_{j=0}^{N} \frac{1}{c_j} \left[\rho_N(y_j) - d_1 - d_2\, S_N(y_j,y_s) - \sum_{\ell=1}^{p<N} b_\ell\, T_\ell\,(y_j)\right]^2 \tag{4.16}$$

$$c_j = \begin{cases} 1 & 1 \leqslant j \leqslant N-1 \\ 2 & j = 0, N \end{cases}. \tag{4.17}$$

Differentiating (4.16) with respect to the parameters d_1, d_2, s and $b_\ell (1 \leqslant \ell \leqslant p < N)$, using the orthogonality relations for the Tchebyshev polynomials and manipulations similar to those used in the previous section, we get p + 3 nonlinear algebraic equations which are completely analogous to (4.5) - (4.8). They are:

$$b_\ell = \hat{\rho}_\ell - d_2\, A_\ell, \qquad \ell = 1,2,\ldots,p. \tag{4.18}$$

$$\hat{\rho}_0 - d_2\, A_0 - d_1 = 0 \tag{4.19}$$

$$\sum_{\ell=p+1}^{N} c_\ell\, A_\ell\, \hat{\rho}_\ell - d_2 \sum_{\ell=p+1}^{N} c_\ell\, A_\ell^2 = 0 \tag{4.20}$$

$$\sum_{\ell=p+1}^{N} c_\ell\, A'_\ell\, \hat{\rho}_\ell - d_2 \sum_{\ell=p+1}^{N} c_\ell\, A_\ell\, A'_\ell = 0 \tag{4.21}$$

where

$$\hat{\rho}_\ell = \frac{2}{Nc_\ell} \sum_{j=0}^{N} \frac{1}{c_j} \rho(y_j) T_\ell(y_j) \qquad (4.22)$$

$$A' = \frac{\partial}{\partial s} A_\ell(s). \qquad (4.23)$$

Again, we combine (4.20) and (4.21) into a single nonlinear equation for the shock location index, s:

$$\Sigma c_\ell A'_\ell \rho_\ell \Sigma c_\ell A^2_\ell - \Sigma c_\ell A_\ell \hat{\rho}_\ell \Sigma c_\ell A_\ell A'_\ell = 0 \qquad (4.24)$$

where all the sums are from $\ell = p+1$ to $\ell = N$.

The procedure for extracting the shock location, jump magnitude and smooth part of the solution from the raw data $\rho(x,y_j)$ (given in Table VI) is exactly the same as described above for the Fourier problem.

For the wedge-flow problem considered here, this procedure applied in the case of a coarse net (N = 8), located the shock with an error only in the fourth significant figure. The smooth part was recovered to within 1% at the worst field point.

5. CONCLUSION

We have demonstrated that the realization of a numerical solution via its grid-point value may be misleading when the true solution has a complicated structure which is not resolved by the grid. We have shown, however, that the numerical solution does contain highly accurate information about the solution and we suggested two ways of extracting this information.

The analysis outlined in this paper is a linear one (though the procedure was applied also to nonlinear problems). However, Lax (1978) has argued that more information about the solution is contained in high resolution schemes even in the nonlinear case. In fact, using notions from information theory, Lax has shown that the ε-capacity of the set of the set of approximate solutions is closer to the ε-capacity of the set of the projections of exact solutions if the numerical scheme is a high-order scheme.

In the area of digital filters one always processes the data in order to overcome the Gibbs phenomenon. If we look at the initial conditions as an input signal and at the final result as the output signal, the idea of filtering is a natural one.

6. ACKNOWLEDGEMENTS

This research was supported in part by the National Aeronautics and Space Administration under NASA Contract No. NASI-17070 while the authors were in residence at ICASE, NASA Langley Research Center, Hampton, VA 23665. Additional support was also provided in part by the US Air Force Office of Scientific Research under Contract No. AFOSR 83-0089 for the first and second author.

7. REFERENCES

Gottlieb, D., Hussaini, M.Y. and Orszag, S.A. (1983) "Introduction to the Proc. of the Symposium on Spectral Methods", SIAM CBMS Series.

Gottlieb, D. and Tadmor, E. (1985) "Recovering pointwise values of discontinuous data within spectral accuracy", ICASE Report No. 85-3.

Lax, P.D. (1978) "Accuracy and resolution in the computation of solutions of linear and nonlinear equations", Recent Advances in Numerical Analysis, Proc. Symp. Mathematical Research Center, University of Wisconsin, Academic Press, pp. 107-117.

Majda, A. and Osher, S. (1978) "The Fourier method for nonsmooth initia data", *Math. Comp.*, Vol. 32, pp. 1041-1081.

Mock, M.S. and Lax, P.D. (1978) "The computation of discontinuous solutions for linear hyperbolic equations", Comm. Pure Appl. Math., Vol. 31, pp. 423-430.

THE NUMERICAL SOLUTION OF BIFURCATION PROBLEMS WITH SYMMETRY WITH APPLICATION TO THE FINITE TAYLOR PROBLEM

K.A. Cliffe
(Theoretical Physics Division, AERE Harwell)

A.D. Jepson
(Department of Computer Science, University of Toronto)

and

A. Spence
(School of Mathematics, University of Bath)

1. INTRODUCTION

The steady-state behaviour of many physical systems can often be modelled by non-linear multiparameter equations of the form

$$F(x,\lambda,\underline{\alpha}) = 0 \quad F : X \times R \times R^p \to X \tag{1.1}$$

where F is a smooth non-linear function, X is a Banach space, $x \varepsilon X$ is a state variable, $\lambda \varepsilon R$ is the bifurcation parameter, and $\underline{\alpha} \varepsilon R^p$ $(p \geqslant 1)$ is a vector of control (or auxiliary) parameters. The distinction of the bifurcation parameter λ is made for two reasons. First it is often the case that one parameter is "dynamical" in character and is varied quasistatically, while other parameters are "geometrical" and are less easily varied. Second the distinction of λ is numerically convenient since it allows the use of the singularity theory of Golubitsky and Schaeffer (1985). (We remark that a good understanding of the solution set of (1.1) is often a prerequisite for understanding the solution behaviour of the time-dependent problem

$$x_t = F(x,\lambda,\underline{\alpha}) ,$$

though we do not discuss this aspect any further in this paper).

As an example (1.1) could be the system (3.13) in Cliffe and Spence (1985) arising from a finite-element discretization of the Navier-Stokes equations for steady axisymmetric flow in the Taylor problem. In this case $X = R^n$, x represents the primitive variables velocity and pressure, λ is the Reynolds number, and there are two control parameters $\underline{\alpha} = (\alpha_1,\alpha_2)$, with α_1 the aspect ratio and α_2 the radius ratio. (N.B. in Cliffe and

Spence $(\lambda,\underline{\alpha}) = (R,\Gamma,\eta)$). The numerical strategy described in this paper was tested on the Taylor problem and numerical results are given in section 5 (see Fig. 4). In the experiment of Benjamin and Mullin (1981) the radius ratio α_2 was fixed at a constant value and so direct comparison with experimental results is not possible. We note however that numerical results for the case of α_2 fixed are given in Cliffe (1983) and Cliffe and Spence (1984) and show excellent agreement with the experimental results. As is standard practice Benjamin and Mullin describe their results in terms of bifurcation diagrams. They give graphs of solutions of the Navier-Stokes equations against R for various values of Γ, with η constant. Three such graphs are given in Fig. 1.

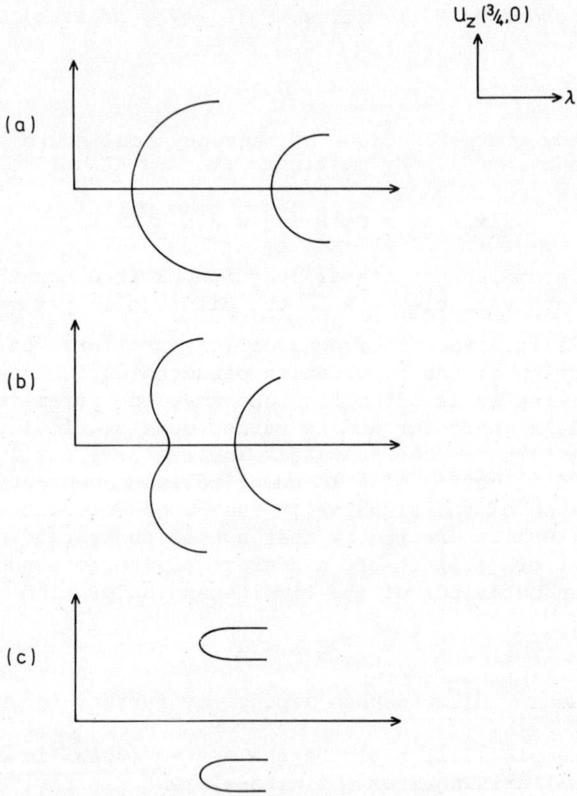

Fig. 1 Bifurcation diagrams for the finite Taylor problem for various values of aspect ratio: a) $\alpha_1 = 1.24$, b) $\alpha_1 = 1.26$, c) $\alpha_1 = 1.27$.

Note that though the solution is $x = (u_r, u_\phi, u_z, p)$ a representative component of the solution (in this case $u_z(3/4, 0)$) is used along the vertical axis. Because of the reflection symmetry in the problem (see Cliffe and Spence (1985), section 4) the two-cell flows have $u_z(3/4,0) = 0$ whereas non-symmetric solutions have $u_z(3/4,0) \neq 0$. The point at which the two-cell (symmetric) flow changes to a non-symmetric flow is called a *symmetry-breaking bifurcation point*. Because of the non-unique nature of the solution set near a bifurcation point it is straightforward to show that

$$F_x(x,\lambda,\underline{\alpha}) \text{ is singular at a bifurcation point } (x,\lambda,\underline{\alpha}), \quad (1.2)$$

since, otherwise, the implicit function theorem would guarantee uniqueness. This is a key property and we develop the theory in detail in section 2, though we note here that if (1.2) holds at $(x_o,\lambda_o,\underline{\alpha}_o)$ we say that $(x_o,\lambda_o,\underline{\alpha}_o)$ is a *singular point*.

The overall objective of this paper is to describe an efficient systematic procedure for the numerical solution of non-linear multiparameter problems. Continuation methods (Keller (1977), Rheinboldt (1980)) have been developed to compute bifurcation diagrams for problems of the form $F(x,\lambda,\underline{\alpha}) = 0$ with $\underline{\alpha}$ given. However these techniques require a starting point on each connected component of the bifurcation diagram and are inefficient when applied to multiparameter problems. Here we describe an approach which alleviates these two difficulties. The approach is based on combining standard continuation methods with the singularity theory of Golubitsky and Schaeffer (1985). This approach has been used successfully on non-symmetric problems, see Jepson and Spence (1985a), and this paper is an extension of that work to problems with (reflection) symmetry. An expanded version of this work containing proofs and further details is in preparation.

The following example, given by Golubitsky and Schaeffer (1985), Ch.VI, illustrates several important concepts for the case $n = 1$, $p = 2$ in (1.1). Consider:

$$F(x,\lambda,\underline{\alpha}) := x^5 - 2m\lambda x^3 + \lambda^2 x + \alpha_1 x - \alpha_2 x^3 , \quad m > 1 , \qquad (1.3a)$$

$$= xa(z,\lambda,\underline{\alpha}) , \quad z = x^2, \; x \in R, \qquad (1.3b)$$

where

$$a : R \times R \times R^2 \to R .$$

In Fig. 2 the varieties $\mathcal{B}_o$, $\mathcal{B}_1$ and $\mathcal{H}_o$ are shown to divide the control parameter space into 5 open regions labelled (1)-(5). These varieties are given by

$$\mathcal{B}_o = \{\underline{\alpha}\varepsilon R^2 : \ a(0,\lambda,\underline{\alpha}) = a_\lambda(0,\lambda,\underline{\alpha}) = 0\} \tag{1.4a}$$

$$\mathcal{B}_1 = \{\underline{\alpha}\varepsilon R^2 : \ \text{(i) } a(0,\lambda,\underline{\alpha}) = 0; \ \text{ and}$$

$$\text{(ii) } F(x,\lambda,\underline{\alpha}) = F_x(x,\lambda,\underline{\alpha}) = F_\lambda(x,\lambda,\underline{\alpha}) = 0, \ x \neq 0\} \tag{1.4b}$$

$$\mathcal{H}_o = \{\underline{\alpha}\varepsilon R^2 : \ a(0,\lambda,\underline{\alpha}) = a_z(0,\lambda,\underline{\alpha}) = 0\} . \tag{1.4c}$$

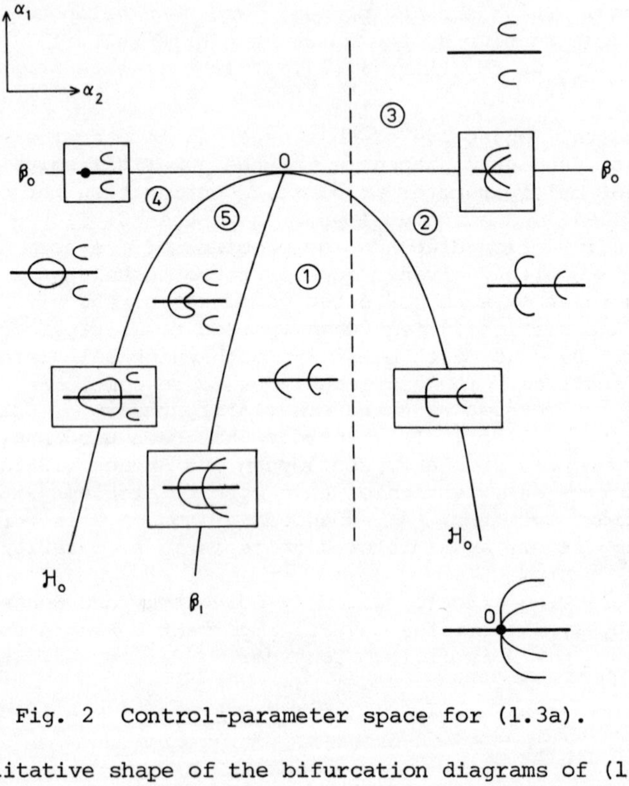

Fig. 2 Control-parameter space for (1.3a).

The qualitative shape of the bifurcation diagrams of (1.3) is the same for all values of $\underline{\alpha}$ in any one region (1)-(5). These shapes are sketched in Fig. 2, and we note that bifurcation diagrams similar to those found in the Taylor problem are recovered if we vary α_1 and $\alpha_2 > 0$. (Follow the dashed line in

Fig. 2 as α_1 increases through zero). Intuitively we say that two bifurcation diagrams are qualitatively similar if they have the same number, type and arrangement of singular points. This notion is made more precise in Golubitsky and Schaeffer (1985).

The two main objectives of our approach to solving (1.1) can now be described. First we aim to completely decompose the control parameter space R^p into regions for which the bifurcation diagrams for (1.1) are qualitatively similar. This means that we need to compute the varieties which divide that space (see, for example, $\mathcal{B}_o$, $\mathcal{B}_1$ and $\mathcal{H}_o$ in Fig. 2). Second we aim to provide initial points on *each* connected component of the bifurcation diagram for any given $\underline{\alpha}\varepsilon R^p$. In this paper we discuss the symmetry-breaking case only and so we do not concern ourselves with varieties like $\mathcal{B}_1$ in Fig. 2.

The plan of the paper is as follows. In section 2 we give the theory of singular points in the presence of symmetry. In section 3 we describe the singularity theory of Golubitsky and Schaeffer which we need for our numerical approach. In section 4 we briefly outline how systems for singular points of (1.1) can be derived from the theoretical results in sections 2 and 3. Finally in section 5 we outline the numerical approach and give a decomposition of control parameter space R^2 for the Taylor problem.

2. SINGULAR POINTS IN THE PRESENCE OF SYMMETRY

In this section we give some theory of bifurcation points. As was mentioned in the introduction a necessary condition for $(x_o,\lambda_o,\underline{\alpha}_o)$ to be a bifurcation point of (1.1) is that $F_x^o: = F_x(x_o,\lambda_o,\underline{\alpha}_o) : X\to X$ is singular. If this holds then $(x_o,\lambda_o,\underline{\alpha}_o)$ is said to be a *singular* point, otherwise it is a *regular* point. Whether or not bifurcation occurs at a singular point depends on certain conditions on the derivatives of F at $(x_o,\lambda_o,\underline{\alpha}_o)$: typically there will be certain *defining conditions* and some associated *non-degeneracy conditions*. Throughout this paper we shall consider only *simple* singular points, i.e. points $(x_o,\lambda_o,\underline{\alpha}_o)$ satisfying (1.1) and such that

$$\text{Null }(F_x^o) = \text{span}\{\phi_o\},\ \text{Range }(F_x^o) = \{y\varepsilon X:\psi_o y = 0,\ \psi_o\varepsilon X'\}\ ,\quad(2.1)$$

where X' is the dual space of X. (If $X = R^n$, then $X' = R^n$ and the usual dual pairing is given by the inner product $\psi_o^T y$). Thus, under (2.1), the defining conditions

$$\begin{aligned} &F^o = 0 \quad , \\ &F_x^o \phi_o = 0 \quad , \quad x_o, \phi_o \in X,\ \lambda_o \in R,\ \underline{\alpha}_o \in R^p \quad , \\ &\ell\phi_o - 1 = 0 \ , \end{aligned} \tag{2.2}$$

(here $\ell \in X'$ and $\ell\phi_o - 1 = 0$ is simply a normalization of the eigenfunction ϕ) with the non-degeneracy conditions

$$\psi_o F_\lambda^o \neq 0 \ , \quad \psi_o F_{xx}^o \phi_o \phi_o \neq 0 \tag{2.3}$$

characterize a simple "quadratic" turning point, which is the most likely type of singular point for a one parameter problem without symmetry (see, for example, Spence and Werner (1982)). In this paper we consider the case when the equations satisfy a reflectional symmetry condition which we represent by the following common assumption:

There exists a linear operator S on X satisfying $S \neq I$, $S^2 = I$, and

$$F(Sx, \lambda, \underline{\alpha}) = SF(x, \lambda\underline{\alpha}) \ . \tag{2.4}$$

For example, in Cliffe and Spence (1985), F represents a variational form of the Navier-Stokes equations, $X = W^{1,2}(D)^3 \times L^2(D)$ and is given by (4.1) in that paper. In the finite-element discretization, $X = R^n$ and S is the matrix given by (4.16).

The symmetry condition (2.4) has several important theoretical and numerical implications, though we discuss mainly the theoretical aspects here. We refer the reader to Cliffe and Spence (1985) for some discussion on the numerical implications.

The following description is based on that in Werner and Spence (1984). The operator S induces a decomposition of X into

$$X = X_s \oplus X_a \tag{2.5a}$$

where

$$X_s = \{x\varepsilon X : Sx = x\} \;, \quad X_a = \{x\varepsilon X : Sx = -x\} \tag{2.5b}$$

consist of the *symmetric* and *anti-symmetric* elements of X respectively. An immediate consequence of (2.5) is that for $x\varepsilon X_s$,

$$\begin{aligned}
&\text{(a)} \quad F, F_\lambda, F_{\alpha_i} \;\varepsilon\; X_s \;,\\
&\text{(b)} \quad X_s \text{ and } X_a \text{ are invariant with regard to } F_x, F_\lambda, F_{\alpha_i} \;,\\
&\text{(c)} \quad \text{for } v, w\varepsilon X_s \text{ or } v, w\varepsilon X_a, \; F_{xx}\, vw\varepsilon X_s \;,\\
&\text{(d)} \quad \text{for } v\varepsilon X_s, \; w\varepsilon X_a, \; F_{xx}\, vw\varepsilon X_a \;,
\end{aligned} \tag{2.6}$$

where F: = F(x,λ,$\underline{\alpha}$) etc. Also it is easy to show, using (2.6b), that if $x_o\varepsilon X_s$ then $\phi_o\varepsilon X_s$ or $\phi_o\varepsilon X_a$. Of most interest is the latter case, the so-called, *symmetry-breaking* case,

$$x_o\varepsilon X_s \;, \; \phi_o\varepsilon X_a \;. \tag{2.7}$$

It follows that

$$\psi_o x = 0 \quad \text{for all } x\varepsilon X_s \tag{2.8}$$

and hence, using (2.6a) and (2.6c),

$$\psi_o F^o_\lambda = 0 \;, \quad \psi_o F^o_{xx}\phi_o\phi_o = 0 \;. \tag{2.9}$$

Clearly, under (2.4) and (2.7), we do not expect to see quadratic turning point singularities since the non-degeneracy conditions (2.3) are automatically violated. In fact the generic type of singular point found when (2.1), (2.4) and (2.7) hold, which is the "quadratic" symmetry-breaking bifurcation point shown in Fig. 1, is characterized by the defining conditions

$$\begin{aligned}
&F(x_o,\lambda_o,\underline{\alpha}_o) = 0 \;,\\
&F_x(x_o,\lambda_o,\underline{\alpha}_o)\phi_o = 0 \;, \quad x\varepsilon X_s, \; \phi\varepsilon X_a, \; \lambda\varepsilon R, \; \underline{\alpha}\varepsilon R^p \;,\\
&\ell\phi_o - 1 = 0 \;,
\end{aligned} \tag{2.10}$$

(note the changes in domains from (2.2.)) and the non-degeneracy conditions

$$\psi_o F^o_{x\lambda}\phi_o + \psi_o F^o_{xx}\phi_o v_o \neq 0 \ , \quad F^o_x v_o = -F^o_\lambda \ , \quad v_o \varepsilon X_s \ , \tag{2.11a}$$

$$\psi_o F^o_{xxx}\phi_o\phi_o\phi_o + \psi_o F^o_{xx}\phi_o v_o \neq 0 \ , \ F^o_x z_o = - F^o_{xx}\phi_o\phi_o \ , \ w_o \varepsilon X_s , \tag{2.11b}$$

where v_o and w_o are well defined because of (2.9). (We remark that we show in section 4 how such systems and non-degeneracy conditions are derived using the theory in section 2.1 and section 3).

An important theoretical tool for analysing the solutions of non-linear problems at and in the neighbourhood of a singular point is the Liapunov-Schmidt reduction process. This is a general process by which a problem like (1.1), of possibly infinite dimension, is reduced to a problem of small dimension, namely, the dimension of $\mathrm{Null}(F^o_x)$. In our case under assumption (2.1) the problem (1.1) will be reduced to an *equivalent scalar* problem, and such problems can then be analysed using the singularity theory in Golubitsky and Schaeffer (1985).

2.1 The Liapunov-Schmidt reduction with symmetry

Under assumptions (2.1) F^o_x is a Fredholm operator of index zero and therefore induces natural decompositions of X into

$$X = X_1 \oplus \mathrm{Null}(F^o_x) \ ; \ X = \mathrm{Range}(F^o_x) \oplus Y_2 \tag{2.12}$$

where X_1 and Y_2 are closed. Let P and Q be the projections

$$P{:}X\to X \quad R(P) = N(F^o_x) \ , \quad N(P) = X_1 \ , \tag{2.13a}$$

$$Q{:}X\to X \quad R(Q) = Y_2 \ , \quad N(Q) = \mathrm{Range}(F^o_x) \ . \tag{2.13b}$$

Note that

$$\dim N(F^o_x) = \dim Y_2 = 1 \ . \tag{2.14}$$

The key result, on which the reduction process depends, is that

$$F^o_x : X_1 \to \mathrm{Range}\ (F^o_x) \tag{2.15a}$$

is invertible (see Schechter (1971)), and hence

$$(I-Q)F_x^o : X_1 \to \text{Range}(F_x^o) \qquad (2.15b)$$

is invertible, since I-Q acts as the identity on Range (F_x^o).
The reduction process proceeds as follows:

In a neighbourhood of $(x_o, \lambda_o, \underline{\alpha}_o)$

(1) Write $x = x_o + (I-P)x + Px$

$$= x_o + x_1 + x_2 \ , \ x_1 \varepsilon X_1, \ x_2 \varepsilon N(F_x^o) \ . \qquad (2.16)$$

(2) Write $F(x, \lambda, \underline{\alpha}) = 0$ as

$$(I-Q)F\ (x_o + x_1 + x_2,\ \lambda,\ \underline{\alpha}) = 0\ \varepsilon\ \text{Range}(F_x^o) \qquad (2.17a)$$

$$QF\ (x_o + x_1 + x_2,\ \lambda,\ \underline{\alpha}) = 0\ \varepsilon\ Y_2 \qquad (2.17b)$$

(3) Using (2.15b), we can solve (2.17a) for x_1 in terms of x_2, λ and $\underline{\alpha}$ using the Implicit Function Theorem, i.e.,

$$x_1 = x_1(x_2, \lambda, \underline{\alpha}) \qquad (2.18)$$

and substituting this into (2.17b) gives

$$QF(x_o + x_1(x_2, \lambda, \underline{\alpha}) + x_2,\ \lambda,\ \underline{\alpha}) = 0\ , \qquad (2.19)$$

which is a mapping from $N(F_x^o) \times R \times R^p$ to Y_2 .

(4) Finally it is convenient to rewrite (2.19) in co-ordinate form using

$$N(F_x^o) = \varepsilon\phi_o \quad ,$$

where ε is real, and

$$Qz = 0 \text{ implies } \psi_o z = 0.$$

Thus (2.19) becomes

$$h(\varepsilon, \lambda, \underline{\alpha}) : \ = \psi F(x_o + x_1\ (\varepsilon\phi_o, \lambda, \underline{\alpha}) + \varepsilon\phi_o, \lambda, \underline{\alpha}) = 0 \qquad (2.20)$$

under the above assumptions.

Thus (1.1) is equivalent to (2.20) and information about the solutions of the scalar problem (2.20) can be converted back, via. (2.17)-(2.19) to information about solutions of (1.1). We merely note the following results which are readily proved by differentiation and evaluation at $(\varepsilon,\lambda,\underline{\alpha}) = (0,\lambda_o,\underline{\alpha}_o)$ (see Golubitsky and Schaeffer (1985))

$$h_{\varepsilon}(0,\lambda_o,\underline{\alpha}_o) = 0 , \tag{2.21a}$$

which follows immediately from $F^o_x\phi_o = 0$,

$$h_{\varepsilon\varepsilon}(0,\lambda_o,\underline{\alpha}_o) = \psi_o F^o_{xx}\phi_o\phi_o \tag{2.21b}$$

$$h_{\varepsilon\varepsilon\varepsilon}(0,\lambda_o,\underline{\alpha}_o) = 3\psi_o F^o_{xx}\phi_o x_{1\varepsilon\varepsilon} + \psi_o F^o_{xxx}\phi_o\phi_o\phi_o \tag{2.21c}$$

$$h_{\lambda}(0,\lambda_o,\underline{\alpha}_o) = \psi_o F^o_{\lambda} \tag{2.21d}$$

$$h_{\lambda\varepsilon}(0,\lambda,\underline{\alpha}_o) = \psi_o F^o_{\lambda x}\phi_o + \psi_o F^o_{xx}\phi_o x_{1\lambda} \tag{2.21e}$$

where

$$x_{1\varepsilon\varepsilon} = \frac{\partial^2}{\partial\varepsilon^2} x_1 \text{ and } x_{1\lambda} = \frac{\partial}{\partial\lambda} x_1 \quad .$$

Finally in this section we note that if F satisfies the symmetry property (2.4) then it is straightforward to show from (2.18)

$$x_1 = x_1(\varepsilon\phi_o,\lambda,\underline{\alpha}) = - x_1(-\varepsilon\phi_o,\lambda,\underline{\alpha})$$

and that

$$h(\varepsilon,\lambda,\underline{\alpha}) = - h(-\varepsilon,\lambda,\underline{\alpha}) \quad , \tag{2.22}$$

i.e. the reduced problem on X = R satisfies the symmetry condition (2.4) with S = -1. Also using (2.9) and (2.21b), (2.21d) we have

$$h_{\varepsilon\varepsilon} = 0 \ , \quad h_{\lambda} = 0 \quad , \tag{2.23}$$

and one readily shows that $x_{1\varepsilon\varepsilon} = z_o$ given by (2.11b) and $x_{1\lambda} = v_o$ given by (2.11a). Hence conditions (2.11a) and (2.11b) are precisely $h_{\varepsilon\lambda} \neq 0$ and $h_{\varepsilon\varepsilon\varepsilon} \neq 0$. We discuss these links further in section 4. First we have to analyse problems like (2.20) which satisfy (2.22) and this is done next in section 3.

Remark 2.24

We note that if the symmetry condition (2.4) does not hold then it can be readily shown using (2.21a), (2.21b) and (2.21d) that conditions (2.2) and (2.3) which characterize a quadratic turning point are equivalent to

$$h(\varepsilon,\lambda,\underline{\alpha}) = 0$$

$$h_\varepsilon(\varepsilon,\lambda,\underline{\alpha}) = 0$$

with $h_\lambda(\varepsilon,\lambda,\alpha) \neq 0$, $h_{\varepsilon\varepsilon}(\varepsilon,\lambda,\underline{\alpha}) \neq 0$. This subject is discussed more fully in Jepson and Spence (1985b).

3. SINGULARITY THEORY WITH Z_2-SYMMETRY

In order to achieve the objectives laid out in the second last paragraph in section 1 it is clear that we need a computational method both for the calculation of the varieties which split up control parameter s ce (for example, H_0 and B_0 in Fig. 2) and for recognizing and calculating any exceptional points on these varieties (for example, O, in Fig. 2). In this section we describe the singularity theory which gives our numerical approach its theoretical basis. The material is to be found in Golubitsky and Schaeffer (1985) and Golubitsky and Langford (1981). We omit all proofs in this section; they will appear in a later publication.

It was shown in Section 2.1 how problems like (1.1) satisfying the symmetry condition (2.4) might be reduced, under appropriate assumptions, to scalar problems of the form

$$h(x,\lambda,\underline{\alpha}) = 0 \ , \quad h: R \times R \times R^p \to R \ , \tag{3.1}$$

$$h(-x,\lambda,\underline{\alpha}) = -h(x,\lambda,\underline{\alpha}). \tag{3.2}$$

(Here we use x instead of ε in section 2.1). Problems satisfying (3.2) are said to have Z_2-*symmetry* and can be rewritten as

$$h(x,\lambda,\underline{\alpha}) = x\, a(z,\lambda,\underline{\alpha}) \ , \ z = x^2 \ , \tag{3.3}$$

i.e. a is *even* in x. It turns out that (3.3) is the most convenient form for study.

First consider problems of the form

$$h(x,\lambda) = 0 \ , \ h = R \times R \rightarrow R \ , \qquad h(-x,\lambda) = -h(x,\lambda) \ , \tag{3.4}$$

with the aim of classifying the Z_2-symmetry-breaking singularities of $h(x,\lambda) = 0$ i.e. the points $(0,\lambda_o)$ where $h_x(0,\lambda_o) = 0$ (recall (2.21a)) and where bifurcation from $x = 0$ takes place. (Note that for the case $X = R$, $S = -1$, then $X_s = \{0\}$ since, using (2.5b), $Sx = x$ implies $x = 0$). In Golubitsky and Langford (1981) the notion of Z_2-equivalence is used to classify the singularities of (3.4) into equivalence classes.

Definition 3.5

Let h, g: $R^2 \rightarrow R$ be two smooth (i.e. C^∞) functions with Z_2-symmetry, and suppose that $h(0,\lambda_o) = 0$, $g(0,0) = 0$. The two bifurcation problems

$$h(x,\lambda) = 0 \ ,$$

$$g(x,\lambda) = 0 \ ,$$

are said to be Z_2 - *equivalent* if

$$g(x,\lambda) = T(x,\lambda)\, h(X(x,\lambda),\Lambda(\lambda)) \tag{3.5a}$$

where T is *even* in x, X is *odd* in x, and

$$T(0,0) \neq 0 \ , \ \frac{\partial X}{\partial x}(0,0) \neq 0 \ , \ X(0,0) = 0 \ , \qquad \frac{\partial \Lambda}{\partial \lambda}(0,0) \neq 0, \ \ \Lambda(0) = \lambda_o \ . \tag{3.5b}$$

Associated with each equivalence class is a number, called the Z_2-codimension, which, if finite, shows that each member of the equivalence class is Z_2- equivalent to a *polynomial* canonical form. Also bifurcation diagrams for problems in the same equivalence class have the same qualitative behaviour as that of the canonical form. The equivalence classes are referred to simply as "singularities". As in Jepson and Spence we introduce a graph, which we call the Z_2-*hierarchy of singularities,* in which are arranged the singularities of

Z_2-codimension less than 4. The graph, Fig. 3, is structured to emphasize the relationships between the singularities and to allow a clear description of our numerical approach. The (q,j)-singularity is defined to be the singularity which the (q,j)-node has at (0,0). Note that q takes the values $0,1,2,3^*,3$ where for rather technical reasons we distinguish one codimension 3 singularity with a superscript '*' and create an extra level between q=2 and q=3 in the graph for this singularity. Throughout this paper $0 \leqslant q \leqslant 3$ is taken to include the case $q = 3^*$.

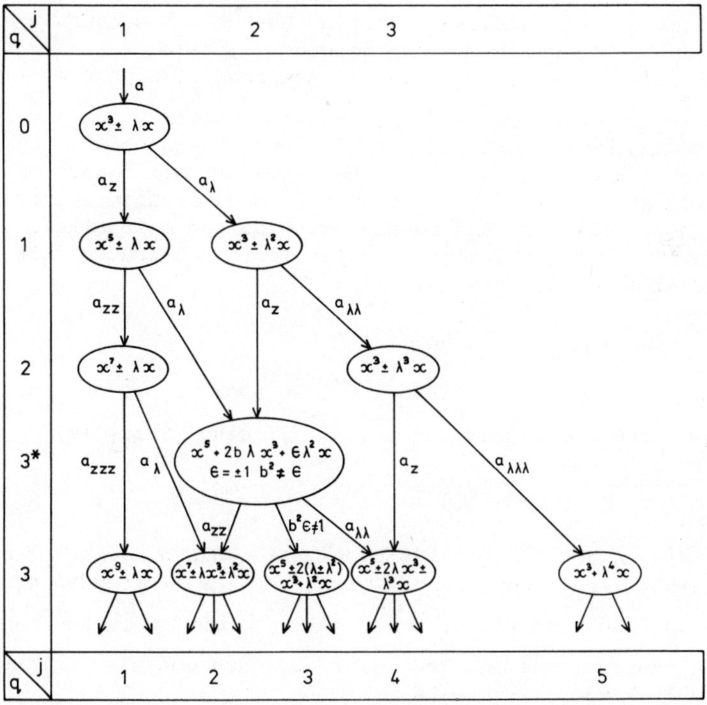

Fig. 3 Hierarchy of Z_2-singularities

An important concept is the *stability* of a singularity with respect to Z_2-symmetry preserving perturbations.

Definition 3.6

A singularity of $h(x,\lambda,\underline{\alpha}) = 0$ satisfying (3.2) at $(0,\lambda_0,\underline{\alpha}_0)$ is said to be *structurally stable* if for any smooth $r(x,\lambda,\underline{\alpha})$

also satisfying (3.2), for any neighbourhood V of $(\lambda_o, \underline{\alpha}_o)$ and for any δ with $|\delta|$ sufficiently small, the perturbed problem

$$\hat{h}(x,\lambda,\underline{\alpha}) = h(x,\lambda,\underline{\alpha}) + \delta r(x,\lambda,\underline{\alpha}) \tag{3.6}$$

has a singularity of the same type at $(0,\lambda(\delta),\underline{\alpha}(\delta))$, $(\lambda(\delta),\underline{\alpha}(\delta)) \varepsilon V$, (i.e. $\hat{h}$ is Z_2-equivalent to h for δ sufficiently small). If in addition $(\lambda(\delta),\underline{\alpha}(\delta)) = (\lambda(0),\underline{\alpha}(0)) + O(\delta)$ the singularity is *structurally linearly stable*. Clearly in numerical calculations on a physical problem one can only expect to compute structurally stable phenomena and this is reflected in Theorem 3.14 below, which states that if a singularity is structurally stable then we can construct an *extended system* which has an isolated solution at the singularity. The Z_2-hierarchy is designed such that a suitable extended system for a (q,j)-singularity can be constructed by following a path down the hierarchy to the (q,j)-node and requiring that the label on each branch of the path vanishes at $(0,\lambda,\underline{\alpha})$. We assume that the path starts above the (0,1)-node and so the equation a = 0 is always included. Thus the extended system for a "quadratic" pitchfork bifurcation point, (q,j) =(0,1) is

$$A_{0,1} := a = 0 , \tag{3.7}$$

(cf. Jepson and Spence (1985a), where symmetry is not assumed), and the extended system for the $(3^*,2)$ singularity is

$$A_{3^*,2} := (a, a_z, a_\lambda) = 0 . \tag{3.8}$$

As in Jepson and Spence (1985a) we associate non-degeneracy conditions (side-constraints) with each (q,j)-extended system. These are the conditions, $C^k_{q,j}$, k = 1,2,...K, that the labels on the branches leaving the (q,j)-node are non-zero. Thus for the (0,1)-singularity we have

$$C^1_{0,1} := a_z \neq 0 , \quad C^2_{0,1} = a_\lambda \neq 0 . \tag{3.9}$$

The following theorem, which is an immediate consequence of proposition 3.47 of Golubitsky and Langford (1981) shows that a (q,j)-extended system with appropriate non-degeneracy conditions completely characterizes a (q,j) - singularity of (3.3).

Theorem 3.10

With $0 \leq q \leq 3$, assume that

$$A_{q,j}(0,\lambda,\underline{\alpha}) = 0 \in R^{q+1} \tag{3.10}$$

is a (q,j)-extended system derived from the Z_2-hierarchy and that

$$C^k_{q,j}(0,\lambda,\underline{\alpha}) \neq 0 \quad k = 1,2,\ldots,K, \tag{3.11}$$

are the appropriate side constraints. Then $(0,\lambda,\underline{\alpha})$ is a solution of (3.10) satisfying (3.11) if and only if $(0,\lambda,\underline{\alpha})$ is a (q,j) singularity of (3.3).

The question arises as to how one might compute such singular points. Obviously one must assume that in a physical problem any singularity is structurally stable, which essentially means that the control parameters in (3.3) enter in such a way as to include all possible Z_2-perturbations of (3.3). These concepts are made precise using the concept of a Z_2-unfolding. Briefly, let $g(x,\lambda)$ be a bifurcation problem with Z_2-symmetry, then $h(x,\lambda,\underline{\alpha})$ is a *p-parameter* Z_2*-unfolding* of $g(x,\lambda)$ near $(0,\lambda_0,\underline{\beta}_0)$ if $g(x,\lambda) = h(x,\lambda,\beta_0)$ and h satisfies (3.2). If h and f are Z_2-unfoldings of g then h *factors through* f if $h(x,\lambda,\underline{\beta}) = T(x,\lambda,\underline{\beta})f(X(x,\lambda,\underline{\beta}),\Lambda(\lambda,\underline{\beta}),M(\underline{\beta}))$ where $M(\underline{\beta}_0) = \underline{\beta}_0$ and the usual Z_2-equivalence conditions (3.5b) hold. h is a *universal* Z_2*-unfolding* of g if all unfoldings of g factor through h. The Z_2*-codimension* of g is the number of parameters in any universal unfolding. The following theorem follows immediately from the definitions.

Theorem 3.13

Let $(0,\lambda_0,\underline{\alpha}_0)$ be a (q,j)-singularity of $h(x,\lambda,\underline{\alpha}) = 0$ and suppose h is a Z_2-universal unfolding. Then this (q,j)-singularity is structurally linearly stable.

We can now discuss briefly the situation about the $(3^*,2)$-singularity. The difficulty is explained in Golubitsky and Langford (1981) after Theorem 3.19. The $(3^*,2)$-singularity, $x^5 + 2m\lambda x^3 \pm \lambda^2 x = 0$, $m^2 \neq 1$ has Z_2-codimension 3 because of the smooth (C^∞) requirement on the equivalence relation (3.5a). If we allow only topological equivalence (i.e. the equivalence only involves continuous functions) then the singularity has (topological) Z_2- codimension 2. This is in fact a deep theoretical problem concerned with *moduli* (which we do not discuss here) but numerically is quite easy to cope with as we see in section 5. Essentially we think of the $(3^*,2)$-singularity having codimension 2.

Finally we have two very useful theoretical results from the computational point of view:

Theorem 3.14

With $0 \leq q \leq 3 \leq p$, assume $(0,\lambda_o,\underline{\alpha}_o)$ is a (q,j)-singularity of $h(x,\lambda,\underline{\alpha}) = a(z,\lambda,\underline{\alpha})x$, $z = x^2$. Let $A_{q,j}(0,\lambda_1,\underline{\alpha})$ be the (q,j)-extended system obtained from the Z_2-hierarchy. If $h(x,\lambda,\underline{\alpha})$ is a universal unfolding of $h(x,\lambda,\underline{\alpha}_o)$ then

$$\text{rank}\left(\frac{\partial A_{q,j}}{\partial(\lambda,\alpha)} \quad (0,\lambda_o,\underline{\alpha}_o)\right) = q+1$$

i.e. $\dfrac{\partial A_{q,j}}{\partial(\lambda,\underline{\alpha})}$ has full rank.

Theorem 3.15

Let $1 \leq q \leq 3$ and suppose $(0,\lambda_o,\underline{\alpha}_o)$ is a linearly stable (q,j) singularity of (3.3). Let j' be chosen so that (q-1,j') is a node in the hierarchy above the (q,j) node. Then $(x_o,\lambda_o,\underline{\alpha}_o)$ is a solution of

$$A_{q-1,j'} = 0$$

with

$$\text{rank}\left(\frac{\partial A_{q-1,j'}}{\partial(\lambda,\alpha)} \quad (x_o,\lambda_o,\underline{\alpha}_o)\right) = q \text{ (i.e. full)}.$$

The importance of Theorems 3.14 and 3.15 is explained in section 5.

4. EXTENDED SYSTEMS FOR (1.1)

In this section we briefly give the extended systems for problems like (1.1) which are derived from the extended systems $A_{q,j}$ in section 3. We only give those needed in the Taylor problem.

Recall that in section 2.1 we showed how for *simple* singular points problems like (1.1) could be converted into scalar problems like (2.20) via the Liapunov-Schmidt reduction process. This process is reversible and it is straightforward to show that the extended system for the (0,1)-singularity (Z_2-equivalent to $x^3-\lambda x$) given by (3.4) with non-degeneracy conditions (3.9) is equivalent to the system (2.10) with non-degeneracy conditions (2.11).

The extended system $A_{1,1} = (a, a_z) = 0$ for (λ, α_1) to compute a (1,1)-singularity (Z_2-equivalent to $x^5-\lambda x$) is equivalent to

$$
\begin{aligned}
&F^o = 0, \\
&F^o_x \phi_o = 0, \\
&\ell\phi - 1 = 0, \qquad\qquad (4.1)\\
&F^o_x z_o + F_{xx}\phi_o\phi_o = 0, \\
&F^o_x q_o + F^o_{xx}\phi_o z_o + 1/3\, F^o_{xxx}\phi_o\phi_o\phi_o = 0, \\
&\ell q_o = 0,
\end{aligned}
$$

with unknown $(x_o, \phi_o, \lambda_o, z_o, q_o, (\alpha_1)_o) \in X_s \times X_a \times R \times X_s \times X_a \times R$,

and the extended system $A_{1,2} = (a, a_\lambda) = 0$ for (λ, α_1) to compute a (1,2)-singularity (Z_2 - equivalent to $x^3-\lambda^2 x$) is equivalent to

$$
\begin{aligned}
&F^o = 0, \\
&F^o_x \phi_o = 0, \\
&\ell\phi - 1 = 0, \qquad\qquad (4.2)\\
&F^o_x v_o + F^o_\lambda = 0, \\
&F^o_x w_o + F^o_{x\lambda}\phi_o + F_{xx}\phi_o v_o = 0, \\
&\ell w_o = 0,
\end{aligned}
$$

with unknown $(x_o,\phi_o,\lambda_o,v_o,w_o,(\alpha_1)_o) \in X_s x X_a x R x X_s x X_a x R$. Systems (4.1) and (4.2) were first given in Cliffe and Spence (1984) where the (1,1)-and (1,2)-singularities were called *quartic pitchfork* and *coalescence points* respectively.

It should be clear now how to derive the extended system for a$(3^*,2)$- singularity in (1.1) which is equivalent to $A_{3^*,2} = (a,a_z,a_\lambda) = 0$. Such a system consists of adding the last three equations of (4.1) to (4.2).

Fuller details of the various possible extended systems for singular points are given in Jepson and Spence (1985b). We merely note that numerically useful theorems like 3.10, 3.13, 3.14 and 3.15 can be proved for the extended systems like (2.10) (4.1) and (4.2).

5. NUMERICAL APPROACH AND RESULTS FOR THE TAYLOR PROBLEM

The numerical approach is essentially the same as that described in Jepson and Spence (1984), (1985a). Here we give a brief summary and present some numerical results for the Taylor problem. We describe the technique for a problem like (3.1). The extension to problems like (1.1) is clear.

Step 1 Descending the hierarchy

We begin by trying to locate the points of highest codimension in a given problem like (3.1). To do this we descend the hierarchy as follows. Assume at some stage we know a solution $(0,\lambda,\underline{\alpha}_o)$ of the (q,j)-extended system

$$A_{q,j}(0,\lambda,\underline{\alpha}_o) = A_{q,j}(z,\beta,\underline{\gamma}_o) \tag{5.1}$$

where $z = (0,\lambda,\underline{a})$, $\beta = \alpha_{q+1}$, $\gamma = (\alpha_{q+2},\ldots\alpha_p)$ with $\underline{a} = (\alpha_1,\ldots,\alpha_q)$ the *unfolding* parameters $\beta \in R$ and $\underline{\gamma} \in R^{p-q-1}$ called the *slice* parameters. For the slice parameters held constant, (5.1) may be viewed as a one parameter bifurcation problem with state variable z, and bifurcation parameter β. From theorem 3.14 we see that $(0,\lambda,\underline{\alpha}_o)$ corresponds to a regular point of (5.1) and hence paths of such points can be computed using standard continuation algorithms. Also it is important to note that Theorem 3.15 implies that the extended systems are well behaved at codimension q+1 singularities. In particular we can expect only simple turning points in (5.1) as β varies.

Singular points of codimension q+1 on this path are recognized by locating a range of values of β at which a non-degeneracy condition $C^k_{q,j}$ changes sign. It may be computed using the appropriate extended system and the process of following paths of such (q+1)-codimension singularities is repeated. In this way the hierarchy is descended until the singularity with the highest codimension is located.

Step 2 Ascending the hierarchy

Given a (q,j)-singularity $(z,\underline{\beta},\gamma)$ that is a regular point of (5.1) the point is certainly a solution of

$$A_{q-1,j'}(z',\beta',\underline{\gamma}') = 0 \tag{5.2}$$

where the (q-1,j')-node is a node in the hierarchy directly above the (q,j)-node and $(z',\beta') = z$, $\underline{\gamma}' = (\beta,\underline{\gamma})$. Then theorem 3.15 shows that (5.2) has, at worst, a simple turning point at the (q,j)-singular point. Standard continuation methods can now be used to compute solution branches of (5.2).

Step 3 Obtaining initial points

By unfolding all the singular points of codimension ≥ 1 we can obtain all the quadratic pitchfork points and turning points and hence obtain starting values on each connected component of the bifurcation diagram of (3.1) which contains at least one singular point.

Calculations were done on the Taylor problem described in section 1, some numerical results for the case of α_2 fixed have already been reported, Cliffe (1983) and Cliffe and Spence (1984), and these numerical results agree well with the experimental results of Benjamin and Mullin (1981). We present here, in Fig. 4, a plot of the control parameter space where α_1 = aspect ratio and α_2 = radius ratio. The dashed line corresponds to the results for $\alpha_2 = 0.615$, the value in the Benjamin and Mullin experiment. In Fig. 4 we only give the varieties $\mathcal{H}_o$ and $\mathcal{B}_o$. Variety $\mathcal{H}_o$ is the path of (1,1)-singularities (quartic bifurcation points) computed using (4.1) and letting α_2 vary, and $\mathcal{B}_o$ is the path of (1.2) - singularities (coalescence points) computed using (4.2) and letting α_2 vary. These paths touch tangentially at O as predicted in Fig. 2, which gives the control parameter space for the unfolding of the canonical form of the (3*,2) singularity. The point O is recognised on H_0 because the side constraint $a_\lambda \neq 0$ (equivalently (2.11a)) is violated. Finally we note that for the Taylor problem we can derive a condition equivalent to m > 1 in (1.3a) and this can be confirmed numerically.

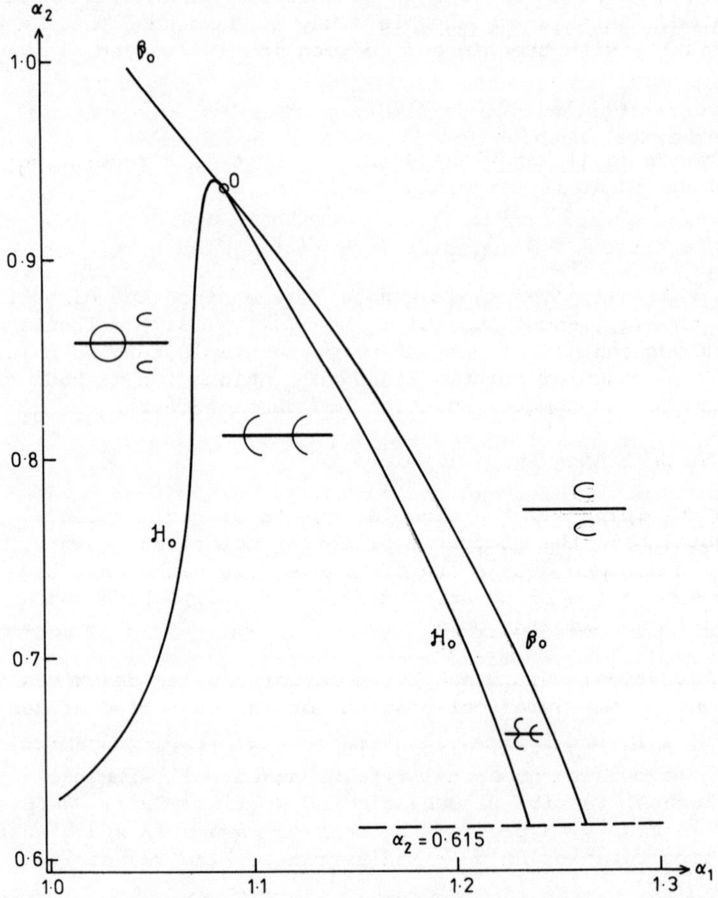

Fig. 4 Control-parameter-space plot of solutions for the finite Taylor problem.

REFERENCES

Benjamin, T.B. and Mullin, T., (1981). Anomalous modes in the Taylor Experiment. *Proc. Roy. Soc. Lond.* **A359**, 24-43.

Cliffe, K.A., (1983). Numerical calculations of two-cell and single-cell Taylor flows. *J. Fluid Mech.* **135**, 219-233.

Cliffe, K.A. and Spence, A., (1984). The calculation of high order singularities in the finite Taylor problem. In "Numerical Methods for Bifurcation Problems." pp 129-144, *ISNM* **70** Birkhauser, Basel.

Cliffe, K.A. and Spence, A., (1985). Numerical calculations of bifurcations in the finite Taylor problem. These proceedings.

Golubitsky, M. and Langford, W.F., (1981). Classification and unfoldings of degenerate Hopf bifurcations. *J. Diff. Equns.* **41**, 375-415.

Golubitsky, M. and Schaeffer, D.G., (1985). Singularities and Groups in Bifurcation Theory. Springer, New York.

Jepson, A.D. and Spence, A., (1984). Singular points and their computation. In "Numerical Methods for Bifurcation Problems". (T. Küpper, H.D. Mittleman and H. Weber, eds.), pp. 195-209, *ISNM* **70**, Birkauser, Basel.

Jepson, A.D. and Spence, A., (1985a). The numerical solution of nonlinear equations having several parameters I: Scalar equations (to appear in *SIAM J. Numer. Analysis*).

Jepson, A.D. and Spence, A., (1985b). On the equivalence of extended systems for singular points (in preparation).

Keller, H.B., (1977). Numerical solution of bifurcation and nonlinear eigenvalue problems. In "Applications of Bifurcation Theory." (P.H. Rabinowitz, ed.), pp359-384, Academic Press, New York.

Rheinboldt, W.C., (1980). Solution fields of nonlinear equations and continuation methods. *SIAM J. Numer. Anal.* **17**, 221-237.

Schechter, M., (1971). Principles of Functional Analysis. Academic Press, New York.

Spence, A. and Werner, B., (1982). Nonsimple turning points and and cusps. *IMA J. of Numer. Anal.* **2**, 413-427.

Werner, B. and Spence, A., (1984). The computation of symmetry-breaking bifurcation points. *SIAM J. Numer. Anal.* **21**, 388-399.

NUMERICAL CALCULATIONS OF BIFURCATIONS IN THE FINITE TAYLOR PROBLEM

K.A. Cliffe
(Theoretical Physics Division, AERE Harwell)

and

A. Spence
(School of Mathematics, University of Bath)

1. INTRODUCTION

The Taylor problem of flow between concentric circular cylinders has been studied for over 60 years. Despite this intensive investigation it is still producing new phenomena which are not yet fully understood. In this paper we shall be concerned with the application of recently developed techniques in numerical bifurcation theory to the steady, axisymmetric flows possible in the Taylor experiment. The experimental situation we have in mind consists of two concentric circular cylinders. The inner cylindrical wall rotates and the outer cylinder and both ends are stationary. The annular gap between the cylinders is filled with a fluid and it is the motion of this fluid that is studied. One of the ends consists of a moveable annular collar so that the length of the annulus may be adjusted. Benjamin and Mullin (Benjamin, 1978; Benjamin and Mullin, 1981; Mullin, 1982; Mullin, Pfister and Lorenzen, 1982) have carried out several experimental investigations of flow in the above apparatus and have discovered an interesting variety of bifurcation phenomena.

From the point of view of bifurcation theory Benjamin and Mullin's apparatus has several salient features. Firstly the equations governing the steady motion of the fluid are non-linear; secondly the problem has two parameters which may be adjusted, namely the speed of the inner cylinder (in non-dimensional form the Reynolds number, R) and the length of the annulus (in non-dimensional form the aspect ratio, Γ); finally the apparatus is mirror symmetric about the midplane of the annulus. (The apparatus also possesses axisymmetry about the axis of the cylinders, however all the flows considered in this paper preserve this symmetry and so it has no effect on the structure of the bifurcations).

The approach we have adopted in studying this problem is to discretize the governing boundary value problem by the finite-element method. This leads to a finite dimensional set of equations with two adjustable parameters R and Γ. The equations also possess Z_2 or mirror symmetry. The solution structure of these equations is determined by applying numerical methods for two-parameter problems with Z_2-symmetry. (Jepson and Spence, 1980; Werner and Spence, 1984; Cliffe, Jepson and Spence, 1985).

The rest of the paper is as follows. In section 2 we present the governing equations for the problem and describe the finite-element discretization of these equations in section 3. In section 4 we discuss the implications of Z_2-symmetry. In section 5 we describe the various systems of equations that can be used to compute singular points in problems with two parameters and Z_2-symmetry. Finally in section 6 we illustrate the use of these techniques on a problem in the Taylor experiment (Cliffe, 1985).

2. GOVERNING EQUATIONS

The equations describing the flow of an incompressible, viscous fluid in the Taylor experiment are the Navier-Stokes equations. In cylindrical polar coordinates (r^*, ϕ, z^*) with origin midway between the ends of the annulus and velocity (u_r^*, u_ϕ^*, u_z^*) the equations for steady, axisymmetric flow are:

$$R\left(\Gamma u_r \frac{\partial u_r}{\partial r} + u_z \frac{\partial u_r}{\partial z} - \frac{\Gamma u_\phi^2}{r+\beta} + \Gamma \frac{\partial p}{\partial r}\right.$$

$$- \frac{\Gamma}{(r+\beta)} \frac{\partial}{\partial r} 2(r+\beta) \frac{\partial u_r}{\partial r} - \frac{\partial}{\partial z}\left(\frac{1}{\Gamma}\frac{\partial u_r}{\partial z} + \frac{\partial u_z}{\partial r}\right) + \frac{2\Gamma u_r}{(r+\beta)^2} = 0, \quad (2.1)$$

$$R\left(\Gamma u_r \frac{\partial u_\phi}{\partial r} + u_z \frac{\partial u_\phi}{\partial z} + \Gamma \frac{u_r u_\phi}{(r+\beta)}\right)$$

$$- \frac{\Gamma}{(r+\beta)} \frac{\partial}{\partial r} (r+\beta) \frac{\partial u_\phi}{\partial r} - \frac{1}{\Gamma} \frac{\partial^2 u_\phi}{\partial z^2} + \frac{\Gamma u_\phi}{(r+\beta)^2} = 0 , \quad (2.2)$$

$$R\left(\Gamma u_r \frac{\partial u_r}{\partial r} + u_z \frac{\partial u_z}{\partial z}\right) + \frac{\partial p}{\partial z}$$

$$- \frac{1}{(r+\beta)} \frac{\partial}{\partial r} (r+\beta) \left(\Gamma \frac{\partial u_r}{\partial r} + \frac{\partial u_r}{\partial z}\right) - \frac{2}{\Gamma} \frac{\partial^2 u_z}{\partial z^2} = 0, \quad (2.3)$$

$$\frac{\Gamma}{(r+\beta)} \frac{\partial}{\partial r} (r+\beta) u_r + \frac{\partial u_z}{\partial z} = 0 . \quad (2.4)$$

In the above equations r, z, (u_r, u_ϕ, u_z) and p are given by:

$$r = \frac{r^*}{d} - \beta, \; z = \frac{z^*}{\ell}, \; (u_r, u_\phi, u_z) = \frac{(u_r^*, u_\phi^*, u_z^*)}{r_1 \Omega}, \; p = \frac{dp^*}{\mu r_1 \Omega},$$

and

$$R = \frac{\Omega r_1 d}{\nu}, \; \Gamma = \frac{\ell}{d},$$

where

$\beta = r_1/d = \eta/(1-\eta)$ and η is the radius ratio; $d = r_2 - r_1$ where r_2 and r_1 are the radii of the outer and inner cylinders respectively; ℓ is the length of the annulus; Ω is the angular velocity of the inner cylinder and ν and μ are the kinematic and dynamic viscosity of the fluid respectively.

Equations (2.1)-(2.4) hold in the region

$$D = \{(r,z) \,|\, 0 \leq r \leq 1 , \; -0.5 \leq z \leq 0.5\} . \quad (2.5)$$

The boundary conditions are that u_r and u_z are zero on the entire boundary of D, and that u_ϕ is zero on the outer cylinder (r=1) and one on the inner cylinder (r=0). On the ends ($z=\pm 0.5$) u_ϕ is zero except near the inner cylinder where it increases smoothly to one over a small distance, ε. The exact value of ε and the variation of u_ϕ will depend on the details of the experiment; however, we have found the results to be insensitive to the value of ε and therefore conclude that any sufficiently small value will be adequate. We note that ε must be positive because when $\varepsilon = 0$ the rate of dissipation of energy in the fluid becomes infinite (see Benjamin and Mullin, 1981, pp. 224, 225).

3. FINITE-ELEMENT DISCRETIZATION

As the starting point for the finite-element discretization, equations (2.1)-(2.4) are converted into a non-linear operator equation in an appropriate Hilbert space. We introduce the following notation: let $L^2(D)$ be the space of functions which are square integrable over D; let $W^{1,2}(D)$ be the space of functions whose generalized first derivatives lie in $L^2(D)$ and let $W_o^{1,2}(D)$ be that subspace of $W^{1,2}(D)$ whose elements vanish (weakly) on the boundary of D. $W^{1,2}(D)^3$ is the space of vector valued functions each component of which is in $W^{1,2}(D)$. Finally let $\underset{\sim}{H} = W_o^{1,2}(D)^3 \times L^2(D)$. We introduce the following functionals:

$$a(\underset{\sim}{U};\underset{\sim}{V},\underset{\sim}{W},R,\Gamma,\eta) = R\int_D [\{(r+\beta)\ (\Gamma u_r \frac{\partial v_r}{\partial r} + u_z \frac{\partial v_r}{\partial z}) - \Gamma\ u_\phi w_\phi\} w_r$$

$$+ \{(r+\beta)\ (\Gamma\ u_r \frac{\partial v_\phi}{\partial r} + u_z \frac{\partial v_\phi}{\partial z}) + \Gamma\ u_r\ v_\phi\} w_\phi$$

$$+ \{(r+\beta)\ (\Gamma\ u_r \frac{\partial v_z}{\partial r} + u_z \frac{\partial v_z}{\partial z}) w_z\}], \tag{3.1}$$

$$b_1(\underset{\sim}{U},\underset{\sim}{W},R,\Gamma,\eta) = \int_D [2\Gamma(r+\beta)\ \frac{\partial u_r}{\partial r}\frac{\partial w_r}{\partial r} + (r+\beta)(\frac{1}{\Gamma}\frac{\partial u_r}{\partial z} + \frac{\partial u_z}{\partial r})\frac{\partial w_r}{\partial z}$$

$$+ 2\Gamma\ \frac{u_r w_r}{r+\beta} + \Gamma(r+\beta)\ \frac{\partial u_\phi}{\partial r}\frac{\partial w_\phi}{\partial r} + \frac{(r+\beta)}{\Gamma}\frac{\partial u_\phi}{\partial z}\frac{\partial w_\phi}{\partial z}$$

$$+ \Gamma\ \frac{u_\phi w_\phi}{(r+\beta)} + (r+\beta)\ (\Gamma\ \frac{\partial u_z}{\partial r} + \frac{\partial u_r}{\partial z})\frac{\partial w_z}{\partial r} + 2\ \frac{(r+\beta)}{\Gamma}\frac{\partial u_z}{\partial z}\frac{\partial w_z}{\partial z}]$$

$$- \int_D [\Gamma\ p\ \frac{\partial}{\partial r}\ (r+\beta) w_r + p(r+\beta)\ \frac{\partial w_z}{\partial z}]$$

$$- \int_D [\Gamma\ s\ \frac{\partial}{\partial r}\ (r+\beta) u_r + s(r+\beta)\ \frac{\partial u_z}{\partial z}], \tag{3.2}$$

where $\underset{\sim}{u} = (u_r, u_\phi, u_z, p)$, $\underset{\sim}{V} = (v_r, v_\phi, v_z, q)$ and $\underset{\sim}{W} = (w_r, w_\phi, w_z, s)$ with $\underset{\sim}{U}, \underset{\sim}{V}, \underset{\sim}{W} \varepsilon \underset{\sim}{H}$. Let $\hat{\underset{\sim}{U}} \equiv (0, \hat{u}_\phi, 0, 0) \varepsilon \underset{\sim}{H}$, where $\hat{u}_\phi$ satisfies the boundary conditions on the azimuthal component of velocity, and let

$$b(\underset{\sim}{U}, \underset{\sim}{W}, R, \Gamma, \eta) = a(\hat{\underset{\sim}{U}}; \underset{\sim}{U}, \underset{\sim}{W}, R, \Gamma, \eta) + a(\underset{\sim}{U}; \hat{\underset{\sim}{U}}, \underset{\sim}{W}, R, \Gamma, \eta) + b_1(\underset{\sim}{U}, \underset{\sim}{W}, R, \Gamma, \eta), \tag{3.3}$$

$$c(\underset{\sim}{W}, R, \Gamma, \eta) = a(\hat{\underset{\sim}{U}}; \hat{\underset{\sim}{U}}, \underset{\sim}{W}, R, \Gamma, \eta) + b_1(\hat{\underset{\sim}{U}}, \underset{\sim}{W}, R, \Gamma, \eta) . \tag{3.4}$$

We now define the operator

$$\underset{\sim}{A} : \underset{\sim}{H} \times R \times R \times R \to \underset{\sim}{H} ,$$

by

$$\underset{\sim}{A}(\underset{\sim}{U}, R, \Gamma, \eta) = \underset{\sim}{W} ,$$

where

$$a(\underset{\sim}{U}; \underset{\sim}{U}, \underset{\sim}{V}, R, \Gamma, \eta) + b(\underset{\sim}{U}, \underset{\sim}{V}, R, \Gamma, \eta) + c(\underset{\sim}{V}, R, \Gamma, \eta) = \langle \underset{\sim}{W}, \underset{\sim}{V} \rangle \text{ for all } \underset{\sim}{V} \varepsilon \underset{\sim}{H}, \tag{3.5}$$

and $\langle .,. \rangle$ denotes the inner product in the Hilbert space $\underset{\sim}{H}$. For fixed $\underset{\sim}{U}$ the left hand side of (3.5) defines a bounded linear functional on $\underset{\sim}{H}$, which by the Reisz representation theorem, means $\underset{\sim}{W}$ exists, and thus $\underset{\sim}{A}$ is well defined. A classical solution of (2.1)-(2.4) satisfies

$$\underset{\sim}{A}(\underset{\sim}{U}, R, \Gamma, \eta) = \underset{\sim}{0} . \tag{3.6}$$

Solutions of (3.6) are called weak or generalized solutions and may, under certain smoothness conditions, be shown to be classical solutions of (2.1)-(2.4).

We turn now to the finite-element approximation of (2.1)-(2.4) for each $h>0$ let W_h and M_h be finite-dimensional spaces such that $W_h \subset W^{1,2}(D)^3$, $M_h \subset L^2(D)$, and let $W_{h,0} = W_h \cap W_0^{1,2}(D)^3$. Let $\hat{\underset{\sim}{U}}_h = (0, \hat{u}_{\phi,h}, 0, 0) \varepsilon \underset{\sim}{H}_h$ $(\equiv W_h \times M_h)$, where $\hat{u}_{\phi,h}$ approximates the boundary conditions on the azimuthal component of velocity, and let

$$a_h(\underset{\sim}{U}_h; \underset{\sim}{V}_h, \underset{\sim}{W}_h, R, \Gamma, \eta) = a(\underset{\sim}{U}_h; \underset{\sim}{V}_h, \underset{\sim}{W}_h, R, \Gamma, \eta) \text{ for } \underset{\sim}{U}_h, \underset{\sim}{V}_h, \underset{\sim}{W}_h \varepsilon \underset{\sim}{H}_h, \tag{3.7}$$

$$b_h(\underset{\sim}{U}_h,\underset{\sim}{W}_h,R,\Gamma,\eta) = a(\hat{\underset{\sim}{U}}_h;\underset{\sim}{U}_h,\underset{\sim}{W}_h,R,\Gamma,\eta) + a(\underset{\sim}{U}_h;\hat{\underset{\sim}{U}}_h,\underset{\sim}{W}_h,R,\Gamma,\eta)$$

$$+ b_1(\underset{\sim}{U}_h,\underset{\sim}{W}_h,R,\Gamma,\eta) \text{ for } \underset{\sim}{U}_h,\underset{\sim}{W}_h \varepsilon \underset{\sim}{H}_h \ , \tag{3.8}$$

$$c_h(\underset{\sim}{W}_h,R,\Gamma,\eta) = a(\hat{\underset{\sim}{U}}_h;\hat{\underset{\sim}{U}}_h,\underset{\sim}{W}_h,R,\Gamma,\eta) + b_1(\hat{\underset{\sim}{U}}_h,\underset{\sim}{W}_h,R,\Gamma,\eta) \text{ for } \underset{\sim}{W}_h \varepsilon \underset{\sim}{H}_h . \tag{3.9}$$

The finite-element aprroximation of $\underset{\sim}{A}$,

$$\underset{\sim}{A}_h : \underset{\sim}{H}_h \times R \text{ X } R \text{ X } R \rightarrow \underset{\sim}{H}_h$$

is defined by

$$\underset{\sim}{A}_h(\underset{\sim}{U}_h,R,\Gamma,\eta) = \underset{\sim}{W}_h$$

where

$$a_h(\underset{\sim}{U}_h;\underset{\sim}{U}_h,\underset{\sim}{V}_h,R,\Gamma,\eta) + b_h(\underset{\sim}{U}_h,\underset{\sim}{V}_h,R,\Gamma,\eta) + c_h(\underset{\sim}{V}_h,R,\Gamma,\eta) =$$

$$\langle \underset{\sim}{W}_h,\underset{\sim}{V}_h \rangle \text{ for all } \underset{\sim}{V}_h \varepsilon \underset{\sim}{H}_h . \tag{3.10}$$

The finite-element space $W_{h,0}$ is generated by none-node isoparametric quadrilateral elements with biquadratic interpolation and the space M_n is generated by piecewise linear interpolation on the same elements, the interpolation being, in general, discontinuous across element boundaries (Engleman et al., 1982). The parameter h is the length of the longest edge in the mesh. There is a natural basis associated with this finite-element approximation in which each coefficient appearing in any linear combination of the basis functions is either the value of the velocity or pressure or one of the first derivatives of the pressure at a node in the grid. We denote this basis by

$$\{\underset{\sim}{N}_i\}_{i=1}^{n},$$

where n is the number of degrees of freedom in the discretization. Any $\underset{\sim}{U}_h \varepsilon \underset{\sim}{H}_h$ has a unique representation of the form

$$\underset{\sim}{U}_h = \sum_{i=1}^{n} u_i \underset{\sim}{N}_i \ . \tag{3.11}$$

We can now write the discretized equations (3.10) in the form of a set of non-linear equations for the coefficients $\underset{\sim}{u}$ in (3.11) as follows. We define $\underset{\sim}{f}:R^n \times R \times R \times R \to R^n$ by

$$f_i(\underset{\sim}{u},R,\Gamma,\eta) = \langle A_h(\underset{\sim}{U}_h,R,\Gamma,\eta),\underset{\sim}{N}_i\rangle , \quad i=1, \ldots n . \tag{3.12}$$

Thus the discretized equations take the form

$$\underset{\sim}{f}(\underset{\sim}{u},R,\Gamma,\eta) = 0 . \tag{3.13}$$

Equations (3.13) are treated by numerical methods for bifurcation problems and we shall describe these in section 5.

4. THE TREATMENT OF SYMMETRY

As we mentioned in the introduction an important feature of the particular version of the Taylor problem we are considering is that the apparatus is mirror symmetric about the mid-plane of the annulus. In this section we describe how this symmetry is represented mathematically and discuss its implications for the discretization.

To begin we define the reflection symmetry operator $S:\underset{\sim}{H} \to \underset{\sim}{H}$ by

$$S\{u_r(r,z),u_\phi(r,z),u_z(r,z),p(r,z)\} = \{u_r(r,-z), u_\phi(r,-z),u_z(r,-z), p(r,-z)\} \tag{4.1}$$

for smooth functions, and use continuity to extend the definition to the whole space. It is easy to see that $S \neq I$ and that $S^2=I$. Since the boundary conditions of the problem are symmetric we may take

$$S\,\hat{\underset{\sim}{U}} = \hat{\underset{\sim}{U}} . \tag{4.2}$$

The change of variables $(r,z) \to (r, -z)$ in the integrals in (3.1) and (3.2) gives

$$a(S\underset{\sim}{U};S\underset{\sim}{V},\underset{\sim}{W},R,\Gamma,\eta) = a(\underset{\sim}{U};\underset{\sim}{V},S\underset{\sim}{W},R,\Gamma,\eta) , \tag{4.3}$$

$$b_1(S\underset{\sim}{U},\underset{\sim}{W},R,\Gamma,\eta) = b_1(\underset{\sim}{U},S\underset{\sim}{W},R,\Gamma,\eta) . \tag{4.4}$$

Similarly we have

$$\langle S\underset{\sim}{U},\underset{\sim}{W}\rangle = \langle\underset{\sim}{U},S\underset{\sim}{W}\rangle . \tag{4.5}$$

Now, using (3.3)-(3.5) and (4.2)-(4.5) it follows that

$$\underset{\sim}{A}(S\underset{\sim}{U},R,\Gamma,\eta) = S\underset{\sim}{A}(\underset{\sim}{U},R,\Gamma,\eta) \ . \tag{4.6}$$

It is clearly important that the discretization process preserves the symmetry properties of the original equations (Brezzi, Rappaz and Raviart, 1981). Since $\underset{\sim}{H}_h \subset \underset{\sim}{H}$ we can use the same operator S in the discrete case. It is easy to see that $S:\underset{\sim}{H}_h \to \underset{\sim}{H}_h$ if and only if the finite element mesh is symmetric about the line z=0. If this condition is satisfied then we may choose $\hat{\underset{\sim}{U}}_h$ such that

$$S\hat{\underset{\sim}{U}}_h = \hat{\underset{\sim}{U}}_h \ , \tag{4.7}$$

and it then follows from (3.7-3.10), (4.4), (4.5) and (4.7) that

$$\underset{\sim}{A}_h(S\underset{\sim}{U}_h,R,\Gamma,\eta) = S\underset{\sim}{A}_h(\underset{\sim}{U}_h,R,\Gamma,\eta) \ . \tag{4.8}$$

Thus, provided the finite-element mesh is symmetric, the discrete problem has the same symmetry property as the original problem.

We define $D^+ \subset D$ by

$$D^+ = \{(r,z) \,|\, 0 \leq r \leq 1,\ 0 \leq z \leq 0.5\} \ , \tag{4.9}$$

and note that

$$a(\underset{\sim}{U};\underset{\sim}{V},\underset{\sim}{W},R,\Gamma,\eta) = a^+(\underset{\sim}{U};\underset{\sim}{V},\underset{\sim}{W},R,\Gamma,\eta) + a^+(S\underset{\sim}{U};S\underset{\sim}{V},S\underset{\sim}{W},R,\Gamma,\eta) , \tag{4.10}$$

$$b_1(\underset{\sim}{U},\underset{\sim}{W},R,\Gamma,\eta) = b_1^+(\underset{\sim}{U},\underset{\sim}{W},R,\Gamma,\eta) + b_1^+(S\underset{\sim}{U},S\underset{\sim}{W},R,\Gamma,\eta) , \tag{4.11}$$

where the superscript + indicates that the integral in (3.1) and (3.2) is over D^+ rather than D. These results are used in the next section.

We now construct a representation of the symmetry operator S on R^n. First we label all the degrees of freedom in the problem. Those associated with nodes on the line z=0 we label from 1 to ℓ. We assume that this line coincides with element boundaries, and since the pressure degrees of freedom for the element we are using are all associated with the central node (see Fig. 1), there are no pressure degrees of freedom on it. We then label all the other degrees of freedom in D^+ from $\ell+1$ to $\ell+m$. Finally the degrees of freedom in $D-D^+$ are labelled from $\ell+m+1$ to $\ell+2m(=n)$ in such a way that a velocity (or pressure) freedom at a node with coordinates (r,z) is labelled i $(\ell+m+1\leqslant i\leqslant \ell+2m)$ where $i-m$ is the label of the corresponding velocity (or pressure) freedom at the node with coordinates $(r,-z)$. It follows from the symmetry of the grid and the above labelling scheme that:

$$S\underset{\sim}{N}_i = \varepsilon(i)\underset{\sim}{N}_j \,, \tag{4.12a}$$

where

$$j = \begin{cases} i & \text{if } 1\leqslant i\leqslant \ell \\ i+m & \text{if } \ell+1\leqslant i\leqslant \ell+m \\ i-m & \text{if } \ell+m+1\leqslant i\leqslant \ell+2m \end{cases} \,, \tag{4.12b}$$

and

$$\varepsilon(i) = \pm 1 \,. \tag{4.12c}$$

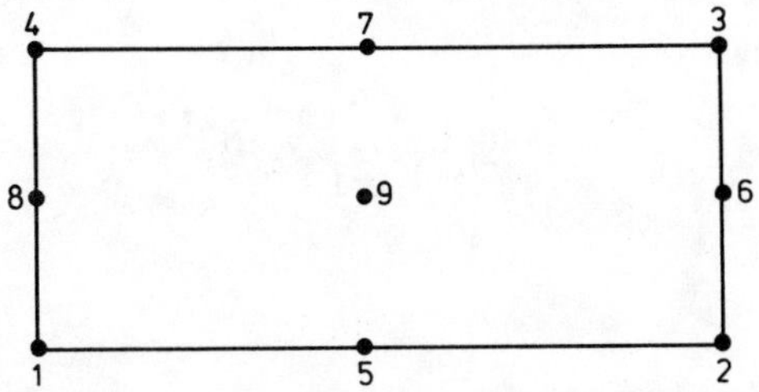

Fig. 1 The nine-node quadrilateral element used for the calculations. Velocity values are associated with nodes 1-9, the pressure and both its first derivatives are associated with node 9 only.

The sign in (4.12c) depends on the particular freedom in question. If it is the r- or ϕ- component of velocity or the pressure or its r-derivative the sign is positive. If it is the z-component of the velocity or the z-derivative of pressure the sign is negative. This follows directly from the definition of S (4.1). Let $\underset{\sim}{U}_h \in \underset{\sim}{H}_h$, then using (4.12) we have

$$SU_{\sim h} = \sum_{i=1}^{n} u_i SN_{\sim i} = \sum_{i=1}^{\ell} u_i \varepsilon(i) N_{\sim i} + \sum_{i=\ell+1}^{\ell+m} u_{i+m}\varepsilon(i+m) N_{\sim i}$$

$$+ \sum_{i=\ell+m+1}^{\ell+2m} u_{i-m}\varepsilon(i-m) N_{\sim i} . \qquad (4.13)$$

Let

$$E = \begin{vmatrix} \varepsilon(1) & & & O \\ & \varepsilon(2) & & \\ & & \ddots & \\ O & & & \varepsilon(n) \end{vmatrix} \qquad (4.14)$$

and

$$P = \begin{vmatrix} I_\ell & O & O \\ O & O & I_m \\ O & I_m & O \end{vmatrix} \qquad (4.15)$$

where I_ℓ is the $\ell x \ell$ identity matrix etc. Equation (4.13) may now be written

$$SU_{\sim h} = \sum_{i=1}^{n} v_i N_{\sim i}$$

where

$$\underset{\sim}{v} = \hat{S}\, \underset{\sim}{u} ,$$

and

$$\hat{S} = PEP . \qquad (4.16)$$

Thus $\hat{S}:R^n \to R^n$ is the required representation of S on R^n, and the function $\underset{\sim}{f}$ defined in (3.12) satisfies

$$\hat{S}\, \underset{\sim}{f}(\underset{\sim}{u},R,\Gamma) = \underset{\sim}{f}(\hat{S}\underset{\sim}{u},R,\Gamma) , \qquad (4.17)$$

which follows immediately from (4.8) and (4.16). Clearly $\hat{S} \neq I$ and $\hat{S}^2 = PEPPEP = I$ since $P^2 = E^2 = I$.

5. SYSTEMS FOR COMPUTING SINGULAR POINTS IN TWO PARAMETER PROBLEMS

Equation (3.13) has three parameters in it, however, in an experiment it is only possible to vary the Reynolds number, R, and the aspect ratio Γ. The radius ratio, η, remains fixed. For the rest of the paper, then, we will omit the explicit dependence on η and treat (3.13) as a two parameter problem. Thus we want to find all $(\underset{\sim}{u},R,\Gamma)$ that satisfy (3.13). This equation defines a "two-dimensional" surface in $R^n xRxR$. The techniques for solving this problem will be described in the next two subsections and are based on Jepson and Spence (1985) and Cliffe, Jepson and Spence (1985). In subsection 5.1 we discuss continuation and extended systems for singularities when the Z_2-symmetry is not broken and in subsection 5.2 we consider singularities where the Z_2-symmetry is broken. In subsection 5.3 we point out the numerical implications of the symmetry.

5.1 Two parameter problems without symmetry

First we consider the problem of solving a one parameter problem:

$$g(\underset{\sim}{x},\lambda) = \underset{\sim}{O} \quad x\varepsilon R, \ \lambda\varepsilon R \ .$$

The basic continuation method (Keller 1977) introduces an arc length parameter s to parametrize the solution. This involves solving the system of equations

$$\underset{\sim}{g}(\underset{\sim}{x},\lambda) = \underset{\sim}{O} \ ,$$

$$N(\underset{\sim}{x},\lambda,s) = 0 \ ,$$

where

$$N(\underset{\sim}{x},\lambda,s) = \frac{\partial \underset{\sim}{x}}{\partial s}(s_o)^T\,(x(s)-x(s_o)) + \frac{\partial\lambda}{\partial s}(s_o)(\lambda(s)-\lambda(s_o))-(s-s_o) \qquad (5.1.1)$$

by Euler-Newton continuation applied to the parameter s (Keller 1977). The reason for introducing s is that when there is a limit point on the solution branch ordinary continuation methods fail whereas the Keller method will succeed in following the branch around the limit point.

System (5.1.1) is used to calculate slices through the "two-dimensional" surface by fixing Γ (or R) at $\Gamma = \Gamma_o$ (or $R = R_o$), with $\underset{\sim}{x}=\underset{\sim}{u}$, $\lambda = R$ (or Γ) and $\underset{\sim}{g}(.,.) = \underset{\sim}{f}(.,.,\Gamma_o)$ (or $\underset{\sim}{f}(.,R_o,.)$).

A limit point may be characterized as an isolated solution of the following extended system (Moore and Spence 1980)

$$\underset{\sim}{F}(\underset{\sim}{y},\Gamma) \equiv \begin{Bmatrix} \underset{\sim}{f}(\underset{\sim}{u},R,\Gamma) \\ \underset{\sim}{f}_{\underset{\sim}{u}}(\underset{\sim}{u},R,\Gamma)\underset{\sim}{\phi} \\ \underset{\sim}{\ell}^T\underset{\sim}{\phi}-1 \end{Bmatrix} = \underset{\sim}{O} \qquad (5.1.2)$$

with $\underset{\sim}{y} = (\underset{\sim}{u},\underset{\sim}{\phi},R)$, $\underset{\sim}{\phi}\varepsilon R^n$ and $\underset{\sim}{\ell}\varepsilon R^n$. The set of points $(\underset{\sim}{u},R,\Gamma)$ such that (5.1.2) is satisfied by $(\underset{\sim}{y},\Gamma)$ defines a curve in $R^n xRxR$ which Jepson and Spence (1985) call a fold curve. Fig. 2 illustrates the reason for this term: the surface is folded back on itself and the fold curve marks the edge of the fold. To calculate this curve we apply the Keller arc length continuation method (5.1.1) to (5.1.2) with $\underset{\sim}{x} = \underset{\sim}{y}$, $\lambda = \Gamma$ and $g(.,.) = \underset{\sim}{F}(.,.)$.

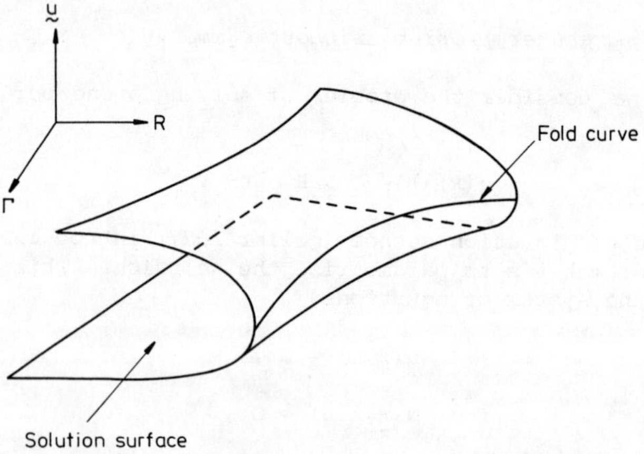

Fig. 2 Part of the solution surface of a two parameter problem (R and Γ) illustrating a fold in the surface and the fold curve.

The fold curve may itself have limit points, these correspond to points at which the solution to (5.1.2) is not isolated, so that in the neighbourhood of the point there will be two or more limit points for each value of Γ. In a two parameter problem these limit points on the fold curve can occur in two ways that are stable to small perturbations. We note that only stable singularities can be computed as there are always small perturbations present due to rounding errors. At these special points an extra equation must be satisfied in addition to (5.1.2).

Let $(\underset{\sim}{u}_o, \underset{\sim}{\phi}_o, R_o, \Gamma_o)$ satisfy (5.1.2) and denote $\underset{\sim}{f}(\underset{\sim}{u}_o, R_o, \Gamma_o)$ by $\underset{\sim}{f}^o$ etc. Let $\underset{\sim}{\psi}_o$ be the left eigenvector of $\underset{\sim}{f}^o_{\underset{\sim}{u}}$. If in addition to (5.1.2) the solution satisfies

$$\underset{\sim}{\psi}_o \underset{\sim}{f}^o_R = 0 , \qquad (5.1.3)$$

where the subscript R denotes differentiation with respect to R, the point $(\underset{\sim}{u}_o, R_o, \Gamma_o)$ is either a transcritical bifurcation point (which we shall denote by T) or else an isola formation point, depending on the sign of a scalar quantity involving higher derivatives of f (Jepson and Spence, 1985). Bifurcation diagrams near T are given in Fig. 3.

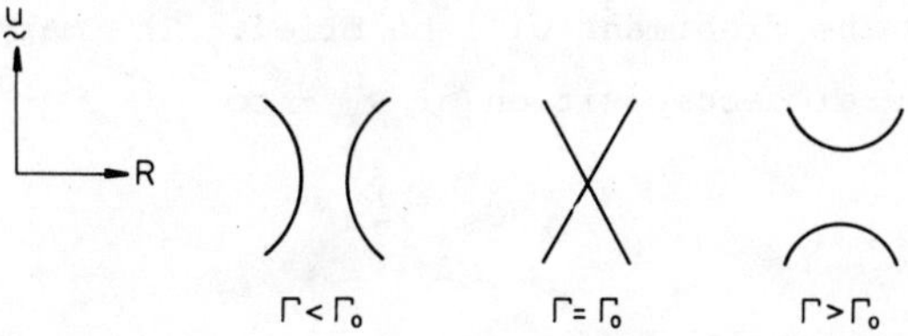

Fig. 3 State diagrams near a transcritical bifurcation point, T. The critical value of the second parameter Γ is Γ_o; the order of the diagrams could be reversed.

If in addition to (5.1.2) the solution satisfies

$$\underset{\sim}{\psi}_o \underset{\sim}{f}_{\underset{\sim}{u}\underset{\sim}{u}} \underset{\sim}{\phi}_o \underset{\sim}{\phi}_o = 0 , \qquad (5.1.4)$$

the point (u_o, R_o, Γ_o) is a non-degenerate hysteresis point which we shall denote by H (Jepson and Spence, 1985; Golubitsky and Schaeffer, 1979). The bifurcation diagrams in this case are shown in Fig. 4. The projection of the fold curve onto the R-Γ plane has a characteristic cusp shape, with the tip of the cusp at the hysteresis point.

$$\Gamma < \Gamma_0 \qquad \Gamma = \Gamma_0 \qquad \Gamma > \Gamma_0$$

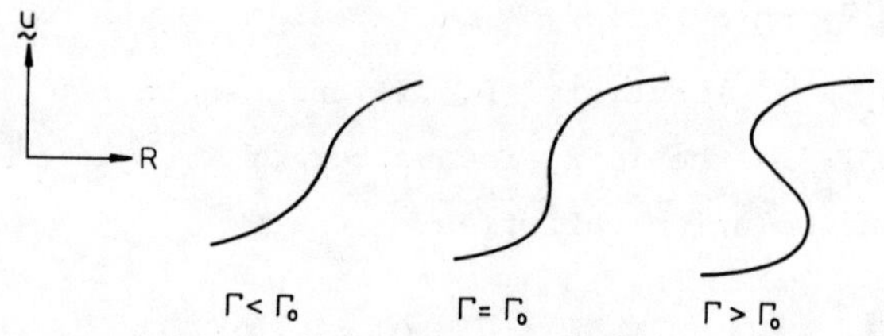

Fig. 4   State diagrams near a non-degenerate hysteresis point,
H. The critical value of the second parameter Γ is
Γ_0; the order of the diagrams could be reversed.

5.2   *Two parameter problems with Z_2-symmetry*

We now consider the implications of the Z_2-symmetry. This
subject is covered in detail by Cliffe, Jepson and Spence
(1985) and so the treatment will be brief. The mapping $\hat{S}$
induces a natural decomposition of R^n into

$$R^n = \underset{\sim}{X}_s \oplus \underset{\sim}{X}_a \qquad (5.2.1)$$

where

$$\underset{\sim}{X}_s = \{\underset{\sim}{x} \in R^n \mid \hat{S}\underset{\sim}{x} = \underset{\sim}{x}\} \;,\quad \underset{\sim}{X}_a = \{\underset{\sim}{x} \in R^n \mid \hat{S}\underset{\sim}{x} = -\underset{\sim}{x}\} \;, \qquad (5.2.2)$$

consist of the *symmetric* and *antisymmetric* elements of R^n
respectively. A symmetry-breaking bifurcation point may be
characterized as an isolated solution of the following
extended system (Werner and Spence 1984)

$$\underset{\sim}{F}(\underset{\sim}{y},\Gamma) = \left\{ \begin{array}{c} \underset{\sim}{f}(\underset{\sim}{u},R,\Gamma) \\ \underset{\sim}{f}_u(\underset{\sim}{u},R,\Gamma)\underset{\sim}{\phi} \\ \ell^T\underset{\sim}{\phi}-1 \end{array} \right\} = \underset{\sim}{O} \quad \begin{array}{l} \underset{\sim}{y} = (\underset{\sim}{u},\underset{\sim}{\phi},R)\in\underset{\sim}{Y} \\ \underset{\sim}{Y} = \underset{\sim}{X}_s x\underset{\sim}{X}_a xR \\ \underset{\sim}{F}:\underset{\sim}{Y}xR\to\underset{\sim}{Y} \;. \end{array} \qquad (5.2.3)$$

It is important to note both the similarity with (5.1.2) and
the essential differences, namely that u must belong to $\underset{\sim}{X}_s$ and
ϕ must belong to $\underset{\sim}{X}_a$. Thus the basic solution is symmetric but
the eigenvector is antisymmetric so that the bifurcating
branch is asymmetric. In direct analogy with a path of limit
points, the set of points $(\underset{\sim}{u},R,\Gamma)$ such that (5.2.3) is
satisfied by $(\underset{\sim}{y},\Gamma)$ defines a curve which lies on the symmetric
solution surface. This curve is the intersection of the
surface of asymmetric solutions with the symmetric ones.

A path of symmetry-breaking bifurcation points may have a turning point and this happens if

$$\underset{\sim}{\psi}_o(\underset{\sim}{f}^o_{R\underset{\sim}{u}}\,\underset{\sim}{\phi}_o + \underset{\sim}{f}^o_{\underset{\sim}{u}\underset{\sim}{u}}\,\underset{\sim}{\phi}_o\,\underset{\sim}{v}_R) = 0\ , \qquad (5.2.4)$$

where

$$\underset{\sim}{f}^o_{\underset{\sim}{u}}\,\underset{\sim}{v}_R + \underset{\sim}{f}^o_R = 0\ ;\quad \underset{\sim}{v}_R \in \underset{\sim}{X}_s\ . \qquad (5.2.5)$$

Cliffe and Spence (1984) call this type of singularity a point of coalescence because two symmetry-breaking bifurcation points coalesce at such a point. The bifurcation diagrams depend on the sign of a scalar quantity involving higher derivatives of $\underset{\sim}{f}$ (we shall not give the details here). We denote these two cases by C^+ and C^- and the relevant diagrams are shown in Fig. 5.

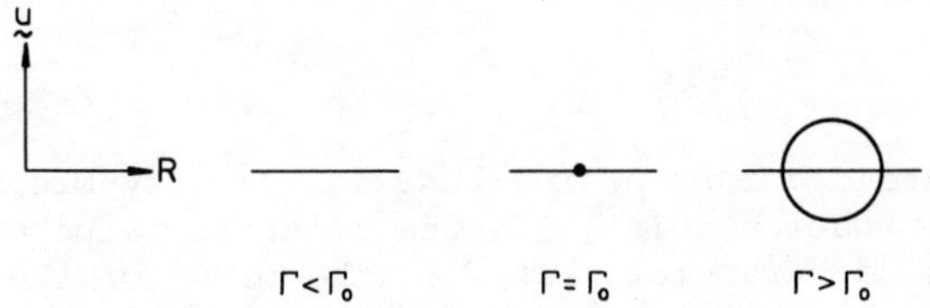

Fig. 5a

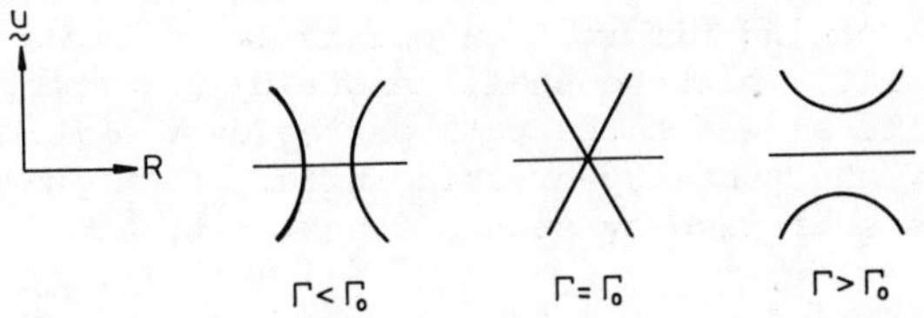

Fig. 5b

Fig. 5 State diagrams near the two types coalescence point, (a) C^+ and (b) C^-. The critical value of the second parameter Γ is Γ_o; the order of the diagrams could be reversed.

Another type of singularity that can occur on a path of symmetry-breaking bifurcation points is a quartic symmetry-breaking bifurcation point which we denote by Q (Cliffe and Spence, 1984). Here the bifurcating asymmetric branch has quartic dependence on the excess Reynolds number near the singularity rather than quadratic dependence. In addition to

(5.2.3) the quartic point satisfies

$$\psi_o \,(f^o_{uuu}\,\phi_o\phi_o\phi_o + 3f^o_{uu}\,\phi_o\, z_o) = 0 \,, \tag{5.2.6}$$

where

$$f^o_u\, z_o + f^o_{uu}\,\phi_o\phi_o = 0 \,, \quad z_o \varepsilon X_s \,. \tag{5.2.7}$$

The bifurcation diagrams near such a point are shown in Fig. 6. Note that a pair of fold curves of the asymmetric solution surface splits off from the symmetric solution at a quartic symmetry-breaking bifurcation point.

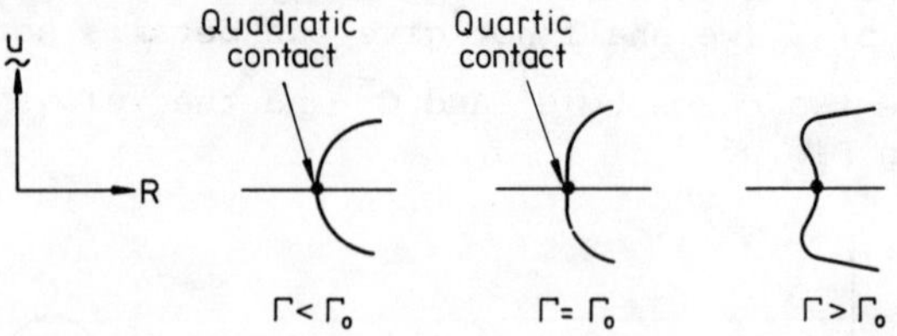

Fig. 6 State diagrams near a quartic symmetry-breaking bifurcation point, Q. The critical value of the second parameter Γ is Γ_o; the order of the diagrams could be reversed.

The final type of singularity which can occur along a path of symmetry-breaking bifurcation points is called a double singular S-point, which we shall denote by D (Werner, 1984). This is a point at which the path passes over a fold so that a limit point and symmetry-breaking bifurcation point coincide. The equations satisfied by such a singularity are

$$\begin{Bmatrix} f \\ f_u\,\phi_s \\ f_u\,\phi_a \\ \ell_s\,\phi_s - 1 \\ \ell_a\,\phi_a - 1 \end{Bmatrix} = 0 \,, \quad \begin{matrix} \phi_s \,\varepsilon\, X_s \,, \\ \phi_a \,\varepsilon\, X_a \,. \end{matrix} \tag{5.2.8}$$

5.3 Numerical implications of symmetry

In this subsection we discuss some of the implications of symmetry for the calculation of symmetry breaking bifurcation points. In particular we shall show that the number of degrees of freedom involved in solving each equation in (5.2.3) is about n/2 or roughly half the total number of degrees of freedom in the problem.

First we define (cf. (5.2.2))

$$\underset{\sim}{H}_h^s = \{\underset{\sim}{U}_h \in \underset{\sim}{H}_h \,|\, S\underset{\sim}{U}_h = \underset{\sim}{U}_h\} ,$$

$$\underset{\sim}{H}_h^a = \{\underset{\sim}{U}_h \in \underset{\sim}{H}_h \,|\, \underset{\sim}{S}\underset{\sim}{U}_h = -\underset{\sim}{U}_h\}. \tag{5.3.1}$$

Now (5.2.3) may be written in the form: find $\underset{\sim}{U}_h \in \underset{\sim}{H}_h^s$, $\underset{\sim}{\Phi}_h \in \underset{\sim}{H}_h^a$ such that

$$a_h(\underset{\sim}{U}_h;\underset{\sim}{U}_h,\underset{\sim}{W}_h,R,\Gamma,\eta) + b_h(\underset{\sim}{U}_h,\underset{\sim}{W}_h,R,\Gamma,\eta)$$

$$+ c_h(\underset{\sim}{W}_h,R,\Gamma,\eta) = 0 \text{ for all } \underset{\sim}{W}_h \in \underset{\sim}{H}_h , \tag{5.3.2a}$$

$$a_h(\underset{\sim}{U}_h;\underset{\sim}{\Phi}_h,\underset{\sim}{W}_h,R,\Gamma,\eta) + a_h(\underset{\sim}{\Phi}_h,\underset{\sim}{U}_h,\underset{\sim}{W}_h,R,\Gamma,\eta)$$

$$+ b_h(\underset{\sim}{\Phi}_h,\underset{\sim}{W}_h,R,\Gamma,\eta) = 0 \text{ for all } \underset{\sim}{W}_h \in \underset{\sim}{H}_h, \tag{5.3.2b}$$

$$\ell\underset{\sim}{\Phi}_h - 1 = 0 \tag{5.3.2c}$$

It follows from $U_h \in H_h^s$, $\underset{\sim}{\Phi}_h \in \underset{\sim}{H}_h^a$ and (3.7)-(3.9), (4.7) and (4.10), (4.11) that (5.3.2) may be written as

$$a_h^+(\underset{\sim}{U}_h;\underset{\sim}{U}_h,\underset{\sim}{W}_h + S\underset{\sim}{W}_h,R,\Gamma,\eta) + b_h^+(\underset{\sim}{U}_h,\underset{\sim}{W}_h + S\underset{\sim}{W}_h,R,\Gamma,\eta)$$

$$+ c_h^+(\underset{\sim}{W}_h + S\underset{\sim}{W}_h,R,\Gamma,\eta) = 0, \text{ for all } \underset{\sim}{W}_h \in \underset{\sim}{H}_h, \tag{5.3.3a}$$

$$a_h^+(\underset{\sim}{U}_h;\underset{\sim}{\Phi}_h,\underset{\sim}{W}_h-S\underset{\sim}{W}_h,R,\Gamma,\eta) + a_h^+(\underset{\sim}{\Phi}_h;\underset{\sim}{U}_h,\underset{\sim}{W}_h-S\underset{\sim}{W}_h,R,\Gamma,\eta)$$

$$+ b_h^+(\underset{\sim}{\Phi}_h, \underset{\sim}{W}_h-S\underset{\sim}{W}_h,R,\Gamma,\eta) = 0, \text{ for all } \underset{\sim}{W}_h \in \underset{\sim}{H}_h, \tag{5.3.3b}$$

$$\underset{\sim}{\ell}\Phi_h - 1 = 0 \,. \qquad (5.3.3c)$$

Equations (5.3.3) involve integrals over the half region D^+, it follows that (5.3.3a) and (5.3.3b) are equations for $\underset{\sim}{U}_h$ and $\underset{\sim}{\Phi}_h$ (equivalently $\underset{\sim}{u}$ and $\underset{\sim}{\phi}$) which involve approximately n/2 degrees of freedom each.

It is informative to regard equations (5.3.3) as the discretization of a boundary value problem for $\underset{\sim}{U}$ $(\equiv u_r, u_\phi, u_z, p)$ and $\underset{\sim}{\Phi}$ $(\equiv \phi_r, \phi_\phi, \phi_z, \pi)$ on the region D^+. The boundary conditions on the line z = 0 are

$$\frac{\partial u_r}{\partial z} = \frac{\partial u_\phi}{\partial z} = u_z = 0 \,,$$

$$\phi_r = \phi_\phi = \frac{\partial \phi_z}{\partial z} = 0 \,.$$

6. RESULTS

In this section we briefly describe some calculations performed using the numerical techniques discussed in the previous sections. The calculations are relevant to the way in which the number of Taylor cells present in the apparatus changes from four to six as the aspect ratio increases. A detailed description of the problem and the results can be found in Cliffe (1985).

Fig. 7 shows the locus of paths of limit points (or fold curve) and symmetry-breaking bifurcation points in the R-Γ plane. The radius ratio was fixed at η=0.6 to correspond to the experiment of Mullin (1982). The fold curve is indicated by the solid line and there is a nondegenerate hysteresis point H and transcritical bifurcation point T on this curve.

The path of symmetry-breaking bifurcation points (dashed curve) is rather interesting. At the bottom right hand corner of the figure the bifurcation is supercritical, that is the asymmetric branches exist for R greater than the critical value. This path moves from one side of the fold surface to the other at the point D. An enlarged diagram of the region near D is shown in the inset in Fig. 7. At the point Q there is a quartic symmetry-breaking bifurcation point and at aspect ratios greater than that at Q the bifurcation is subcritical. The solution branch bends back round as indicated in Fig. 6, giving rise to a path of limit points in the asymmetric solutions which are indicated by the chained line in Fig. 7. At C^- there is a turning point in the path of symmetry-breaking bifurcation points and the solution diagrams near this point look like those in Fig. 5b. Finally there is a second turning point in the path at C^+ where the solution diagrams look like those in Fig. 5a.

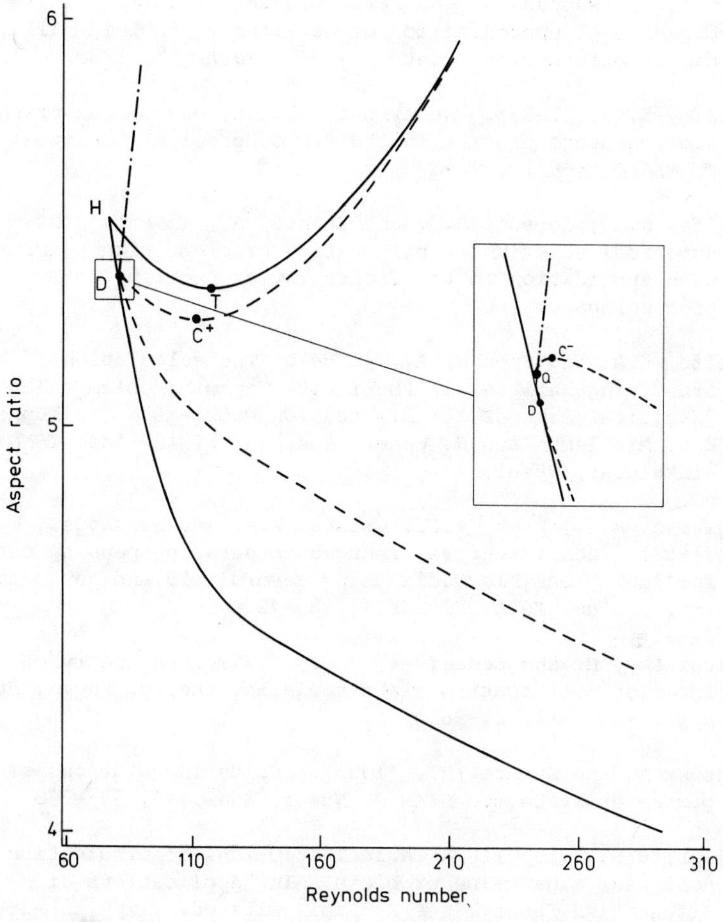

Fig. 7 Critical loci for four- and six-cell modes with radius ratio 0.6: —— numerically calculated fold curve, ---- numerically calculated path of symmetry-breaking bifurcation points, -.-.- numerically calculated fold curve for asymmetric solution surface.

REFERENCES

Benjamin, T.B., (1978). Bifurcation phenomena in steady flows of a viscous liquid. II. Experiments. *Proc. R. Soc. Lond.* **A359**, 27-43.

Benjamin, T.B. and Mullin, T., (1981). Anomalous modes in the Taylor experiment. *Proc. R. Soc. Lond.* **A377**, 221-249.

Brezzi, F., Rappaz, J. and Raviart, P.A., (1981). Finite dimensional approximation of non-linear problems. III. Simple bifurcation points. *Numer. Math.* **38**, 1-30.

Cliffe, K.A., (1985). Numerical calculations of the primary flow exchange process in the Taylor problem (to appear in *J. Fluid Mech.*)

Cliffe, K.A., Jepson, A.D. and Spence, A., (1985). The numerical solution of bifurcation problems with symmetry with application to the finite Taylor problem (these proceedings).

Cliffe, K.A. and Spence, A., (1984). The calculation of high order singularities in the finite Taylor problem. In "Numerical Methods for Bifurcation Problems". (T. Küpper, H.D. Mittleman and H. Weber, eds.), pp. 129-144, *ISNM* **70**, Birkhauser, Basel.

Engleman, M.S., Sani, R.L., Gresho, P.M. and Bercovier, M., (1982). Consistent vs. reduced integration penalty methods for incompressible media using several old and new elements. *Int. J. Num. Meth. Fluids,* **2**, 25-42.

Golubitsky, M. and Schaeffer, D.G., (1979). A theory of imperfect bifurcation via singularity theory. *Comm. Pure Appl. Math.* **32**, 21-98.

Jepson, A. and Spence, A., (1985). Folds in solutions of two parameter systems. *SIAM J. Numer. Anal.* **22**, 347-368.

Keller, H.B., (1977). Numerical solutions of bifurcation and nonlinear eigenvalue problems. In "Applications of Bifurcation Theory. (P.H. Rabinowitz ed.). pp 359-384. Academic Press, New York.

Moore, G. and Spence, A., (1980). The calculation of turning points of nonlinear equations. *SIAM J. Numer. Anal.* **17**, 567-576.

Mullin, T., (1982). Mutations of steady cellular flows in the Taylor experiment. *J. Fluid Mech.* **121**, 207-218.

Mullin, T., Pfister, G. and Lorenzen, A., (1982). New observations on hysteresis effects in Taylor-Couette flow. *Phys. Fluids* **25**, 1134-1136.

Werner, B., (1984). Regular systems for bifurcation points with underlying symmetries. In "Numerical Methods for Bifurcation Problems" (T. Küpper, H.D. Mittleman and H. Weber, eds.), pp. 562-574, *ISNM* **70**, Birkhauser, Basel.

Werner, B. and Spence, A., (1984). The computation of symmetry-breaking bifurcation points. *SIAM J. Numer. Anal.* **21**, 388-399.

HYPERBOLIC PRECONDITIONING TO SOLVE THE INCOMPRESSIBLE EULER EQUATIONS FOR VORTEX FLOW

A. Rizzi*
(FFA The Aeronautical Research Institute of Sweden, S-161 11, Bromma, Sweden)

1. INTRODUCTION

The ability to compute detailed information about incompressible flowfields by the numerical solution of a set of governing equations offers great utility to industrial designers of turbomachinery, internal ducts, and pumps as well as transport vehicles, be they ocean-going ships, road vehicles, or airplanes. Most of these applications involve flows having very high Reynolds numbers, and the potential equation is what first comes to mind as the suitable flow model. Many of the flows, however, possess substantial regions with rotation, one very common feature being the shedding of vorticity from sharp edges. Working with the potential model, one faces the cumbersome task of setting up vortex-sheet discontinuities in the field and then adjusting or fitting them to the surrounding flow. This requires prior knowledge of where the sheets begin and becomes very complicated for all but the simplest situations. The alternative, especially when the vortex topology is complex or unknown, is to adopt the Euler equations as the flow model because they allow rotational flow everywhere and vortical regions are "captured" implicitly as an integral part of the solution.

The recent discovery that a vortex sheet can be generated and captured in the solution of a <u>compressible</u> flowfield with separation from the leading edge of a delta wing at high angle of attack has stirred up a lot of activity in the development of numerical methods to solve the compressible Euler equations. Unfortunately these methods do not work very well for an <u>incompressible</u> problem simply by setting the freestream Mach number to a very small value, zero in the limit, because with

---

*also Adjunct Professor, Royal Institute of Technology, Stockholm

decreasing Mach number the speed of the sound waves becomes much larger than the speed of convection. This increasing disparity in wave speeds causes the governing system of hyperbolic equations to be poorly conditioned, and the stability of the computation is greatly impaired. The conventional approach to solve the equations is to recast them into the vorticity-streamfunction formulation, but while effective in two-dimensional problems, it turns out to be impractical in three. If, however, the interest is only in steady three-dimensional flow, artificial compressibility is one way around the difficulty because this approach removes the sound waves from the system by prescribing a pseudotime evolution for the pressure which is hyperbolic and which converges to the true steady-state value. In a theoretical study Viviand (1983) recently has generalized this concept of a pseudo-unsteady approach, and he has defined a broad class of such systems that includes the one of artificial compressibility. He did not, however, demonstrate any of them in an actual computation. The purpose here is to investigate in detail two particular systems from the class of Viviand and compare them to artificial compressibility (Rizzi and Eriksson, 1985). I show how the pseudo-unsteady approach leads to a hyperbolic system, carry out a numerical study of its condition, set forth the CFL stability limit for the time integration, and examine the types of discontinuities that it admits. The method can be seen as a hyperbolic preconditioning technique because it modifies the wave character of the system to be optimal for numerical solution by a hyperbolic integration scheme.

2. FLOW SEPARATION WITH EULER EQUATIONS

The central issues then become physical rather than numerical, i.e. to understand how vorticity is created in the flow, to calibrate the diffusion of the vorticity, and to gauge to what extent errors in total pressure degrade the overall accuracy of the solution. The first perhaps is most controversial. If an oncoming stream of fluid is initially irrotational, how, in the absence of irreversible processes, can vorticity appear in the flow? Consider now a solid body that presents an obstacle to the flow. The fluid wets the surface and at some point, or along some line, it detaches from the body and continues downstream. If the flow is potential, its velocity nowhere can be in shear. But if it obeys the Euler equations, the flow can separate along a line on the surface, across which the velocity may be in shear. The discontinuity in velocity direction then extends in a surface out into the flow which is called a vortex sheet. The total pressure on each side of the sheet is the same as that upstream of the body. No losses have been inflicted and vorticity exists only inside the vortex sheet, an infinitesimally thin surface of discontinuity, which is also a stream surface. The circulation along every closed fluid circuit therefore remains the same. The sheet, however, is linearly unstable and begins to deform according to the Kelvin-Helmholtz instability. But in some cases, like a stream sweeping past a sharp edge at some angle (Fig. 1) the sheet can, under the influence of its own vorticity, coil up to form a steady but infinite spiral.

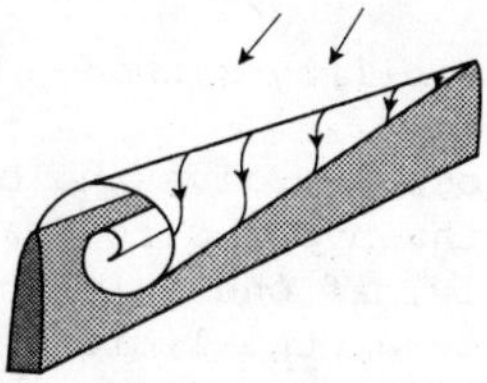

Fig. 1 Steady tip vortex formed by flow sweeping obliquely past a sharp edge.

But allowed just the slightest amount of diffusion, vorticity begins to seep out of the sheet and forms a vortex core after a finite number of turns in the spiral. Do we observe such phenomena in a numerical simulation? To try to answer this let us look at a numerical solution to this problem in its simplest possible setting, incompressible flow past a flat delta wing in which a stable vortex sheet is shed from the leading edge and then coils up into a steady vortex over the wing (Fig. 2).

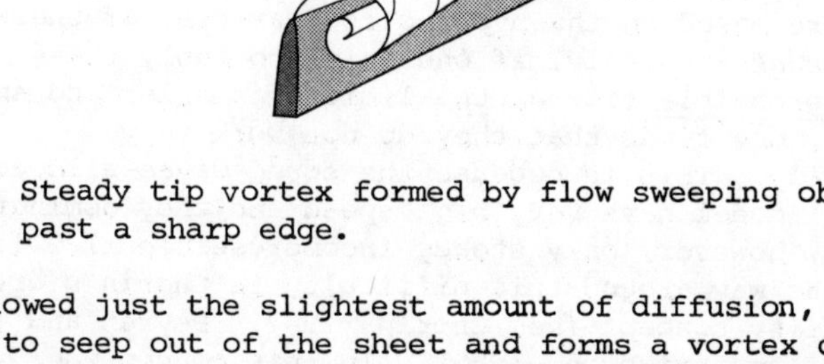

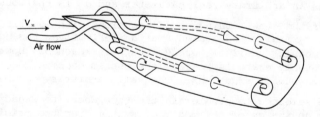

Fig. 2 Leading-edge vortices on lee side of a delta wing.

I analyze the qualitative as well as quantitative aspects of this flow, computed by hyperbolic preconditioning, in comparison to the results of a 3D panel method that fits the vortex sheet to the surrounding potential flowfield. The fine-mesh solution with over 639 000 grid points confirms, but in greater detail, the presence in a medium-mesh calculation of a torsional wave superposed upon the vortex core as it approaches the trailing edge of the wing. The shearing of the flow by the trailing edge causes this wave-like disturbance.

3. PSEUDO-UNSTEADY METHODS FOR INCOMPRESSIBLE FLOW

This section describes the pseudo-unsteady methods I use to solve the Euler equations for incompressible flow. It starts with a discussion of the artificial compressibility method, shows its generalization to the optimal system of Viviand, followed by a description of the computational method.

3.1 Artificial Compressibility Method

Practically all methods to solve the compressible Euler equations are based on the hyperbolic character of these equations. Unfortunately, if one tries to apply these methods to an incompressible flow as the limit of smaller and smaller Mach number, one finds that they do not work very satisfactorily. This is because the sound waves allowed in the mathematical model have very high speed and they dominate the system. If, however, only steady incompressible flow is of interest, the way around this difficulty is Chorin's artificial compressibility concept (see Chorin, 1967; Peyret and Taylor, 1983; Rizzi and Eriksson, 1985). In this concept an artificial time-dependent term is added to the continuity equation so that the governing equations become

$$\frac{1}{\rho_o}\ \frac{\partial p}{\partial t} + c^2 \operatorname{div} \underset{\sim}{V} = 0 \ ; \quad \frac{\partial \underset{\sim}{V}}{\partial t} + (\underset{\sim}{V}.\operatorname{grad})\underset{\sim}{V} + \frac{1}{\rho_o} \operatorname{grad} p = 0 \tag{3.1}$$

where c is an arbitrary real parameter and ρ_o is the constant density of the flow. This set of equations with modified dependent variables and scaled spatial derivatives has no physical meaning until the steady state is reached. However, when the steady state is reached the system becomes identical to the true steady equations. The still arbitrary parameter c^2 can be selected to accelerate the time decay to steady state. The major advantage of the modified system over the original one is that the high-speed sound waves have been eliminated, rendering the modified system much better conditioned for numerical solution. The advantage here is very similar to the removal of gravity waves by the geostrophic approximation in the equations of meterology.

3.2 Well Posedness

In three-dimensional Cartesian coordinates the conservation equations for the artificial compressibility approach to steady incompressible flow are

$$\frac{\partial}{\partial t} q + M.[\frac{\partial f}{\partial x} + \frac{\partial g}{\partial y} + \frac{\partial h}{\partial z}] = 0 \qquad (3.2)$$

where

$$q=\begin{bmatrix} p/\rho_o \\ u \\ v \\ w \end{bmatrix} \quad M=\begin{bmatrix} c^2 & 0 & 0 & 0 \\ 0 & 1 & 0 & 0 \\ 0 & 0 & 1 & 0 \\ 0 & 0 & 0 & 1 \end{bmatrix} \quad f=\begin{bmatrix} u \\ u^2+p/\rho_o \\ uv \\ uw \end{bmatrix} \quad g=\begin{bmatrix} v \\ uv \\ v^2+p/\rho_o \\ vw \end{bmatrix} \quad h=\begin{bmatrix} w \\ uw \\ vw \\ w^2+p/\rho_o \end{bmatrix} .$$

The equivalent quasilinear form of system (3.2) is

$$\frac{\partial}{\partial t} q + A \frac{\partial}{\partial x} q + B \frac{\partial}{\partial y} q + C \frac{\partial}{\partial z} q = 0 \qquad (3.3)$$

with

$$A=M.\frac{\partial f}{\partial q} = \begin{bmatrix} 0 & c^2 & 0 & 0 \\ 1 & 2u & 0 & 0 \\ 0 & v & u & 0 \\ 0 & w & 0 & u \end{bmatrix} \quad B=M.\frac{\partial g}{\partial q} = \begin{bmatrix} 0 & 0 & c^2 & 0 \\ 0 & v & u & 0 \\ 1 & 0 & 2v & 0 \\ 0 & 0 & w & v \end{bmatrix}$$

$$C=M.\frac{\partial h}{\partial q} = \begin{bmatrix} 0 & 0 & 0 & c^2 \\ 0 & w & 0 & u \\ 0 & 0 & w & v \\ 1 & 0 & 0 & 2w \end{bmatrix} .$$

It is called hyperbolic at the point (x,y,z,t,q) if there exists a nonsingular matrix $T(\alpha,\beta,\varepsilon)$ that diagonalizes the linear combination $D= \alpha A+\beta B+\varepsilon C$

$$T^{-1}DT= \text{ diag. } \{\lambda^{(1)}, \lambda^{(2)}, \lambda^{(3)}, \lambda^{(4)}\}$$

where the eigenvalues λ of D are real and the norms of T and T^{-1} are uniformly bounded for arbitrary real α,β and ε. The eigenvalues of D are found to be with the definitions

$U = \alpha u+\beta v+\varepsilon w$ and $a^2 = U^2+c^2(\alpha^2+\beta^2+\varepsilon^2)$, $\lambda^{(1)} = \lambda^{(2)} = U$, $\lambda^{(3)} = U+a$, $\lambda^{(4)} = U-a$ and are always real. The range between the smallest and largest eigenvalues can be adjusted according to the value of the parameter c. Using matrix D's complete set of linearly independent right eigenvectors as the columns of matrix T we find

$$T = \begin{bmatrix} 0 & 0 & c^2a & -c^2a \\ -\varepsilon & -\beta & u(U+a)+\alpha c^2 & u(U-a)+\alpha c^2 \\ 0 & \alpha & v(U+a)+\beta c^2 & v(U-a)+\beta c^2 \\ \alpha & 0 & w(U+a)+\varepsilon c^2 & w(U-a)+\varepsilon c^2 \end{bmatrix}. \tag{3.4}$$

Its inverse is formed from the left eigenvectors of D

$$T^{-1} = \begin{bmatrix} \dfrac{\varepsilon U-(\alpha^2+\beta^2+\varepsilon^2)w}{\alpha a^2} & -\dfrac{wU+\varepsilon c^2}{a^2} & -\dfrac{\beta(wU+\varepsilon c^2)}{\alpha a^2} & \dfrac{(\alpha u+\beta v)U+(\alpha^2+\beta^2)c^2}{\alpha a^2} \\ \dfrac{\beta U-(\alpha^2+\beta^2+\varepsilon^2)v}{\alpha a^2} & -\dfrac{vU+\beta c^2}{a^2} & \dfrac{(\alpha u+\varepsilon w)U+(\alpha^2+\varepsilon^2)c^2}{\alpha a^2} & -\dfrac{\varepsilon(vU+\beta c^2)}{\alpha a^2} \\ -\dfrac{U-a}{2a^2c^2} & \dfrac{\alpha}{2a^2} & \dfrac{\beta}{2a^2} & \dfrac{\varepsilon}{2a^2} \\ -\dfrac{U+a}{2a^2c^2} & \dfrac{\alpha}{2a^2} & \dfrac{\beta}{2a^2} & \dfrac{\varepsilon}{2a^2} \end{bmatrix}. \tag{3.5}$$

Working with the maximum norm one can then go on to complete the demonstration of hyperbolicity by making reasonable estimates to show that these last two matrices are uniformly bounded if $|u|$, $|v|$, and $|w|$ are bounded.

But how can we choose a numerical value for the parameter c in order to obtain a mix of wave speeds beneficial for numerical solution? To do so we need to look at a measure more quantitative than just the boundedness of the transformation matrices. In fact what we want to know is how the bound varies with c. This can be determined numerically for given values of $\underset{\sim}{V}$ and c by computing the eigenvalues σ of T^*T since the

L_2 norms are $\|T\|=(\sigma_{max})^{1/2}$ and $\|T^{-1}\|=(\sigma_{min})^{-1/2}$. It is less cumbersome but sufficient for insight to consider only the two-dimensional problem, i.e. $w=\varepsilon=0$. For any specified values of $\underset{\sim}{V}$ and c a good measure of the condition of the system is the number $K=\|T\|\ \|T^{-1}\| = (\sigma_{max}/\sigma_{min})^{1/2}$. One can surmise, and we have verified it in an actual computation, that this condition number K depends only on the ratio $c^2/\underset{\sim}{V}\cdot\underset{\sim}{V}$. Computed numerically and plotted in Fig. 3 as the radial coordinate of the polar diagram (K,θ) where the wave angle θ defines $\alpha=\sin\theta$ and $\beta=\cos\theta$ for $0\leqslant\theta\leqslant 2\Pi$, the condition number K is displayed for three different values of $c^2/\underset{\sim}{V}\cdot\underset{\sim}{V}$. When the ratio $r=c^2/\underset{\sim}{V}\cdot\underset{\sim}{V}$ is greater than unity, the pressure waves dominate over the convection waves and the system is less directionally dependent and better conditioned.

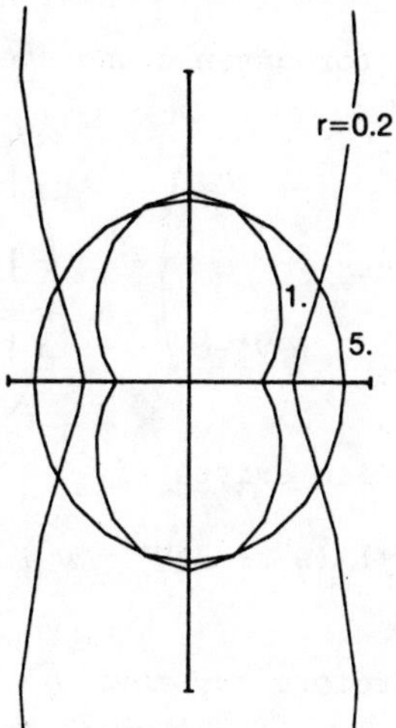

Fig. 3 Polar diagram of condition number K of the hyperbolic system (3.2) as a function of the plane wave angle θ.

3.3 Discontinuities

In steady flow the true incompressible equations admit of course only the tangential discontinuity with jump conditions $[p]=0$ and $\underset{\sim}{V}_1\cdot\underset{\sim}{n} = \underset{\sim}{V}_2\cdot\underset{\sim}{n} = 0$. The artificial-compressibility however, approaches steady flow only asymptotically in time so we must investigate what transient discontinuities, if any, are allowed in this pseudo-system of equations. This aspect of the method has been overlooked, it seems, by Chorin (1967) and Peyret and Taylor (1983). We follow the standard analysis of discontinuities for conservation laws (see Rizzi and Eriksson, 1985) to obtain the jump relations across the discontinuity surface

$$\begin{aligned} s[p] &= c^2[V_n] \\ s[V_n] &= [V_n^2+p] = 2V_n^*[V_n] + [p] \\ s[V_t] &= [V_nV_t] = V_n^*[V_t] + V_t^*[V_n] \end{aligned} \qquad (3.6)$$

where s is the speed of the discontinuity in the direction of its normal $\tilde{n}$. Parallel to the surface is the tangential direction $\tilde{t}$, the asterisk indicates the average of the value on each side of the discontinuity e.g. $V_n^* = (V_{n_1}+V_{n_2})/2$, and the square brackets their difference $[V_n] = V_{n_2} - V_{n_1}$. System (3.6) is linear homogeneous for given s and average velocities, viz.

$$\begin{pmatrix} -s & c^2 & 0 \\ 1 & 2V_n^*-s & 0 \\ 0 & V_t^* & V_n^*-s \end{pmatrix} \begin{pmatrix} [p] \\ [V_n] \\ [V_t] \end{pmatrix} = 0 \qquad (3.7)$$

and a nontrivial solution exists if

$$(s-V_n^*)\,(s^2 - 2sV_n^* - c^2) = 0.$$

The discontinuity therefore may move with any of three different speeds $s_1 = V_n^*$, $s_2 = V_n^*+(V_n^{*2}+c^2)^{1/2}$, or $s_3 = V_n^*-(V_n^{*2}+c^2)^{1/2}$. The jump relations for the first speed then are eigenvectors of (3.6) which work out to be

$$\begin{pmatrix} [p] \\ [V_n] \\ [V_t] \end{pmatrix} = \kappa \begin{pmatrix} 0 \\ 0 \\ 1 \end{pmatrix} \quad \text{with } s = V_{n_1} = V_{n_2} = V_n^* \qquad (3.8)$$

and κ is an arbitrary constant. We immediately recognize this solution as the unsteady tangential discontinuity that corresponds directly to the steady one. The second and third eigenvectors associated with the speeds $s=s_2$ and $s=s_3$ respectively

$$\begin{pmatrix} [p] \\ [V_n] \\ [V_t] \end{pmatrix} = \kappa \begin{pmatrix} c^2 \\ s \\ sV_t^*(V_n^{*\,2}+c^2)^{-1/2} \end{pmatrix} \tag{3.9}$$

are more unexpected since they allow jumps in both pressure and velocity across a discontinuity. With $s=s_2$ the shock travels downstream, and the pressure and normal velocity of a fluid particle passing through it rises or falls together depending on the sign of κ. With $s=s_3$ the shock moves upstream and the jumps in pressure and normal velocity take opposite signs. These last two (shock) discontinuities are of course nonphysical, but it appears that they do not alter the accuracy of the stationary solution because they are not allowed to remain part of any steady flowfield (Rizzi and Eriksson, 1985).

4. OPTIMAL HYPERBOLIC PRECONDITIONING

Is artificial compressibility the only hyperbolic approach for incompressible flow? In an exhaustive study of this question, Viviand (1983) examines arbitrary time transformations represented by the matrix M^{-1} that leads to the pseudo-unsteady systems

$$M^{-1}\frac{\partial q}{\partial t} + \frac{\partial}{\partial x}f + \frac{\partial}{\partial y}g + \frac{\partial}{\partial z}h = 0 \tag{4.1}$$

where the dependent variables q are velocity and either static or total pressure or some function of all of these. Two essential constraints are set on the elements of M^{-1}; 1) system (4.1) must be hyperbolic, and 2) it must possess the same number of positive and negative eigenvalues as the real equations do in order to ensure that the physically relevant information is brought in by the boundary conditions. If this last constraint is not met, the possibility arises that the solution may satisfy the steady equations with incorrect boundary conditions. Viviand then finds that there are still a number of elements of M free to specify, and he goes on to define a class of systems that are optimal in the sense that their eigenvalues are all as nearly equal in magnitude as possible. Waves then all travel at about the same speed, and this may lead to more rapid convergence. In this sense M can be considered a hyperbolic preconditioning matrix.

Of this class of optimal systems I choose to look in detail at two specific systems (Viviand, private communication) which, to clarify the concepts, are presented here for two dimensions. The results do, however, carry over to three dimensions. They are

$$\frac{\partial}{\partial t} p/\rho_o + 2V^2 \operatorname{div} \underset{\sim}{V} - \underset{\sim}{V}\cdot\operatorname{div}(\underset{\sim\sim}{VV} + \underset{\sim}{I}\, p/\rho_o) = 0$$

$$\frac{\partial}{\partial t} \underset{\sim}{V} + \operatorname{div}(\underset{\sim\sim}{VV} + \underset{\sim}{I}\, p/\rho_o) - \underset{\sim}{V} \operatorname{div} \underset{\sim}{V} = 0 \tag{4.2}$$

and

$$\frac{\partial}{\partial t} p/\rho_o + V^2 \operatorname{div} V = 0$$

$$\frac{\partial}{\partial t} \underset{\sim}{V} + \operatorname{div}(\underset{\sim\sim}{VV} + \underset{\sim}{I}\, p/\rho_o) - 2\underset{\sim}{V} \operatorname{div} \underset{\sim}{V} = 0. \tag{4.3}$$

They differ from each other in the way the pseudotransients of the pressure are driven. In Eq. (4.2) both the divergence of mass and momentum directly influence the pressure transient, while only mass does so in Eq. (4.3). Neither is strictly conservative, Viviand calls them semiconservative, because of the variable coefficients on the divergence terms. Adopting the total pressure $P=p/\rho_o+1/2\ V^2$ insteady of the static pressure p as the dependent variable, Eq. (4.2) simplifies to

$$\frac{\partial}{\partial t} P + V^2 \operatorname{div} \underset{\sim}{V} = 0$$

$$\frac{\partial}{\partial t} \underset{\sim}{V} + \operatorname{div}(\underset{\sim\sim}{VV} + \underset{\sim}{I}\, p/\rho_o) - \underset{\sim}{V} \operatorname{div} \underset{\sim}{V} = 0 \tag{4.4}$$

and is more convenient for numerical computation.

In Cartesian coordinates these become

$$\frac{\partial}{\partial t} q + M\cdot\left[\frac{\partial f}{\partial x} + \frac{\partial g}{\partial y}\right] = 0 \tag{4.5}$$

where f and g are defined in Eq. (3.2) and only q and M may differ. The quasilinear form with $A=M\cdot(\partial f)/(\partial q)$ and $B=M\cdot(\partial g)/(\partial q)$ is

$$\frac{\partial}{\partial t} q + A\frac{\partial q}{\partial x} + B\frac{\partial q}{\partial y} = 0\,. \tag{4.6}$$

4.1 Analysis of System (4.3)

Equations (4.3) take the values

$$q=\begin{bmatrix} p/\rho_o \\ u \\ v \end{bmatrix} \quad M=\begin{bmatrix} V^2 & 0 & 0 \\ -2u & 1 & 0 \\ -2v & 0 & 1 \end{bmatrix} \quad A=\begin{bmatrix} 0 & V^2 & 0 \\ 1 & 0 & 0 \\ 0 & -v & u \end{bmatrix} \quad B=\begin{bmatrix} 0 & 0 & V^2 \\ 0 & v & -u \\ 1 & 0 & 0 \end{bmatrix} \qquad (4.7)$$

and an analysis of the matrix $D=\alpha A+\beta B$ determines its eigenvalues to be $\lambda^{(1)}=\alpha u+\beta v=U$ and $\lambda^{(2)}, \lambda^{(3)}=\pm\sqrt{\alpha^2+\beta^2}\sqrt{u^2+v^2} = \pm\gamma V$ which are always real. In comparison to those of artificial compressibility, the range in their magnitudes is smaller, and the largest one is substantially smaller. The transformation matrices T and T^{-1} work out to

$$T = \begin{bmatrix} 0 & V(U-\gamma V) & V(U+\gamma V) \\ -\beta & u\gamma -\alpha V & -u\gamma -\alpha V \\ \alpha & v\gamma -\beta V & -v\gamma -\beta V \end{bmatrix} \qquad (4.8)$$

and

$$T^{-1}=\begin{bmatrix} \beta u - \alpha v & u(\beta u-\alpha v) & v(\beta u-\alpha v) \\ \dfrac{\gamma V+U}{2V} & \dfrac{\alpha(U+\gamma V)}{2\gamma} & \dfrac{\beta(U+\gamma V)}{2\gamma} \\ \dfrac{-\gamma V+U}{2V} & \dfrac{\alpha(-U+\gamma V)}{2\gamma} & \dfrac{\beta(-U+\gamma V)}{2\gamma} \end{bmatrix}.$$

Where the flow stagnates, V goes to zero and some elements of T^{-1} become infinite. But the matrix can be bounded by redefining V to be $V^2=\max(\varepsilon_o^2, V^2)$ in the coefficient matrices A and B as demonstrated by the plot of the condition number K in Fig. 4.

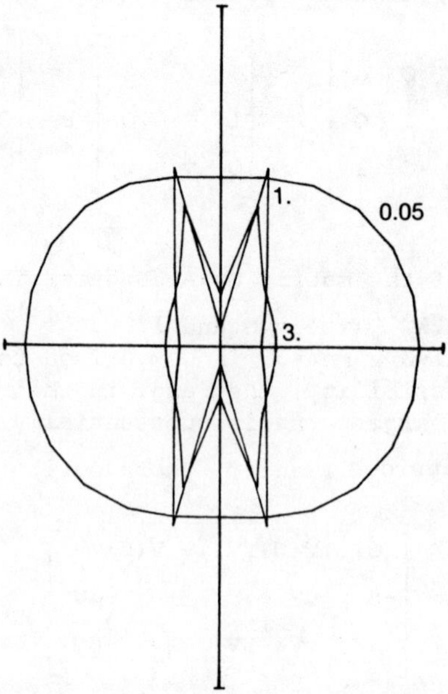

Fig. 4 Polar diagram of condition number K of the hyperbolic system (4.3) as a function of the plane wave angle θ.

4.2 Analysis of System (4.4)

If we carry out the same analysis of Eq. (4.4) with

$$q=\begin{bmatrix}P\\u\\v\end{bmatrix}\quad M=\begin{bmatrix}V^2 & 0 & 0\\-u & 1 & 0\\-v & 0 & 1\end{bmatrix}\quad A=\begin{bmatrix}0 & V^2 & 0\\1 & 0 & -v\\0 & 0 & u\end{bmatrix}\quad B=\begin{bmatrix}0 & 0 & V^2\\0 & v & 0\\1 & -u & 0\end{bmatrix} \tag{4.9}$$

we find that this system has exactly the same eigenvalues as system (4.3): $U, \pm\gamma V$ but a different eigenvector structure

$$T=\begin{bmatrix}V^2 & \gamma V & \gamma V\\u & \alpha & -\alpha\\v & \beta & -\beta\end{bmatrix} \tag{4.10}$$

and

$$T^{-1} = \frac{1}{2\gamma(\alpha v-\beta u)} \begin{bmatrix} 0 & -2\beta\gamma & 2\alpha\gamma \\ \frac{\alpha v-\beta u}{V} & \beta V+v\gamma & -\alpha V-u\gamma \\ \frac{\alpha v-\beta u}{V} & \beta V-v\gamma & -\alpha V+u\gamma \end{bmatrix}$$

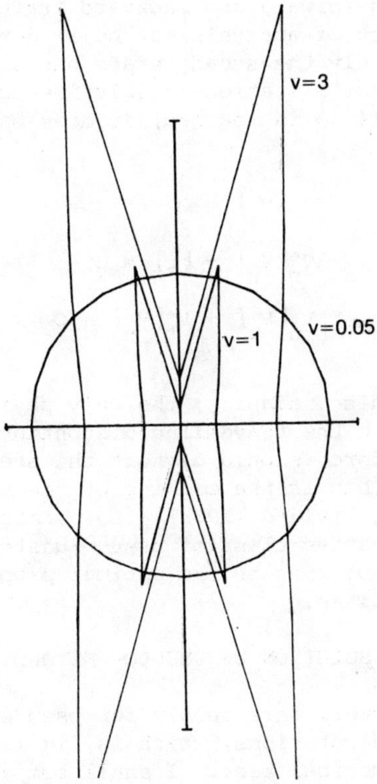

Fig. 5 Polar diagram of condition number K of the hyperbolic system (4.4) as a function of the plane wave angle θ.

Figure 5 displays its boundness by the plot of condition number K. It is interesting to note that while all three systems (4.2), (4.3) and (4.4) possess the same optimal-size eigenvalues, they all have different eigenvectors. Even the simple change of dependent variable p to total pressure P in going from Eq. (4.2) to Eq. (4.4) results in a different eigenvector structure. But whether the wave patterns of one system offer more benefit to the numerical solution than either of the other two awaits discovery by actual computation.

4.3 Discontinuities

Because Eqs. (4.2) and (4.3) are not strictly conservative, we cannot formally derive unsteady jump relations as we did for the artificial-compressibility method. If one supposes these variable coefficients to be locally constant, say, at their average values on each side of the discontinuity, then relations can be derived analogous to those of Eqs. (3.6) and (3.9) for exactly the same types, namely a moving tangential discontinuity, and forward and backward travelling shocks. How valid this sort of analysis may be is not of great concern, however, because only the steady state has any physical relevance. And once a stationary solution to any of Eqs. (4.2), (4.3), or (4.4) is reached, it does satisfy the steady jump relations

$$\begin{aligned} [V_n] &= 0 \\ 2V_n^*[V_n] + [p] &= 0 \\ V_n^*[V_t] + V_t^*[V_n] &= 0 \end{aligned} \tag{4.11}$$

for a tangential discontinuity, the only physical one allowed. That aberrations in the travelling discontinuities, due to the semiconservative form, should distort the steady state then seems unlikely. This is the meaning of the term semiconservative. Viviand (1983) also points out that, although a strictly conservative class of pseudo-unsteady systems can be derived, it does not possess the optimal property of the semiconservative class.

5. FINITE VOLUME SOLUTION OF PSEUDO-UNSTEADY SYSTEMS

Let us see how well this theory for pseudo-unsteady systems works in actual computations, which is, in fact, the first time it is put to a numerical test. I shall compare two solutions for incompressible flow past a 70 deg. swept flat delta wing, the first is by the usual artificial compressibility system (3.2), and the second by the optimal system (4.4). These are solved numerically in the general finite-volume form (Rizzi and Eriksson, 1984)

$$\frac{\partial}{\partial t} \int q \, dvol + M \cdot \iint \underset{\sim}{H} \cdot \underset{\sim}{n} \, ds = 0 \tag{5.1}$$

where $\underset{\sim}{H} \cdot \underset{\sim}{n} = [\underset{\sim}{V} \cdot \underset{\sim}{n},\ u\underset{\sim}{V} \cdot \underset{\sim}{n} + p/\rho_o\, \underset{\sim}{n} \cdot \underset{\sim}{e}_x,\ v\underset{\sim}{V} \cdot \underset{\sim}{n} + p/\rho_o\, \underset{\sim}{n} \cdot \underset{\sim}{e}_y,\ w\underset{\sim}{V} \cdot \underset{\sim}{n} + p/\rho_o\, \underset{\sim}{n} \cdot \underset{\sim}{e}_z]$ is the vector flux of q across the surrounding grid of quadrilateral cells with volume VOL and vector areas

$\underset{\sim}{S}_I = SIX\ \underset{\sim}{e}_x + SIY\ \underset{\sim}{e}_y + SIZ\ \underset{\sim}{e}_z$, $\underset{\sim}{S}_J = SJX\ \underset{\sim}{e}_x + SJY\ \underset{\sim}{e}_y + SJZ\ \underset{\sim}{e}_z$, and $\underset{\sim}{S}_K = SKX\ \underset{\sim}{e}_x + SKY\ \underset{\sim}{e}_y + SKZ\ \underset{\sim}{e}_z$. To analyze Eq. (5.1) locally after semi-discretization using centred space differences, hold the metrics of the cell ijk constant to obtain

$$\frac{d}{dt}(q\ \mathrm{vol})_{ijk} + (\tilde{A}\,\delta_I + \tilde{B}\,\delta_J + \tilde{C}\,\delta_K)q_{ijk} = 0 \qquad (5.2)$$

where $\tilde{A} = M\cdot\frac{\partial(\underset{\sim}{H}\cdot\underset{\sim}{S}_I)}{\partial q}$, $\tilde{B} = M\cdot\frac{\partial(\underset{\sim}{H}\cdot\underset{\sim}{S}_J)}{\partial q}$, and $\tilde{C} = M\cdot\frac{\partial(\underset{\sim}{H}\cdot\underset{\sim}{S}_K)}{\partial q}$.

5.1 CFL Condition

In order to solve the hyperbolic system (5.1), assuming that under appropriate boundary conditions it does converge to a steady state, I straight-forwardly apply the time-marching finite-volume procedure developed originally for the compressibly Euler equations which uses an explicit three-stage Runge-Kutta-type time integration scheme (see Rizzi and Eriksson, 1984). In the absence of boundaries the usual linearized Fourier analysis of Eq. (5.2) specifies the limit on the time step for which the integration locally is stable, i.e. the CFL condition, is $\Delta t \leqslant CFL/|\lambda|max$ where the constant CFL depends on the particular multistage method that is used. A conservative estimate for the maximum eigenvalue $|\lambda|max$ of the 3D spatial difference operator for Eqs. (3.2) leads to the local-time-step condition

$$\Delta t_{ijk} \leqslant CFL\left[\frac{VOL}{\tilde{U} + (\tilde{U}^2 + c^2 S^2)^{1/2}}\right]_{ijk} \qquad (5.3)$$

where

$$\begin{aligned}\tilde{U} = &\ |uSIX + vSIY + wSIZ| \\ &+ |uSJX + vSJY + wSJZ| \\ &+ |uSKX + vSKY + wSKZ|\end{aligned}$$

$$\begin{aligned}S^2 = &\ (|SIX| + |SJX| + |SKX|)^2 \\ &+ (|SIY| + |SJY| + |SKY|)^2 \\ &+ (|SIZ| + |SJZ| + |SKZ|)^2 .\end{aligned}$$

For the optimal system (4.4) a similar estimate gives

$$\Delta t_{ijk} \leq CFL\left[\frac{VOL}{\tilde{V}\cdot S}\right]_{ijk}$$

where $\tilde{V}=(u^2+v^2+w^2)^{1/2}$. Thus we see that optimal preconditioning offers a time step that is 2 to 3 times larger than the standard pseudo-unsteady system. The computed results presented here have all been carried out using these step sizes in the local-time-step integration scheme.

5.2 Boundary conditions

Boundary conditions of course specify the particular problem and two different types are of concern here: flow conditions on a solid wall and at the farfield boundary of the mesh. Since the mesh is aligned to the wall, for the first type I set the velocity flux through the wall to zero and determine the pressure p on it from the normal component of the incompressible momentum equation $\tilde{V}\cdot(\tilde{V}\cdot \text{grad})\ \tilde{n} = \tilde{n}\cdot \text{grad}\ p/\rho_o$ which is exactly analogous to the conditions used for compressible flow (Rizzi, 1978). When it is differenced to formally first-order accuracy, the pressure on the surface is deduced from the interior values.

The farfield conditions are specified as a form of Engquist's hierarchical series of absorbing boundary conditions (see Engquist and Majda, 1978). The First Approximation of this theory reads that setting conditions on the characteristic variables, instead of the flow variables q, of the component of Eq. (5.2) normal to the boundary is maximally dissipative and therefore absorbs more energy. Selecting α,β and ε as the components of the unit vector normal to the boundary means that the eigenvalues λ of D are the slopes in time of the characteristic surfaces in the direction normal to the boundary. The transformation matrices T^{-1} and T diagonalize this one-dimensional equation so that

$$\frac{\partial}{\partial t}\phi+\Lambda\frac{\partial}{\partial \eta}\phi=0 \text{ where } \phi=T^{-1}q \text{ and } \Lambda=T^{-1}D\,T=\text{diag}\{\lambda^{(1)},\lambda^{(2)},\lambda^{(3)},\lambda^{(4)}\},$$

The combination of boundary conditions determined from outside the domain and auxiliary conditions set from inside follows in the now-standard way according to whether the associated characteristic directions enter or leave the domain. When $U<0$ set the three (corresponding to negative eigenvalues) ingoing characteristic variables $\phi^{(1)}$, $\phi^{(2)}$, and $\phi^{(4)}$ to their freestream values, linearly extrapolate the third $\phi^{(3)}$ from the computational field, and then solve for the original unknowns $q = T\phi$. At outflow it is $\phi^{(4)}$ that is given the values of undisturbed flow, and $\phi^{(1)}$, $\phi^{(2)}$ and $\phi^{(3)}$ are extrapolated.

5.3 Artificial Viscosity Model

The rationale for introducing an artificial viscosity operator into the numerical procedure has been put forward before (Eriksson and Rizzi, 1984). To counteract nonlinear aliasing associated with centred differences, to control the cascading of energy from large scale to small scale, and perhaps even to ensure the existence of a steady state are among the reasons that bespeak such a model. The dissipative operator Dq that I add to the convective differencing of the incompressible Euler equations contain a linear fourth-difference term for each of the three spatial directions (see Rizzi and Eriksson, 1984). They are implemented with boundary conditions so that the complete operator has the negative semi-definite property $(q^T Dq) \leq 0$ which has been found important for minimizing the occurrence of spurious errors at solid boundaries (see Eriksson, 1984).

6. COMPARISON OF COMPUTED RESULTS

The mesh around the delta wing is an O-O type constructed by Eriksson's (1982) interpolation method that places a polar singular line at the apex and a parabolic singular line at the tip of the trailing edge. The computation begins with the freestream flow as initial conditions on a coarse mesh that has 40 cells around the half span, 20 each on the upper and lower chord, and 12 outwards, (see Fig. 6). This particular grid topology focusses cells along the leading and trailing edges, as well as the apex where the flow changes most rapidly. It requires, however, a slight rounding of the wing tip.
Figure 7 presents an overall comparison of the time evolutions of the two solutions to system (3.2) and (4.4), respectively, given by the root-mean-square and maximum residuals of the time differences of p, and by the lift and drag coefficients. The high-frequency changes in the residuals of the optimal system (4.4) suggest more rapid wave propagation than for the artificial-compressibility solution (3.2). However, the comparison of pressure coefficients computed on the wing surface (Fig. 7) reveals that the vortex is not as fully developed in the optimal-system solution because its suction peaks are not as large. The low pressure near the leading edge indicates that the vortex sheet may not be developing properly. A preliminary explanation for this may be that the semi-

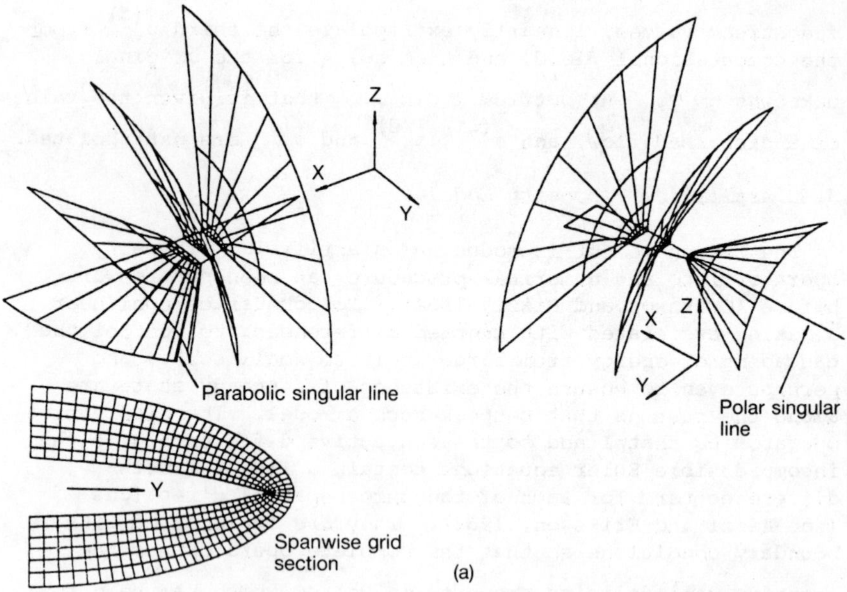

Fig. 6a Main features of an O-O grid generated around a delta-shaped small aspect ratio wing. The polar singular line produces a dense and nearly conical distribution of points at the apex which is needed to resolve the rapidly varying flow there.

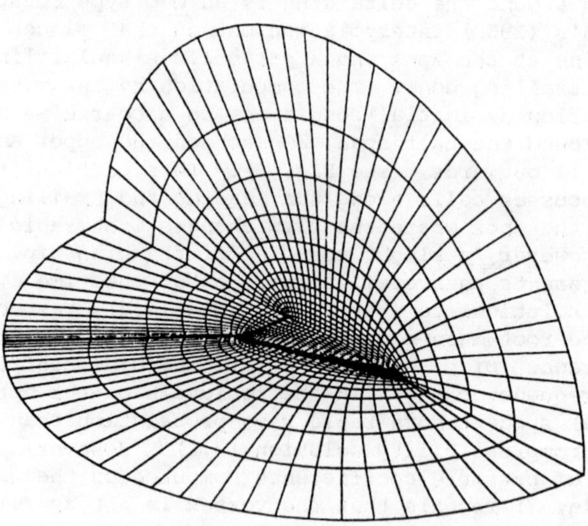

Fig. 6b Three-dimensional view of the delta wing mesh.

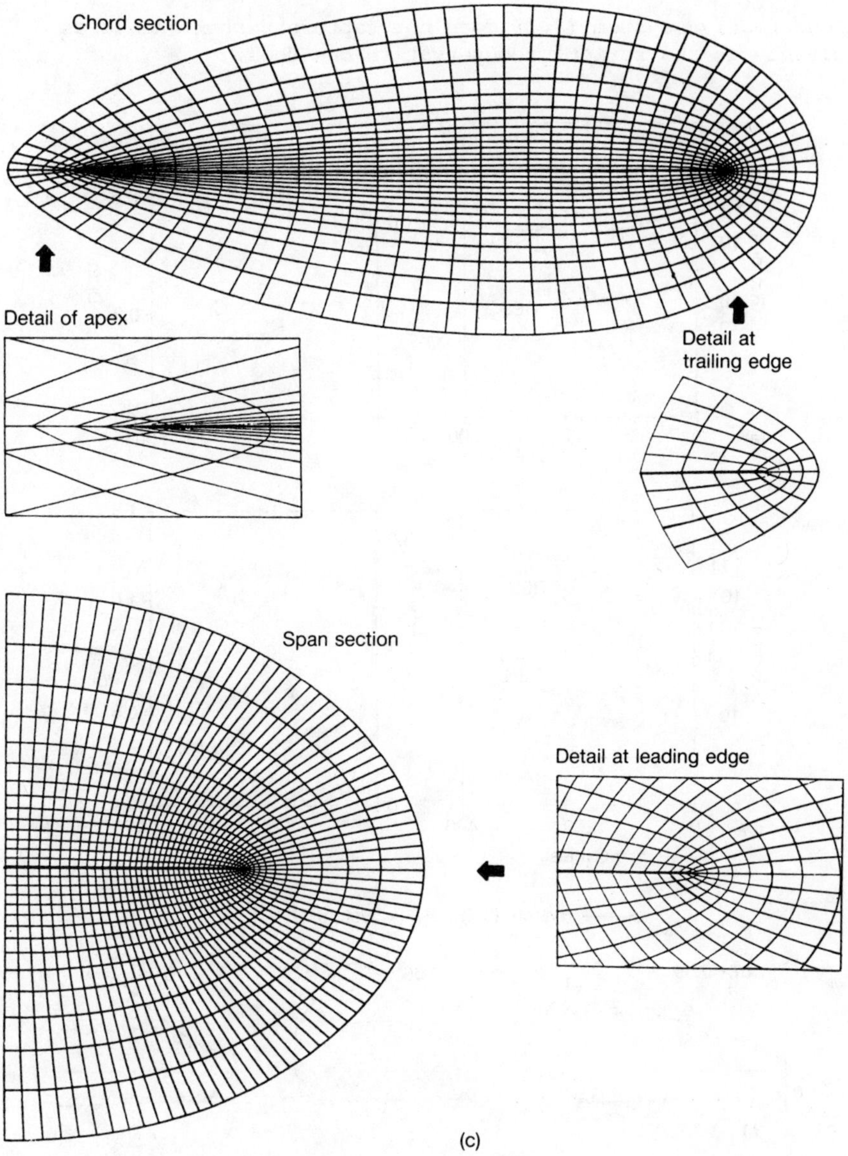

Fig. 6c Partial chordwise and spanwise views of the actual medium (80x24x40) mesh used for the computation of the 70 deg. swept flat plate.

conservative system (4.4) does not capture vortex sheets as cleanly as the strictly conservative one does.

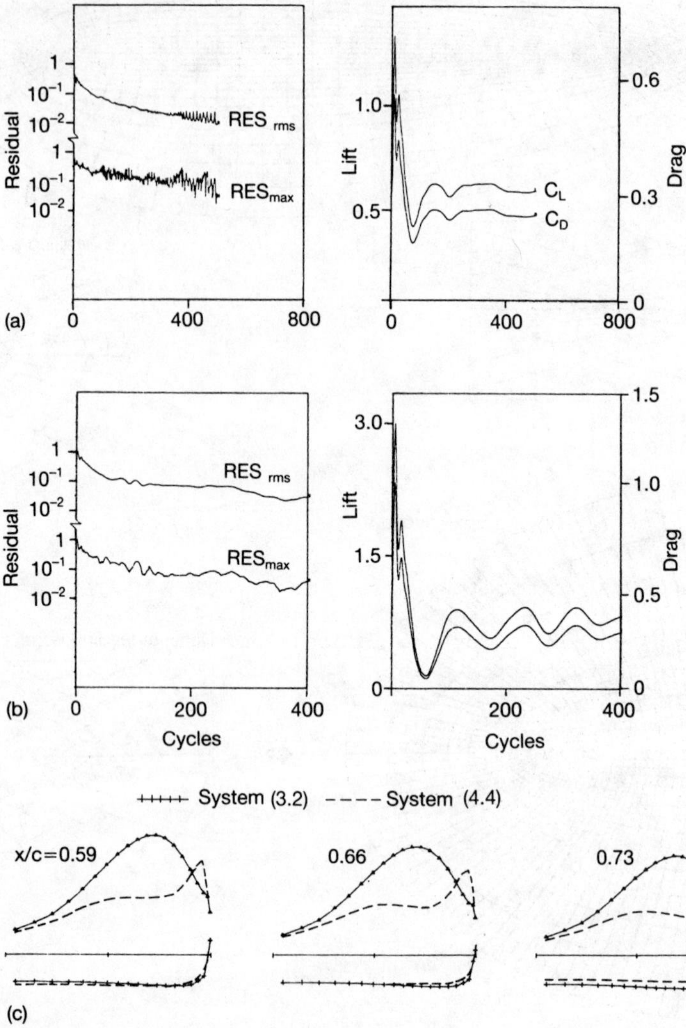

Fig. 7 Comparison of time evolution of the coarse-mesh solutions to (a) system (4.4) and (b) system (3.2), and (c) comparison of the two computed surface pressure distribution.

Very little is known right now about the capturing of vortex sheets in numerical grid solutions, so it is difficult to come to any definite conclusion. Hoeijmakers and Rizzi (1984) carried out one of the first comparisons between a vortex sheet captured in a solution to the Euler equations and one fitted to a potential flow. That comparison, using a solution to Eq. (3.2), demonstrates that a stable vortex sheet can be captured which does agree well with the fitted potential one. The Euler solution, however, indicates a wave-like structure superposed upon the vortex core as it nears the trailing edge. A further question then arises, does the stability of the captured sheet, its close agreement in size and position with the fitted one, and the observed wave structure all change as the mesh size is refined? The fine-mesh solution presented in the next section attempts to answer it.

7. VORTEX SHEET CAPTURED IN FINE MESH SOLUTION

This section presents the conservative solution (3.2) of incompressible flow around the delta wing of zero thickness and unit length at 20 deg. angle of attack obtained upon two mesh densities. The first, the medium density, contains 80x24x40 cells, and is the same one used by Hoeijmakers and Rizzi (1984). The second, the fine density has 160x48x80 cells. Steady flow separates from the leading edge in a vortex sheet which then, under the influence of its own vorticity, rolls up to form a vortex over the wing. Due to the lack of experimentally measured turbulent flow data for this case, a comparison with measurements is not carried out. Instead the results are compared with those from Hoeijmakers's potential boundary-integral (panel) method which inserts a vortex sheet, adjusts it to the surrounding irritational flowfield, and allows it to rollup under its own influence for several turns, and then models the remaining core by an isolated line vortex (see Hoeijmakers et al, 1983, and Hoeijmakers and Rizzi, 1984). The position and strength of the vortex sheet and isolated vortex are determined as part of the solution, sometimes termed "fitting" the rotational flow features. They are true discontinuities, infinitesimally thin, and for this reason a very good choice for comparison because the sheet and vortex in my solutions are not infinitesimally thin but smeared or "captured" over a number of computational mesh cells. The comparison therefore offers a good control on the position of the computed vortex and the diffusion of the sheet as well as how these change with mesh size. Furthermore such panel-method results have been found to agree reasonably well with measurements made in turbulent flow (see e.g. Hoeijmakers et al., 1983).

The thickness of the rotational flow features captured in the solution to the Euler equations varies directly with the size of the mesh cells. The simplest way, therefore, to minimize the diffusion of vorticity is to use as dense a mesh as possible. The fine mesh here consists of over 600 000 cells. The computations were carried out on the CYBER 205 vector computer in 32-bit precision at the rate of 6 microsec. per cell per iteration which translates to over 125 mflops. The medium-mesh solution was advanced an additional 2000 steps beyond the solution discussed by Rizzi and Eriksson (1985) with no discernible change found with time. That medium-mesh solution, therefore, is steady. It is interpolated to the fine mesh and then iterated for 2500 steps.

Global features of the flow in the medium and the fine mesh are compared in Fig. 8 by isograms, drawn in plane projection, of the computed solution in three nonplanar mesh surfaces x/c= 0.3, 0.6 and 0.9 over the wing, one surface in the wake at x/c= 1.15 and one cutting axially through the core of the vortex. The isograms of C_p, vorticity magnitude $|\tilde{\Omega}|$, and total pressure coefficient $(p_t - p_\infty)/(p_{t_\infty} - p_\infty)$ in Figs. 8(a-c) agree for the two solutions and reveal qualitatively the leading-edge vortex over the wing, as well as the trailing-edge vortex that develops from the trailing-edge sheet, interacts, and counter rotates with the leading-edge vortex, and produces the double-vortex pattern in the wake. This wake phenomenon has been observed in the wind tunnel (Hummel, 1979). The comparison shows that overall the broad features of the flow are represented in both grids. The isograms viewed axially through the core indicate the approximately conical nature of the flow starting at the apex. In both solutions at about the 80% chord position, however, the leading-edge vortex lifts up slightly and an abrupt change takes place which might be interpreted as an additional, and unexpected, vortex phenomenon. Although the cause of this feature may still be numerical, it is known that the shear layer separating from the leading edge does show small-scale wave-like structure near the trailing edge (Hoeijmakers and Vaatstra, 1983), which could perhaps evolve into a second vortex that wraps up and intertwines with the primary vortex (Hoeijmakers, private communication). It may also be a precursory effect of the development of the trailing edge vortex. In any event more of the details are brought out in the fine mesh. The contours of total pressure (Fig. 8c) display qualitative agreement with those observed in wind tunnel measurements. The computed losses in the core, however, are high, nearly 80% of the total pressure, and can be attributed to the numerical effect of capturing the vortex sheet. Theoretically the loss

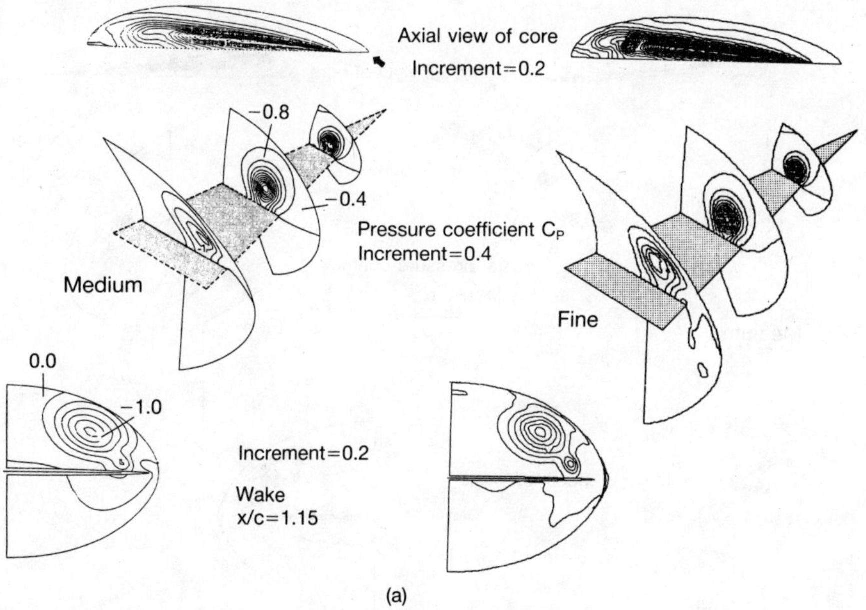

Fig. 8a Isobars of pressure coefficient C_p.

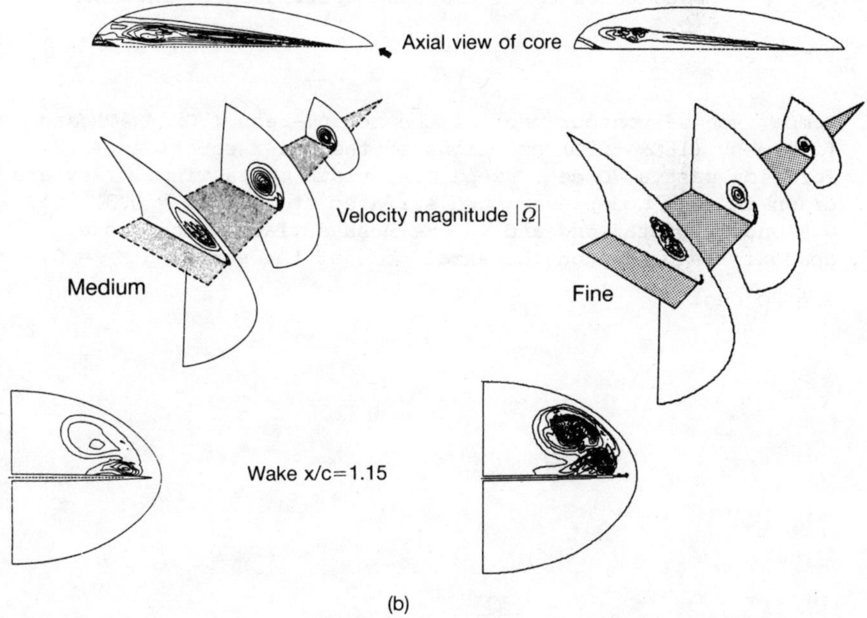

Fig. 8b Vorticity magnitude (not same increment for medium and fine contours).

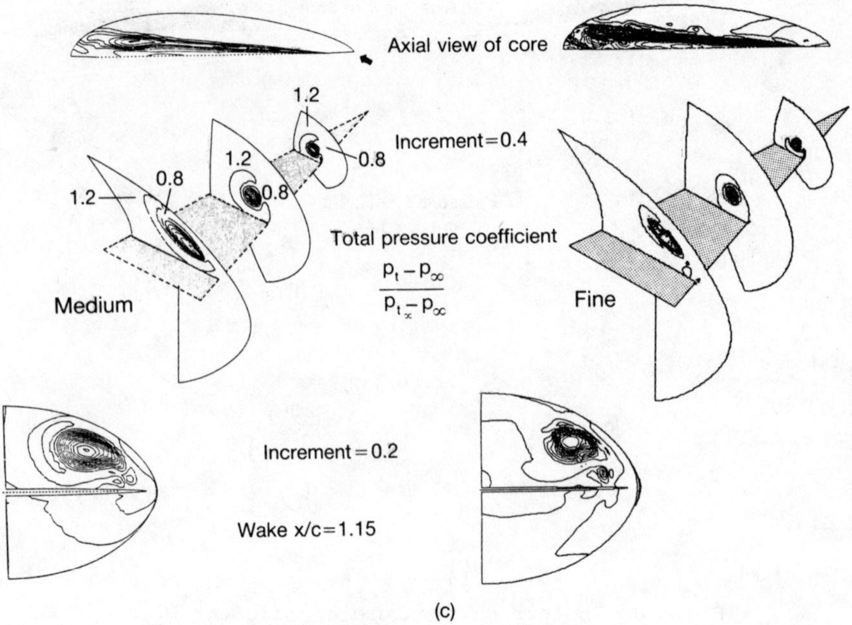

Fig. 8c Contours of total pressure coefficient. Increment = 0.4.

Comparison of contour maps of the medium-mesh (80x24x40) and fine-mesh (160x48x80) solutions to the Euler equations (3.2) for flow past a 70 deg. swept flap plate delta wing. They are drawn in four non-planar mesh surfaces at the x/c = 0.3, 0.6, 0.9 and 1.15 stations and in one mesh surface which passes approximately through the axial core of the vortex. $M_\infty = 0$, $\alpha = 20$ deg.

should be zero (total pressure coefficient =1) on each side of the sheet, even though the velocity is in shear. The numerical solution has to support this shear with a continuous profile over several mesh cells through the sheet, and any sort of reasonable profile (say a linear one) connecting the velocity vector on one side with the one on the other side immediately implies a total pressure loss for the profile even if the velocities at both sides are correct. And the level of loss quickly mounts as the number of cells available to support the profile diminishes. As the vortex sheet wraps up tighter and tighter there will be only one or two cells located between the coils, and ultimately it will disappear off the mesh completely. At that point it is not unreasonable to expect the amount of loss of total pressure that we see in Fig. 8c. The fine mesh supports more coils, but the sheet must eventually disappear in the same way off this mesh too, producing about the same loss at the centre of a now smaller diameter core. It is the size therefore, of the contour rings, but not its level, that varies with mesh spacing.

Let us now inspect the solution more quantitatively by comparison with the results of the potential method. Figure 9 presents the shape of the fitted vortex sheet (dashed lines) from the potential method superimposed upon the vorticity magnitude contours of the Euler-equation solution in three cross-flow planes. Bear in mind that the dashed lines are in the planes x/c= constant while the full lines are projections onto these planes of the vorticity contours in the non-planar mesh surfaces that intersect the wing at the corresponding value of x/c. We see that the vorticity captured in the field is diffused over 5 or 6 cells in both the medium and fine mesh solutions*, and that, in general, the vorticial flow region occupies a larger volume than that enclosed by the vortex sheet fitted to the potential solution. But the positions of the vortex cores in the comparisons and even the curvature of the sheets agree remarkably well, except at the first station x/c= 0.3 where the vorticity contours are somewhat larger and more inboard of the fitted vortex sheet. The vorticity in the fitted sheet is largest near the leading edge where the curvature of the sheet is singular, and the Euler-equation solutions indicate the same trend. The sheet appears to depart tangentially from the lower surface of the leading edge. This comparison with mesh refinement confirms that a stable vortex sheet separating from a swept leading edge can be captured in the vorticity field of the Euler-

* the increment between isovorticity contours is not the same for the medium and fine mesh solutions.

equation solution with a reasonable degree of realism. The curious distortion of the contours in the x/c= 0.9 station and the associated islands of vorticity are another indication that a torsional wave may be standing on the core and giving rise to subsidiary vortices.

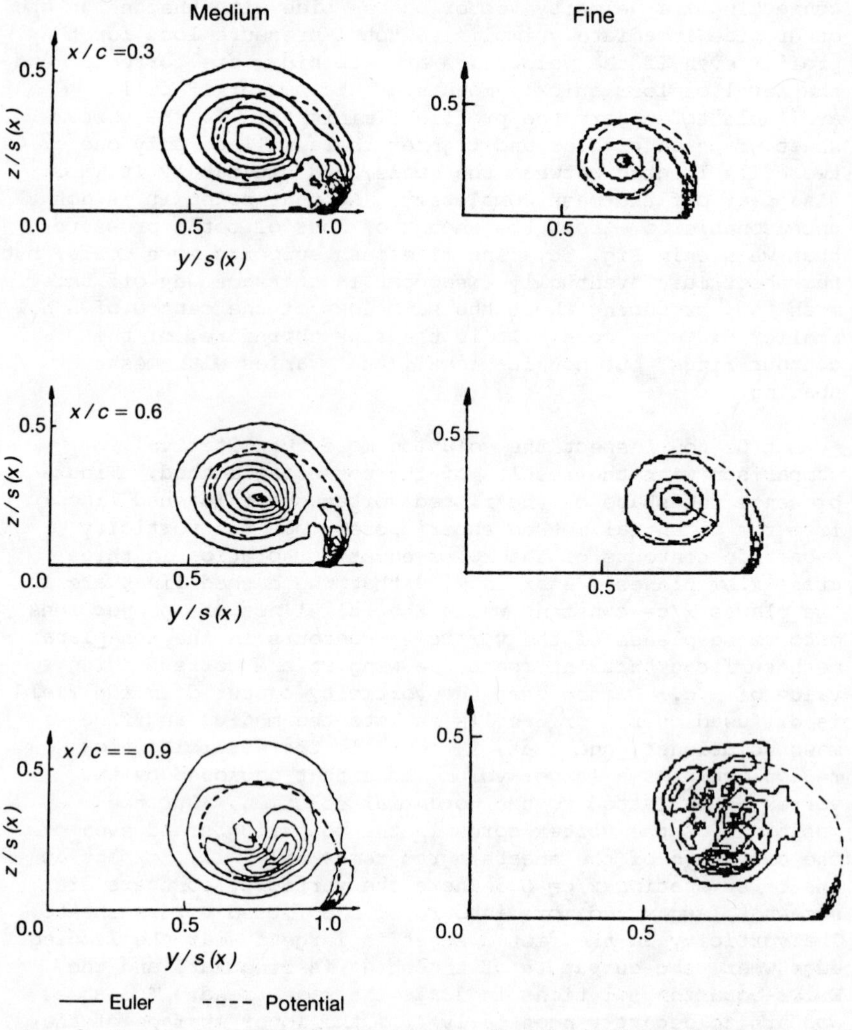

Fig. 9 Comparison of the vorticity fields indicated by vorticity magnitude contours (solid lines) computed with the Euler equations (3.2), using the medium and fine meshes, and the shed vortex (dashed lines) that is fitted as a discontinuity to the surrounding potential solution obtained by the 3D panel method.

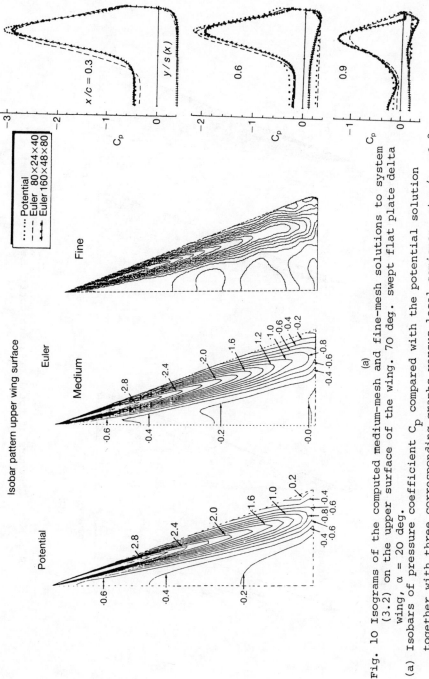

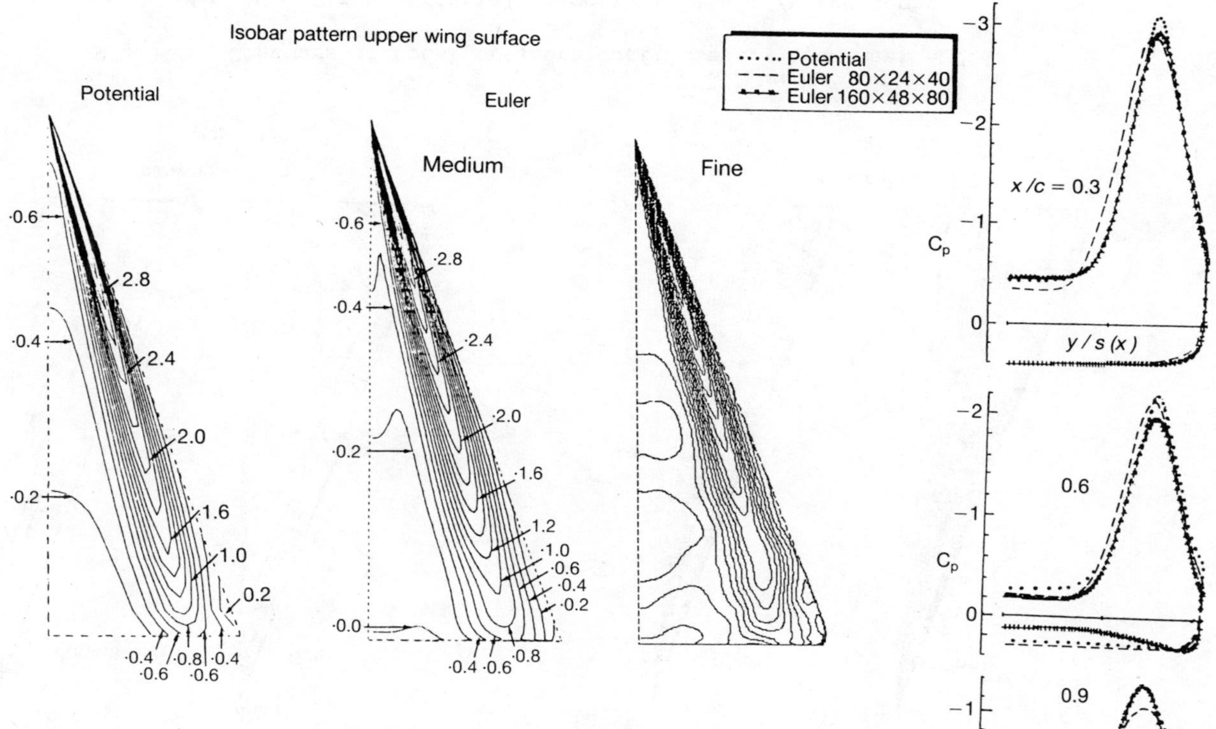

(a)

Fig. 10 Isograms of the computed medium-mesh and fine-mesh solutions to system (3.2) on the upper surface of the wing. 70 deg. swept flat plate delta wing, $\alpha = 20$ deg.

(a) Isobars of pressure coefficient C_p compared with the potential solution together with three corresponding graphs versus local semispan at $x/c = 0.3$, 0.6 and 0.9. Increment = 0.2.

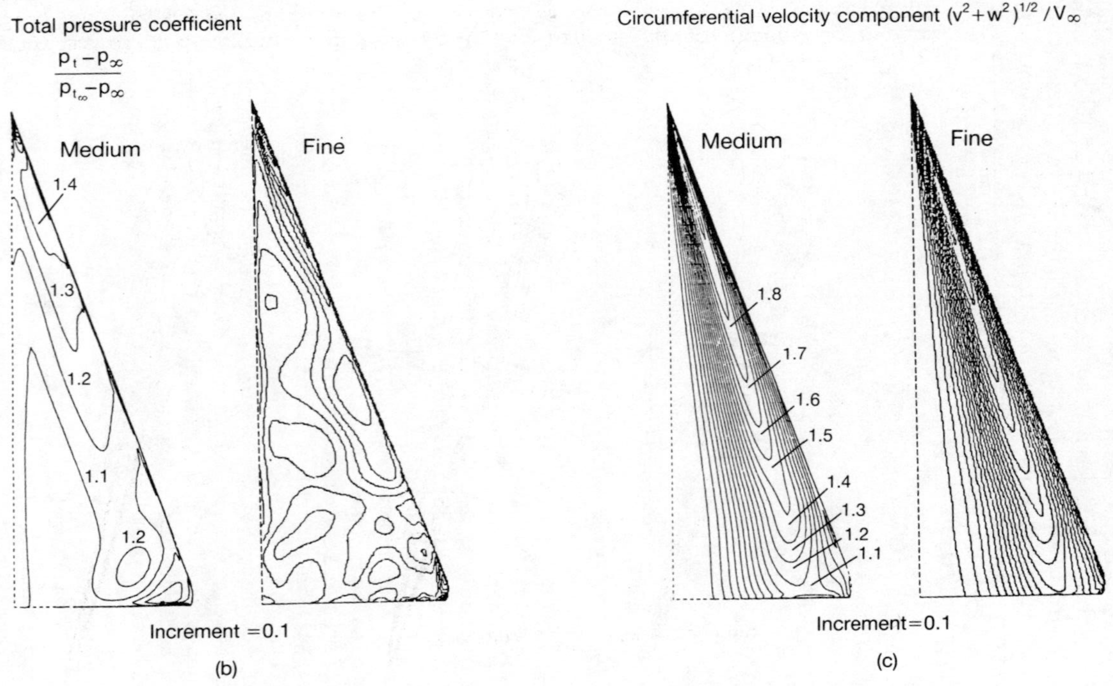

Fig. 10b Contours of total pressure coefficient. Increment = 0.4.

Fig. 10c Circumferential velocity $(v^2 + w^2)^{\frac{1}{2}}/V_\infty$ contours.

An informative, and usually qualitative, manner to examine a flow over a wing is to look at its isograms on the wing surface. Figures 10(a-c) present such views together with the more quantitative graphs of spanwise distributions at three x/c= constant stations and compare them with the potential values. In the sets of computed isobars (Fig. 10a) the pressure trough under the leading-edge vortex has about the same shape, position, and width, and the three agree rather well. The peak level of the suction along the entire trough on the upper surface is somewhat lower in the medium-mesh Euler results, and shifted slightly inboard at x/c= 0.3, but the fine-mesh results show a trend towards the potential solution. The largest losses in total pressure coefficient (Fig. 10b), even negative values, are found near the apex. The fine-mesh results show a pronounced waviness that may be a reflection of the character of the vortex sheet as it approaches the trailing edge. This waviness, however, is not present in the circumferential $(v^2+w^2)^{1/2}/V_\infty$ velocity components on the upper surface (Fig. 10c).

8. CONCLUDING REMARKS

The pseudo-unsteady approach is an interesting one for solving the incompressible Euler equations because it is equally applicable in two or three dimensional problems and because its solutions provide a lot of insight into the types of flows that can occur. Whether the resulting hyperbolic system is strictly conservative, or only semiconservative, may influence how well a vortex sheet is captured. As a demonstration of the method, and as a benchmark solution for the testing of such methods a steady 3D flowfield with a free-shear layer has also been computed with a conservative system, and its features have been discussed under grid refinement. Comparison with an accepted solution was reasonable, even large (but local) errors in total pressure do not seriously degrade the global accuracy, and a curious vortex-like phenomenon coupled to a wave structure in the flow properties was seen to develop just ahead of the trailing edge. This feature may arise from a disturbance of the leading-edge sheet or it may be associated with the rollup of the trailing-edge sheet.

ACKNOWLEDGEMENT

I wish to thank H. Viviand at ONERA for may helpful discussions and insight on his development of pseudo-unsteady systems, ETA Systems, Inc. for providing computer time for the further development of this method, and Charles Purcell for his ever-present help in the running of the program.

REFERENCES

Chorin, A.J., (1967) A Numerical Method for Solving Incompressible Viscous Flow Problems, *J. Comp. Phys.*, 2, pp. 12-26.

Engquist, B. and Majda, A., (1977) Absorbing Boundary Conditions for the Numerical Simulation of Waves, *Math. Comp.*, Vol. 31, pp. 629-651.

Eriksson, L.-E., (1975) Calculation of Two-Dimensional Potential Flow Wall Interference for Multicomponent Airfoils in Closed Low Speed Wind Tunnels, *FFA TN* AU-1116, Part 1, Stockholm.

Eriksson, L.-E., (1982) Generation of Boundary-Conforming Grids Around Wing-Body Configurations Using Transfinite Interpolation, *AIAA J.*, Vol. 20, pp. 1313-1320.

Eriksson, L.-E. and Rizzi, A., (1983) Computer-Aided Analysis of the Convergence to Steady State of a Discrete Approximation to the Euler Equations, *AIAA Paper* No. 83-1951 presented at the 6th CFD Conf., Danvers, MA, (also *J. Comp. Phys.*, in press).

Eriksson, L.-E., Rizzi, A. and Therre, J.P., (1984) Numerical Solutions of the Steady Incompressible Euler Equations Applied to Water Turbines, *AIAA Paper* No. 84-2145.

Eriksson, L.-E., (1984) Boundary Conditions for Artificial Dissipation Operators, *FFA TN* 1984-53, Stockholm.

Hoeijmakers, H.W.M., and Vaatstra, W., (1973) A Higher-Order Panel Method Applied to Vortex-Sheet Roll-Up. *AIAA Journal*, Vol. 21, pp. 516-523.

Hoeijmakers, H.W.M., Vaatstra, W., and Verhaagen, N.G., (1983) Vortex Flow Over Delta and Double-Delta Wings, *J. Aircraft*, Vol. 21, No. 9.

Hoeijmakers, H.W.M. and Rizzi, A., (1984) Vortex-Fitted Potential Solution Compared with Vortex-Captured Euler Solution for Delta Wing with Leading Edge Vortex Separation, *AIAA Paper* No. 84-2144.

Hummel, D., (1979) On the Vortex Formation Over a Slender Wing at Large Incidence, *AGARD* CP-247.

Peyret, R. and Taylor, T., (1983) *Computational Methods for Fluid Flow,* Springer, New York.

Rizzi, A.W., (1978) Numerical Implementation of Solid-Body Boundary Conditions for the Euler Equations, *ZAMM*, Vol. 58, pp. 301-T304.

Rizzi, A., and Eriksson, L.E., (1984) Computation of Flow Around Wings Based on the Euler Equations, *J. Fluid Mech.*, Vol. 148, pp. 45-71.

Rizzi, A. and Eriksson, L.-E., (1985) Vortex-Sheet Capturing in Numerical Solutions of the Incompressible Euler Equations, *SIAM Sci. Stat. Comp.* in press.

Rizzi, A. and Eriksson, L.-E., (1985) Computation of Inviscid Incompressible Flow with Rotation, *J. Fluid Mechanics,* Vol. 153, pp. 275-312.

Viviand, H., (1983) Systems Pseudo-Instationannairès Pour Les Ecoulements Stationnairès de Fluide Parfait, *ONERA Publication* No. 1983-4, Chatillon, (In French).

PARTICLE NUMERICAL MODELS IN FLUID DYNAMICS

P.A. Raviart
(Analyse Numérique, Université Pierre et Marie, Curie, Paris)

1. INTRODUCTION

By particle methods of solution of time-dependent problems we mean numerical methods where some of the dependent variables are approximated at each time t by a linear combination of Dirac measures in the space variables x. Although particle methods are far less popular than classical numerical methods such as finite difference or finite element methods, they play an increasing role in the numerical simulation of very complex flows where convection phenomena are dominant. In fact, vortex methods are now commonly used in the simulation of incompressible fluid flows at high Reynolds numbers. On the other hand, particle methods seem to be well suited to the computation of compressible multifluid flows.

The purpose of this paper is to provide a fairly general introduction to the particle method which can be also presented as a generalized finite difference method using Lagrangian coordinates. The paper is organized as follows: Section 2 is devoted to the problem of approximation of continuous functions by linear combinations of Dirac measures (the particles). In Section 3, we consider general convection-diffusion problems. First, we show that the particle method provides a simple and effective way of dealing with the convection terms. Next, we introduce a deterministic method for approximating the diffusion term. In Section 4, we consider the Navier-Stokes equations for an incompressible viscous fluid. By applying the particle method to the vorticity equation, we introduce in a very natural manner the two-dimensional and three-dimensional point-vortex methods with a new treatment of the viscous term. In Section 5, we show how to adapt the particle method to the nonlinear hyperbolic system of gas dynamics. An extension of the method to a detonation problem is then discussed.

2. PARTICLE APPROXIMATION OF FUNCTIONS

In order to introduce the particle method, we first need to discuss the representation of functions in Lagrangian coordinates. Given a velocity field $\underset{\sim}{u} = \underset{\sim}{u}(\underset{\sim}{x},t) = (u_1(\underset{\sim}{x},t),\ldots, u_d(\underset{\sim}{x},t))$, $\underset{\sim}{x} = (x_1,\ldots,x_d) \in R^d$, $t \geq 0$, we define the integral curves of the vector-field $\underset{\sim}{u}$, or equivalently the characteristic curves of the first-order differential operator

$$\frac{\partial}{\partial t} + \underset{\sim}{u}.\underset{\sim}{\nabla} ,$$

as the solutions of the differential system

$$\frac{d}{dt}\underset{\sim}{x} = u(\underset{\sim}{x},t). \tag{2.1}$$

For all $\xi \in R^d$, we denote by $t \to \underset{\sim}{x}(\xi,t)$ the solution of (2.1) which satisfies the initial condition

$$\underset{\sim}{x}(0) = \underset{\sim}{\xi} . \tag{2.2}$$

Setting

$$J(\underset{\sim}{\xi},t) = \det \left(\frac{\partial x_i}{\partial \xi_j} (\underset{\sim}{\xi},t)\right), \tag{2.3}$$

a standard calculation shows that

$$\frac{\partial J}{\partial t} (\underset{\sim}{\xi},t) = J(\underset{\sim}{\xi},t) (\underset{\sim}{\nabla}.\underset{\sim}{u}) (\underset{\sim}{x}(\underset{\sim}{\xi},t),t). \tag{2.4}$$

Now, let $g = g(\underset{\sim}{x},t)$, $\underset{\sim}{x} \in R^d$, $t \geq 0$, be a continuous function. For deriving a particle approximation of the function g, we are given a quadrature formula over the space R^d

$$\int_{R^d} \psi(\underset{\sim}{\xi})d\underset{\sim}{\xi} \simeq \sum_{j\in J} \alpha_j \, \psi(\underset{\sim}{\xi}_j) \tag{2.5}$$

where $(\underset{\sim}{\xi}_j,\alpha_j)_{j\in J}$ is some set of points $\underset{\sim}{\xi}_j \in R^d$ and weights $\alpha_j > 0$. By using the change of variables $\underset{\sim}{x} = \underset{\sim}{x}(\underset{\sim}{\xi},t)$, we obtain for all continuous function $\phi = \phi(\underset{\sim}{x})$ with compact support

$$\int_{R^d} g(\underset{\sim}{x},t)\phi(\underset{\sim}{x})dx = \int_{R^d} g(\underset{\sim}{x}(\underset{\sim}{\xi},t),t)\phi(\underset{\sim}{x}(\underset{\sim}{\xi},t))J(\underset{\sim}{\xi},t)d\underset{\sim}{\xi} .$$

Hence, setting

$$\underset{\sim}{x}_j(t) = \underset{\sim}{x}(\underset{\sim}{\xi}_j, t) \qquad (2.6)$$

and

$$a_j(t) = \alpha_j \, J(\underset{\sim}{\xi}_j, t) \, , \qquad (2.7)$$

we have by (2.5)

$$\int_{R^d} g(\underset{\sim}{x}, t)\phi(\underset{\sim}{x})\,d\underset{\sim}{x} \simeq \underset{j\in J}{\Sigma} a_j(t)g(\underset{\sim}{x}_j(t), t)\phi(\underset{\sim}{x}_j(t)).$$

In other words, we find that the function g is approximated by the following measure

$$\pi^h g(\underset{\sim}{x}, t) = \underset{j\in J}{\Sigma} a_j(t)g(\underset{\sim}{x}_j(t), t)\delta(\underset{\sim}{x}-\underset{\sim}{x}_j(t)) \qquad (2.8)$$

where $\delta(x)$ is the "delta-function". Observe that (2.8) is a Lagrangian representation of the function g: at each time t, $g(\cdot, t)$ is approximated by a linear combination of Dirac measures, or <u>particles</u>, which have fixed Lagrangian coordinates $\underset{\sim}{\xi}_j$.

On the other hand, given a set of particles $j \in J$ with positions $\underset{\sim}{x}_j(t)$ and weights $b_j(t)$, or equivalently the measure

$$g^h(\underset{\sim}{x}, t) = \underset{j\in J}{\Sigma} b_j(t)\,\delta(\underset{\sim}{x}-\underset{\sim}{x}_j(t)),$$

we construct a smooth function in the following way. We introduce a sufficiently smooth cut-off function ζ such that

$$\int_{R^d} \zeta\,d\underset{\sim}{x} = 1.$$

Then, for all $\epsilon > 0$, we set

$$\zeta_\epsilon(\underset{\sim}{x}) = \frac{1}{\epsilon^d}\,\zeta(\underset{\sim}{x}/\epsilon)$$

and

$$g_\epsilon^h(\underset{\sim}{x}, t) = \underset{j\in J}{\Sigma} b_j(t)\zeta_\epsilon(\underset{\sim}{x}-\underset{\sim}{x}_j(t)).$$

This amounts to replacing each point-particle j by a finite-size particle whose shape is defined by the function ζ_ε.

By applying this procedure to the measure $\pi^h g$, we obtain the function

$$\pi_\varepsilon^h g(\underset{\sim}{x},t) = \sum_{j\in J} a_j(t) g(\underset{\sim}{x}_j(t),t)\zeta_\varepsilon(\underset{\sim}{x}-\underset{\sim}{x}_j(t)). \qquad (2.9)$$

In fact, $g - \pi_\varepsilon^h g$ may be viewed as some measure of the error between the function g and its particle approximation $\pi^h g$. Moreover, it is possible to derive bounds for $g - \pi_\varepsilon^h g$ in terms of the properties of the cut-off function ζ. Let us describe a typical result in that direction. We suppose for simplicity that

$$J = Z^d,\ \underset{\sim}{\xi}_j = (j_i h)_{1\leqslant i\leqslant d},\ \alpha_j = h^d \quad \forall j = (j_1,\ldots,j_d) \in Z^d. \qquad (2.10)$$

Then, we have (see(Raviart, 1983), (Gallic and Raviart, 1985a) for more details):-

<u>Proposition 1.</u> <u>Assume that the velocity field</u> $\underset{\sim}{u}$ <u>and the function</u> g <u>are smooth enough and that</u> (2.10) <u>holds. Assume in addition that the cut-off function</u> ζ <u>satisfies the following properties:</u>

(i) <u>there exists an integer</u> $k \geqslant 1$ <u>such that</u>

$$\left|\begin{array}{l} \displaystyle\int_{R^d} \zeta\, d\underset{\sim}{x} = 1, \\ \displaystyle\int_{R^d} x_1^{\beta_1}\ldots x_d^{\beta_d}\, \zeta\, d\underset{\sim}{x} = 0 \quad \forall \beta = (\beta_1,\ldots,\beta_d) \in N^d,\ 1\leqslant|\beta|\leqslant k-1; \end{array}\right. \qquad (2.11)$$

(ii) <u>the function</u> ζ <u>belongs to the space</u> $W^{m+s,1}(R^d)$ <u>for some integers</u> $m > d$ <u>and</u> $s \geqslant 0$.
<u>Then we find for all</u> $\beta \in N^d$ <u>with</u> $|\beta| = s$

$$\partial^\beta (g-\pi_\varepsilon^h g)(x,t) = O(\varepsilon^k + h^m/\varepsilon^{m+s} \qquad (2.12)$$

In the above result, $W^{\mu,1}(R^d)$ denotes the Sobolev space

$$W^{\mu,1}(R^d) = \{\psi \in L^1(R^d);\ \partial^\beta\psi \in L^1(R^d),\ 1 \leq |\beta| \leq \mu\}$$

and we have set, if $\beta = (\beta_1,\ldots,\beta_d) \in N^d$,

$$|\beta| = \beta_1 + \ldots + \beta_d\ ,\ \partial^\beta = \frac{\partial^{|\beta|}}{\partial x_1^{\beta_1} \ldots \partial x_d^{\beta_d}}\ .$$

It follows from (2.12) that the interparticle distance h must be chosen much smaller than the regularization parameter ε in order to obtain satisfactory approximation results.

3. PARTICLE SOLUTION OF CONVECTION-DIFFUSION EQUATIONS.

Let us apply the above considerations to the numerical solution of convection-diffusion equations. We begin by considering convection problems of the form

$$\begin{cases} \dfrac{\partial}{\partial t} \underset{\sim}{\phi} + \underset{\sim}{\nabla}\cdot(\underset{\sim}{u}\ \underset{\sim}{\phi}) = \underset{\sim}{g}(\underset{\sim}{x},t,\underset{\sim}{\phi}),\ \underset{\sim}{x} \in R^d,\ t > 0 \\ \underset{\sim}{\phi}(\underset{\sim}{x},0) = \underset{\sim}{\phi}_o(\underset{\sim}{x}) \end{cases} \tag{3.1}$$

where $\underset{\sim}{\phi} = (\phi_1,\ldots,\phi_p)$ is a p-vector, $\underset{\sim}{g}$ is a continuous function from $R^d \times R_+ \times R^p$ into R^p and $\underset{\sim}{\nabla}\cdot(\underset{\sim}{u}\ \underset{\sim}{\phi})$ stands for

$$(\underset{\sim}{\nabla}\cdot(\underset{\sim}{u}\ \phi_1),\ldots,\underset{\sim}{\nabla}\cdot(\underset{\sim}{u}\ \phi_p)).$$

We are looking for an approximate particle solution of (3.1) of the form

$$\underset{\sim}{\phi}^h(\underset{\sim}{x},t) = \sum_{j\in J} a_j(t)\underset{\sim}{\phi}_j(t)\delta(\underset{\sim}{x}-\underset{\sim}{x}_j(t)), \tag{3.2}$$

where $\underset{\sim}{\phi}_j(t)$ is some approximation of $\underset{\sim}{\phi}(\underset{\sim}{x}_j(t),t)$. We first notice that we have

$$\frac{\partial}{\partial t} \underset{\sim}{\phi}^h + \nabla\cdot(\underset{\sim}{u}\ \underset{\sim}{\phi}^h) = \sum_{j\in J} \frac{d}{dt}(a_j(t)\underset{\sim}{\phi}_j(t))\delta(\underset{\sim}{x}-\underset{\sim}{x}_j(t)) \tag{3.3}$$

in the sense of distributions on $R^d \times]0,\infty[$. Indeed, let ψ be a C^∞ function with compact support in $R^d \times]0,\infty[$, we have by (2.1)

$$\frac{d}{dt}\psi(\underset{\sim}{x}(\underset{\sim}{\xi},t),t) = (\frac{\partial\psi}{\partial t} + \underset{\sim}{u}.\underset{\sim}{\nabla}\psi)(\underset{\sim}{x}(\underset{\sim}{\xi},t),t)$$

so that

$$< \frac{\partial}{\partial t}\underset{\sim}{\phi}^h + \underset{\sim}{\nabla}.(\underset{\sim}{u}\,\underset{\sim}{\phi}^h),\psi > =$$

$$= -\int_0^\infty \sum_{j\in J} a_j(t)\underset{\sim}{\phi}_j(t)(\frac{\partial\psi}{\partial t} + \underset{\sim}{u}.\underset{\sim}{\nabla}\psi)(\underset{\sim}{x}_j(t),t)dt =$$

$$= -\int_0^\infty \sum_{j\in J} a_j(t)\underset{\sim}{\phi}_j(t)\frac{d}{dt}\psi(\underset{\sim}{x}_j(t),t)dt$$

and (3.3) follows. Next, by the results of Section 2, we may write

$$\underset{\sim}{g}(\underset{\sim}{x},t,\underset{\sim}{\phi}(\underset{\sim}{x},t)) = \sum_{j\in J} a_j(t)\underset{\sim}{g}(\underset{\sim}{x}_j(t),t,\underset{\sim}{\phi}(\underset{\sim}{x}_j(t)\delta(\underset{\sim}{x}-\underset{\sim}{x}_j(t)). \tag{3.4}$$

Therefore, using (3.3) and (3.4), a (semi-) discrete form of Problem (3.1) consists in finding the functions $t \to \underset{\sim}{\phi}_j(t)$, $j\in J$, solutions of

$$\begin{cases} \frac{d}{dt}(a_j\,\underset{\sim}{\phi}_j) = a_j\,\underset{\sim}{g}(\underset{\sim}{x}_j,t,\underset{\sim}{\phi}_j), \\ \underset{\sim}{\phi}_j(0) = \underset{\sim}{\phi}_0(\underset{\sim}{\xi}_j), \quad j\in J. \end{cases} \tag{3.5}$$

On the other hand, using (2.1), (2.2), (2.4), (2.6) and (2.7), we note that the functions $t \to \underset{\sim}{x}_j(t)$ and $t \to a_j(t)$ are the solutions of the differential equations

$$\begin{cases} \frac{d}{dt}\underset{\sim}{x}_j = \underset{\sim}{u}(\underset{\sim}{x}_j,t), \\ \underset{\sim}{x}_j(0) = \underset{\sim}{\xi}_j \end{cases} \tag{3.6}$$

and

$$\begin{cases} \frac{d}{dt} a_j = a_j (\underset{\sim}{\nabla} . \underset{\sim}{u})(\underset{\sim}{x}_j, t) \\ a_j(0) = \alpha_j . \end{cases} \tag{3.7}$$

The particle method is thus defined by the equations (3.5), (3.6) and (3.7). It remains however to introduce suitable time-stepping in order to obtain a practically implementable scheme.

Remark 3.1 Note that we need not solve (3.7) in the case of a purely convective problem ($\underset{\sim}{g} \equiv \underset{\sim}{0}$) since the first equation (3.5) reduces to

$$\frac{d}{dt}(a_j \underset{\sim}{\phi}_j) = \underset{\sim}{0}.$$

This means that Problem (3.1) is solved by moving the particles along the characteristic curves without changing their weights $\alpha_j \underset{\sim}{\phi}_o(\underset{\sim}{\xi}_j)$.

On the other hand, in the case $\underset{\sim}{\nabla} \cdot \underset{\sim}{u} = 0$ which occurs frequently in practice, (3.7) is trivially solved by setting $a_j(t) = \alpha_j$.

Remark 3.2 An equivalent way of deriving our scheme consists in noticing that (2.4) and (2.7) imply

$$\frac{1}{a_j}\frac{d}{dt}(a_j \underset{\sim}{\phi}(\underset{\sim}{x}_j, t)) = (\frac{\partial}{\partial t}\underset{\sim}{\phi} + \nabla \cdot (\underset{\sim}{u}\, \underset{\sim}{\phi}))(\underset{\sim}{x}_j, t) . \tag{3.8}$$

This amounts to writing first equation (3.1) in the Lagrangian coordinates $(\underset{\sim}{\xi}, t)$ and using collocation method at the points $\underset{\sim}{\xi}_j$, $j \in J$.

Remark 3.3 Consider the linear case where

$$\underset{\sim}{g}(\underset{\sim}{x}, t, \underset{\sim}{\phi}) = \underset{\approx}{G}(\underset{\sim}{x}, t)\underset{\sim}{\phi}$$

and $\underset{\approx}{G}(\underset{\sim}{x}, t)$ is a p x p matrix. Then, it is a simple matter to check that our particle approximation $\underset{\sim}{\phi}^h$ is indeed the unique measure solution of

$$\begin{cases} \dfrac{\partial}{\partial t} \underset{\sim}{\phi}^h + \underset{\sim}{\nabla} \cdot (\underset{\sim}{u}\, \underset{\sim}{\phi}^h) = \underset{\approx}{G}(\underset{\sim}{x},t) \cdot \underset{\sim}{\phi}^h , \\ \underset{\sim}{\phi}^h(\underset{\sim}{x},0) = \sum_{j \in J} \alpha_j\, \underset{\sim}{\phi}_0(\underset{\sim}{\xi}_j)\,\delta(\underset{\sim}{x}-\underset{\sim}{\xi}_j). \end{cases}$$

Hence, in the linear case, the particle approximation of Problem (3.1) is an exact solution of the convection equation corresponding to a perturbed initial condition.

As we have noticed above, the particle method provides an easy and natural way of solving convection problems. We now turn to convection-diffusion problems of the form

$$\begin{cases} \dfrac{\partial}{\partial t} \underset{\sim}{\phi} + \underset{\sim}{\nabla} \cdot (\underset{\sim}{u}\, \underset{\sim}{\phi}) - \nu\, \Delta \underset{\sim}{\phi} = \underset{\sim}{g}(\underset{\sim}{x},t,\underset{\sim}{\phi}),\ \underset{\sim}{x} \in R^d,\ t > 0. \\ \underset{\sim}{\phi}(\underset{\sim}{x},0) = \underset{\sim}{\phi}_0(\underset{\sim}{x}) , \end{cases} \tag{3.9}$$

where the diffusion coefficient $\nu > 0$ is assumed to be small. Clearly, the difficulty stems from the particle treatment of the diffusion term. A first procedure proposed by Chorin (1973) uses a random walk method for solving the diffusion equation

$$\frac{\partial}{\partial t} \underset{\sim}{\phi} - \nu \Delta \phi = 0$$

and a splitting algorithm. We propose in this paper a deterministic method which is based on the notion of generalized finite-difference.

Given a sufficiently smooth function $R^d \to R$, we want to find an approximation of $\Delta\psi(\underset{\sim}{x}_j)$ in terms of the values $\psi(\underset{\sim}{x}_k)$ of ψ at the particle positions $\underset{\sim}{x}_k$. In fact, using a Taylor expansion at the point $\underset{\sim}{x}_j$, one can prove

Proposition 2. (Gallic and Raviart, 1985b). Assume the hypotheses of Proposition 1 with $s = 1$. Assume in addition that the cut-off function ζ is radially symmetric, i.e., there exists a function $\bar{\zeta} : R_+ \to R$ such that

$$\zeta(\underset{\sim}{x}) = \bar{\zeta}(|\underset{\sim}{x}|). \tag{3.10}$$

Then, if the function ψ is sufficiently smooth, we have

$$\begin{cases} 2 \sum_{k\in J} a_k(\psi(\underset{\sim}{x}_k) - \psi(\underset{\sim}{x}_j)) \dfrac{D\zeta_\varepsilon(\underset{\sim}{x}_j-\underset{\sim}{x}_k)\cdot(\underset{\sim}{x}_k - \underset{\sim}{x}_j)}{|\underset{\sim}{x}_k - \underset{\sim}{x}_j|^2} = \\ = \Delta\psi(x_j) + O(\varepsilon^k + h^m/\varepsilon^{m+1} \end{cases} \tag{3.11}$$

In (3.11), $D\zeta_\varepsilon(\underset{\sim}{x})$ is the total derivative of the function ζ_ε at the point $\underset{\sim}{x}$ so that

$$D\zeta_\varepsilon(\underset{\sim}{x})\cdot\underset{\sim}{y} = \sum_{i=1}^{d} \frac{\partial\zeta_\varepsilon}{\partial x_i}(\underset{\sim}{x})\, y_i, \quad \underset{\sim}{y} \in R^d.$$

Remark 3.4 One can slightly extend the generalized difference operator appearing in (3.11) in order to approximate any second order differential operator of the form $\nabla\cdot(v\ \nabla)$. Assuming the hypotheses of Proposition 2, we obtain

$$\sum_{j\in J} (b(\underset{\sim}{x}_k) + b(\underset{\sim}{x}_j))(\psi(\underset{\sim}{x}_k) - \psi(\underset{\sim}{x}_j)) \frac{D\zeta_\varepsilon(\underset{\sim}{x}_j-\underset{\sim}{x}_k)\cdot(\underset{\sim}{x}_k-\underset{\sim}{x}_j)}{|\underset{\sim}{x}_k - \underset{\sim}{x}_j|^2} =$$

$$= \underset{\sim}{\nabla}\cdot(b\ \underset{\sim}{\nabla}\psi)(\underset{\sim}{x}_j) + O(\varepsilon^k + h^m/\varepsilon^{m+1}).$$

Now, using (3.8) and Proposition 2, it is a simple matter to adapt the particle method to the convection-diffusion problem (3.9). Looking again for an approximate solution $\underset{\sim}{\phi}^h$ of the form (3.2), we define the functions $t \to \underset{\sim}{\phi}_j(t)$, $j \in J$, to be the solutions of the differential system

$$\begin{cases} \dfrac{d}{dt}(a_j\, \underset{\sim}{\phi}_j) - 2\nu\, a_j \sum_{k\in J} a_k(\underset{\sim}{\phi}_k - \underset{\sim}{\phi}_j) \dfrac{D\zeta_\varepsilon(\underset{\sim}{x}_j-\underset{\sim}{x}_k)(x_k-x_j)}{|\underset{\sim}{x}_k - \underset{\sim}{x}_j|^2} \\ \qquad = a_j\, \underset{\sim}{g}(\underset{\sim}{x}_j,t,\underset{\sim}{\phi}_j), \\ \underset{\sim}{\phi}_j(0) = \underset{\sim}{\phi}_0(\underset{\sim}{\xi}_j), \quad j \in J. \end{cases} \tag{3.12}$$

The numerical simulation of the diffusion term is thus obtained by introducing interactions of particles.

If we assume $\underset{\sim}{g} \equiv 0$, we note again that the convection terms are taken into account by moving the particles without changing their weights $a_j \underset{\sim}{\phi}_j$. On the contrary, the particle treatment of the diffusion term consists in modifying the weights of the particles without changing their positions $\underset{\sim}{x}_j$.

Remark 3.5. It may seem natural to choose for the cut-off function ζ the Gaussian function

$$\zeta(\underset{\sim}{x}) = \frac{1}{(2\pi)^{d/2}} \exp\left(-\frac{|\underset{\sim}{x}|^2}{2}\right). \tag{3.13}$$

This leads to a simple form for the first equation (3.12). Indeed, since in that case

$$D\zeta_\varepsilon(\underset{\sim}{x})\cdot\underset{\sim}{y} = \frac{1}{(2\pi)^{d/2}\,\varepsilon^{d+2}}\; \underset{\sim}{x}\cdot\underset{\sim}{y} \exp\left(-\frac{|\underset{\sim}{x}|^2}{2\varepsilon^2}\right),$$

we obtain

$$\begin{cases} \dfrac{d}{dt}(a_j \underset{\sim}{\phi}_j) - \dfrac{2\nu}{(2\pi)^{d/2}\,\varepsilon^{d+2}}\, a_j \displaystyle\sum_{k\in J} a_k(\underset{\sim}{\phi}_k - \underset{\sim}{\phi}_j) \exp\left(-\dfrac{|\underset{\sim}{x}_k - \underset{\sim}{x}_j|^2}{2\varepsilon^2}\right) = \\ = g(x_j, t, \phi_j). \end{cases}$$

Concerning the convergence of this particle method of approximation of the convection-diffusion problem (3.9), we refer again to (Gallic and Raviart, 1985a). Using Proposition 2, the convergence is proved under the additional hypothesis

$$\frac{d}{dr}\bar{\zeta}(r) \leqslant 0 \qquad r \geqslant 0$$

together with the error estimate

$$(\underset{\sim}{\phi} - \underset{\sim}{\phi}_\varepsilon^h)(\underset{\sim}{x}, t) = O(\varepsilon^2 + h^m/\varepsilon^{m+1}).$$

See also (Cottet and Gallic, 1985a) for a closely related result when the cut-off function ζ is given by (3.13).

4. APPLICATIONS TO VORTEX METHODS

Let us now apply the particle methodology to fluid dynamics problems. In this Section, we consider the two-dimensional and three-dimensional Navier-Stokes equations for an incompressible fluid flow

$$\frac{\partial}{\partial t}\underset{\sim}{u} + (\underset{\sim}{u}\cdot\underset{\sim}{\nabla})\underset{\sim}{u} + \underset{\sim}{\nabla}p - \nu\,\Delta\underset{\sim}{u} = \underset{\sim}{f}\,, \quad x \in R^d,\ d = 2,3,\ t > 0 \tag{4.1}$$

$$\underset{\sim}{\nabla}\cdot\underset{\sim}{u} = 0. \tag{4.2}$$

We supplement these equations with the following condition which means that the flow is uniform at infinity

$$\underset{\sim}{u}(\underset{\sim}{x},t) \to \underset{\sim}{u}_\infty(t)\,, \quad |\underset{\sim}{x}| \to \infty \tag{4.3}$$

and an initial condition

$$\underset{\sim}{u}(\underset{\sim}{x},0) = \underset{\sim}{u}_o(\underset{\sim}{x}). \tag{4.4}$$

We begin by analyzing the two-dimensional case. We set

$$\omega = \operatorname{curl}\underset{\sim}{u} = \frac{\partial u_2}{\partial x_1} - \frac{\partial u_1}{\partial x_2} \tag{4.5}$$

and we assume for simplicity that curl $\underset{\sim}{f} = 0$. By taking the curl of Equation (4.1), we obtain the following equivalent formulation of Problem (4.1), (4.2), (4.3), (4.4):

$$\begin{cases} \dfrac{\partial\omega}{\partial t} + \underset{\sim}{\nabla}\cdot(\underset{\sim}{u}\,\omega) - \nu\,\Delta\omega = 0\,, & x \in R^2\,,\ t > 0 \\[2ex] \omega(\underset{\sim}{x},0) = \omega_0(\underset{\sim}{x})\,, & \omega_0 = \operatorname{curl}\underset{\sim}{u}_o\,, \end{cases} \tag{4.6}$$

$$\begin{cases} \operatorname{curl}\underset{\sim}{u} = \omega & \\ & x \in R^2\,,\ t > 0 \\ \underset{\sim}{\nabla}\cdot\underset{\sim}{u} = 0 & \\ \underset{\sim}{u}(\underset{\sim}{x},t) \to \underset{\sim}{u}_\infty(t)\,, & |\underset{\sim}{x}| \to \infty. \end{cases} \tag{4.7}$$

Let us introduce the kernel

$$\underset{\sim}{K}(\underset{\sim}{x}) = \frac{1}{2\pi|\underset{\sim}{x}|^2}(-x_2, x_1).$$

Then, it is well known that the equations (4.7) may be equivalently written in the form

$$\underset{\sim}{u}(\cdot,t) = \underset{\sim}{u}_\infty(t) + \underset{\sim}{K} * \omega(\cdot,t). \tag{4.8}$$

Hence, we obtain a convection-diffusion problem (4.6) for the vorticity ω but here the velocity field $\underset{\sim}{u}$ is a function of ω itself through the equations (4.7) or (4.8).

The vortex method is then obtained by applying the particle method developed in Section 3 to the convection-diffusion problem (4.6). We are looking for an approximate vorticity of the form

$$\omega^h(\underset{\sim}{x},t) = \sum_{j\in J} a_j(t)\omega_j(t)\delta(\underset{\sim}{x}-\underset{\sim}{x}_j(t)). \tag{4.9}$$

Since the convolution operator $\underset{\sim}{K}*$ does not operate on measures and therefore on $\omega^h(\cdot,t)$, we need first to regularize it before using (4.8). Setting

$$\underset{\sim}{K}_\varepsilon = \underset{\sim}{K} * \zeta_\varepsilon\ , \tag{4.10}$$

we define an approximate velocity field $\underset{\sim}{u}^h_\varepsilon$ by

$$\underset{\sim}{u}^h_\varepsilon(\cdot,t) = \underset{\sim}{u}_\infty(t) + \underset{\sim}{K}_\varepsilon * \omega^h(\cdot,t). \tag{4.11}$$

Since $\underset{\sim}{\nabla}\cdot\underset{\sim}{u}^h_\varepsilon = 0$, we have

$$a_j(t) = \alpha_j \tag{4.12}$$

and the vortex method is defined by the equations

$$\frac{d}{dt}\underset{\sim}{x}_j = \underset{\sim}{u}^h_\varepsilon(\underset{\sim}{x}_j,t) = \underset{\sim}{u}_\infty + \sum_{j\in J}\alpha_k\,\omega_k\,\underset{\sim}{K}_\varepsilon(\underset{\sim}{x}_j - \underset{\sim}{x}_k) \tag{4.13}$$

and

$$\frac{d}{dt}\omega_j - 2\nu\sum_{k\in J}\alpha_k(\omega_k - \omega_j)\frac{D\zeta_\varepsilon(\underset{\sim}{x}_j - \underset{\sim}{x}_k)\cdot(\underset{\sim}{x}_k - \underset{\sim}{x}_j)}{|\underset{\sim}{x}_k - \underset{\sim}{x}_j|^2} = 0\ . \tag{4.14}$$

For the convergence of the two-dimensional vortex method in the case $\nu = 0$, i.e., for the incompressible Euler equations, we refer to (Beale and Majda, 1982a and 1982b), (Cottet, 1982) and (Raviart, 1983).

<u>Remark 4.1</u>. There are other ways of dealing with the diffusion term in the vortex method. For a discussion of this topic, see the survey of Leonard (1980). See also (Beale and Majda, 1981) for a mathematical analysis of Chorin's random walk method.

<u>Remark 4.2</u>. One can choose other forms than (4.10) for the regularized kernel $\mathbf{K}_\varepsilon$ which may be simpler to use in practice. For instance, one can take

$$\mathbf{K}_\varepsilon(\mathbf{x}) = \frac{\quad}{2\pi(|\mathbf{x}|^2 + \varepsilon^2)} (-x_2, x_1).$$

On the other hand, instead of using (4.11), one can solve directly the equations (4.7) in order to define an approximate velocity field. If we introduce a stream function ψ such that

$$\mathbf{u} = \mathbf{curl}\ \psi = \left(\frac{\partial \psi}{\partial x_2}, -\frac{\partial \psi}{\partial x_1}\right),$$

the equations (4.7) become

$$\begin{cases} -\Delta\psi = \omega, & x \in R^2, \quad t > 0, \\ \mathbf{curl}\ \psi(\mathbf{x}, t) \to \mathbf{u}_\infty(t), & |\mathbf{x}| \to \infty. \end{cases} \tag{4.15}$$

Roughly speaking, the vortex in cell method consists in solving numerically (4.6) by the particle method and (4.15) by a classical finite-difference or finite-element method. For a precise description of the vortex in cell method, see again (Leonard, 1980). A proof of convergence of the method may be found in (Cottet, 1984).

Let us give a simple interpretation of the particle approximation of the diffusion term in the vortex method (4.13), (4.14). Setting $t_n = n\,\Delta t$, we consider the following time-stepping

$$\mathbf{x}_j^{n+1} = \mathbf{x}_j^n + \Delta t(\mathbf{u}_\infty(t_n) + \sum_{k \in J} \alpha_k\, \omega_k^n\, \mathbf{K}_\varepsilon(\mathbf{x}_j^n - \mathbf{x}_k^n)), \tag{4.16}$$

$$\omega_j^{n+1} = \omega_j^n + 2\nu\ \Delta t \sum_{k\in J} \alpha_k (\omega_k^n - \omega_j^n) \frac{D\zeta_\varepsilon(\underset{\sim}{x}_j^{n+1} - \underset{\sim}{x}_k^{n+1}).(\underset{\sim}{x}_k^{n+1} - \underset{\sim}{x}_j^{n+1})}{|\underset{\sim}{x}_k^{n+1} - \underset{\sim}{x}_j^{n+1}|^2} . \tag{4.17}$$

By choosing as in Remark 3.4

$$\zeta(\underset{\sim}{x}) = \frac{1}{2\pi} \exp\left(-\frac{1}{2}|\underset{\sim}{x}|^2\right)$$

the equation (4.17) becomes

$$\omega_j^{n+1} = \omega_j^n + \frac{\nu\ \Delta t}{\pi\varepsilon^4} \sum_{k\in J} \alpha_j (\omega_k^n - \omega_j^n) \exp\left(-\frac{|\underset{\sim}{x}_j^{n+1} - \underset{\sim}{x}_k^{n+1}|^2}{2\varepsilon^2}\right) .$$

Now, if we take

$$\varepsilon = \sqrt{2\nu\ \Delta t} , \tag{4.18}$$

we obtain

$$\begin{aligned} \omega_j^{n+1} = \frac{1}{4\pi\ \nu\ \Delta t} \sum_{k\in J} \alpha_k\ \omega_k^n \exp\left(-\frac{|\underset{\sim}{x}_j^{n+1} - \underset{\sim}{x}_k^{n+1}|^2}{4\nu\ \Delta t}\right) + \\ + \left(1 - \frac{1}{4\pi\ \nu\ \Delta t} \sum_{k\in J} \alpha_k \exp\left(-\frac{|\underset{\sim}{x}_j^{n+1} - \underset{\sim}{x}_k^{n+1}|^2}{4\nu\ \Delta t}\right)\right)\omega_j^n . \end{aligned} \tag{4.19}$$

The fully discrete vortex method (4.16), (4.18) may be viewed as a very natural splitting algorithm which can be described as follows. We start from the particle approximation

$$\sum_{j\in J} \alpha_j\ \omega_j^n\ \delta(\underset{\sim}{x} - \underset{\sim}{x}_j^n)$$

of $\omega(\cdot, t_n)$. Then:

(i) the vortex particles are moved during the time interval (t_n, t_{n+1}) according to the equations (4.16). This amounts to solving the incompressible Euler equations

$$\begin{cases} \dfrac{\partial \omega}{\partial t} + \nabla \cdot (\underset{\sim}{u}\, \omega) = 0 \\[2ex] \underset{\sim}{u}(\cdot,t) = \underset{\sim}{u}_{\infty}(t) + \underset{\sim}{K} * \omega(\cdot,t) \end{cases}$$

by using the vortex method and a forward Euler time-stepping. We obtain a new vortex dirstribution

$$\sum_{j \in J} \alpha_j \, \omega_j^n \, \delta(\underset{\sim}{x} - \underset{\sim}{x}_j^{n+1}).$$

(ii) We solve exactly the diffusion problem

$$\begin{cases} \dfrac{\partial \tilde{\omega}}{\partial t} - \nu \, \Delta \tilde{\omega} = 0 \,, \\[2ex] \tilde{\omega}(\underset{\sim}{x}, t_n) = \displaystyle\sum_{j \in J} \alpha_j \, \omega_j^n \, \delta(\underset{\sim}{x} - \underset{\sim}{x}_n^{n+1}) \end{cases}$$

on the interval (t_n , t_{n+1}). We find

$$\tilde{\omega}(\underset{\sim}{x}, t_{n+1}) = \frac{1}{4\pi \, \nu \, \Delta t} \sum_{k \in J} \alpha_k \, \omega_k^n \exp\left(- \frac{|\underset{\sim}{x} - \underset{\sim}{x}_k^{n+1}|^2}{4\nu \, \Delta t}\right).$$

(ii) we approximate the continuous function $\tilde{\omega}(\cdot, t_{n+1})$ by a particle distribution

$$\sum_{j \in J} \alpha_j \, \omega_j^{n+1} \, \delta(\underset{\sim}{x} - \underset{\sim}{x}_j^{n+1})$$

using the new positions $\underset{\sim}{x}_j^{n+1}$ of the vortex particles. Instead of choosing

$$\omega_j^{n+1} = \tilde{\omega}(\underset{\sim}{x}_j^{n+1}, t_{n+1}) = \frac{1}{4\pi \, \nu \, \Delta t} \sum_{k \in J} \alpha_k \, \omega_k^n \exp\left(- \frac{|\underset{\sim}{x}_j^{n+1} - \underset{\sim}{x}_k^{n+1}|^2}{4\nu \, \Delta t}\right),$$

we define ω_j^{n+1} by the formula (4.19) which includes a small correction term ensuring that the highly desirable conservation property

$$\sum_{j\in J} \alpha_j \, \omega_j^{n+1} = \sum_{j\in J} \alpha_j \, \omega_j^n$$

holds.

(iv) Time is advanced by Δt and one goes back to the step (i) of the procedure.

For the convergence of this splitting algorithm, we refer to (Cottet and Gallic, 1985b).

We now turn to the three-dimensional case. We set

$$\underset{\sim}{\omega} = \underset{\sim}{\nabla} \times \underset{\sim}{u} \tag{4.20}$$

and we assume again for simplicity that $\underset{\sim}{\nabla} \times \underset{\sim}{f} = \underset{\sim}{0}$. By taking the curl of Equation (4.1), we obtain the equivalent formulation of the Navier-Stokes problem (4.1) - (4.4):

$$\begin{cases} \dfrac{\partial}{\partial t} \underset{\sim}{\omega} + \nabla\cdot(\underset{\sim}{u}\,\underset{\sim}{\omega}) - (\underset{\sim}{\nabla}\,\underset{\sim}{u})\underset{\sim}{\omega} - \nu\,\Delta\underset{\sim}{\omega} = \underset{\sim}{0}\ , & \underset{\sim}{x} \in R^3\ , \quad t > 0 \\ \underset{\sim}{\omega}(\underset{\sim}{x},0) = \underset{\sim}{\omega}_0(\underset{\sim}{x})\ , \quad \underset{\sim}{\omega}_0 = \underset{\sim}{\nabla} \times \underset{\sim}{U}_0\ , & \end{cases} \tag{4.21}$$

$$\begin{cases} \underset{\sim}{\nabla} \times \underset{\sim}{u} = \underset{\sim}{\omega} & \\ & \underset{\sim}{x} \in R^3\ , \quad t > 0 \\ \underset{\sim}{\nabla}\cdot\underset{\sim}{u} = 0 & \\ \underset{\sim}{u}(\underset{\sim}{x},t) \to \underset{\sim}{u}_\infty(t)\ , & |\underset{\sim}{x}| \to \infty\ . \end{cases} \tag{4.22}$$

In (4.21), $\underset{\sim}{\nabla}\,\underset{\sim}{u}$ is the 3 a 3 matrix-valued function $\left(\dfrac{\partial u_i}{\partial x_j}\right)_{1\leq i,j\leq 3}$.

If we introduce the matrix-kernel

$$\underset{\approx}{K}(\underset{\sim}{x}) = \frac{1}{4\pi}\,\frac{\underset{\sim}{x}}{|\underset{\sim}{x}|^3}\,\underset{\sim}{x}\ ,$$

we obtain that the equations (4.22) may be replaced by

$$\underset{\sim}{u}(\cdot,t) = \underset{\sim}{u}_\infty(t) + \underset{\approx}{K} * \underset{\sim}{\omega}(\cdot,t). \tag{4.23}$$

The three-dimensional vortex method is based on the formulation (4.21), (4.23). Note again that (4.21) is a convection-diffusion problem that we have already considered in Section 3. The numerical method is then defined as follows: We look for an approximate vortex distribution of the form

$$\underset{\sim}{\omega}^h(\underset{\sim}{x},t) = \sum_{j\in J} \alpha_j \, \underset{\sim}{\omega}_j(t) \, \delta(\underset{\sim}{x} - \underset{\sim}{x}_j(t)). \tag{4.24}$$

Setting

$$\underset{\approx}{K}_\varepsilon = \underset{\approx}{K} * \zeta_\varepsilon \tag{4.25}$$

and

$$\underset{\sim}{u}_\varepsilon^h = \underset{\sim}{u}_\infty + \underset{\approx}{K}_\varepsilon * \underset{\sim}{\omega}^h = \underset{\sim}{u}_\infty + \sum_{j\in J} \alpha_j \, \underset{\approx}{K}_\varepsilon(\underset{\sim}{x} - \underset{\sim}{x}_j).\underset{\sim}{\omega}_j \, , \tag{4.26}$$

we define the vector-valued functions $t \to \underset{\sim}{x}_j(t)$ and $t \to \underset{\sim}{\omega}_j(t)$ to be solutions of the differential systems

$$\frac{d}{dt} \underset{\sim}{x}_j = \underset{\sim}{u}_\varepsilon^h(\underset{\sim}{x}_j, t) \tag{4.27}$$

and

$$\begin{cases} \dfrac{d}{dt} \underset{\sim}{\omega}_j - (\underset{\sim}{\nabla} \, \underset{\sim}{u}_h^\varepsilon)(\underset{\sim}{x}_j,t).\underset{\sim}{\omega}_j - \\ - 2\nu \displaystyle\sum_{k\in J} \alpha_k \frac{D\zeta_\varepsilon(\underset{\sim}{x}_j - \underset{\sim}{x}_k)\cdot(\underset{\sim}{x}_k - \underset{\sim}{x}_j)}{|\underset{\sim}{x}_k - \underset{\sim}{x}_j|^2} (\underset{\sim}{\omega}_k - \underset{\sim}{\omega}_j) = \underset{\sim}{0}. \end{cases}$$

The convergence of this three-dimensional vortex method has been recently established independently by Beale (1985) and Cottet (1985) again in the case $\nu = 0$.

5. APPLICATION TO COMPRESSIBLE FLUID DYNAMICS

The particle methodology can now be extended to the compressible gas dynamics equations

$$\left.\begin{aligned} &\frac{\partial\rho}{\partial t} + \underset{\sim}{\nabla}\cdot(\rho \, \underset{\sim}{u}) = 0 \\ &\frac{\partial}{\partial t}(\rho \, \underset{\sim}{u}) + \underset{\sim}{\nabla}\cdot(\rho \, \underset{\sim}{u} \, \underset{\sim}{u}) + \underset{\sim}{\nabla} p = 0 \qquad x \in R^d \, , \quad t > 0 \\ &\frac{\partial}{\partial t}(\rho \, e) + \underset{\sim}{\nabla}\cdot((\rho \, e + p)\underset{\sim}{u}) = 0 \, . \end{aligned}\right\} \tag{5.1}$$

In (5.1), $e = \varepsilon + \frac{1}{2}|\underset{\sim}{u}|^2$ where ε is the specific internal energy and e is the specific total energy. In order to close the equations (5.1), we need to add an equation of state of the form

$$p = p(\rho,\varepsilon). \tag{5.2}$$

A particle-in-cell method of solution of the equations (5.1), (5.2) has been introduced by Harlow (1964) for the numerical computations of compressible multifluid flows. In this method, particles are used in order to simulate the convection terms and a grid is used for computing the pressure forces acting on the particles. Recently, Gingold and Monaghan (1983) have proposed a pure particle scheme, the smoothed particle hydrodynamics (S.P.H.) method where the grid is eliminated and the pressure forces are easily put into our particle framework. We refer to (Ovadia and Raviart, 1985) for a fairly general presentation.

First setting

$$\underset{\sim}{\phi} = \begin{pmatrix} \rho \\ \rho\underset{\sim}{u} \\ \rho e \end{pmatrix} \tag{5.3}$$

and

$$\underset{\sim}{f}_i(\underset{\sim}{\phi}) = \begin{pmatrix} 0 \\ p\delta_{i1} \\ \vdots \\ p\delta_{id} \\ pu_i \end{pmatrix}, \quad 1 \leqslant i \leqslant d, \quad \underset{\approx}{f}(\underset{\sim}{\phi}) = (\underset{\sim}{f}_1(\underset{\sim}{\phi}),\ldots,\underset{\sim}{f}_d(\underset{\sim}{\phi})), \tag{5.4}$$

the gas dynamics equations (5.1) can be written in the form

$$\frac{\partial}{\partial t}\underset{\sim}{\phi} + \underset{\sim}{\nabla}\cdot(\underset{\sim}{u}\,\underset{\sim}{\phi}) + \underset{\sim}{\nabla}\cdot\underset{\approx}{f}(\underset{\sim}{\phi}) = \underset{\sim}{0} \tag{5.5}$$

with

$$\underset{\sim}{u} = \underset{\sim}{u}(\underset{\sim}{\phi}) = (\rho\underset{\sim}{u})/\rho. \tag{5.6}$$

Again, we look for an approximate particle solution of the form

$$\underset{\sim}{\phi}^h(\underset{\sim}{x},t) = \sum_{j\in J} a_j(t)\underset{\sim}{\phi}_j(t)\delta(\underset{\sim}{x} - \underset{\sim}{x}_j(t)). \tag{5.7}$$

The difficulty lies in the particle discretisation of the pressure term $\underset{\sim}{\nabla}\cdot\underset{\approx}{f}(\underset{\sim}{\phi})$. Given a smooth function $\psi : R^d \to R$, we observe that, under the hypotheses of Proposition 1, we have

$$\begin{cases} \sum_{k\in J} a_j(\psi(\underset{\sim}{x}_k) + \psi(\underset{\sim}{x}_j)) \dfrac{\partial\zeta_\varepsilon}{\partial x_i}(\underset{\sim}{x}_k - \underset{\sim}{x}_j) = \\ \quad = \dfrac{\partial\psi}{\partial x_i}(x_j) + O(\varepsilon^k + \dfrac{h^m}{\varepsilon^{m+1}}) . \end{cases} \tag{5.8}$$

Now, using (5.8), we define the functions $t \to \underset{\sim}{\phi}_j(t)$, $j \in J$, to be solutions of the differential system

$$\frac{d}{dt}(a_j \underset{\sim}{\phi}_j) + a_j \sum_{k\in J} a_k(\underset{\approx}{f}(\underset{\sim}{\phi}_k) + \underset{\approx}{f}(\underset{\sim}{\phi}_j))\cdot\underset{\sim}{\nabla}\zeta_\varepsilon(\underset{\sim}{x}_j - \underset{\sim}{x}_k) = \underset{\sim}{0},\ j \in J. \tag{5.9}$$

<u>Remark 5.1</u>. Note that the form of the particle approximation of the pressure terms leads to a conservative method. In fact, assuming that

$$\zeta(-\underset{\sim}{x}) = \zeta(\underset{\sim}{x}) \quad , \quad \underset{\sim}{x} \in R^d ,$$

we obtain

$$\sum_{j,k\in J} a_j a_k(\underset{\approx}{f}(\underset{\sim}{\phi}_k) + \underset{\approx}{f}(\underset{\sim}{\phi}_j))\cdot\underset{\sim}{\nabla}\zeta_\varepsilon(\underset{\sim}{x}_j - \underset{\sim}{x}_k) = \underset{\sim}{0}$$

and therefore by (5.9)

$$\frac{d}{dt}\Big(\sum_{j\in J} a_j \underset{\sim}{\phi}_j\Big) = \underset{\sim}{0}$$

which is precisely the required conservation property.

If we replace in (5.9) $\underset{\sim}{\phi}$ by its value (5.3) and if we set

$$(\rho \underset{\sim}{u})_j = \rho_j \underset{\sim}{u}_j \quad , \quad (\rho e)_j = \rho_j e_j \ ,$$

the system (5.9) gives

$$\begin{cases} \dfrac{d}{dt}(a_j \rho_j) = 0, \\[2ex] \dfrac{d}{dt}(a_j \rho_j \underset{\sim}{u}_j) + a_j \sum\limits_{k \in J} a_k (p_k + p_j) \underset{\sim}{\nabla} \zeta_\varepsilon (\underset{\sim}{x}_j - \underset{\sim}{x}_k) = \underset{\sim}{0}, \\[2ex] \dfrac{d}{dt}(a_j \rho_j e_j) + a_j \sum\limits_{k \in J} a_k (p_k \underset{\sim}{u}_k - p_j \underset{\sim}{u}_j) \cdot \underset{\sim}{\nabla} \zeta_\varepsilon (\underset{\sim}{x}_j - \underset{\sim}{x}_k) = 0. \end{cases} \tag{5.10}$$

The first equation (5.10) simply means that the mass of the particle j

$$m_j = a_j \rho_j \tag{5.11}$$

is independent of t. Then, the other equations (5.10) can be equivalently written in the form

$$\frac{d}{dt} \underset{\sim}{u}_j + \sum_{k \in J} \frac{m_k}{\rho_j \rho_k} (p_k + p_j) \underset{\sim}{\nabla} \zeta_\varepsilon (\underset{\sim}{x}_j - \underset{\sim}{x}_k) = \underset{\sim}{0}, \tag{5.12}$$

$$\frac{d}{dt} e_j + \sum_{k \in J} \frac{m_k}{\rho_j \rho_k} (p_k \underset{\sim}{u}_k + p_j \underset{\sim}{u}_j) \cdot \underset{\sim}{\nabla} \zeta_\varepsilon (\underset{\sim}{x}_j - \underset{\sim}{x}_k) = 0. \tag{5.13}$$

Hence each particle j has a mass m_j, a velocity u_j and a specific total energy e_j. Its motion is defined by

$$\frac{d}{dt} \underset{\sim}{x}_j = \underset{\sim}{u}_j \tag{5.14}$$

and it satisfies the equations (5.12), (5.13) together with

$$p_j = p(\rho_j, \varepsilon_j) \quad , \quad \varepsilon_j = e_j - \frac{1}{2} |\underset{\sim}{u}_j|^2. \tag{5.15}$$

It remains only to specify a_j or equivalently ρ_j; we take

$$\rho_j = \sum_{k \in J} m_k \, \zeta_\varepsilon(\underset{\sim}{x}_j - \underset{\sim}{x}_k). \tag{5.16}$$

Remark 5.2. One can also derive slightly different particle approximations of the momentum equations and the energy equation which can be easier to use in practical computations. For instance, one can replace (5.13) by

$$\frac{d}{dt} e_j + \sum_{k \in J} \frac{m_k}{\rho_j \, \rho_k} (p_k \underset{\sim}{u}_j + p_j \underset{\sim}{u}_k) \cdot \underset{\sim}{\nabla} \zeta_\varepsilon (\underset{\sim}{x}_j - \underset{\sim}{x}_k) = 0.$$

Together with (5.12), this gives a very simple equation for the internal energy

$$\frac{d}{dt} \varepsilon_j + \frac{p_j}{\rho_j} \sum_{k \in J} \frac{m_k}{\rho_k} (\underset{\sim}{u}_k - \underset{\sim}{u}_j) \cdot \underset{\sim}{\nabla} \zeta_\varepsilon (\underset{\sim}{x}_j - \underset{\sim}{x}_k) = 0.$$

The numerical method is completed by introducing a leap-frog time-stepping and an artificial viscosity technique to avoid nonlinear instabilities. For details, see again (Gingold and Monaghan, 1983) and (Ovadia and Raviart, 1985). Numerical experiments show that the method is robust and accuract for a reasonable number of particles.

Let us point out that putting more physics in the numerical model is generally an easy matter with the particle method. As an example, following (Bourgeade, 1985), we consider a combustion problem where, for simplicity, we neglect heat conduction. This amounts to add to the equations (5.1) a chemical reaction equation of the form

$$\frac{\partial}{\partial t} (\rho Z) + \underset{\sim}{\nabla} \cdot (\rho Z \underset{\sim}{u}) = \rho r , \tag{5.17}$$

where Z is the mass fraction of unburnt gas and $r = r(\rho, p, Z)$ is the reaction rate. In order to close the system of equations (5.1), (5.17), we usually need an equation of state of the form

$$\varepsilon = \varepsilon(\rho, p, Z).$$

In practice, such an equation of state is frequently unknown. In fact, the knowledge of the equations of state of the unburnt gas and the burnt gas alone enable us to use the particle method.

We consider the fluid as a mixture of unburnt ($Z_j = 1$) and burnt particles ($Z_j = 0$) and we set

$$Z^h(\underset{\sim}{x},t) = \sum_{j\in J} \frac{m_j}{\rho_j} Z_j \, \zeta_\varepsilon(\underset{\sim}{x} - \underset{\sim}{x}_j(t)). \tag{5.18}$$

Next, we use a random method for describing the combustion process. At time t_n, the probability for an unburnt particle j to burn between t_n and t_{n+1} is approximated by

$$\Delta t \; r(\rho_j^n, p_j^n, Z^h(x_j^n, t_n)).$$

Using this "probability law", we decide at random if the particle j burns, we add to its internal energy the energy liberated by the combustion process. We thus obtain a numerical method which is easy to implement.

6. REFERENCES

Beale, J.T., (1985). A convergent 3-D vortex method with grid-free stretching. Preprint.

Beale, J.T. and Majda, A., (1981). Rate of convergence for viscous splitting of the Navier-Stokes equations. *Math. Comp.*, **37**, 243-259.

Beale, J.T. and Majda, A., (1982a). Vortex methods I: Convergence in three dimensions. *Math. Comp.*, **39**, 1-27.

Beale, J.T. and Majda, A., (1982b). Vortex methods II: Higher order accuracy in two and three dimensions. *Math. Comp.*, **39**, 29-52.

Bourgeade, A., (1985). Particle approximation of denotation wave. Preprint.

Chorin, A.J., (1973). Numerical study of slightly viscous flows. *J. Fluid Mech.*, **57**, 785-796.

Cottet, G.H., (1982). Méthodes particulaires pour l'équation d'Euler dans le plan. Thèse de 3ème cycle, Université Pierre et Marie Curie, Paris.

Cottet, G.H., (1984). A vortex in cell method for the two-dimensional Euler equations. Rapport interne n° 108, Centre de Mathématiques Appliquées, Ecole Polytechnique.

Cottet, G.H., (1985). On the convergence of the vortex method in two and three dimensions. (In preparation).

Cottet, G.H. and Gallic, S., (1985a). A particle method to solve transport-diffusion equations I: The linear case. Rapport interne n° 115, Centre de Mathematiques Appliquées, Ecole Polytechnique.

Cottet, G.H. and Gallic, S., (1985b). A particle method to solve transport-diffusion equations II: The Navier-Stokes system. (In preparation).

Gallic, S. and Raviart, P.A., (1985a). A particle method for first order symmetric systems. (to appear).

Gallic, S. and Raviart, P.A., (1985b). Particle approximation of convection-diffusion problems. (to appear).

Gingold R.A. and Monaghan, J.J., (1983). Shock simulation by the particle method S.P.H., *J. Comput. Phys.*, **52**, 374-389.

Harlow, F.H., (1964). The particle in cell computing method for fluid dynamics in Methods in Computational Physics (B. Alder, S. Fernbach, M. Rotenberg ed.) vol. 3, Academic Press, New York.

Leonard, A., (1980). Vortex methods for flow simulations. *J. Comput. Phys.*, **37**, 283-335.

Ovadia, J. and Raviart, P.A., (1985). Particle simulation of nonlinear systems of conservation laws. (In preparation).

Raviart, P.A., (1983). An analysis of particle methods. CIME Course, Como, Italy.

COMPRESSIBLE LAGRANGIAN HYDRODYNAMICS WITHOUT LAGRANGIAN CELLS

R.A. Clark
(Computational Physics Group X-7, Los Alamos National Laboratory)

1. INTRODUCTION

Traditional Lagrangian hydrodynamic codes for time dependent, compressible, multimaterial problems in two dimensions use the same general method. A Lagrangian mesh is defined which moves with the fluid and this mesh defines a set of Lagrangian cells. The mass in each cell remains fixed and the motion of the mesh determines the volume and hence the density of each cell. These methods work well until the mesh becomes distorted due to shear or turbulence. Large distortions cause computer codes to quickly grind to a halt.

The usual solution to distortion is to "rezone" the mesh. Here we move the mesh points artificially so as to reduce distortions and then map the quantities from the old to the new. This results in unwanted diffusion of mass, momentum and energy throughout the mesh. Even with rezoning, few Lagrangian codes can handle more than limited distortions. Recently, what we call "Free-Lagrangian" codes have been developed specifically to handle large distortions. These codes, in addition to adjusting the mesh points, can reconnect mesh points, thus creating new cells. While Free-Lagrangian codes can handle virtually any distortion, they are even more diffusive than rezoners.

We are trying a different approach to the problem. We abandon the idea of Lagrangian cells entirely. In the next section we will discuss how the conservation equations can be solved directly without resorting to Lagrangian cells. Next we will give some examples of calculations using this method. Finally, we will give details of the calculational method presently being used.

2. SOLVING THE CONSERVATION EQUATIONS

The equations we are trying to solve can be written

$$\frac{D}{Dt}\rho = -\rho\,\vec{\nabla}\cdot\vec{U} \qquad (2.1)$$

$$\frac{D}{Dt}\vec{U} = -\frac{1}{\rho}\vec{\nabla}P \qquad (2.2)$$

$$\frac{De}{Dt} = -\frac{P}{\rho}\vec{\nabla}\cdot\vec{U} \qquad (2.3)$$

$$P = P(\rho, e) \qquad (2.4)$$

where $\vec{U}$ represents the vector velocity, ρ the density, e the specific internal energy and P the pressure of the fluid. Equation (2.1) expresses conservation of mass, (2.2) conservation of momentum and (2.3) conservation of energy. The Lagrangian time derivative, i.e., the derivative following the fluid, is indicated by $\frac{D}{Dt}$.

In a standard Lagrangian calculation only equation (2.2), the momentum equation is solved directly. The procedure is to integrate (2.2) over some region of space to arrive at the acceleration of each mesh point. The mesh points are then moved and the new cell volumes along with the fixed cell mass determine the new density, hence indirectly solving equation (2.1). The associated PdV work term updates the cell energy and indirectly solves equation (2.3) and the new pressure is obtained from the equation of state (2.4).

We propose the following: Instead of Lagrangian cells, we think of a set of Lagrangian points which are embedded in and move with the fluid. There is no mass associated with these points. They are just moving tracer points at which we will attempt to keep track of the velocity, density, energy, and pressure of the fluid. In our later example calculations we will show point positions at various times in the calculation. At each of these points, we know the density, energy and velocity of the fluid, but we do not associate any particular mass with the point.

Looking now at equation (2.1), we note that to approximate the time integral of the density change from time t to time t + δt we need an approximation to $\vec{\nabla}\cdot\vec{U}$ at that point. To solve equation (2.2) we need an approximation for $\vec{\nabla}P$, and for (2.3) we again need $\vec{\nabla}\cdot\vec{U}$. To obtain these we select a set of "representative" neighbours. We then make a finite difference approximation to $\vec{\nabla}P$ and $\vec{\nabla} \cdot \vec{U}$ using these neighbours, and update ρ, $\vec{U}$ and e at each point. Each point is then moved the distance $\vec{U}\ \delta t$ and one time step is completed.

At the next time step the selection of a set of "representative" neighbours may change, but this does not require any sort of re-mapping of variables. It only means that a different set of points will be used in the next finite difference approximation to $\vec{\nabla} \cdot \vec{U}$ and $\vec{\nabla}P$. Large distortions in the flow will produce frequent changes in neighbour selection, but since there are no cells to distort and no re-mapping to be done the calculation proceeds from cycle to cycle with no difficulty.

3. SOME EXAMPLE CALCULATIONS

3.1 Spherical example

Here we will give three examples of calculations performed by the code HOBO using the free Lagrangian method described herein. In the first test problem, the initial condition is a sphere of perfect gas with a gamma of 5/3. The gas is divided into four regions as seen in Fig. 3.1. Pressures are in megabars, density in gm/cc and dimension in cm.

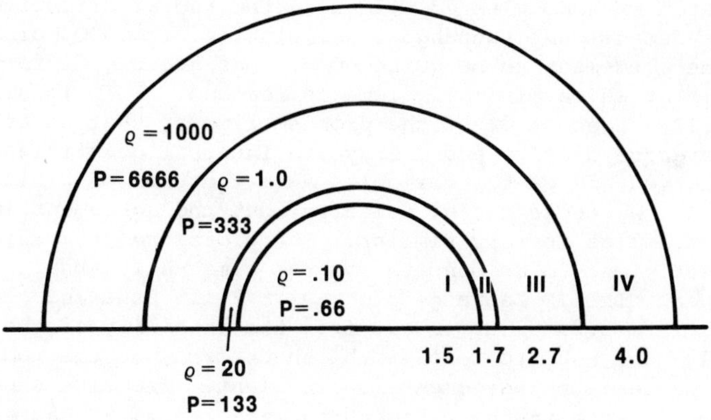

Fig. 3.1

The high pressure in region IV will drive a spherical implosion which will greatly compress region III, II, and particularly I. There are two challenges to this problem, the first is to maintain a spherical ball while running the calculation in cylindrical (r,Z) geometry. Six snapshots of region II are shown in Fig. 3.2. Region I is interior to region II. The minimum volume of region I occurs in the fifth snapshot after which region I begins to expand. We ran 1000 calculational cycles with 73 points in the radial direction and 64 points covering 180° of angle. The left half of the snapshot is a reflection of the right half which was calculated.

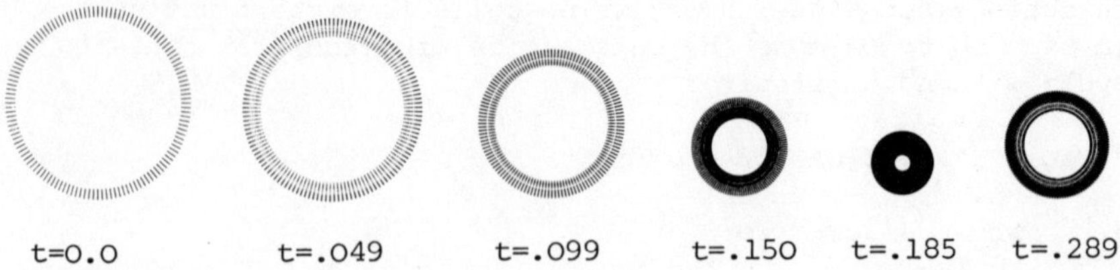

Fig. 3.2

The second challenge is the accuracy of the solution. For comparison purposes we ran a standard one-dimension Lagrangian code using 800 zones, 100 zones in each region. In Figs. 3.3a, b, c, and d we have plotted the average density and average specific internal energy in regions I and II as calculated by HOBO with 73 points in the radial direction and the one-dimensional Lagrangian calculation with 800 points. We feel the agreement to be quite good. One notable difference is the time at which minimum volume is reached. HOBO is slow by about .0075 μsec or 4% of the problem time at that point. Since average density and energy are integral quantities we have plotted one of the variables as a function of radius in Fig. 3.4. We chose radial velocity, but the agreement in all other variables is very similar. The plots are from slightly different times to compensate for the time shift just mentioned. The 1D Lagrange plot is at 2.125 μsec and the HOBO plot is from 2.25 μsec. Apart from the inability of the more coarsely zone HOBO to resolve the shock front at the radius 1.2 cm we feel the agreement is excellent. The time chosen for the plot is late in the calculation when region II has expanded almost back to its original volume.

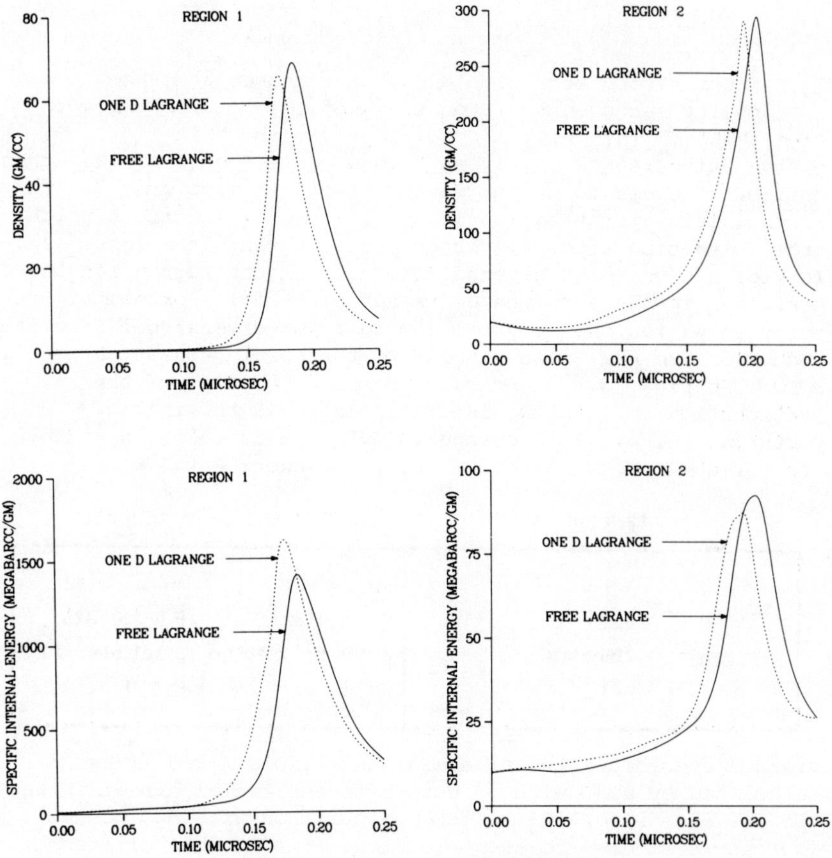

Figs. 3.3 a, b, c and d

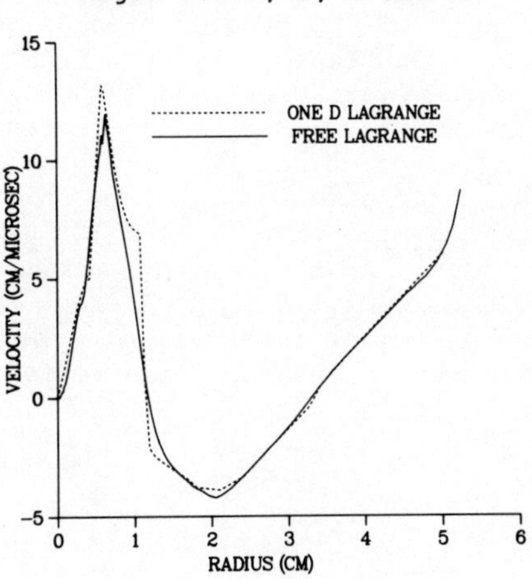

Fig. 3.4

3.2 Meshkov Instability

For our second test problem we have chosen a Meshkov
instability based on the geometry used in one of Meshkov's
experiments. The initial conditions are shown in Fig. 3.5. A
piston driven shock is driven through a region of air and then
helium. The air to helium density ratio is just over 7. There
is an initial perturbation in the air -- He interface which
grows with time after the shock passes through the interface.
In Fig. 3.6 we plot several snapshots of the Lagrangian point
positions in the air (the He is not plotted). For comparison
purposes we ran the same problem in a two-dimensional Eulerian
code with the cell size similar to the point separation used in
HOBO. In Figs. 3.7 a and b, we compare the size of the
perturbation as it grows in time. In 3.7 a the initial
perturbation, δ, is .2 cm and in 3.7.b it is .4 cm in width.
The agreement between the two codes is excellent.

33.8 cm	12.675 cm	16.9 cm
air $\gamma = 1.4$ $P = 1.8289$ $\varrho = 1.795 \times 10^{-3}$ $U = 15.31$	air $\gamma = 1.4$ $P = 1.01325$ $\varrho = 1.184 \times 10^{-3}$ $U = 0$	He $\gamma = 1.63$ $P = 1.01325$ $\varrho = 1.664 \times 10^{-4}$ $U = 0$

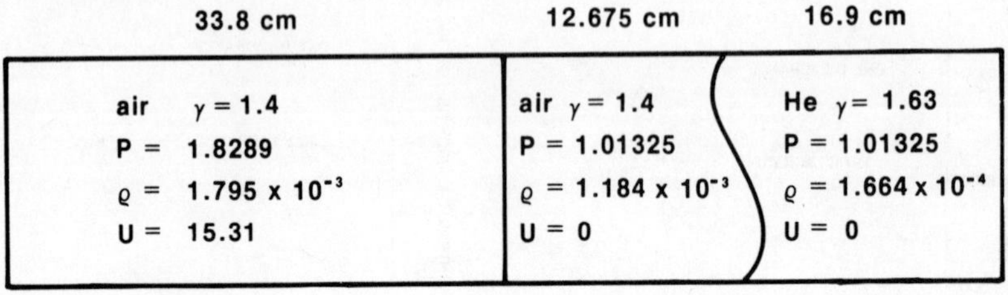

Fig. 3.5

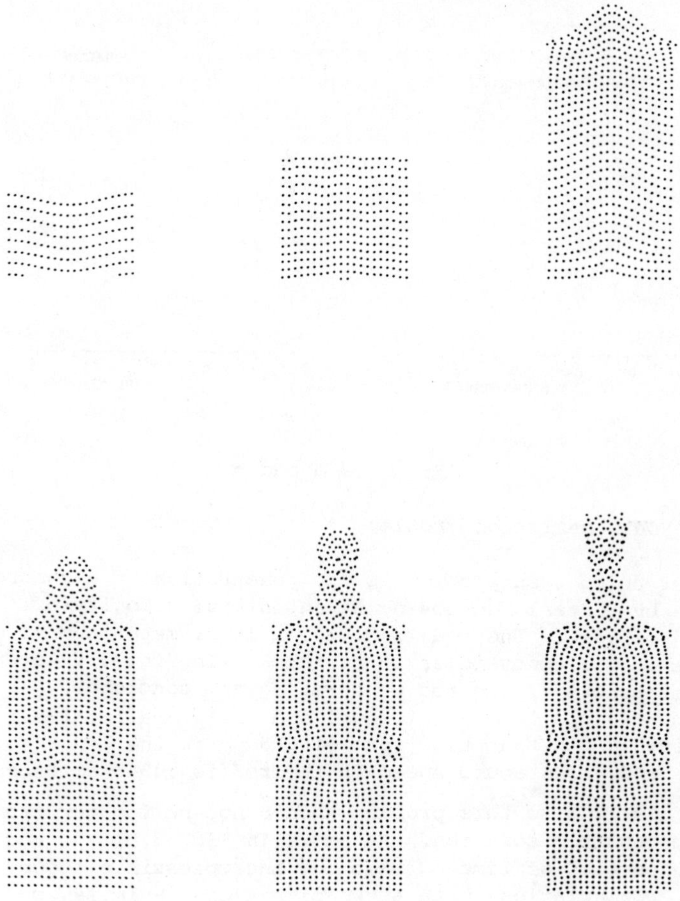

Fig. 3.6

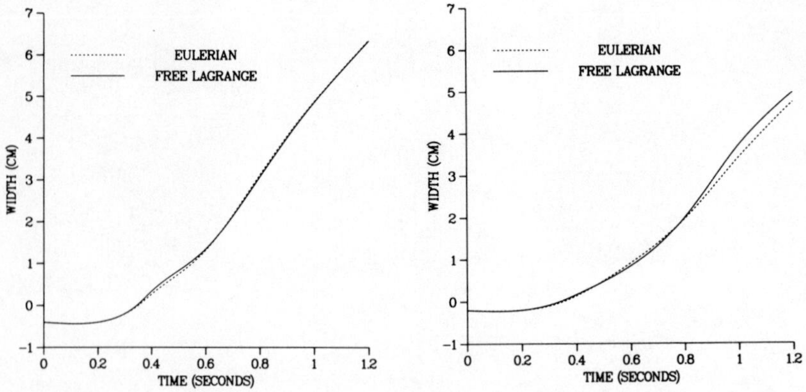

(a) (b)

Fig. 3.7 a and b

3.3 Rod Penetration Problem

Our third test problem is the penetration of a concrete plate by a steel rod moving at an initial velocity of .2134 cm/μsec. The rod is 9.066 cm in diameter and 45 cm in length. The concrete is 50 cm thick. In Fig. 3.8 we show six snapshots of the rod penetrating the concrete. Incompressible theory (Birkhoff and Farantonello, 1957) predicts a constant time rate of change in the length of the steel rod. The sound speed in the rod is .4545 cm/μsec and $(v/c)^2 = .22$, so this problem should not be too far from the incompressible solution. As shown in Fig. 3.9, the rod length as a function of time matches the incompressible theory very well. Calculations with a two-dimensional Eulerian code produced an almost identical result.

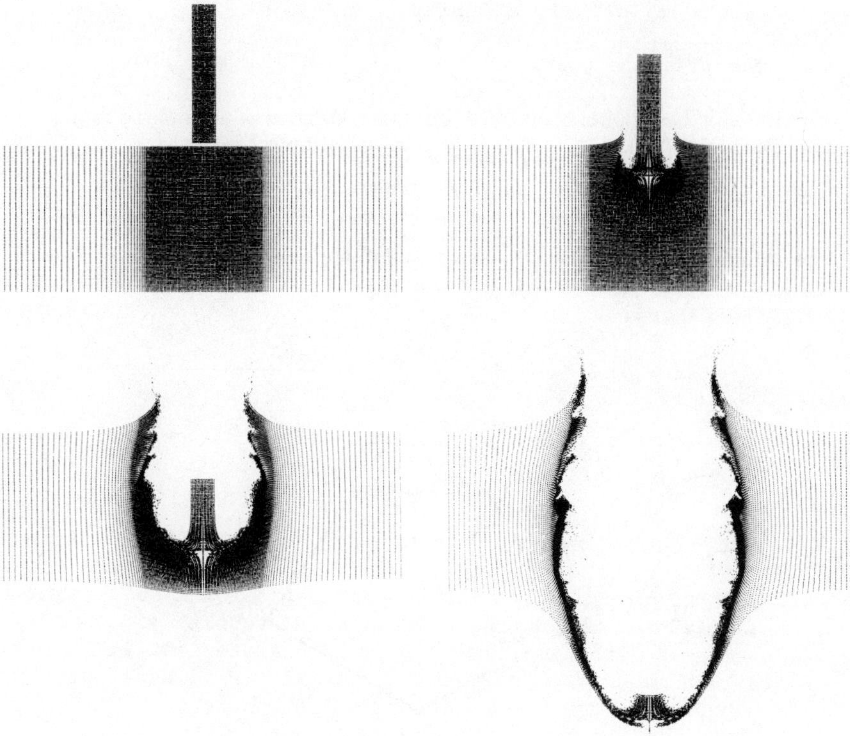

Fig. 3.8

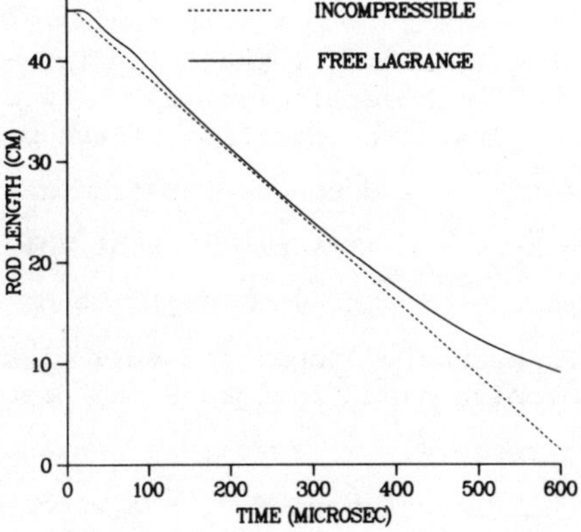
INCOMPRESSIBLE
FREE LAGRANGE
ROD LENGTH (CM)
TIME (MICROSEC)
0
10
20
30
40
50
0
100
200
300
400
500
600

Fig. 3.9

4. THE FINITE DIFFERENCE SCHEME

4.1 The pressure gradient

We want to approximate $\vec{\nabla}P$ at the point k where neighbours are the points $k_1, k_2 \ldots k_{nmax}$. Our neighbour selection guarantees at least three neighbours for each point, the average is six and there is no maximum number. Clearly there are many methods that could be used to approximate $\vec{\nabla}P$. The following was arrived at through trial and error and appears to work very well.

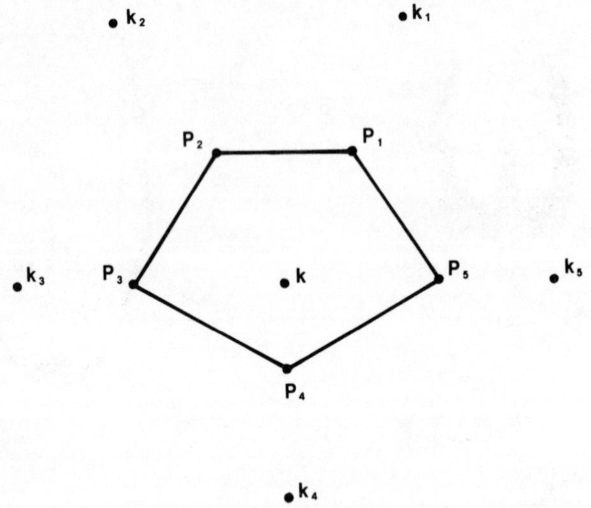

Fig. 4.1

Consider point k in Fig. 4.1 that has five neighbours. We construct a polygon with vertices midway between the point and each of its neighbours. The position of the nth vertex is $\vec{x}_n = 1/2\ [\vec{x}(k) + \vec{x}(k_n)]$ and the vector from $\vec{x}_k$ to $\vec{x}_n$ is denoted by $\delta\vec{x}_n = \vec{x}_n - \vec{x}_k$. The pressure at the nth vertex, P_n, is a weighted average of P(k) and $P(k_n)$ (to be described in section 4.3). We assume a linear pressure distribution along each edge of the polygon and integrate the pressure over the surface to get a force $\vec{F}$. We assume a constant density ρ_k over the polygon to calculate a mass M. Then we have $\frac{D}{Dt}\vec{U} = \frac{\vec{F}}{M}$. Now let $\vec{x}'_n = \vec{x}_R + \varepsilon\partial\vec{x}_n$ and the pressure at the new vertex is $P'_n = P_k + \varepsilon(P_n - P_k)$. Now $\vec{F}$ and M are functions of ε and we compute

$$\lim_{\varepsilon \to 0} \frac{\vec{F}(\varepsilon)}{M(\varepsilon)} \; .$$

The resulting expression for the pressure gradient is

$$\vec{\nabla}P_k = \frac{\hat{x}\sum_n P_n\,(\delta y_{n-1} - \delta y_{n+1}) + \hat{y}\sum_n P_n\,(\delta x_{n+1} - \delta x_{n-1})}{\sum_n (\delta x_{n+1}\,\delta y_n - \delta y_{n+1}\,\delta x_n)} \qquad (4.1.1)$$

where $\hat{x}$ and $\hat{y}$ are respectively the unit vectors in the x and y directions and $\vec{\delta x}_n = \delta x_n\,\hat{x} + \delta y_n\,\hat{y}$.

If the preceding is done in cylindrical geometry, the result is identical for $\vec{\nabla}P$ with x and y replaced by r and z. It is of interest to note that if the $\lim_{\varepsilon \to 0}$ is not taken the result does not give a spherically symmetric pressure gradient in a spherically symmetric problem using cylindrical coordinates.

There is an easier way to arrive at equation (4.1), although the method just described is how we originally derived it. Since it takes only three points to describe a plane surface, each consecutive pair of neighbours along with the point k defines a pressure plane to first order. If we assign a weight to each of these approximations we have an approximation for $\vec{\nabla}P$. If the weighting function is the area of the triangle formed by the three points, the result is the same as Eq. (4.1). We have tried other weighting functions, Θ and $\sin\Theta$ where Θ is the angle between $\vec{\delta x}_n$ and $\vec{\delta x}_{n+1}$ and both work fairly well, but area weighting appears to be best at this time.

4.2 The divergence of the velocity field

In cartesian coordinates we represent the velocity at the point k by $\vec{U}_k = u_k\,\hat{x} + v_k\,\hat{y}$. The divergence of the velocity field can be expressed as $\vec{\nabla} \cdot \vec{U} = \frac{1}{V}\frac{\partial V}{\partial t}$ where V is the specific volume of the fluid. Referring back to Fig. 4.1 the specific volume of the constructed polygon is proportional to the area of the polygon given by

$$A = 1/2\ \Sigma_n\ (x_{n+1} + x_n)\ (y_{n+1} - y_n)\ .$$

Hence we can write

$$\vec{\nabla}\ .\ \vec{U} = \frac{1}{A}\frac{\partial A}{\partial t}\ \frac{\Sigma_n (u_{n+1} + u_n)\ (y_{n+1} - y_n) + (x_{n+1} + x_n)(v_{n+1} - v_n)}{\Sigma_n\ x_{n+1}\ y_n - y_{n+1}\ x_n}\ . \qquad (4.2.1)$$

Equation (4.2.1) can be derived directly from equation (4.1.1) by noting that (4.1.1) implies a definition for the operators $\frac{\partial}{\partial x}$ and $\frac{\partial}{\partial y}$ and when these are applied to $\vec{\nabla}\ .\ \vec{U} = \frac{\partial u}{\partial x} + \frac{\partial v}{\partial y}$ equation 4.2.1 is obtained. Thus we have in effect three ways of deriving the same finite difference approximation to the operators $\frac{\partial}{\partial x}$ and $\frac{\partial}{\partial y}$. In cylindrical coordinates we express the divergence of the velocity field as

$$\vec{\nabla}\ .\ \vec{U} = \frac{1}{r}\frac{\partial}{\partial r}\ (ru) + \frac{\partial v}{\partial z} = \frac{u}{r} + \frac{\partial u}{\partial r} + \frac{\partial v}{\partial z}$$

where $\frac{\partial u}{\partial r} + \frac{\partial v}{\partial z}$ is calculated by equation (4.2.1) with x,y replaced by r,z.

4.3 The midpoint pressure and velocity

In equation 4.1.1 we use a pressure P_n which is midway between points k and k_n. This is not a numerical average. Consider the one-dimensional problem depicted in Fig. 4.2.a.

What pressure should we use for $P_i^+ = P_{i+1}^-$? If we use the average, 1/2 $(P_1 + P_2)$ the acceleration at i+1 will be much greater than at i. However, we know that the velocity should be continuous across the discontinuity. Given equal zoning the boundary pressure which gives equal acceleration to points i and i+1 is $P_i^+ = (P_i\ \rho_{i+1} + P_{i+1}\rho_i)/(\rho_i + \rho_{i+1})$.

It can be shown that the resulting finite difference approximation $P_x = (P_i^+ - P_i^-)/\delta x$ is second order accurate when the density is continuous.

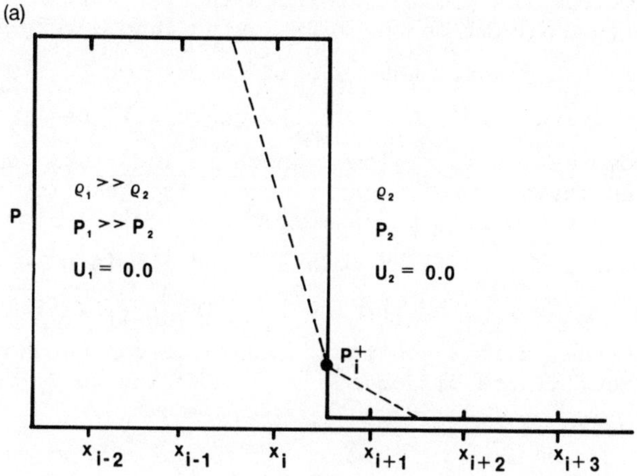

$$P_i^+ = \frac{\varrho_{i+1} P_i + \varrho_i P_{i+1}}{\varrho_i + \varrho_{i+1}}$$

Fig. 4.2.a

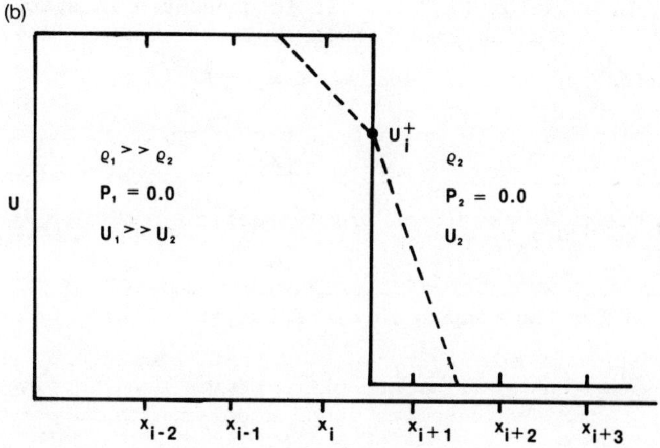

$$U_i^+ = \frac{(\varrho c^2)_i U_i + (\varrho c^2)_{i+1} U_{i+1}}{(\varrho c^2)_i + (\varrho c^2)_{i+1}}$$

Fig. 4.2.b

Now consider the problem depicted in Fig. 4.2.b. Here we have a heavy material on the left moving into a very light material on the right. What should we use for $U_i^+ = U_{i+1}^-$? If we use the average $1/2\ (U_i + U_{i+1})$, there will be a very large rate of compression in region 2 which is incorrect because region 1 is moving into a near vaccum. The quantity that should be continuous is pressure. The velocity which causes equal pressure increases at points i and i+1 is $U^+ =$ $[(\rho c^2)_i\, u_i + (\rho c^2)_{i+1}\, u_{i+1}]/[(\rho c^2)_i + (\rho c^2)_{i+1}]$. This assumes the sound speed c is a constant. Again it can be shown that the resultant finite difference approximation to U_x is second order accurate if ρc^2 is continuous.

The midpoint pressures used in equation (4.1.1) are inverse density weighted and the midpoint velocities in equation (4.2.1) are ρc^2 weighted.

4.4 The artificial viscosity

An artificial visocsity q is added to the midpoint pressure in equation (4.1.1). It is quadratic in space form. Let U_C be the closing rate between points k and k_n, i.e.

$$U_C = (\vec{U}_k - \vec{U}_{k_n}) \cdot \frac{(\vec{X}_r - \vec{X}_{k_n})}{|\vec{X}_k - \vec{X}_{k_n}|} .$$

Then let $q_k = a^2 \rho_k U_C^2$ and $q_{k_n} = a^2 \rho_{k_n} U_C^2$. In the spirit of paragraph 4.3, we inverse density weight the two to get our expression for the midpoint q, i.e.

$$q_n = 2a^2\, U_C^{\ 2} \ /\ (1/\rho_k + 1/\rho_{k_n}) . \tag{4.4.1}$$

In all of our example calculations in section 2 we used $a^2 = 5.76$. Now we must fold q into the internal energy equation in which we need to evaluate $(P + q)\vec{\nabla}.\vec{U}$. Our approximation for $\vec{\nabla}.\vec{U}$ is given by equation (4.2.1). The q term is brought inside the summation so that

$$(\rho+q)\vec{\nabla}.\vec{U} = \frac{\sum_n (P_k + q_n)u_n(y_{n-1} - y_{n+1}) + \sum_n (P_k + q_n)v_n(x_{n+1} - x_{n-1})}{\sum_n x_{n+1}\, y_n - y_{n+1}\, x_n} \tag{4.4.2}$$

4.5 Prevention of density striations

The method so far described has one remaining difficulty. By having all of the variables centred in space it becomes impossible to detect a sawtooth type wave as depicted in one dimension in Fig. 4.3.

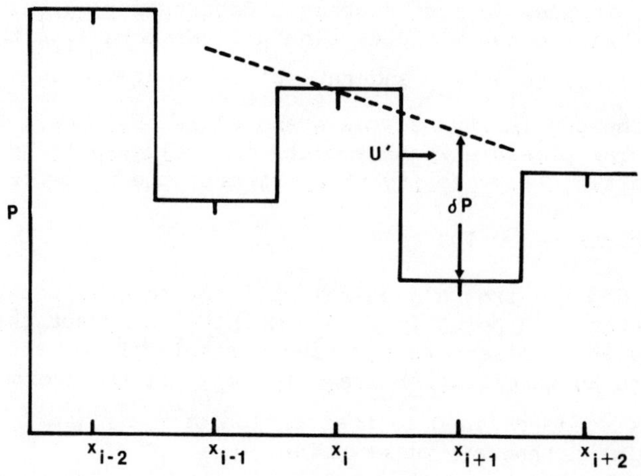

Fig. 4.3

If such a wave develops it cannot be detected by a centred difference scheme. To correct for this, we define an artificial velocity u' as depicted in Fig. 4.3. We use our calculated $\vec{\nabla}$ P to extrapolate from point k to point k_n giving $P^{ext}_{k_n} = P_k + (\vec{X}_{k_n} - \vec{X}_k)\ .\ \vec{\nabla} P_k$. If the pressure field is linear then $P^{ext}_{k_n} = P_{k_n}$. If they are not equal, there is a second derivative in the pressure field which we attempt to reduce. Physically what should happen is a velocity would be produced at the midpoint as indicated in 4.3., which would be produced point i and compress point i+l. This velocity must be proportioned to $\delta P = P^{ext}_{k_n} - P_{k_n}$. We chose to use $u' = b^2\delta P/\rho c$.

We then use ρc^2 weighting between points k and k_n to arrive at

$$u'_n = \frac{b^2 \, \delta P \, (c_k + c_{k_n})}{\rho_k \, c_k^2 + \rho_{k_n} \, c_{k_n}^2} \, . \tag{4.4}$$

u'_n is added to u_n in calculating $\vec{\nabla}.\vec{U}$.

In our present calculations $b^2 = 1.44$. We further limit $|u'_n|$ to be less than 20% of the maximum of (c_k, c_{k_n}). In practice u' is a very small term but an absolutely necessary one. For example, in test problem 1 density striations of around 50% will occur without using u'. We note also that δP is proportional to $\delta x^2 P_{xx}$ and thus u' is quadratic in nature. The similarity between q and u' is striking. The q is an artificial pressure which smooths the velocity field while u' is an artificial velocity which smooths the pressure field.

4.6 Neighbour selection

The method requires a good selection of representative neighbours at each point in time. We have found out that the neighbours whose bisectors form the Voronoi polygon around the point k are an excellent choice. The k_{th} Voronoi (Voronoi, 1908) polygon is defined as that region of space which is nearer point k than any other point.

5. SUMMARY

The partial differential equations (2.1), (2.2) and (2.3) along with the equation of state (2.4), which describe the time evolution of compressible fluid flow can be solved without use of a Lagrangian mesh. The method follows embedded fluid points and uses finite difference approximations to $\vec{\nabla}P$ and $\vec{\nabla} \cdot \vec{u}$ to update ρ, $\vec{u}$ and e. We have demonstrated that the method can accurately calculate highly distorted flows without difficulty. The finite difference approximations are not unique and improvements may be found in the near future. The neighbour selection is not unique, but the one being used at present appears to do an excellent job. The method could be directly extended to three dimensions. One drawback to the method is its failure to explicitly conserve mass, momentum and energy. In fact, at any given time the mass is not defined. We must perform an auxiliary calculation by integrating the density field over space to obtain mass, energy and momentum.

However, in all cases where we have done this, we have found the drift in these quantities to be no more than a few present.

6. REFERENCES

Birkhoff, G. and Farantonello, E.H., (1957) Jets, Wakes and Cavities, Academic Press, New York and London.

Voronoi, G.J., (1908) *Journal Reine Angew. Math,* **134**, 198.

APPLICATION OF DISCRETE VORTEX METHODS TO THE COMPUTATION OF SEPARATED FLOWS

J.M.R. Graham
(Department of Aeronautics, Imperial College, London)

1. INTRODUCTION

In its simplest form the discrete vortex method uses sequences of discrete vortices to represent a cross-section of a two-dimensional vortex sheet which in turn may be viewed as the infinite Reynolds number limit of a free shear layer. The velocity field of these inviscid "point" vortices is used to convect the vortices and hence the evolution of the sheet is calculated by a step by step forward time integration. A great deal has been published on this subject and the present paper will concentrate on only one aspect, the modelling of separated low speed flows around two-dimensional bodies. In this case the shed vortex sheets are usually assumed to be two-dimensional also, although in many cases that is only true in the mean. A comprehensive review of the simulation of three-dimensional vortex flows has been given recently by Leonard (1985). These simulations have been mainly concerned with isolated free vorticity (eg. the evolution of vortex rings or transitional boundary-layer disturbances where major wall separation is not involved). Three-dimensional vortex sheet separation from bluff bodies has been studied much less than two-dimensional because of the greatly increased computing requirement. Rehbach's (1978) work and Billet's (1980) calculation of the flow over a swept and tapered step are examples.

Application of the discrete vortex method to the simulation of flow around two-dimensional bluff bodies was largely initiated by Clements' (1973) modelling of separated flow past a semi-infinite bluff body and Gerrard's (1967) more approximate model for the wake of a circular cylinder. In addition to open separations where two shear layers on either side of a wake interact, the discrete vortex method has been used to study closed separations in which a separated shear layer reattaches to the body surface (eg. Kiya et al. (1982), Ashurst (1979a)). Simulation of this type of flow has much in common with the former but will be considered in less detail in this paper.

In the case of two-dimensional (planar) low speed flows the vorticity convection equation:

$$\frac{\partial \omega}{\partial t} + U_j \frac{\partial \omega}{\partial x_j} = \nu \frac{\partial^2 \omega}{\partial x_j \partial x_j} \tag{1.1}$$

shows that the vorticity $\omega(=\nabla \wedge U)$ is convected with the fluid particles. In an inviscid flow (ν=0) the vorticity of each fluid particle therefore remains unchanged with time, and for an irrotational free stream can only be generated by pressure gradients acting at the boundaries of the fluid. Whether the flow is treated as inviscid or viscous, vortex methods replace the convective left-hand side of equation (1.1) by the Lagrangian equivalent:

$$\frac{\partial z(\omega)}{\partial t} = U(z) \tag{1.2}$$

where z is the position vector, (x+iy) in complex notation, of a given particle having vorticity ω; U is the corresponding (complex) velocity. In a two-dimensional flow the vorticity is related to the stream-function by the Poisson equation

$$\nabla^2 \psi = - \omega. \tag{1.3}$$

2. REPRESENTATION OF VORTEX SHEETS

In the absence of three-dimensional disturbances and viscous diffusion all the vorticity in the fluid lies on infinitesimally thin sheets. Some parts of these sheets may roll up forming core regions containing spirals of infinite length. But, these regions apart, the Biot-Savart integral solution of equation (1.3) can be evaluated over sheets ΣS_j of finite length so that the velocity field may be written:

$$U(z') = \frac{i}{2\pi} \int_{\Sigma S_j} \frac{d\Gamma(z)}{z'^* - z^*} \tag{2.1}$$

where $\Gamma(z)$ is the circulation of a sheet from its end up to the point z, and * denotes complex conjugate.

The evolution of the vortex sheets is obtained by numerical integration of equations (1.2) and (2.1). The simplest formula uses first order Euler integration of 1.2:

$$z'_j = z_j + U(z_j)\delta t \tag{2.2}$$

$$\text{with } U(z_j) = \frac{i}{2\pi} \sum_{\substack{k=1 \\ k\neq j}}^{N} \frac{\Delta\Gamma_k}{z_j^* - z_k^*} \tag{2.3}$$

z'_j is the new value of z_j after a time step δt. The z_j, j=1 to N, are a set of discrete points covering the vortex sheets, each at the centroid of a small section of sheet having circulation $\Delta\Gamma_k$. The $\Delta\Gamma_k$ are conventionally thought of and depicted as discrete point vortices attached to fluid particles, hence the name of the method.

Equation (2.2) is first order accurate in δt. Greater accuracy may be obtained by using, for example, a Runge-Kutta multistep integration procedure. But the accuracy with which (2.3) represents (2.1) depends on the distribution of the z_j along the sheets.

Rosenhead (1931) and later Westwater (1935), for the roll-up of a trailing vortex sheet, used a small number N of the z_j points on the sheet. They found that in each case the sheet rolled up in a fairly orderly way. Later work by Moore (1974) using a closer spacing of the z_j led to rapidly developing chaos in the structure of the sheet. Fink and Soh (1978) suggested that this could be prevented by rediscretising the sheet at each time step to retain a uniform distribution of the convected z_j and therefore a more accurate representation of the principal value integral. However Moore (1979) has shown that although rediscretization slows the development of instability it does not remove it since all such discrete representations of a vortex sheet of zero thickness are pathologically susceptible to the Helmholz instability. Van de Vooren (1980) has given an analysis of the singular integral in (2.1) as a discrete sum including the effect of curvature of the vortex sheet.

The Helmholz instability may also be reduced by damping the rate of growth of the smallest wavelengths. Chorin (1973) considered each discrete vortex in the discretization of the sheet as having an independently diffusing viscous core. This scheme, which Chorin has shown to give a representation of viscous diffusion, does also stabilise discrete vortex models of vortex sheets. Other workers have used constant finite core vortices for the latter purpose only. Leonard (1980) gives a detailed discussion of the representation of diffused regions of vorticity.

It is also possible to carry out the integration of (2.1) along the sheet by considering piecewise constant (or linear) distributions of vorticity over short (usually straight) elements of sheet, eg. Faltinsen and Pettersen (1983), Basu and Hancock (1978), Higdon and Pozrikidis (1985) (arc elements) and Pullin (1978). The last of these used a circulation parameter as the variable of integration along the sheet following Smith's (1968) technique for the integration of the steady sheet from the leading edge of a slender wing. It appears from these results that the higher the order of the representation and hence the smoother the resulting integral function along the sheet the smaller are the instabilities which develop in a given time interval.

In the case of the representation of three-dimensional vortex sheets by line elements of infinitesimal core size a further problem arises due to the singularity in the self-induced velocity. The problem may be avoided by considering the sheet locally as a strip of curved surface rather than as a line, see eg. de Bernardinis et al. (1981). This will not be considered further here.

3. IMPULSIVELY STARTED FLOW PAST A SHARP EDGE

The local details of vortex shedding from a sharp edged body, where separation is observed to be fixed at the edge, can be investigated by considering the separating shear layer from a semi-infinite wedge (figure 1) of the same internal angle δ. It is convenient to transform the region exterior to the wedge into the right half (ζ) plane by the conformal transformation: $\zeta = c^{1-1/k} z^{1/k}$ where c is a length scale and $k=2 - \delta/\pi$. The complex potential in the ζ plane of a general unseparated flow at the edge has the form:

$$W_\infty(\zeta) = i\,V(t)\zeta + U(t)\zeta^2 + \ldots.$$

where only the first two terms corresponding to flow round and symmetrically off the edge are relevant here. As a result of the onset at t = 0 of V velocity round the edge, neglecting viscous diffusion, separation occurs and a vortex sheet is shed into the fluid.

The equation for the subsequent development of the sheet, $z=\tilde{z}(\Gamma,t)$ is

$$\frac{\partial \tilde{z}^*}{\partial t} = \left.\frac{\partial W}{\partial z}\right|_{z=\tilde{z}} . \tag{3.1}$$

In addition a condition of smooth separation from the edge must be applied. The correct application of this Kutta-Joukowski condition has been discussed by Giesing (1969). He showed that provided the sheet is modelled by a continuous sheet of vorticity in the vicinity of the edge, it must leave the edge tangentially to the windward side, but with curvature which may be infinite at the point of separation. The surface velocity Q is therefore finite, continuous and non-zero on the windward side and goes to zero as $z \to 0$ on the lee side, except when the wedge angle $\delta=0$. In order to achieve this at the singular point $z = 0$, $\left.\frac{\partial W}{\partial \zeta}\right|_{\zeta=0} = 0$ in the ζ-plane. The total complex potential for the separated flow is $W=W_\infty + W_\Gamma$, where W_Γ is the part due to the vortex sheet. Replacing the interaction of the vortex sheet with the body by an image sheet in the transformed plane which makes the imaginary axis a streamline of the flow:

$$\frac{\partial W}{\partial z} = \frac{\partial W}{\partial z}_\infty + \frac{1}{2\pi i}\,\frac{\partial \zeta}{\partial z}\left\{\int_0^{\Gamma_0} \frac{d\Gamma'}{\zeta-\tilde{\zeta}(\Gamma',t)} + \int_0^{-\Gamma_0} \frac{d\Gamma'}{\zeta+\tilde{\zeta}^*(\Gamma',t)}\right\} \tag{3.2}$$

The convection equation for the vortex sheet becomes

$$\frac{\partial \tilde{z}^*}{\partial t} = -\frac{i}{k}\, c^{1-1/k}\, V(t)\,\tilde{z}^{1/k-1} + \frac{2U(t)}{k}\, c^{2-2/k}\, \tilde{z}^{2/k-1} +$$

$$\frac{\tilde{z}^{\,1/k-1}}{2\pi i k}\left\{\int_0^{\Gamma_0} \frac{d\Gamma'}{\tilde{z}^{1/k} - \tilde{z}(\Gamma')^{1/k}} + \int_0^{-\Gamma_0} \frac{d\Gamma'}{\tilde{z}^{1/k} + \tilde{z}^*(\Gamma')^{1/k}}\right\} \tag{3.3}$$

and the Kutta-Joukowski condition gives

$$\int_0^{\Gamma_0} \frac{d\Gamma'}{\tilde{z}(\Gamma')^{1/k}} - \int_0^{-\Gamma_0} \frac{d\Gamma'}{\tilde{z}^*(\Gamma')^{1/k}} = \frac{c^{1/k-1}}{4\pi k} V(t) . \qquad (3.4)$$

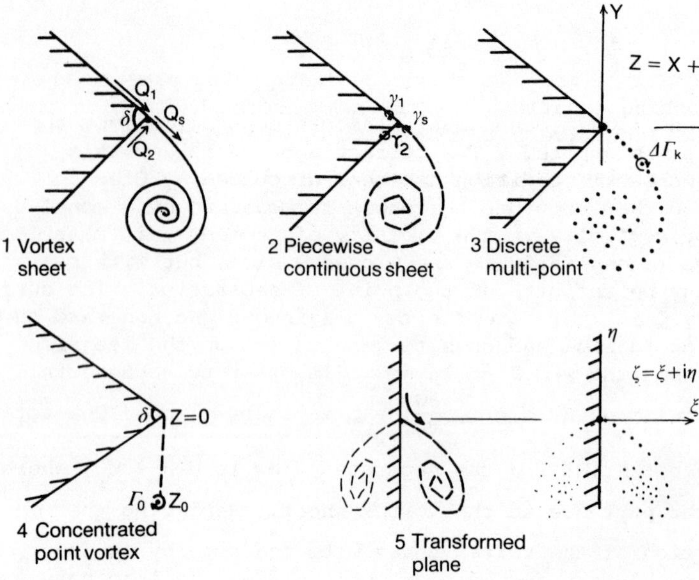

Fig. 1. Vortex sheet shed from a sharp edge.

4. CONCENTRATED POINT VORTEX REΚRESENTATION (FIG. 1)

In the most simplified model of a spiral vortex sheet, the sheet is replaced by a concentrated point vortex Γ_o at z_o at the approximate 'centre' of the spiral joined to the shedding edge by a cut in the plane across which the velocity potential is discontinuous. The cut represents the effect of the sheet feeding vorticity from the edge to the growing vortex core. The equations governing the growth and motion of this vortex have been derived by Rott (1956) and Graham (1977) for unsteady separated flow, based on the original theory of Brown and Michael (1954) for steady separated flow over a slender wing.

The zero total force (complex) equation

$$\frac{\partial z_o^*}{\partial t} + \frac{z_o^*}{\Gamma_o} \frac{\partial \Gamma_o}{\partial t} = \left. \frac{\partial W}{\partial z} \right|_{z=z_o} \qquad (4.1)$$

represents in integrated form the condition of continuity of pressure across the sheet. The Kutta-Joukowski condition at the edge is:

$$\left.\frac{\partial W}{\partial \zeta}\right|_{\zeta=0} = 0. \qquad (4.2)$$

These equations are dimensionally exactly the same as the corresponding equations for the continuous sheet which they replace. Impulsively started flow of the form U=0, $V=\hat{V}t^{\gamma}$ gives rise to a self-similar development of the vortex sheet, first investigated by Blendermann (1969). Equation 4.1 requires evaluation of the complex velocity $\frac{\partial W}{\partial z}$ at z_o, less the singular part $\frac{i\Gamma_o}{2\pi(z-z_o)}$.

Thus

$$\left.\frac{\partial W}{\partial z}\right|_{z=z_o} = \frac{i}{k}\, c^{-1/k}\, V\, z_o^{1/k-1} - \frac{i\Gamma_o}{2\pi k}\left\{\frac{k-1}{2z_o} + \frac{z_o^{1/k\,-1}}{z_o^{1/k} + z_o^{*1/k}}\right\} \qquad (4.3)$$

The second term is Routh's correction, arising from the distorting effects of the transformation on the vortex core. This correction is required in all discrete vortex calculations for which the velocity field is calculated in a transformed plane.

Equations (4.1) and (4.2) have the solution

$$\Gamma_o = \frac{2\pi}{\sqrt{k}}\,\chi^{1/(2k-1)}\, L^2\, t^{-1}, \qquad (4.4)$$

$$z_o = \chi_1^{k/(2k-1)}\, L\, e^{i\theta}, \qquad (4.5)$$

where

$$\theta = -k\cos^{-1}\left(\frac{1}{2}\sqrt{k}\right) \qquad (4.6)$$

and

$$\chi_1 = \frac{\sqrt{4-k}\,(k-1)(2k-1)}{2k^2\,(3\gamma k + k + 1)}. \qquad (4.7)$$

The length scale (the only one for this infinite wedge flow) is

$$L = c^{(k-1)/(2k-1)}(Vt)^{k/(2k-1)}.$$

The constant of proportionality for the circulation Γ_o is overestimated in comparison with more detailed representations, see Table 1.

TABLE 1

Impulsively started, constant velocity (γ=0)

Angle	K	Blendermann	Fink and Soh	Pullin	Multi-vortex	Brown and Michael
0°	2.0	2.321	2.418	2.398	2.38	2.493
90°	1.5	1.983	1.794	1.684	1.70	1.923

When the V-velocity dominates the flow as above the vortex is shed at an angle which is greater than $\frac{\delta}{2}$ (the tangent to the windward edge) for all edge angles less than 138.5°. But in cases for which a large U-velocity is present it is shed along the external bisector.

A defect in this model is the necessity to enforce a full stagnation (Kutta) condition in the transformed plane. This leads to an unrepresentative full stagnation point at separation in the physical plane for non-zero edge angles and departure of the separating streamline along the bisector rather than tangentially to the windward edge. However the great merit of the model is its simplicity which allows one to examine the behaviour of some vortex shedding problems approximately without extensive computation.

If the vorticity from a given edge changes sign with time it may become necessary to identify new centres of vortex roll-up. Maskell's (1972) condition for a separating vortex sheet specifies that it will change direction at the same time at which its strength changes sign, from being parallel to one side of the edge to being parallel to the other. This, suggests that $\frac{\partial \Gamma_o}{\partial t}=0$ is a suitable criterion for the shedding of a new point vortex core. Other criteria, such as $\frac{\partial \Gamma_o}{\partial t}$ passing through a minimum may also be used.

5. MULTI-VORTEX REPRESENTATION OF THE VORTEX SHEET SHED FROM AN EDGE

Pullin (1978) analysed the development of the sheet described by equations (3.3) and (3.4) for impulsive flows round a sharp edge. He expressed (3.4) in self-similar form and then evaluated the integral numerically over the main part of the sheet, by a trapezoidal scheme. To avoid the need for an infinite set of elements the inner spiral part of the sheet was represented by a single concentrated core vortex, treated in exactly the same way as the isolated Brown and Michael vortex with a cut joining it to the end of the continuous part of the sheet.

Most calculations of separated flow past bluff bodies have tended to use the simplest multi-point vortex rather than continuous element representation of the sheet. This is more convenient and faster to compute but gives a less accurate and less smooth velocity field. It is also, like the continuous representation, limited to the outer part of a spiral sheet if an orderly roll-up is to be obtained. In the inner part of a spiral it is essential for stability to maintain a smaller separation between succeeding vortices along the sheet than the gap between turns of the spiral (Moore, 1974). Therefore a concentrated inner core vortex should again be used. This is often not done because of the difficulty in general calculations of identifying new centres of roll-up as they form. The result is structureless clouds of point vortices which are or may be justified by the turbulent diffusion which would be present in reality at high Reynolds numbers.

The vortex sheet γ_s leaves the edge tangentially to the wetted surface as shown in Fig. 1, and except in the case of zero internal edge angle $\gamma_1 = \gamma_s$ and $\gamma_2 \to 0$ at the edge. γ_2 vorticity cannot be shed in inviscid flow at the edge at all but secondary separation on surface 2 may occur ahead of the edge releasing vorticity of opposite sign into the primary sheet. Giesing (1969) shows that a shed vortex sheet of this strength satisfies the usual stagnation point Kutta condition in the transformed plane in which the edge has been opened out into a plane surface. This removes the infinite part of the velocity in the physical plane.

If however the vortex sheet is replaced by a finite number of points and the Kutta condition is satisfied at the edge, a full stagnation point must exist in the physical plane with both Q_1 and $Q_2 \rightarrow 0$ at the edge, except for the case $\delta=0$. Many calculation methods therefore fix the strength and position of the last vortex shed by satisfying the Kutta condition and then obtaining an estimate of Q_1 either from the velocity near to but not at the edge (at which it is either 0 or ∞) or from the convection velocity of the last vortex or vortices to have been shed, eg. Clements (1973). Phrases such as boundary layer thickness are often used in this context to define a point at which to evaluate Q_1, but it is clear that no such length scale can be consistently defined for what is basically an inviscid model. The rapid variation of Q_1, near the separation point (Figure 2) makes its direct estimation point vortex representation likely to be inaccurate. Use of the convection velocity (Sarpkaya (1975), Evans and Bloor (1976), and in effect Kiya et al (1982)) which varies less rapidly just off the edge is likely to be more accurate, but must be based on a correct assumption for the velocity of the flow on the separated side of the sheet.

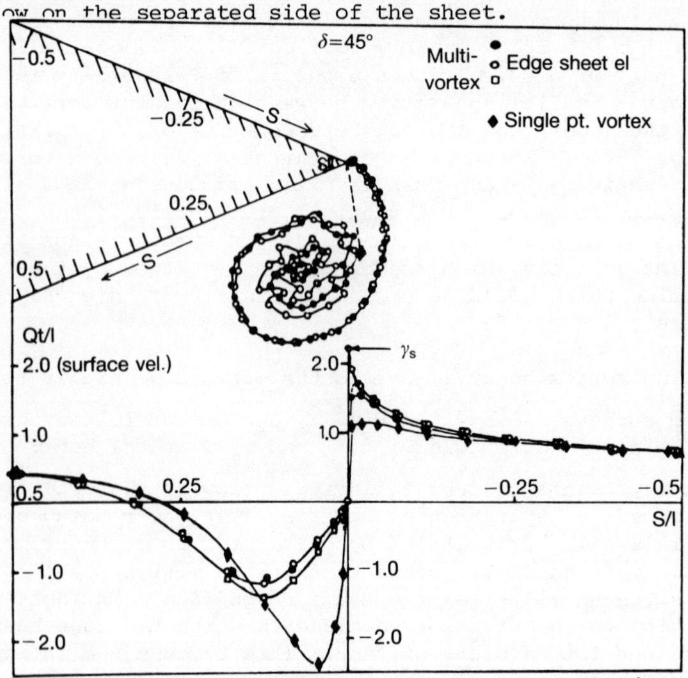

Fig. 2. Vortex sheet and surface velocity distribution for uniformly accelerating flow past a 45° edge.

Alternatively the release point may be specified, usually just above the desired separation point, and the vortex strength calculated by satisfying the Kutta condition at the separation point. Kamamoto and Bearman (1980) systematically studied the effect of such an artificially specified release point for the case of a normal flat plate in steady flow. They concluded that satisfactory results could be obtained for the shape and strength of the shed vortex sheet provided the non-dimensional ratio of the release height, free stream velocity and time step lay within a certain range. However this technique has been criticised for introducing an empirical parameter and for eliminating some of the feed back in the vortex shedding process. It is also possible, Graham (1977), to use the Brown and Michael equations to calculate the position and strength of each new vortex. Since the V velocity is suppressed by the previously shed vortices of the sheet, U-velocity dominates and the new vortex leaves along the bisector of the edge ($\delta<90^{\circ}$) or at a more acute angle for larger edge angles. The predicted initial trajectories of the vortices leaving the edge are therefore incorrect and the local details of the flow field are wrong. However comparison with more accurate representations indicates that the strength of the sheet is given quite accurately and the shape of the sheet and velocity distribution on the body surface a small distance away from the edge are also accurately predicted (Figures 2 and 3).

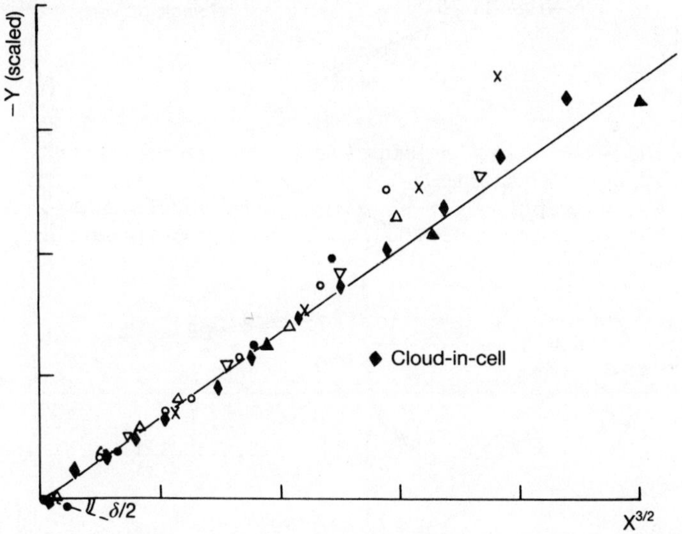

Fig. 3. Local vortex sheet shape from various computations.

Some of the difficulties associated with shedding point vortices to represent a continuous sheet may be avoided by using a continuous plane sheet element for the vorticity shed during the latest timestep. In reality this part of the sheet is curved with infinite curvature at the edge but a plane element is nevertheless sufficient to give the main components of the local flow field. Giesing (1969) has derived the equations relating the strength γ_s of the sheet at the edge to the V velocity component round the edge. Assuming constant γ_s over the short plane element of length s, the Kutta condition in the transformed plane gives:

$$V = \frac{k}{\pi(k-1)} \cos\left(\frac{\delta}{2k}\right) \gamma_s \left(\delta s\right)^{1-1/k} \tag{5.1}$$

In this case V is due both to the external imposed velocity field and to the previously shed point vortices representing the rest of the sheet. At the same time, for a non-zero edge angle the vortex sheet grows to the length $\delta s(t) = \frac{1}{2}\gamma_s(t - t_n)$ since $\frac{1}{2}\gamma_s$ is its convection velocity. If the circulation $\gamma_s \delta s$ is now considered to be centred on a point on the extension of the wetted surface at a distance from the edge:

$$r_n = (1-1/k)^k \delta s, \tag{5.2}$$

closer to the edge than the midpoint of the element, the Kutta condition is satisfied as if by a point vortex. The result is a prescription for shedding point vortices which satisfies the correct condition for a continuous separating sheet with:

$$\Gamma_n = \gamma_s \delta s = 2\left(\frac{\pi(k-1)}{2k\sin\pi/k}\right)^{2k/(2k-1)} L^2\, \delta t^{-1} \tag{5.3}$$

$$z_n = \frac{1}{2}\delta s e^{-i\delta/2} = 2^{-1/(2k-1)}\left(\frac{\pi(k-1)}{2k\sin\pi/k}\right)^{k/(2k-1)} L\, e^{-i\delta/2} \tag{5.4}$$

L is the length scale for the wedge flow $(=(V\,\delta t)^{k/(2k-1)})$. The local velocity field is calculated assuming that the element adjacent to the edge is a sheet element but that all succeeding elements are points. This gives satisfactory results for the surface velocity as the calculation of a uniformly accelerating flow round a 45° edge as shown in Figure 2.

Clapworthy and Mangler (1974) have shown that the shape of a vortex sheet leaving a sharp edge conforms to a curve

$$y' = a_o x'^{3/2} + \dots$$

where the local x' and y' axes are parallel and normal to the wetted surface. Figure 3 shows a plot of the vortex sheet shape adjacent to the edge computed by a number of different point vortex schemes including the 'cloud-in-cell' method discussed later. They all follow this shape closely after the first elements of the sheet.

An altogether more accurate but more expensive method is to represent the whole of the sheet by piecewise continuous sheet elements. Basu and Hancock (1978) for an unsteady aerofoil flow and Faltinsen and Pettersen (1983) for a number of bluff sections used straight elements. The evolution of the sheet is computed in these cases from the velocities at the element midpoints reinterpolated to the node points. Faltinsen and Pettersen found that because of the large distortions induced in a rolling up sheet behind a bluff body it was necessary to continuously rediscretise the sheet to keep the element lengths in sensible proportion. With large rates of strain on the sheet even rediscretisation eventually becomes ineffective and the sheet must be severed. This operation leads to some problems in selection of general criteria for interfering with the sheet in this way and also for the selection of new core vortices representing new centres of roll-up. However the results are generally superior to point vortex representation. Higdon and Pozrikidis (1985) using circular arc elements to represent the sheet similarly obtained very good results.

If the body surface is modelled by the surface vortex method rather than by transformation and images it becomes even more important to use (higher order) panels for sheet and wall elements in close proximity in order to avoid leakage. Ashurst (1979b) avoided this problem by using local images with far field correction by the surface singularity method.

6. SEPARATION FROM A CONTINUOUS SURFACE

Unfortunately many practical bodies, not least the circular cylinder, have continuous curved surfaces without specific edges fixing separation. The prediction of separation for such cases cannot strictly be obtained from a straightforward boundary layer calculation but requires a local Navier-Stokes solution such as Smith's (1977) application of the triple-deck technique. Many discrete vortex methods avoid this by either specifying separation empirically, or using an approximate boundary-layer separation condition or replacing the viscous diffusion in the shear layers by a viscous vortex model. Stansby (1977) in some of his earlier work with a circular cylinder, fixed separation at 95° for supercritical flow. Katz (1981) and Basuki (1983) similarly fixed the forward separation point on stalled aerofoils. In the case of oscillatory incident flows past circular cylinders in which the separation points are known to move through large angles, and also in the case of secondary separation, Stansby (1981) has suggested specifying the separation points empirically at points where the velocity has dropped by 5% from its peak value. With a specified separation point, vortices are usually released from a point at a fixed height h above the separation point. This strategy is required so that the Kutta condition can be satisfied by the point vortex model at the separation point. The point vortex model does not however give a very good representation of the pressure field just ahead of separation and this is a drawback in its use with a separation prediction model. The true shape of the separating sheet should induce ahead of itself an adverse pressure gradient with a square-root singularity at the separation point. As with the sharp edge, this behaviour is generally better modelled by a sheet of piecewise continuous elements in the neighbourhood of separation.

Deffenbaugh and Marshall (1976) and Sarpkaya and Shoaff (1979) and others have included a separation prediction based on laminar boundary layer theory in their discrete vortex models. This technique has been successful because the point vortex model of the sheet smeared out the singularity in the adverse pressure gradient sufficiently for convergence to a stable and realistic separation point. If however a more accurate representation of the sheet is used the square root singularity will tend to drive the separation point predicted by laminar boundary layer theory ahead of it until the smooth separation point far upstream of the true one is reached. Smith's application of triple deck theory allows a proper viscous-inviscid interaction to be carried out giving a better prediction of separation for steady flows. Fiddes (1980) has also used it successfully to predict the laminar separation line on a slender cone at incidence. Faltinsen and Pettersen (see Aarsnes (1984)) used it in a similar way, assuming it can be applied quasi-steadily at the separation points on a circular cylinder. The technique in the latter case involved estimating the strength of the pressure-gradient singularity from the numerical prediction of surface pressure. This was matched to a skin-friction parameter given by an upstream laminar boundary layer calculation. The actual separation point was then solved for iteratively. An average separation angle of about 75° was obtained for the first few periods of development of a Karman vortex street. Extension of the technique to turbulent boundary layers is not yet established.

Stansby and Dixon (1983) following Chorin's (1978) work have as an alternative represented the laminar boundary layer on the cylinder by a distribution of discrete vortices with an added random displacement at each time-step to simulate viscous diffusion. Separation should then, in theory, occur naturally when vorticity of opposite sign to the majority of the wall vorticity is created at the wall faster than it can be absorbed by diffusion. They used this technique with point vortices to predict separation on circular cylinders in both steady and oscillatory streams. This procedure is correct in the limit of very large vortex number density, but its accuracy when relatively few vortices are used is not yet established. Millinazzo and Saffman (1977) have shown that the random walk method only represents viscous diffusion accurately when the vortex number density is of the order of the Reynolds number in magnitude. In practical calculations possible at present the number of vortices across the depth of the boundary layer is very few and therefore the velocity profile through the boundary layer is crude. However Stansby and Dixon obtain good predictions for some of the global parameters such as drag and Strouhal number. Spalart et al. (1983) have in a different way coupled a boundary layer calculation to an outer discrete vortex simulation. The boundary layer vorticity is replaced by discrete vortices ahead of the separation point (where boundary layer theory is still valid). These vortices, initially lying on the displacement surface of the boundary layer are shed into the outer flow region at (separation) points where a rapid increase in the normal flux of vorticity occurs.

7. SECONDARY SEPARATION

Secondary separation beneath the primary separated sheet is known to have a considerable influence on the behaviour of vortex sheets (for example on a slender wing, Smith (1968)). It is observed to be present during starting flow past a circular cylinder. Both Stansby and Dixon (1982) and Faltinsen and Pettersen (Aarsnes)(1984) have incorporated it into their models for flow past circular cylinders and have observed that the release of opposite signed vorticity which weakens the primary sheet leads to a longer vortex formation region. Excessively short formation regions are a frequent, unrealistic, outcome of discrete vortex calculations also associated with overpredicted force coefficients. Significant opposite signed vorticity is only shed during a small part of the main shedding cycle when the latter is close to the surface and induces high reversed flow. It may be therefore that the roll up of the main shear layer is very sensitive to a weakening effect at that time. The prediction of the secondary separation points is difficult because of the large changes in the instantaneous velocity field in the separated region. For that reason the empirical 'velocity drop' rule is often resorted to.

8. THE CLOUD-IN-CELL METHOD

The fact that the number of operations required by the Biot-Savart integration increases as N_v^2 where N_v is the total number of vortex elements in the flow at a given time puts a strong constraint on the number of time-steps for which a developing flow can be computed. For this reason alternative less expensive techniques of computing the velocity field have been sought. If the vorticity distribution is assumed to have a finite support, each vortex element occupying a blob of space (Chorin (1973), Leonard (1980)) and therefore representing a diffused sheet it is possible to solve the Poisson equation (1.3) linking the stream-function to the vorticity field by differential methods. This can be done either by direct reduction methods or by fast Fourier transforms. However the resulting equations still involve $O(N_v^2)$ operations per time-step, due to the irregular distribution of the vortex elements in space. Van der Vegt and Huijsmanns (1984) used a cubic spline interpolation for the Fourier functions which considerably increased the efficiency. This is a grid free process but it is not clear that the interpolation errors are of lower order than those arising from the use of an Eulerian grid.

More usually, methods which calculate the velocity field directly from the Poisson equation do so by finite difference solution on a regular Eulerian grid. This is the basis of the hybrid Langrangian-Eulerian method known as the Cloud-in-cell or

Vortex-in-cell method, (Christiansen (1973)).

The usual procedure for a two-dimensional flow is as follows. The vortex sheets shed into the fluid are represented at time t by arrays of point vortices $\Delta\Gamma_k$, k=1...N, at position $z_k(t)$. The individual circulations $\Delta\Gamma_k$ of these vortices are distributed as vorticity ω_{ij} over a regular mesh, usually by area weighting (bilinear distribution) to the four surrounding node points. The finite difference form of equation (1.3):

$$\frac{\Delta^2\ \psi_{ij}}{\Delta x^2} + \frac{\Delta^2\psi_{ij}}{\Delta y^2} = -\ \omega_{ij} \tag{8.1}$$

may now be solved using a fast elliptic solver. A number of methods are available ranging from calculating the double finite Fourier transform of (8.1) in both x and y directions to the direct block elimination methods.

The velocity field on the mesh u_{ij}, v_{ij} is obtained by central differencing the stream-function on the mesh. The same area weighting as for the vorticity is then used to derive the velocity field at the vortex points themselves which is used to convect the vortices.

This distribution method on a uniform rectangular mesh satisfies conservation of vorticity and of its two first moments $\Sigma x_k \Delta\Gamma_k$ and $\Sigma y_k \Delta\Gamma_k$. The velocity field induced by a point vortex on the mesh falls to zero at the position of the vortex so that no self-induced velocity is generated. Point vortices thus appear within the mesh as if spread over a finite core comparable to the local mesh size, but not perfectly isotropic and depending on the position in the cell. The numerical viscosity introduced by the mesh is non-diffusive, but has the effect of preventing instabilities of sub-mesh length-scale developing, without removing all previously created detail at this scale, see for example Baker (1979).

As with the discrete vortex method either transformation or singularity methods may be used to represent a solid body in the mesh. In the former case we transform a single closed body into a circle which is surrounded by a concentric polar mesh.

The regular mesh $\theta_i = 2\pi i/N \qquad i = 1...N$

$$r_j = R(1+\delta)^{j-1} \qquad j = 1...M$$

where R is the radius of the circle has the advantages of a constant mesh aspect ratio and of concentrating mesh points next to the body. The polar form of equation (1.3), Fourier transformed with respect to θ, is solved in finite difference form:

$$\frac{4(\Psi_{kj+i} - 2\Psi_{kj} + \Psi_{kj-1})}{(r_{j+1} - r_{j-1})^2} - \frac{K^2}{r_j^2}\Psi_{kj} = -\Omega_{kj} \tag{8.2}$$

where Ψ_{kj} and Ω_{kj} are Fourier transforms of ψ_{ij} and ω_{ij}. Gaussian elimination is used for the resulting tridiagonal set of equations.

As with all transformation based methods it is more efficient to convect the vortices in the circle (ζ) plane using a modified convection equation:

$$\frac{\partial\zeta_k^*}{\partial t} = \left\{\left(\frac{\partial W}{\partial\zeta_k} - \frac{i\Delta\Gamma_k}{4\pi}\frac{\partial^2\zeta_k}{\partial z_k^2}\right)\Bigg/\left(\frac{\partial\zeta_k}{\partial z_k}\right)^2\right\}\left|\frac{\partial\zeta_k}{\partial z_k}\right|^2 \tag{8.3}$$

The transformation modulus and Routh correction are calculated at the mesh nodes and prestored. Transformation usually produces a highly distorted mesh in the physical plane with small meshes in the separation regions which is likely to be advantageous.

Alternatively the body may be overlayed by a regular, usually rectangular, mesh. The zero normal velocity boundary condition is then satisfied at N_b points on the body surface thus specifying N_b unknown singularities at the N_b nodes of the grid which are interior to the body. The capacity matrix of influence functions relating these points to the boundary collocation points can be calculated at the start, inverted and stored. A variation of the method which offers improved accuracy for thin bodies, replaces the surface of the body by a sheet of dipole panels. In two dimensions the method of Kennedy and Marsden (1976) can be used to obtain the velocity on the panels in order to satisfy a condition of constant streamfunction on the surface of the body.

Use of Fourier transforms to solve the Poisson equation is consistent with a periodicity in the direction on the polar grid. For other grids there is some error or extra work to avoid the implied periodicity. There is generally a problem with the velocity field associated with any vortex lying within one mesh of a solid boundary. For that case the vortex and its image in the boundary have implied cores which overlap leading to considerable reduction in the calculated velocity. It is therefore necessary to include a short range correction to the velocity field particularly for vortices close to the separation point.

9. FLOW PAST A NORMAL FLAT PLATE CALCULATED BY THE CLOUD-IN-CELL METHOD

This flow has the advantages of fixed separation points and that transformation to a circle can be carried out analytically. If an impulsively started flow past a normal plate is allowed to develop naturally asymmetry grows very slowly from the level of the rounding errors. In order to obtain a Karman vortex street in a practical length of computation time some artificial asymmetric stimulus is required at the start. Examples are the addition of one strong vortex downstream of the plate (Naylor (1982)), weakening of one of the two separating vortex sheets, (Sarpkaya and Shoaff (1979)) or application of a temporarily asymmetric incident velocity (Graham and Naylor (1982), Faltinsen and Pettersen (1983)). In the calculation shown here, the shedding process releases a number n of unit strength vortices at each time step, where n is adjusted to the nearest integer satisfying the Kutta condition. This leads to some jitter in the sheet but gives a denser representation of the sheet.

If the vortex shedding due to a steady stream is given an initial asymmetric perturbation, a Karman vortex street rapidly develops (Figure 4). The Strouhal number for this calculation is S=0.146 which compares reasonably with measured values. But the vortex formation region appears to be too short and the vortices in the street too concentrated and distinct compared with observations of high Reynolds number flows. It is usually reckoned that about 40% of the vorticity shed from the edges of a bluff body is lost in the near wake by mixing with vorticity of opposite sign (Fage and

Johansen (1927)). In the above flow simulation very little mixing occurs across the wake and there is no scope for other forms of cancellation. This in turn leads to drag forces on the body considerably higher than the mean value which is usually measured. However the simulation does compare well with the flow visualisation of a flat plate of low aspect-ratio between end-plates taken by Perry et al. (1982). Reduced aspect-ratio between end-plates is known to increase the spanwise correlation of the vortex shedding, giving approximate two-dimensionality when the aspect-ratio is less than about 4. Locking of the vortex shedding by oscillation of the body has a similar effect and results in a reduced formation region, increased drag and more concentrated vortices.

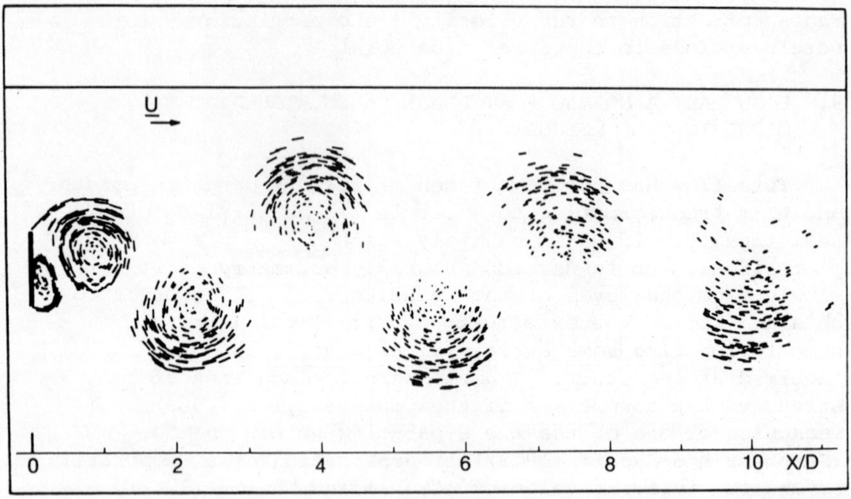

Fig. 4. Cloud-in-cell computation of the vortex sheet behind a flat plate in steady flow.

If, however, the vortex shedding is triggered two-dimensionally by the flow, as is the case for an in-line oscillation of the free stream, much better agreement with experimental measurement is obtained. Figure 5 shows the vortex pairs shed from a flat plate of width D during the first cycle of a reversing oscillatory flow $U=U_o \sin(2\pi t/T)$ compared with flow visualisations taken in a similar case by Oshima and Oshima (1980). For longer periods of flow the predicted vortex shedding patterns are also in substantial agreement with observation for values of Keulegan-Carpenter numbers, $K_c = U_o T/D$, up to about 30. The vortex shedding is generally asymmetric for $K_c > 5$ and the resulting force history periodic.

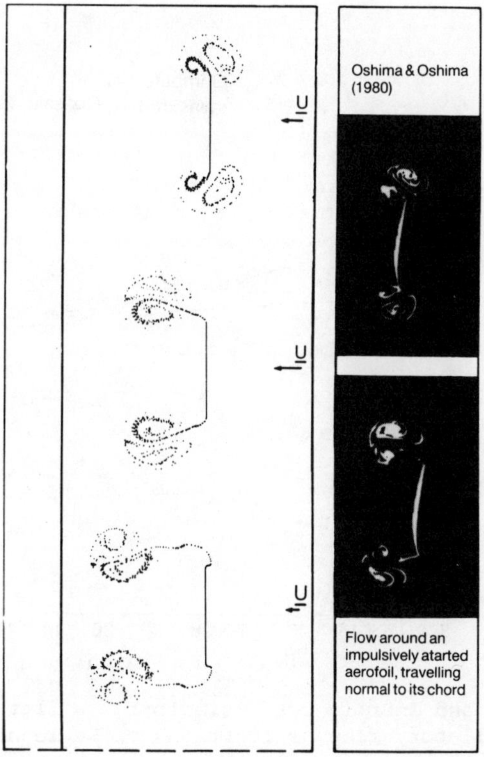

Fig. 5. Symmetric vortex shedding from a flat plate in oscillatory flow, $K_c = 2$.

The predicted drag and inertia coefficients agree well with experimental measurements, Figure 6. Visualisation of these oscillatory flows does show that with some exceptions the vortex shedding is substantially correlated across long lengths of span.

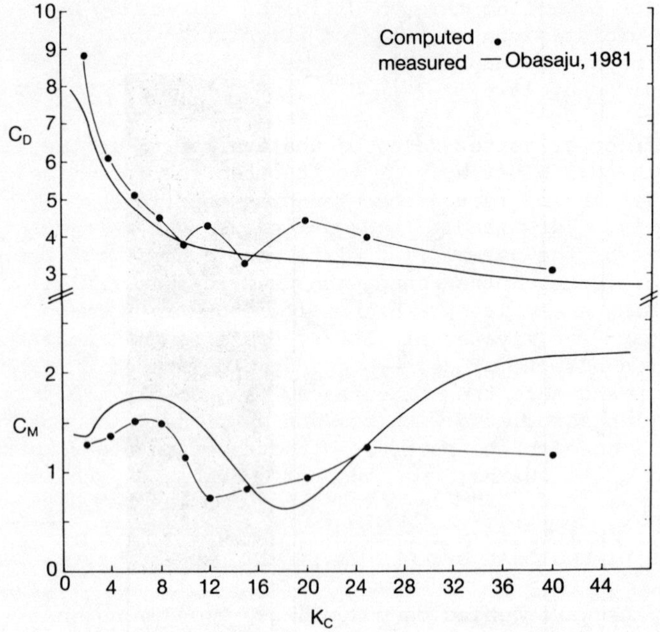

Fig. 6. Drag and Inertia coefficients for a flat plate in oscillatory flow as functions of Keulegan Carpenter number.

10. VORTEX DECAY

In order to obtain more realistic predictions for the forces and vortex strength in the wake, particularly for steady incident flows, an artificial vortex reduction or decay has frequently been incorporated into discrete vortex calculations, see eg. Sarpkaya and Shoaff (1979). Katz (1981) also, reduced the strength of vorticity shed from the leading edge separation point on a stalled aerofoil by an empirical factor at separation. A continuous decay factor, based on the diffusion of a viscous vortex, is very often applied. The circulation within a fixed radius r_c of the centre of such a vortex 'decays' with time as:

$$\Gamma(t) = \Gamma_o \left(1 - \exp\left(-r_c^2/4\nu t\right)\right) \qquad (10.1)$$

This apparent rate of decay is then applied to every point vortex, based on its residence time in the flow (eg. Kiya et al

(1982)). This technique may model some short range cancellation occurring through diffusion and mixing but generally violates conservation of circulation in the far field. Others, eg. Naylor (1982), have used simpler exponential decay laws.

The effect of vortex decay of whatever sort, in the case of flow behind a bluff body is to lengthen the formation region a little and reduce the mean drag coefficient considerably. For example in Naylor's case the mean drag coefficient of the normal flat plate was reduced from above 3.0 to 2.3. On the other hand when used in the simulation of a reattaching shear layer, the bubble length is reduced by vortex decay, see Kiya et al (1982). This is in line with the observed effects of decreasing Reynolds number on flow down a step and here the vortex decay is probably mimicking the reduction associated with production of negative vorticity on the wall beneath the bubble. Without vortex decay there is some doubt as to whether the asymptotic bubble length is finite.

11. THREE DIMENSIONAL EFFECTS

Since the wake behind many nominally two-dimensional bluff bodies is known to be instantaneously three-dimensional, it has been suggested that the observed vortex reduction may be associated with three-dimensional perturbation of initially spanwise vortex lines. This does not reduce the overall amount of circulation unless like diffusion it leads to mixing of vorticity of opposite sign. But it does change the velocity near field induced by a vortex line and it is possible that numerical vortex decay simulates this effect.

A numerical experiment is described in Graham and Naylor (1982) to see the effect on a discrete vortex calculation of a small amount of sinusoidal, spanwise waviness. The results showed a significant effect on the vortex formation region and a corresponding reduction in the drag coefficient. Three-dimensional effects may also be important in modelling the reattachment of a shear layer. Kiya et al (1982), for example, found that the mean velocity profile, 'turbulence' intensity and shear stress are badly predicted in the reattachment region by two-dimensional calculations. Faltinsen and Pettersen (Aarsnes (1984)) and Stansby and Dixon (1982) however have succeeded in obtaining quite long vortex formation regions in the base of a circular cylinder when secondary separation alone was incorporated.

12. APPLICATION TO SEPARATED FLOWS ON AEROFOILS

The Cloud-in-Cell method may be used to calculate a wide range of bluff body flows by making use of numerical, if necessary, transformation techniques. Another important area of application is to unsteady and separated flow over aerofoils. Katz (1981) has applied the discrete vortex method to impulsively started separated flow past an aerofoil. Spalart et al (1983) have applied the same method to this problem together with a boundary layer representation and Lewis and Porthouse (1983) used the discrete vortex method with a random walk method of predicting separation similar to Stansby and Dixon's (1983). Katz used a vortex reduction factor for the sheet separating at the leading edge and Basuki (1983) using the Cloud-in-cell method, found similarly that the upper surface pressure distribution was very sensitive to the strength of the separated sheet above it. He found that without any form of vortex decay the strength of this shed sheet caused the sheet coming off the trailing edge to roll up very strongly above the trailing edge and induce strong suction pressures in that region.

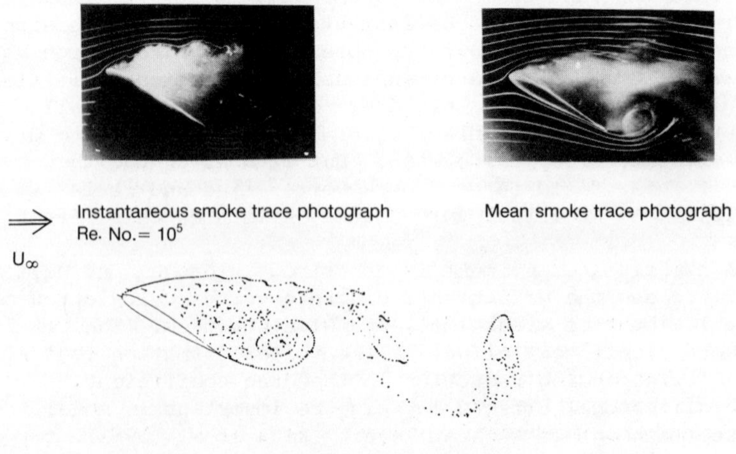

Fig. 7. Vortex wake from separated flow over 12% thick Joukowski aerofoil at 30° incidence.

However the use of a similar rate of decay to that used by Kiya et al (1982) gave a more realistic wake development (Figure 7) and mean pressure distribution (Figure 8). The need to introduce this artificial aid may be associated with the difficulty of modelling a separated sheet in close proximity to a solid surface.

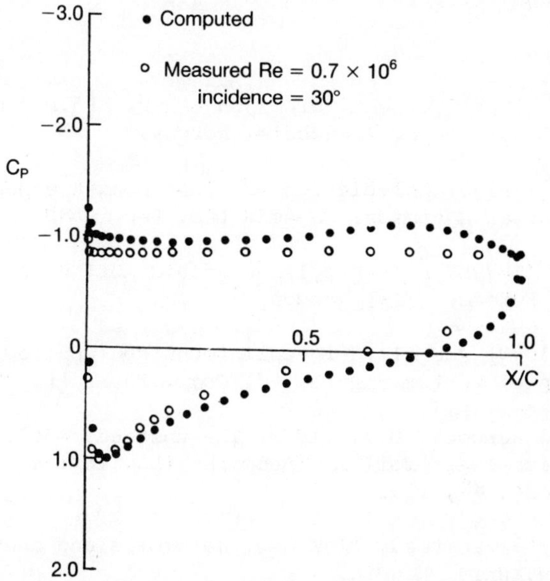

Fig. 8. Mean separated flow pressure distribution on the surface of a 12% thick Joukowski aerofoil at 30° incidence.

13. CONCLUSIONS

This paper has concentrated on the use of the discrete vortex method to model vortex sheets shed from two-dimensional bluff bodies. The method clearly also has important applications in areas of three dimensional wake modelling (as behind road vehicles for example) and for compressible flows.

Discrete vortex calculations of separated flow have perhaps not yet reached the status of conventional Eulerian mesh solutions of the Navier-Stokes or compressible Euler equations for routine purposes. Flow simulations by means of this general method are however becoming increasingly reliable and therefore may offer more hope in the short term of an accurate method of simulating high Reynolds number flows in which large scale separations and large eddy structures dominate. The main problem areas, from a practical point of view, seem to be the prediction of separation points for the shedding of vortex sheets from continuous surfaces and the resolution of the problem of apparently excessive sheet strength. The latter points to the importance of modelling secondary separation and three-dimensional effects as well as diffusion.

14. REFERENCES

Aarsnes, J.V. (1984) Current forces on ships. *Ph.D Thesis.* Norwegian Inst. Tech. Trondheim, Norway.

Ashurst, W.T. (1979a) Calculation of plane sudden expansion flow via vortex dynamics. Sandia Lab. Rep. SAND 79-8679.

Ashurst, W.T. (1979b) Piston cylinder fluid motion via vortex dynamics. Euromech 119, London.

Baker, G.R. (1979) The cloud-in-cell technique applied to the roll up of vortex sheets. *J. Comp. Phys.* **31**, 76.

Basu, B.C. and Hancock, G.J. (1978) The unsteady motion of a two-dimensional aerofoil in incompressible inviscid flow. *J. Fluid Mech.* **87**, 159.

Basuki, J. (1983) Unsteady flow over aerofoils and cascades. *Ph.D Thesis,* Univ. London.

de Bernardinis, B., Graham, J.M.R. and Parker, K.H. (1981) Oscillatory flow around disks and through orifices. *J. Fluid Mech.* **102**, 279.

Billet, G. (1980) Numerical simulation of a three-dimensional wall separation. Rech Aerosp. 1980-4, 11.

Blendermann, W. (1969) Der Spiralwirbel am translatorisch bewegten kreisbogenprofilen Struktur, Bewegung und Reaktion, Schiffsteck. 16, 3.

Brown, C.E. and Michael, W.H. (1954) The effect of leading edge separation on the lift of a delta wing. *J. Aero Sci.* **21**, 690.

Chorin, A.J. (1973) Numerical study of slightly viscous flow. *J. Fluid Mech.* **57**, 785.

Chorin, A.J. (1978) Vortex sheet approximation of boundary layers. *J. Comp. Phys.* **27**, 428.

Christiensen, J.P. (1973) Numerical simulation of hydrodynamics by a method of point vortices. *J. Comp. Phys.* **13**, 363.

Clapworthy, G.J. and Mangler, K.W. (1974) The behaviour of a conical vortex sheet on a slender delta wing near the leading edge. RAE Tech. Rep. No. 74150.

Clements, R.R. (1973) An inviscid model of two-dimensional vortex shedding. *J. Fluid Mech.* **57**, 321.

Deffenbaugh, F.D. and Marshall, F.J. (1976) Time development of the flow about an impulsively started cylinder. *A.I.A.A. J.* **14**, 908.

Evans, R.A. and Bloor, M.I.G. (1976) The starting mechanism of wave-induced flow through a sharp-edged orifice. *J. Fluid Mech.* **82**, 115.

Fage, A. and Johansen, F.C. (1927) The structure of the vortex sheet. *Phil. Mag.* **5**, 417.

Faltinsen O. and Pettersen, B. (1983) Separated flow around marine structures. Proc. S.S.P.A. Ocean Engg. Symp., Gothenborg.

Fiddes, S.P. (1980) A theory of the separated flow past a slender elliptic cone at incidence. RAE Tech. Memo. Aero 1858.

Fink, P.T. and Soh, W.K. (1978) A new approach to roll up calculations of vortex sheets. *Proc. Roy. Soc.* A362, 195.

Gerrard, J.H. (1967) Numerical computation of the magnitude and frequency of the lift on a circular cylinder. *Phil. Trans. Roy. Soc.* **261**, 137.

Giesing, J.P. (1969) Vorticity and the Kutta condition for unsteady flows. Trans. A.S.M.E.(E), *J. App. Mech.* **36**, 608.

Graham, J.M.R. (1977) Vortex shedding from sharp edges. Imperial Coll. Aero Rep. 77-06.

Graham, J.M.R. and Naylor P.J. (1982) The vortex wake of a flat plate in steady and oscillatory flow. Euromech 160, W. Berlin.

Higdon, J.J.L. and Pozrikidis, C. (1985) The self-induced motion of vortex sheets. *J. Fluid Mech.* **150**, 203.

Kamamoto, K. and Bearman, P.W. (1980) An inviscid model of interactive vortex shedding behind a pair of flat plates arranged side by side to the approaching flow. Trans. Japan. S.M.E. 46, 1299.

Katz, J. (1981) A discrete vortex method for the non-steady

separated flow over an aerofoil. *J. Fluid Mech.* **102**, 315.

Kennedy, J.L. and Marsden, D.J. (1976) Potential flow velocity distributions on multi-component airfoil sections. *Canadian Aero. and Space. J.* *22*, 243.

Kiya, M., Sasaki, K. and Arie, M. (1982) Discrete vortex simulation of a turbulent separation bubble. *J. Fluid Mech.* **120**, 219.

Leonard, A. (1980) Vortex methods for flow simulation. *J. Comp. Phys.* **37**, 289.

Leonard, A. (1985) Computing three-dimensional incompressible flows with vortex elements. *Ann. Rev. Fluid Mech.* **17**, 523.

Lewis, R.I. and Porthouse, D.T.C. (1983) Recent advances in theoretical simulation of real fluid flows. Proc. North East Coast Inst. Engrs. and Shpbldrs. Conf. Newcastle-upon-Tyne.

Maskell, E.C. (1972) On the Kutta Joukowski condition in two-dimensional unsteady flow. RAE Aero tech. note (unpublished).

Millinazzo, F. and Saffman, P.G. (1977) The calculation of large Reynolds number two-dimensional flow using discrete vortices with random walk. *J. Comp. Phys.* *23*, 380.

Moore, D.W. (1974) A numerical study of the roll-up of a finite vortex sheet. *J. Fluid Mech.* **63**, 225.

Moore, D.W. (1979) The spontaneous appearance of a singularity in the shape of an evolving vortex sheet. *Proc. Roy. Soc.* A365, 105.

Naylor, P.J. (1982) A discrete vortex model for bluff bodies in oscillatory flow. *Ph.D Thesis*, Univ. London.

Obasaju, E.D. (1981) Unpublished data. Aero. Dept. Imperial Coll.

Oshima, Y. and Oshima, K. (1980) Vortical flow behind an oscillating airfoil. Proc 15th I.C.T.A.M., Toronto, 357.

Perry, A.E., Chong, M.S. and Lim, T.T. (1982) The vortex shedding process behind two-dimensional bluff bodies. *J. Fluid Mech.* **116**, 77.

Pullin, D.I. (1978) The large-scale structure of unsteady self-similar rolled-up vortex sheets. *J. Fluid Mech.* **88**, 401.

Rehbach, C. (1978) Numerical calculation of three-dimensional unsteady flows with vortex sheets. A.I.A.A. paper no. 78-111.

Rosenhead, L. (1931) The formation of vortices from a surface of discontinuity. *Proc. Roy. Soc.* A134, 170.

Rott, N. (1956) Diffraction of a weak shock with vortex generation. *J. Fluid Mech.* **1**, 111.

Sarpkaya, T. (1975) An inviscid model of two-dimensional vortex shedding for transient and asymptotically steady separated flow over an inclined plate. *J. Fluid Mech.* **68**, 109.

Sarpkaya, T. and Shoaff, R.L. (1979) Inviscid model of two-dimensional vortex shedding by a circular cylinder. *A.I.A.A. J.* 17, 1193.

Smith, F.T. (1977) The laminar separation of an incompressible fluid streaming past a smooth surface. *Proc. Roy. Soc.* A356, 443.

Smith, J.H.B. (1968) Improved calculations of leading edge separation from slender delta wings. *Proc. Roy. Soc.* A306, 67.

Spalart P.R., Leonard, A. and Baganoff, D. (1983) Numerical simulation of separated flows. N.A.S.A. Tech. Memo. 84328.

Stansby, P.K. (1977) An inviscid model of vortex shedding from a circular cylinder in steady and oscillatory far flows. *Proc. Inst. Civ. Engrs.* 63, 865.

Stansby, P.K. (1981) A numerical study of vortex shedding from one and two cylinders. *Aero Quarterly* 32, 48.

Stansby P.K. and Dixon, A.G. (1982) The importance of secondary shedding in two-dimensional wake formation at very high Reynolds numbers. *Aero Quarterly* 33, 105.

Stansby, P.K. and Dixon, A.G. (1983) Simulation of flows around cylinders by a Lagrangian vortex scheme. App. Ocean Res. 5, 167.

Van Der Vegt, J.J. and Huijsmans, R.H.M. (1984) Numerical simulation of flow around bluff bodies at high Reynolds numbers. 15th Symp. Naval Hydrodynamics, Hamburg.

Van De Vooren, A.I. (1980) A numerical investigation of the rolling-up of vortex sheets. *Proc. Roy. Soc.* A373, 67.

Westwater, F.L. (1935) Rolling-up of the surface of discontinuity behind an aerofoil of finite span. A.R.C. Rep. and Memo. 1692.

CELL-VERTEX MULTIGRID SCHEMES FOR SOLUTION OF THE EULER EQUATIONS

M.G. Hall
(Royal Aircraft Establishment, Farnborough)

1. INTRODUCTION

The inspiration for the work presented here is a tantalizing paper, "A multiple grid scheme for solving the Euler equations", given at a conference by Ni (1981). Ni showed results that seemed very impressive in accuracy as well as speed, but he gave only an indication of how they were obtained. Such speed and accuracy might enable three-dimensional solutions of the Euler equations to be a matter of routine in practical aerodynamic design. However, other researchers have had difficulties in attempting to reproduce Ni's speed. For example, there were four papers given, on various extensions to Ni's scheme, at the corresponding conference in 1983, and in only one of these, by Ni's fellow-researcher Davis (1983), was the speed of Ni's scheme reproduced. Also presented in 1981, at the same conference, was a paper (Jameson et al, 1981) on numerical (non-multigrid) solution of the Euler equations using multi-stage time-stepping schemes of Runge-Kutta type. Subsequently, Jameson (1983) presented a multigrid method of his own that incorporated such multi-stage time-stepping. The schemes to be described here are developments suggested by the work of Ni and Jameson. The mathematical treatment is restricted to flow in two space dimensions, and the illustrative numerical results are for aerofoils, but the schemes are essentially applicable to a wide tange of configurations.

A basic feature of the present schemes is the form of the approximation for the flux balance for a computational cell, where the values of the dependent variables are specified at cell vertices rather than at cell centres or as cell averages. In this we choose to follow Ni (and Denton, 1981) rather than other researchers who, like Jameson, adopt cell-average approximations. To reflect the roles of cell and vertex we call the schemes 'cell-vertex' schemes. The advantage of cell-vertex schemes is that second-order spatial accuracy can, in principle, be obtained for the flux balance, even at boundaries and with non-smooth grids.

However, to achieve such accuracy new developments, beyond the details given by Ni (1981), have been required. For practical shapes the grid will typically be far from uniform, with cells near the leading edge having dimensions smaller than the leading-edge radius and cell dimensions in the far field of the order of the wing chord. It is found that when the scheme suggested by Ni is applied to such highly stretched grids heavy smoothing is needed to ensure numerical stability, and a substantial loss of accuracy results. This defect is remedied here by ensuring that mass and momentum changes are conserved in a key step of the numerical process. Another development has been required for the treatment of the flow-tangency boundary condition at the body surface - a topic that is not discussed by Ni. Here second-order accuracy is retained by a simple modification of the numerical scheme that involves an image principle but does not involve any extrapolation.

In addition to a cell-vertex formulation Ni (1981) adopted a time-stepping algorithm of Lax-Wendroff type for advancing to the steady state. The process was accelerated by the use of multiple grids, but Ni did not provide enough information, either in guiding principle or in detail, to enable others to reconstruct his multigrid time-stepping algorithm without considerable improvisation. For the present schemes some principles for multigrid acceleration have been formulated and possibilities for time-stepping in a multigrid framework have been explored. It is found helpful to take a viewpoint suggested by the uniqueness theorem for Cauchy's initial value problem for hyperbolic equations. Thus, to reduce the number of grid points involved towards the Cauchy minimum we march in cycles of steps, with each cycle consisting of successively larger steps associated with successively coarser grids; in effect, the time-step is greatly extended at small cost. To ensure convergence to the fine-grid solution the problem is formulated so that coarse-grid changes vanish when corresponding changes in the previous step vanish. To advance the solution efficiently boundary quantities are updated on the coarser grids as well as on the first, finest, grid of a cycle. Finally, for further economy in computing time, the calculation begins with a reduced number of steps or grid-levels per cycle and additional, finer, grids are added as the calculation proceeds. It is found that when the solution is advanced by Lax-Wendroff time-stepping in a multigrid framework constructed on these principles the rate of convergence is indeed greatly improved. Yet the number of multigrid cycles required for a given level of convergence remains significantly higher than the number required in the multigrid method of Jameson (1983).

In contrast to the above, Jameson adopts a time-stepping algorithm of the multi-stage Runge-Kutta type. This admits the use of time steps much larger than those admitted by a Lax-Wendroff algorithm and would be expected to give a reduced number of multigrid cycles for convergence. Jameson's use of residual averaging and enthalpy damping as accelerative devices would contribute further, but it seemed worthwile at this stage to explore the advantages offered solely by such multi-stage* time-stepping. Jameson's multigrid framework seems to be very similar in practice to that outlined above, although it may be viewed as a member of the class of multigrid schemes advocated by Brandt (1977) and conventionally associated with elliptic equations and iteration processes for boundary-value problems. Thus we wish to combine multi-stage time-stepping with the established cell-vertex formulation and multigrid framework and then to compare the overall performance of the scheme with that of the corresponding Lax-Wendroff scheme. The significant factors for comparison are the computing time required for convergence (not the number of cycles), and the robustness and ease of use of the scheme.

The task would have been relatively straightforward if we simply adopted Jameson's multi-stage algorithm as it stood, because the choice of time-stepping algorithm is independent of the form of the approximation for the flux balance. However, one feature of Jameson's algorithm is the inclusion of a fourth-order spatial dissipation as a stabilising device. The flux balance for a particular cell thus involves cells beyond the immediate neighbours, which complicates the scheme considerably, especially when applied to complex configurations with adjoining grids. A further objective, therefore, is to find a stabilizing device that is free from this complication. The device we adopt, after some trial, is essentially a second-order time-change rather than a spatial dissipation. It turns out, when the performances of various multigrid schemes are compared, that the computing times for multi-stage time-stepping are no less than those for Lax-Wendroff time-stepping, even though far fewer cycles are required for convergence with the former. It is found, however, that a composite Lax-Wendroff and multi-stage scheme gives faster convergence, with significantly less computing time, than either Lax-Wendroff or multi-stage time-stepping on its own.

---

* For brevity, the term 'multi-stage' is taken to mean 'multi-stage of Runge-Kutta type', so that a multi-stage scheme is distinct from a Lax-Wendroff scheme here.

This paper begins with derivations of the cell-vertex schemes for Lax-Wendroff and multi-stage time-stepping without multiple grids. The treatment of the flow-tangency boundary condition, and of stability and smoothing, are discussed in turn. Then the multigrid acceleration schemes, for Lax-Wendroff time-stepping and for composite Lax-Wendroff and multi-stage time-stepping, are described. Finally, numerical results are shown for the RAE 2822 and NACA 0012 aerofoils to demonstrate accuracy and speed. Surface distributions of Mach number, and convergence histories for lift, drag and a typical residual are shown. Comparisons with results from other methods are included.

2. CELL-VERTEX FORMULATIONS

2.1 Basic Euler equations

Since we are seeking steady-state solutions of the Euler equations it is sufficient to consider the equations

$$\frac{\partial U}{\partial t} = -\left(\frac{\partial F}{\partial x} + \frac{\partial G}{\partial y}\right) , \qquad (2.1)$$

where U, F and G are three-component vectors given by

$$U \equiv \begin{vmatrix} \rho \\ \rho u \\ \rho v \end{vmatrix} , \quad F \equiv \begin{vmatrix} \rho u \\ \rho u^2 + p \\ \rho uv \end{vmatrix} , \quad G \equiv \begin{vmatrix} \rho v \\ \rho vu \\ \rho v^2 + p \end{vmatrix} ,$$

and p is given by the Bernoulli equation

$$p = \frac{\rho}{\gamma}\left(1 - \frac{\gamma - 1}{2}(u^2 + v^2)\right) .$$

These equations describe a model flow in which the total enthalpy is constant, Bernoulli's equation having been substituted for Euler's energy equation and the equation of state. The model flow varies with a time t in a rectangular Cartesian space x,y. The quantities u and v are the velocity components in the x and y directions respectively, and ρ and p denote the density and pressure. All quantities are normalized, with a typical body dimension, the speed of sound at the free-stream stagnation condition and the free-stream total pressure respectively taken as the units of length, speed and pressure. To derive the required cell-vertex schemes from (2.1) an arbitrary curvilinear grid defined only by the coordinates x, y of its grid points in Cartesian space is adopted. No transformation of independent variables is made and ρ, u and v are retained as dependent variables.

2.2 Lax-Wendroff scheme

The Taylor-series expansion for the component U in (2.1) can be written

$$\delta U = \left(\frac{\partial U}{\partial t}\right)^n \Delta t + \left(\frac{\partial}{\partial t}\left(\frac{\partial U}{\partial t}\right)\right)^n \frac{\Delta t^2}{2} + \text{higher-order terms},$$

where the superscript n denotes the value of a variable at the time level n, $\delta U \equiv U^{n+1} - U^n$ and $\Delta t \equiv t^{n+1} - t^n$. Substitution from (2.1) and change in order of differentiation yields

$$\delta U = -\left(\frac{\partial F}{\partial x} + \frac{\partial G}{\partial y}\right)^n \Delta t - \left(\frac{\partial}{\partial x}\left(\frac{\partial F}{\partial U}\frac{\partial U}{\partial t}\right) + \frac{\partial}{\partial y}\left(\frac{\partial G}{\partial U}\frac{\partial U}{\partial t}\right)\right)^n \frac{\Delta t^2}{2}. \quad (2.2)$$

We denote the first order change in U and ΔU, that is,

$$\Delta U = -\left(\frac{\partial F}{\partial x} + \frac{\partial G}{\partial y}\right)^n \Delta t. \quad (2.3)$$

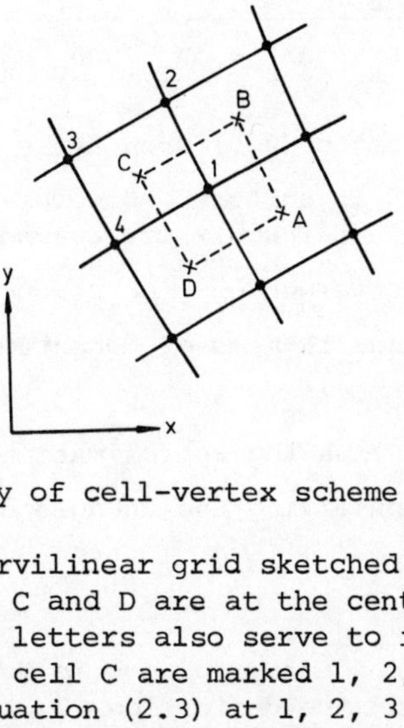

Fig. 1 Geometry of cell-vertex scheme for single grid

Now consider the curvilinear grid sketched in Fig. 1. The points marked A, B, C and D are at the centres of the four cells shown and the letters also serve to identify the cells. The vertices of the cell C are marked 1, 2, 3 and 4. Then, given F and G in equation (2.3) at 1, 2, 3 and 4, the average first-order change in U for the whole cell C in the time-step Δt is, by the integral theorem of Gauss and the trapezoid rule,

$$\Delta U_C = \Big((F_3 - F_1)(y_2 - y_4) + (F_4 - F_2)(y_3 - y_1)$$
$$+ (G_1 - G_3)(x_2 - x_4) + (G_2 - G_4)(x_3 - x_1)\Big)\frac{\Delta t}{2\Delta A_C}, \quad (2.4)$$

where ΔA_C is the area of the cell C and the values of F and G are specified at the time-level n. Note that equation (2.4) is an expression of the flux balance for the cell C. There are similar expressions for the first-order changes ΔU_A, ΔU_B and ΔU_D in the cells A, B and D respectively.

Now it seems that in Ni's scheme (1981), the first-order change at the grid point 1 is taken to be

$$\Delta U_1 = \frac{1}{4}(\Delta U_A + \Delta U_B + \Delta U_C + \Delta U_D). \tag{2.5}$$

It is found that use of the approximation (2.5) for aerofoils, with typical highly-stretched grids, tends to make the numerical solution divergent - the change in U from one time step to the next increases unless large amounts of numerical smoothing are applied. Here we adopt a more general formula: instead of (2.5), the first-order change in U in a time-step Δt at the grid point 1 is taken to be

$$\Delta U_1 = \frac{\Delta U_A \cdot \Delta A_A + \Delta U_B \cdot \Delta A_B + \Delta U_C \cdot \Delta A_C + \Delta U_D \cdot \Delta A_D}{\Delta A_A + \Delta A_B + \Delta A_C + \Delta A_D}. \tag{2.6}$$

Note that (2.6) reduces to (2.5) when $\Delta A_A = \Delta A_B = \Delta A_C = \Delta A_D$, that is, when the grid is uniform. However the approximation (2.6) is regarded not a a formula for averaging but rather as an aid to numerical convergence. The product $\Delta U_1 \cdot \frac{1}{4}(\Delta A_A + \Delta A_B + \Delta A_C + \Delta A_D)$ is the change in mass or momentum for an element of area $\frac{1}{4}(\Delta A_A + \Delta A_B + \Delta A_C + \Delta A_D)$, and ΔU_1 is the average change in U over this element. Thus the approximation (2.6) relates ΔU_1 to ΔU_A, ΔU_B, etc by conserving the changes $\Delta U_A \cdot \frac{1}{4}\Delta A_A$, $\Delta U_B \cdot \frac{1}{4}\Delta A_B$, etc.

To estimate the second-order change in U we note first that, from equation (2.1) and the definition (2.3), we can replace $(\partial U/\partial t)^n$ in equation (2.2) by $\Delta U/\Delta t$. Denote $(\partial F/\partial U)\Delta U$ and $(\partial G/\partial U)\Delta U$ in (2.2) by ΔF and ΔG respectively and assume that their values at the points A, B, C and D are given by approximations of the form

$$\left(\frac{\partial F}{\partial \rho}\Delta\rho\right)_C = \frac{1}{4}\left(\left(\frac{\partial F}{\partial \rho}\right)_1 + \left(\frac{\partial F}{\partial \rho}\right)_2 + \left(\frac{\partial F}{\partial \rho}\right)_3 + \left(\frac{\partial F}{\partial \rho}\right)_4\right) \cdot (\Delta\rho)_C .$$

Then, for the element of area ΔA_1 with vertices at A, B, C and D an expression for the second-order part of the change in U is obtained by integration; it is similar to the right-hand side of (4) with F and G replaced by ΔF and ΔG, respectively. It is assumed that $\Delta A_1 = \frac{1}{4}(\Delta A_A + \Delta A_B + \Delta A_C + \Delta A_D)$ approximately, so that the corresponding first-order change is given by (2.6). Hence the total change for the element of area ΔA_1, which we take to be the total change at the grid point 1, is

$$\delta U_1 = \Delta U_1 + \Big((\Delta F_C - \Delta F_A)(y_B - y_D) + (\Delta F_D - \Delta F_B)(y_C - y_A) + (\Delta G_A - \Delta G_C)(x_B - x_D) + (\Delta G_B - \Delta G_D)(x_C - x_A)\Big)\frac{\Delta t}{4\Delta A_1}\,. \tag{2.7}$$

The above is a two-stage scheme:

(i) flow conditions at the vertices of each cell are used to calculate the first-order changes within the cell - equation (2.4), and

(ii) the first-order changes in four neighbouring cells are used to calculate the total changes at the common, central, vertex - equations (2.6) and (2.7).

Although (2.6) differs significantly from the corresponding equation of Ni, we have followed Ni in adopting a scheme with the dependent variables specified at cell vertices. One advantage over cell-average finite-volume schemes such as that of Jameson et al (1981) is that the resulting steady-state solution is formally second-order accurate spatially even for non-uniform grids that are not smooth. This advantage may be obscured by the fact that equations (2.6) and (2.7) are not second-order accurate for arbitrary grids. However it is observed in numerical calculations that as the steady state is approached, that is when $\delta U \to 0$, the changes given by equation (2.4) behave likewise, that is $\Delta U \to 0$. This would be expected from the form of the right-hand side of equation (2.7), for $\Delta F = \Delta G = 0$ if all $\Delta U = 0$. Equation (2.4) expresses the flux balance for a single cell. The only approximation made in deriving (2.4) is the trapezoid rule, which has an error of third order. The flux balance is, therefore, correct to second order. In contrast it is assumed in typical cell-average schemes that the conditions on a cell boundary required for the balance of fluxes are given by arithmetic means of conditions in the pair of adjacent cells. Any such mean is second-order accurate only for uniform or suitably smooth grids. In practical problems it is difficult to satisfy this condition, especially when treating the more complex configurations, and the advantage of the present scheme could be marked. However, the overall advantages and disadvantages of the two types of scheme have apparently not yet been assessed systematically, It is possible that cell-vertex schemes may require more labour for implementation.

2.3 Multi-stage scheme

In this section we consider 4-stage schemes of Runge-Kutta type. Other schemes, with three stages, have also been considered, but these were found to be less effective and they are not discussed here. If a quantity U satisfies a differential equation

$$\frac{\partial U}{\partial t} + R(U) = 0, \tag{2.8}$$

and we have $U = U^n$ at time $t = t^n$, then to advance the solution by a time-step $\nu\Delta t$, where ν is the Courant number, we may follow Jameson et al (1981) and set

$$\left.\begin{aligned} U^{(0)} &= U^n \\ U^{(1)} &= U^{(0)} - \nu\alpha_1 \Delta t R^{(0)} \\ U^{(2)} &= U^{(0)} - \nu\alpha_2 \Delta t R^{(1)} \\ U^{(3)} &= U^{(0)} - \nu\alpha_2 \Delta t R^{(2)} \\ U^{(4)} &= U^{(0)} - \nu\Delta t R^{(3)} \\ U^{n+1} &= U^{(4)}, \end{aligned}\right\} \tag{2.9}$$

where α_1, α_2 and α_3 are constants. Here we let U be the solution U_1 at the vertex 1, and R(U) be the residual $[R(U)]_1$ in equation (2.1) at that vertex, expressed in some discrete form.

This scheme has been implemented in several ways in attempts to calculate the flow past a lifting aerofoil with Courant numbers as large as possible. A key question has been what to choose for the form of the residual $[R(U)]_1$. The obvious choice

$$[R(U)]_1 = -\frac{\Delta U_1}{\Delta t},$$

where ΔU_1 is given by equations (2.4) and (2.6), resulted in instability for all values of the constants α_1, α_2 and α_3 tried, unless the Courant number was reduced below unity. Presumably the process could be stabilized by the introduction of a fourth-order dissipation as Jameson et al (1981) have done. However, an alternative that is free from the complication of fourth-order dissipation has been found in the second-order part of the total change δU_1 in equation (2.7). Equation (2.7) can be written

$$\delta U_1 = \Delta U_1 + \Delta t[\Delta R(\Delta F,\Delta G)]_1 \,,$$

where $[\Delta R(\Delta F,\Delta G)]_1 \equiv \Big((\Delta F_C - \Delta F_A)(y_B - y_D) + (\Delta F_D - \Delta F_B)(y_C - y_A)$

$$+ (\Delta G_A - \Delta G_C)(x_B - x_D) + (\Delta G_B - \Delta G_D)(x_C - x_A)\Big)/(4\Delta A_1)\,. \qquad (2.10)$$

Note that when $\Delta U \rightarrow 0$ we have $\Delta R \rightarrow 0$. One suitable choice of residual is

$$[R(U)]_1 = -\frac{\Delta U_1}{\Delta t} - k_1 \Delta R_1 \,, \qquad (2.11)$$

where k_1 is a constant. Rather than calculate ΔR at every stage in a time-step it has been found sufficient to freeze it at the value calculated for the first stage. Thus a viable scheme consists of the stages

$$\left.\begin{aligned}
U^{(0)} &= U^n \\
U^{(1)} &= U^{(0)} + \nu\alpha_1\Big(\Delta U^{(0)} + k_1 \Delta t \Delta R^{(0)}\Big) \\
U^{(2)} &= U^{(0)} + \nu\alpha_2\Big(\Delta U^{(1)} + k_1 \Delta t \Delta R^{(0)}\Big) \\
U^{(3)} &= U^{(0)} + \nu\alpha_3\Big(\Delta U^{(2)} + k_1 \Delta t \Delta R^{(0)}\Big) \\
U^{(4)} &= U^{(0)} + \nu\Big(\Delta U^{(3)} + k_1 \Delta t \Delta R^{(0)}\Big) \\
U^{n+1} &= U^{(4)}.
\end{aligned}\right\} \qquad (2.12)$$

Another, equally successful, device is to add the correction ΔR only at the end of the step. Then, instead of (2.12), we have

$$\left.\begin{aligned} U^{(0)} &= U^{n} \\ U^{(1)} &= U^{(0)} + \nu\alpha_1 \Delta U^{(0)} \\ U^{(2)} &= U^{(0)} + \nu\alpha_2 \Delta U^{(1)} \\ U^{(3)} &= U^{(0)} + \nu\alpha_3 \Delta U^{(2)} \\ U^{(4)} &= U^{(0)} + \nu \Delta U^{(3)} \\ U^{n+1} &= U^{(4)} + k_2 \Delta t \Delta R^{(4)} \quad , \end{aligned}\right\} \tag{2.13}$$

where k_2 is another constant. The above schemes have been used to advance to the steady state both without and with the use of multiple grids. By suitable choices of the constants the local Courant number can be made greater than 2 before the onset of instability.

2.4 Flow-tangency boundary condition

The condition that the flow be tangential to the body surface is implemented here by a modification of the scheme used for the interior of the field and involves an image principle. Fig. 2 shows the body surface passing through a cell vertex 1, with the cells B and C in the interior of the field and the slope of the surface at 1 given by the angle α between the local tangent and the x-axis. Given values of U at cell vertices along the body surface and within the field at some time, it is required to determine the total change δU_1 at 1 in the next time step, subject to the tangency condition

$$\frac{v}{u} = \tan \alpha \ . \tag{2.14}$$

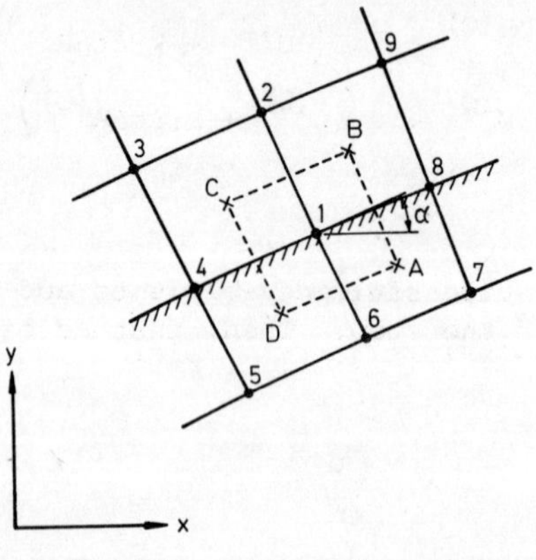

Fig. 2 Geometry of cell-vertex scheme at aerofoil surface

Firstly, a 'non-reflective' or 'incident' change is calculated without reference to the tangency condition or the image principle, the procedure being as follows for the Lax-Wendroff cell-vertex scheme:

(i) The first order changes ΔU_B and ΔU_C for the cells B and C are obtained from equation (2.4).

(ii) The 'non-reflective' or 'incident' total change at 1, δU_1^i, is obtained from equations (2.6) and (2.7), with ΔU_A and ΔU_D taken to be zero but with extrapolated values for ΔA_A and ΔA_D.

The extension for multi-stage cell-vertex schemes is straightforward. For the scheme described by equations (2.12), for example, we have $\delta U_1^i = U_1^{n+1} - U_1^n$, where U_1^{n+1} is given by (2.12) with ΔU_1 given by equations (2.4) and (2.6) and ΔR_1 given by (2.10); while ΔU_A and ΔU_D are again taken to be zero.

We then take, for the 'predicted' change at 1,

$$\delta U_1^p = 2\delta U_1^i \,. \tag{2.15}$$

Finally, the predicted changes in ρ, ρu and ρv are 'corrected' by setting

$$\left.\begin{aligned} \delta\rho_1 &= \delta\rho_1^p \\ \delta(\rho u)_1 &= (\rho q_t)_1^* \cos\alpha - (\rho u)_1 \\ \delta(\rho v)_1 &= (\rho q_t)_1^* \sin\alpha - (\rho v)_1 \,, \end{aligned}\right\} \tag{2.16}$$

where $(\rho q_t)_1^* = \left((\rho u)_1 + \delta(\rho u)_1^p\right)\cos\alpha + \left((\rho v)_1 + \delta(\rho v)_1^p\right)\sin\alpha$. Equations (16) imply that the corrected change in ρq_t, where q_t is the velocity component tangential to the surface, is equal to the predicted value, and that the change in ρq_n, where q_n is the normal velocity component, is cancelled by its image.

Note that the choice of formulae, in particular the choice for obtaining the above predicted quantities, has been decided on essentially by trial. Formulae based on the theory of characteristics were tried and discarded. Physical notions have provided some guidance but the over-riding criterion has been rapidity and ease of convergence to the state where the Euler equations and the boundary conditions are satisfied simultaneously. Thus in the steady state, when $\Delta U_B = \Delta U_C = 0$, the flux balance for each cell bounded by the surface will be correct to second-order, and the tangency condition will be satisfied exactly at the cell vertices on the surface. No extrapolation is involved in the flux balance; the only approximation is the trapezoid rule. It is only in the advance to the converged state that extrapolated values for ΔA_A and ΔA_D are required, and the choice is not critical. Since second-order accuracy is obtained for arbitrary grids, both non-smooth and smooth, there again appears to be an advantage over cell-average schemes.

2.5 Numerical stability and smoothing

The onset of numerical instability places an upper limit on the step sizes Δt that can be used in advancing to the steady state. The permissible maximum varies from cell to cell, depending on the geometry of the cell and the local flow conditions. For the Lax-Wendroff scheme the appropriate condition for numerical stability for the cell C (Fig. 1) is taken to be

$$\Delta t \leqslant \min\left(\frac{\Delta A_C}{|u\Delta y^{\ell} - v\Delta x^{\ell}| + a\Delta\ell}\,,\ \frac{\Delta A_C}{|u\Delta y^{m} - v\Delta x^{m}| + a\Delta m}\right), \tag{2.17}$$

where

$$\Delta x^{\ell} = \tfrac{1}{2}(x_2 + x_3 - x_1 - x_4)\ ,\quad \Delta x^{m} = \tfrac{1}{2}(x_1 + x_2 - x_4 - x_3)\ ,$$

$$\Delta y^{\ell} = \tfrac{1}{2}(y_2 + y_3 - y_1 - y_4)\ ,\quad \Delta y^{m} = \tfrac{1}{2}(y_1 + y_2 - y_4 - y_3)\ ,$$

$$\Delta\ell = \left((\Delta x^{\ell})^2 + (\Delta y^{\ell})^2\right)^{\frac{1}{2}}\ ,\qquad \Delta m = \left((\Delta x^{m})^2 + (\Delta y^{m})^2\right)^{\frac{1}{2}}\ ,$$

and $a = (\gamma p/\rho)^{\frac{1}{2}}$. The condition (2.17) seems satisfactory for present purposes although, strictly, it applies not to our basic equations (2.1) but to the full Euler equations which include an energy equation and in which variations in time are correctly represented. For the three-component system (2.1) the condition corresponding to (2.17) is much more complicated: the speed of propagation of small disturbances differs from that for the full Euler equations. Fortunately the difference is small at moderate Mach numbers. To evaluate ΔU_C from equation

(2.4) we set $(\Delta t)_C = \Delta t_{max}$ where Δt_{max} is given by (2.17). For the evaluation of δU_1 from equation (2.7) we set

$$(\Delta t)_1 = \min\Big((\Delta t)_A, (\Delta t)_B, (\Delta t)_C, (\Delta t)_D\Big). \tag{2.18}$$

For the multi-stage schemes Δt is taken to be a multiple of the value given by equation (2.18). This multiple, the Courant number ν, is kept fixed over the entire field.

Numerical smoothing is needed, to keep spatial oscillations within acceptable bounds. Several forms of smoothing scheme are feasible. The scheme adopted here is derived from numerical experiments and is an elaboration of that suggested by Ni (1981). The smoothing change $\delta^S U_1$ applied at the vertex 1 (Fig. 1) is

$$\delta^S U_1 = \mu(\bar{U}_A + \bar{U}_B + \bar{U}_C + \bar{U}_D - 4U_1), \tag{2.19}$$

where $\bar{U}_C = \frac{1}{4}(U_1 + U_2 + U_3 + U_4)$,
and so on, and the smoothing coefficient μ is given by

$$\mu = \mu_0 \Delta s + f(\rho, \bar{R}), \tag{2.20}$$

where μ_0 is a constant, Δs is an average cell dimension, and $\bar{R}$ is an average residual given by

$$\bar{R} = \left|\frac{\Delta\rho}{\Delta t}\right|_A + \left|\frac{\Delta\rho}{\Delta t}\right|_B + \left|\frac{\Delta\rho}{\Delta t}\right|_C + \left|\frac{\Delta\rho}{\Delta t}\right|_D .$$

The term $f(\rho, \bar{R})$ is introduced to provide strong smoothing in the vicinity of a shock but much less elsewhere, and has a form such that $f \to 0$ when $\Delta\rho \to 0$.

The change (2.19) is implemented to yield new smoothed values before calculating the changes given in sections 2.2 to 2.4. For smooth grids the flux balance for a cell remains second-order accurate in the steady state, except of course in the neighbourhood of a shock wave. For non-smooth grids second-order accuracy can be maintained by setting $\mu_0 \sim \Delta s$.

3. MULTIGRID SCHEMES

3.1 Principles for multigrid acceleration

When a multigrid scheme is adopted in the iterative solution of elliptic equations the role of the calculations on the coarser grids of a cycle is to damp errors of wavelengths of the order of the dimensions of the grid elements or cells. Since the iterative process can be interpreted as a marching process it seems plausible that the advance to the steady state by solution of hyperbolic equations can also be accelerated by the use of multigrids, with the same role for the coarse-grid calculations. This seems to be the view taken by Jameson (1983), who has demonstrated the marked power of multigrid acceleration in hyperbolic systems. (Ni (1981), also demonstrated such a power but he gave little indication of his guiding principles.) In the course of developing the present multigrid schemes it has been helpful to take a different view.

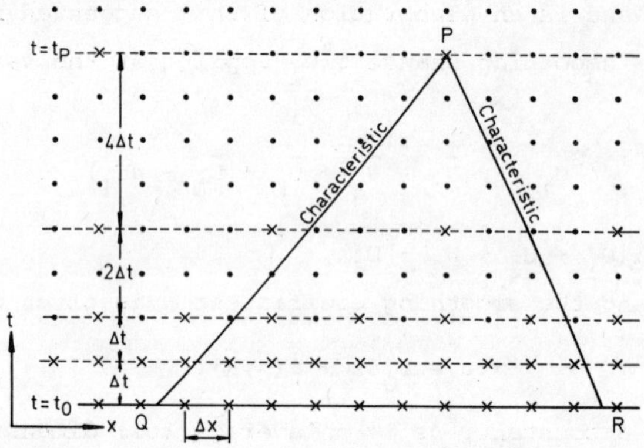

Fig. 3 Multigrid solution of Cauchy's initial-value problem for $f(x,t) = 0$.

We begin by recalling the solution of Cauchy's initial-value problem for the simple hyperbolic equation $f(x,t) = 0$. Fig 3 is a sketch of a grid in the x,t plane showing a line $t = t_O$ on which initial conditions are supposedly given, a point P at the time t_P, at which the solution is required, and the pair of characteristics through P which, together with $t = t_O$, define the domain of dependence of P. The uniqueness theorem states that conditions at P depend only on conditions on the segment of $t = t_O$ that is bounded by the characteristics through P. Let us suppose that the time interval $t_P - t_O$ is much larger than the interval Δt given by the stability condition for the spatial interval Δx. We could, of course, advance from t_O to t_P to determine conditions at P by one of the established explicit or implicit methods. In such an explicit method we would take several steps Δt in each of which we would calculate the conditions at grid points Δx apart on a line t = const.

For each grid point the calculation would be simple but there would be a large number of calculations. In an implicit method we might take only one large step but the calculation of conditions at P would be more complex, involving the simultaneous solution of a set of equations for conditions at the other points along $t = t_P$. But we can observe, in the light of the uniqueness theorem, that both types of method would be unnecessarily laborious: the conditions at P are completely determined by conditions at the grid points from Q to R on $t = t_O$, and it is not theoretically necessary to deal with grid points lying between t_O and t_p, or with points other than P itself on t_p. Thus the multigrid scheme can be regarded as a particular way of reducing the effort required to calculate the solution at P, in which the number of grid points involved is reduced toward the minimum required by the uniqueness theorem. A plausible scheme might involve an advance from t_O to t_P in successively larger time-steps that are associated with successively coarser grids, as indicated on the Figure. In an advance to the steady state we would repeat time-step sequences of the type shown in Fig. 3. Thus we would advance from t_O to t_p, interpolate between the widely-spaced calculated results at t_p to obtain a new set of initial conditions specified at closely-spaced intervals along $t = t_p$, and advance beyond t_p in the same way as we advanced from t_O to t_p. For the process to succeed we require that calculations on a coarse grid

(i) preserve the initially specified information that is needed to determine the solution at P, and

(ii) advance the solution efficiently towards the required state.

Some guiding principles for the construction of a multigrid scheme follow. To preserve information from a previous, finer, grid a coarse-grid calculation should itself satisfy the Euler equations and be formulated so that changes δU vanish when corresponding changes in the previous step on the finer grid vanish. Thus in interpolating coarse-grid results, only changes should be interpolated, not the dependent variables themselves. To advance the solution efficiently boundary quantities should be updated on the coarse grids as well as on the first, finest, grid of a cycle. Note that if a coarse-grid calculation were intended only to damp errors of the appropriate wavelength it would be natural to set instead $\delta U = 0$ on the boundaries; this is found greatly to reduce the rate of convergence to the steady state. Note also that while implicit algorithms may be considered in the search for computational efficiency, such algorithms do not readily fit the view we are taking of the multigrid process; in a multigrid framework explicit algorithms may be as efficient as implicit ones. Only explicit algorithms are considered here.

3.2 A multigrid scheme with Lax-Wendroff time-stepping

To advance the solution on a coarse grid in the required direction the coarse-grid changes could be obtained from equations (2.6) and (2.7) but, if the first-order changes corresponding to ΔU_A, ΔU_B, etc. are required for (2.6) were evaluated directly from (2.4) on the coarse grid, the results would be unacceptably inaccurate and, moreover, the changes on the coarse grid would not vanish as the changes on the previous, finer, grid vanish. Ni (1981) proposes that the first-order changes required for (2.6) be replaced by quantities derived from the overall changes on the previous grid but he does not elaborate further. The present scheme is a development of Ni's proposal that follows the guidelines of the previous section.

Suppose that the coarse-grid calculation under consideration involves an advance $\Delta t^{n+1} = t^{n+1} - t^n$. The Taylor-series expansion for the corresponding change in U can be written

$$(\delta U)^{n+1} \equiv U^{n+1} - U^n = \left(\frac{\partial U}{\partial t}\right)^n \Delta t^{n+1} + \left(\frac{\partial}{\partial t}\left(\frac{\partial U}{\partial t}\right)\right)^n \frac{(\Delta t^{n+1})^2}{2} + \dots ,$$

while for the previous time-step, on the next finer grid, there is another expansion

$$(\delta U)^n \equiv U^n - U^{n-1} = \left(\frac{\partial U}{\partial t}\right)^n \Delta t^n - \left(\frac{\partial}{\partial t}\left(\frac{\partial U}{\partial t}\right)\right)^n \frac{(\Delta t^n)^2}{2} + \dots ,$$

so that elimination of $(\partial U/\partial t)^n$ yields

$$(\delta U)^{n+1} = (\delta U)^n \frac{\Delta t^{n+1}}{\Delta t^n} + \frac{1}{2}\left(\frac{\partial}{\partial t}\left(\frac{\partial U}{\partial t}\right)\right)^n \left((\Delta t^{n+1})^2 + \Delta t^n \Delta t^{n+1}\right) .$$

Substitution from equations (2.1) and change in the order of differentiation yields

$$(\delta U)^{n+1} = (\delta U)^n \frac{\Delta t^{n+1}}{\Delta t^n} - \frac{1}{2}\left(\frac{\partial}{\partial x}\left(\frac{\partial F}{\partial U}\frac{\partial U}{\partial t}\right) + \frac{\partial}{\partial y}\left(\frac{\partial G}{\partial U}\frac{\partial U}{\partial t}\right)\right)^n \times$$

$$. \left((\Delta t^{n+1})^2 + \Delta t^n \Delta t^{n+1}\right) ,$$

and substitution of

$$\frac{\partial U}{\partial t} = \frac{(\delta U)^n}{\Delta t^n} + O(\Delta t)$$

finally yields

$$(\delta U)^{n+1} = (\delta U)^n \frac{\Delta t^{n+1}}{\Delta t^n} - \frac{1}{2}\left(\frac{\partial}{\partial x}\left(\frac{\partial F}{\partial U}\,\delta U\right) + \frac{\partial}{\partial y}\left(\frac{\partial G}{\partial U}\,\delta U\right)\right)^n \frac{\Delta t^{n+1}}{\Delta t^n} \times$$

$$. \ (\Delta t^{n+1} + \Delta t^n) + \dots \ . \qquad (3.1)$$

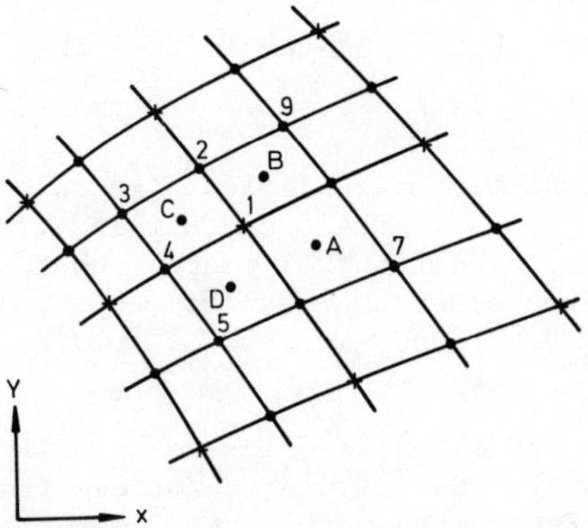

Fig. 4 Geometry of finite-volume scheme for multiple grids

Fig. 4 shows a coarse grid, with the cell vertices marked by crosses, together with the finer grid used for the previous time step. The former is obtained by omitting every second line of the latter. The total change at the vertex 1 on the coarse grid is taken to be the change for the element 3579. The first-order part of the change ΔU_1 is an average of the values of $(\delta U)^n(\Delta t^{n+1}/\Delta t^n)$ over the element, namely

$$\Delta U_1^{n+1} = \left(\frac{(\delta U)_A^n.\ \Delta A_A + (\delta U)_B^n.\ \Delta A_B + (\delta U)_C^n.\ \Delta A_C + (\delta U)_D^n.\ \Delta A_D}{\Delta A_A + \Delta A_B + \Delta A_C + \Delta A_D}\right)\frac{\Delta t^{n+1}}{\Delta t^n}, \qquad (3.2)$$

where $(\delta U)_C^n = \frac{1}{4}\left((\delta U)_1^n + (\delta U)_2^n + (\delta U)_3^n + (\delta U)_4^n\right)$.

Note that this average conserves changes in mass and momentum.

The second-order part of the change is obtained from equation (3.1) as before, by application of the integral theorem of Gauss and the trapezoid rule. Again, for convenience, let $\Delta F \equiv (\partial F/\partial U)\,\delta U$ and $\Delta G \equiv (\partial G/\partial U)\,\delta U$. The final result for the total change at 1 is

$$\delta U_1^{n+1} = \Delta U_1^{n+1} + \Big((\Delta F_3 - \Delta F_7)(y_9 - y_5) + (\Delta F_5 - \Delta F_9)(y_3 - y_7)$$

$$+ (\Delta G_7 - \Delta G_3)(x_9 - x_5) + (\Delta G_9 - \Delta G_5)(x_3 - x_7)\Big) \frac{\Delta t^{n+1}}{\Delta t^n}\left(\frac{\Delta t^{n+1} + \Delta t^n}{4\Delta A_1}\right), \tag{3.3}$$

where ΔA_1 is the area of the element 3579.

Equations (3.2) and (3.3) are evaluated on all but the finest grid of each cycle. It can be seen that if $(\delta U)^n \to 0$ then $(\delta U)^{n+1} \to 0$, which is a necessary feature if the scheme is to be convergent.

The boundary conditions are implemented on each coarse grid as well as on the finest. The procedure is straightforward and follows that given in section 2.4 except that, instead of evaluating the first-order changes from equation (2.4), we take averages given by (3.2) and, instead of solving (2.7) we solve (3.3).

The time steps for a coarse grid are specified by making use of the time steps already calculated for the finest grid of the cycle. In the first coarse grid in the cycle it can be deduced, from inspection of Fig. 4 together with equations (3.2) and (3.3), that $(\Delta t_1)^{n+1} = (\Delta t_1)^n$. For subsequent coarse grids we take (Fig. 4)

$$(\Delta t_1)^{n+1} = 2 \cdot \min\left((\Delta t)^n_A, (\Delta t)^n_B, (\Delta t)^n_C, (\Delta t)^n_D\right), \tag{3.4}$$

where A, B, C and D are points on the grid that is two levels finer than the current grid. Thus if the time steps were all equal on the first, fine, grid the complete cycle of time steps would be Δt, Δt, $2\Delta t$, $4\Delta t$, $8\Delta t$, etc.

Numerical smoothing of the type defined by equation (2.19) cannot be applied on the coarser grids, because the resulting changes would not tend to zero when $(\delta U)^n \to 0$ and the cycles would not converge. The only smoothing applied on coarse grids here is that implied by the averaging of values of $(\delta U)^n$ in equation (3.2).

Interpolation for intermediate vertices is necessary at the end of calculations on the coarsest grid of a cycle, to provide a complete set of initial conditions for calculation on the first, fine, grid of the next cycle. Here we simply interpolate the changes linearly. This has no effect on the accuracy of the steady-state results, because only changes in time are interpolated.

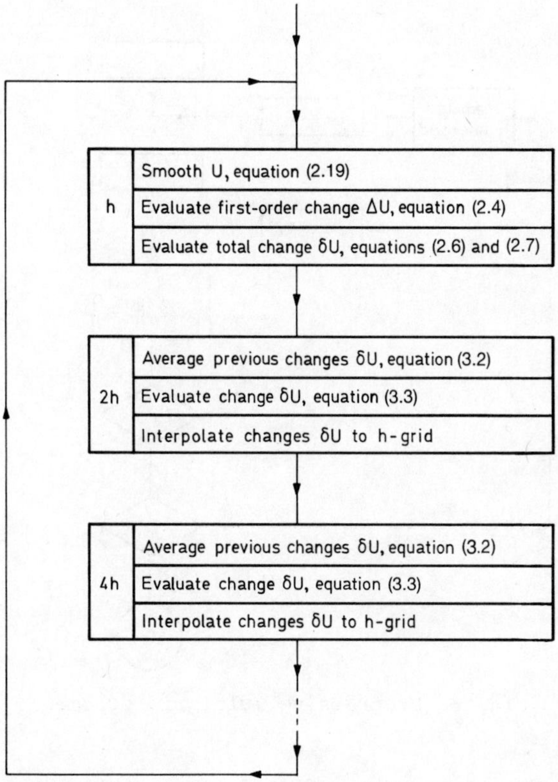

Fig. 5 Cyclic procedure in multigrid scheme with Lax-Wendroff time-stepping.

A multigrid cycle is summarized in the computational flow diagram sketched in Fig. 5. The grids in a cycle are denoted by h, 2h, etc, where h is a typical cell dimension on the finest grid. The overall multigrid scheme has a progressive character, as indicated in the flow diagram sketched in Fig. 6; the number of grid levels in a cycle is denoted by ℓ, and the number of cycles at a fixed number of levels by i. The calculation begins with relatively large cells on the finest grid in each cycle, together with relatively few grid levels per cycle. Extra levels are added progressively in the course of the calculation, with each additional level having a finer grid than the finest hitherto. Linear interpolation is used when adding an extra level.

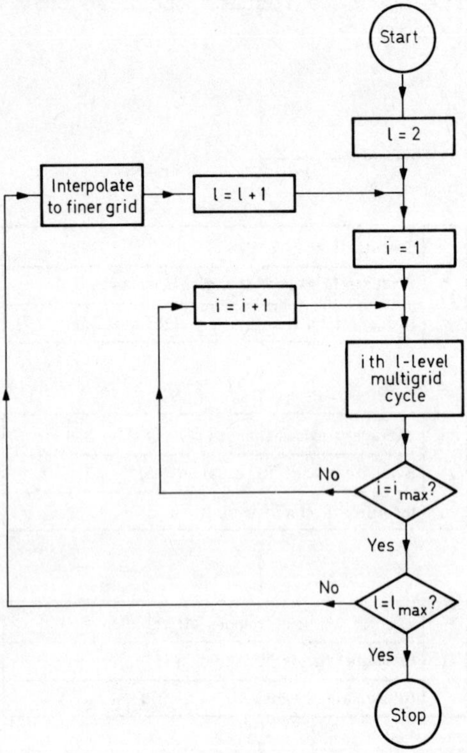

Fig. 6 Progressive multigrid scheme

3.3 Multigrid with Lax-Wendroff and multi-stage time-stepping

In the previous section a multigrid scheme was described in which Lax-Wendroff time-stepping is used at every step of a multigrid cycle. It is possible also to use multi-stage time-stepping at every step of the cycle but, as noted in the Introduction, it is found that a composite multigrid scheme, with both Lax-Wendroff and multi-stage time-stepping, gives converged results in less computing time than either Lax-Wendroff or multi-stage time-stepping on its own.

For the composite scheme, the Lax-Wendroff time-stepping of section 2.2 is retained for the calculation on the first, finest grid of the multigrid cycle, but for each of the coarse grids a four-stage time-stepping algorithm similar to (2.12) is used. The algorithm includes a forcing function P and consists of the stages

$$\left.\begin{aligned}
U_{2h}^{(0)} &= U_h^n \\
U_{2h}^{(1)} &= U_{2h}^{(0)} + \nu\alpha_1[\Delta U_{2h}^{(0)} + \Delta tP] \\
U_{2h}^{(2)} &= U_{2h}^{(0)} + \nu\alpha_2[\Delta U_{2h}^{(1)} + \Delta tP] \\
U_{2h}^{(3)} &= U_{2h}^{(0)} + \nu\alpha_3[\Delta U_{2h}^{(2)} + \Delta tP] \\
U_{2h}^{(4)} &= U_{2h}^{(0)} + \nu[\Delta U_{2h}^{(3)} + \Delta tP] \\
U_{2h}^{n+1} &= U_{2h}^{(4)} \\
\delta U_{2h}^{n+1} &= U_{2h}^{n+1} - U_h^n ,
\end{aligned}\right\} \quad (3.5)$$

where α_1, α_2 and α_3 are constants and ν is the Courant number.

The residual $\Delta U_{2h}^{(0)}$ at the vertex 1 in Fig. 4 is given by a formula analogous to (2.6), namely

$$(\Delta U_{2h})_1 = \frac{\Delta U_7 \, . \, \Delta A_7 + \Delta U_9 \, . \, \Delta A_9 + \Delta U_3 \, . \, \Delta A_3 + \Delta U_5 \, . \, \Delta A_5}{\Delta A_7 + \Delta A_9 + \Delta A_3 + \Delta A_5} , \quad (3.6)$$

where the subscripts 7, 9, 3 and 5 refer to coarse-grid cells with 'centres' at 7, 9, 3 and 5 respectively, and the first-order changes ΔU_7, ΔU_9, etc, are given by the formula analogous to (2.4). The forcing function P is given by

$$P = \left(\frac{\Delta U_h}{\nu\Delta t}\right)^n - \frac{\Delta U_{2h}^{(0)}}{\Delta t} + \Delta R_{2h}, \quad (3.7)$$

where $\Delta U_{2h}^{(0)}$ is given by the formula (3.6); ΔU_h is given by (*cf* equation (3.2))

$$(\Delta U_h)_1 = \frac{(\delta U)_A \, . \, \Delta A_A + (\delta U)_B \, . \, \Delta A_B + (\delta U)_C \, . \, \Delta A_C + (\delta U)_D \, . \, \Delta A_D}{\Delta A_A + \Delta A_B + \Delta A_C + \Delta A_D} , \quad (3.8)$$

where $(\delta U)_C = \frac{1}{4}\left((\delta U_h)_1 + (\delta U_h)_2 + (\delta U_h)_3 + (\delta U_h)_4\right)$;

and ΔR_{2h} is a stabilizing term given by (Fig. 4, *cf* equation (2.10)

$$(\Delta R_{2h})_1 = \{(\Delta F_3 - \Delta F_7)(y_9 - y_5) + (\Delta F_5 - \Delta F_9)(y_3 - y_7)$$

$$+ (\Delta G_7 - \Delta G_3)(x_9 - x_5) + (\Delta G_9 - \Delta G_5)(x_3 - x_7)\}/(4\Delta A_1), \qquad (3.9)$$

where $\Delta F_3 \equiv [(\partial F/\partial U).\delta U_h]_3$, etc, and ΔA_1 is the area of the cell 3579. Note that when $\delta U_h \to 0$

$$P \to -\frac{\Delta U_{2h}^{(0)}}{\Delta t}$$

and hence

$$U_{2h}^{n+1} \to U_h^n$$

as required. In addition, the forcing function is intended to transmit changes from the previous h grid; substitution for P as given by (3.7) into the expression for $U_{2h}^{(1)}$ in (3.5) yields

$$U_{2h}^{(1)} = U_{2h}^{(0)} + \nu\alpha_1\Delta t\left[\left(\frac{\Delta U_h}{\nu\Delta t}\right)^n + \Delta R_{2h}\right],$$

and it can be seen from (3.8) and (3.9) that the terms within the square brackets constitute a residual given by changes on the previous grid.

The other features of the multigrid scheme are straightforward. The time steps for a coarse grid are now $\nu\Delta t$ where ν is the Courant number and Δt is the Lax-Wendroff time-step, as given by equation (3.4) in section 3.2. As with Lax-Wendroff multigrid, no numerical smoothing of the type defined by equation (2.19) is applied on any coarse grid, and coarse-grid changes are interpolated linearly to obtain a complete set of initial conditions for calculations on the first, fine, grid of the next cycle. The overall multigrid scheme is identical to that sketched in Fig. 6.

Of course, the procedure followed in each cycle differs from that sketched in Fig. 5 for the Lax-Wendroff multigrid; a multigrid cycle of the composite scheme is summarized in the flow diagram sketched in Fig. 7.

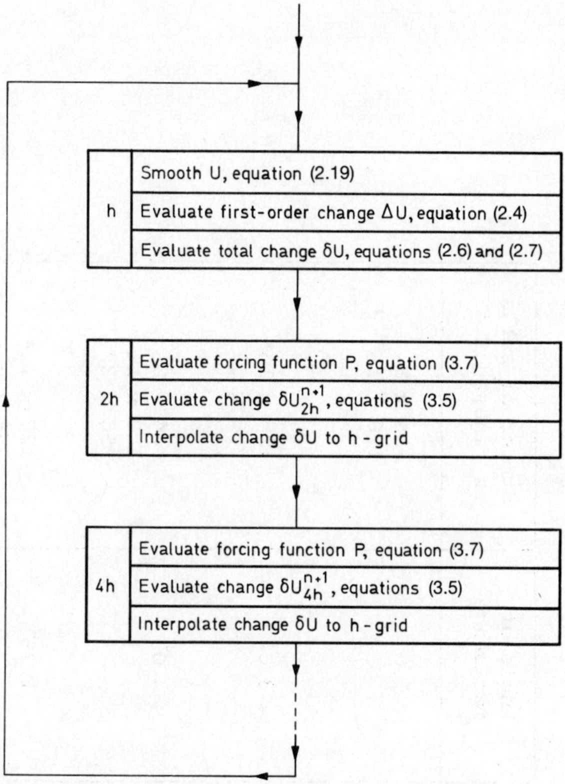

Fig. 7 Cyclic procedure with Lax-Wendroff and multi-stage time-stepping

4. NUMERICAL RESULTS

4.1 Scope of calculations

The present Lax-Wendroff method has been used to calculate the flow of a perfect gas, with $\gamma = 1.4$, past the RAE 2822 and NACA 0012 aerofoils for a variety of conditions. These are listed in Table 1, which shows that for each free-stream condition (M_∞ and incidence) at least two calculations were performed. For the cases marked by an asterisk the composite method was also used to calculate the flow. For all the listed calculations the far-field boundary was located at a distance of approximately eight chord-lengths from the aerofoil. Other calculations not reported here have been performed for distances ranging from 4-16 chords; no significant changes in the results were found for distances beyond eight chords and,

Table 1

List of calculations of flow past RAE 2822 and NACA 0012 aerofoils

Aerofoil	M_∞	Angle of incidence	Number of cells on finest grid
RAE 2822	0.696	1.0°	64 x 8 128 x 16 256 x 32
	0.75	3.0°	128 x 16* 256 x 32
NACA 0012	0.63	2.0°	64 x 8 128 x 16 256 x 32
	0.80	1.25°	128 x 16* 256 x 32

indeed, for subcritical flows no significant changes were found over the entire range.

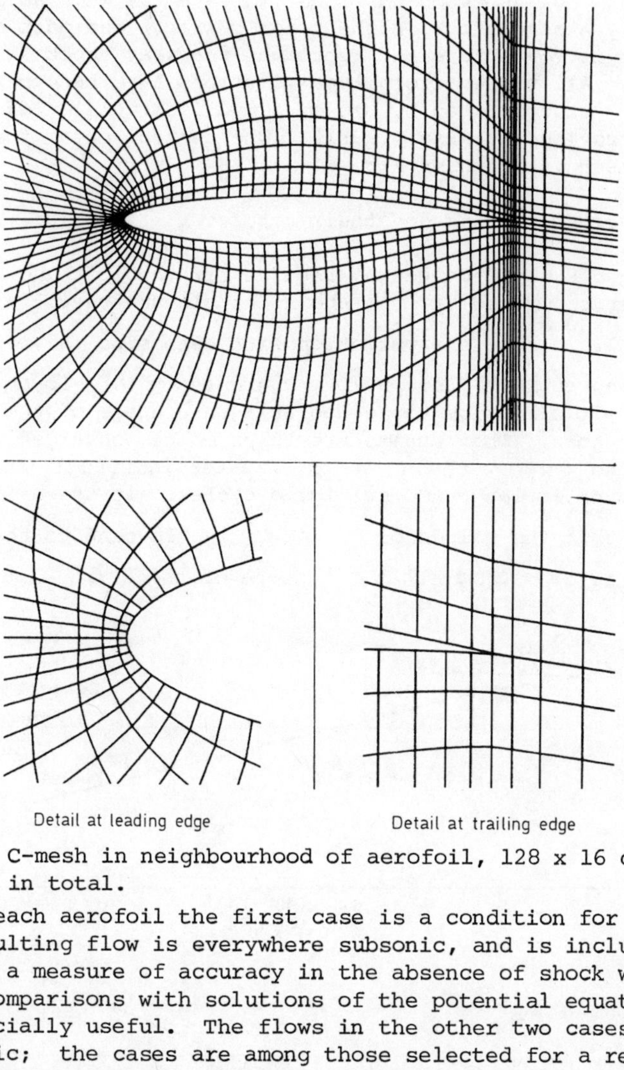

Fig. 8 C-mesh in neighbourhood of aerofoil, 128 x 16 cells in total.

For each aerofoil the first case is a condition for which the resulting flow is everywhere subsonic, and is included to provide a measure of accuracy in the absence of shock waves, where comparisons with solutions of the potential equation can be especially useful. The flows in the other two cases are transonic; the cases are among those selected for a recent AGARD test (FDP WG-07. Report in preparation), and each presents a challenge. The RAE 2822 is an aerofoil with a leading-edge shape that rapidly accelerates the flow over the nose and a rear camber that induces substantial flow gradients locally. At the chosen free-stream condition there is in addition a large region of supersonic flow over the upper surface terminated by a strong shock wave. Thus this case provides a test of capability for dealing with strong shock waves and with substantial flow gradients at nose and rear.

The transonic case chosen for the NACA 0012 aerofoil tests another aspect of the capability for dealing with shock waves. There is a weak shock on the lower surface (as well as a moderately strong shock on the upper surface) and part of the challenge is to capture and reproduce the weak shock as well as possible. All these aerofoil calculations have been performed on a 'C'-mesh generated by use of a modified version of a program provided by Rizzi (1981). The central part of a typical mesh is shown in Fig. 8.

4.2 Converged results and their accuracy

In Figs. 9 to 12 converged results for the lift and drag coefficients, C_L and C_D, are shown for all the cases listed in Table 1. To show the variation with cell size the results are plotted against $(\Delta s)^2 \times 10^4$, where $(\Delta s \times 10^2)$ is an average cell dimension expressed as a percentage of the aerofoil chord. The results are taken to be converged when there is no further change in the fourth significant figure, for C_L, over at least 100 multigrid cycles. It is desirable not only that the values of C_L and C_D should tend to the exact solution as $\Delta s \to 0$ but that the slope of the curve should be small.

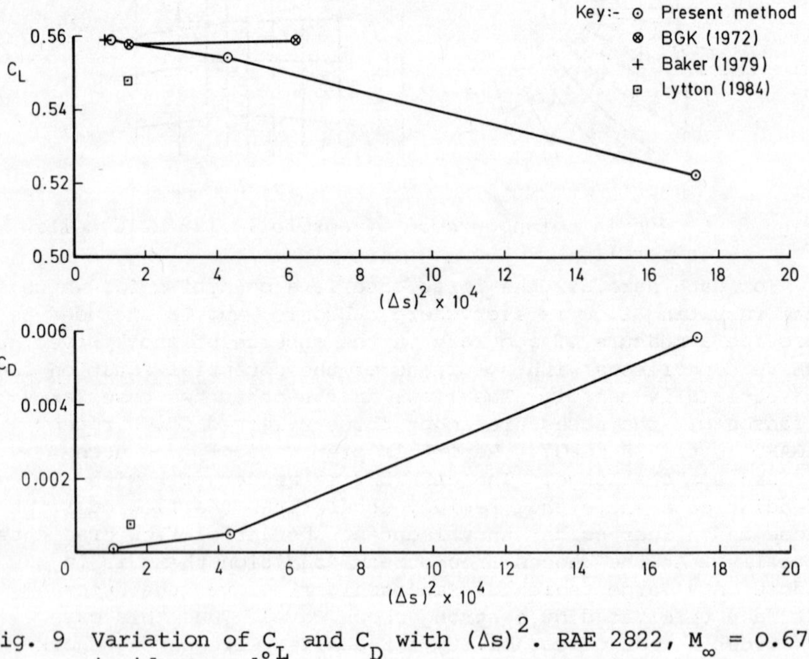

Fig. 9 Variation of C_L and C_D with $(\Delta s)^2$. RAE 2822, $M_\infty = 0.676$, incidence = 1°.

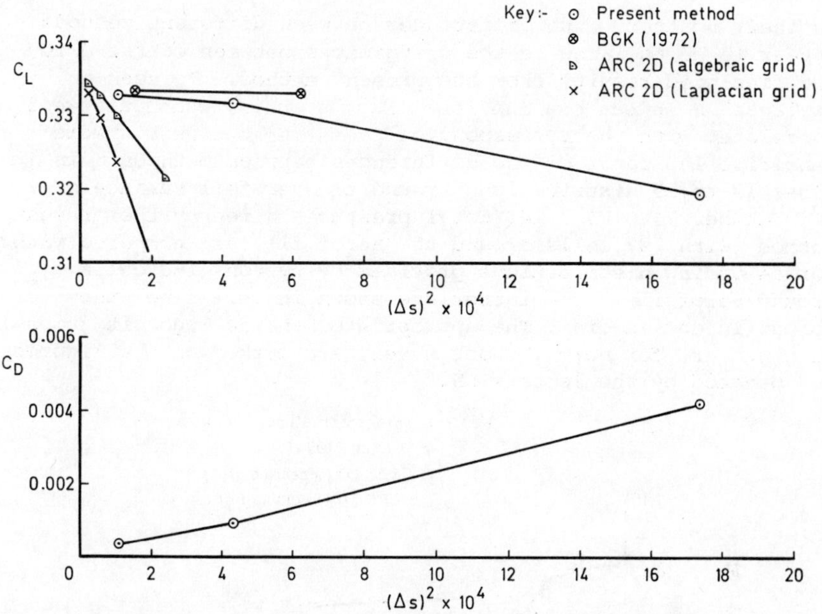

Fig. 10 Variation of C_L and C_D with $(\Delta s)^2$. NACA 0012, $M_\infty = 0.63$, incidence = 2°.

Figs 9 and 10 show results for the two subcritical cases, for the RAE 2822 and the NACA 0012 aerofoils respectively. Included for comparison are values of C_L from solution of the potential equation by the well-established methods of Bauer, Garabedian and Korn (1972) (BGK) and Baker (1979) (Fig. 9 only), and from a solution of the Euler equations by a recent method of Lytton (1984) (Fig. 9 only). Also included in Fig. 10 are the only available variations with Δs of the solution from an alternative method (Flores et al, 1984) (ARC 2D), due to Steger and Pulliam, for the Euler equations. It can be seen from the figures that the present method gives results that closely approximate the solutions of the potential equation when a fine grid ($\Delta s \doteq 0.01$, 192 cells around the aerofoil) is used. Even with 96 cells the method may prove to be accurate enough for some purposes, and this possibility is discussed when we deal with the practical calculation of supercritical flows.

Figs. 11 and 12 show results for the two supercritical cases listed in Table 1. Included are more results obtained by the use of the ARC 2D method (Fig. 11) and the method of Lytton (1984) and a recent solution of the Euler equations by Jameson (1984), (Fig. 12). A feature of these supercritical results is their scatter; the differences between different methods appear to be as large as the differences between coarse-grid and fine-grid results from the present method. To seek an explanation we can examine local Mach numbers, which are well correlated with the corresponding pressures but seem a more sensitive indicator of the differences between methods. In Figs. 13 to 15 distributions around the aerofoil surface of Mach number and (Fig. 14) total pressure, given by the present method (with 192 cells around the aerofoil), are compared with corresponding distributions provided by Lytton (160 cells around aerofoil). The latter are shown in full line, the former in dotted line. The appropriate Rankine-Hugonoit jumps in Mach number, for normal shock waves, are marked on the Figures and denoted by the letters RH.

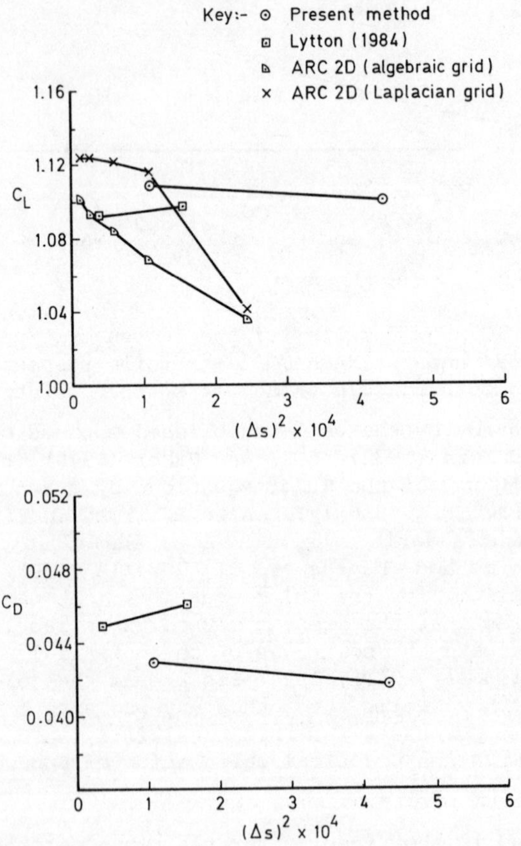

Fig. 11 Variation of C_L and C_D with $(\Delta s)^2$. RAE 2822, $M_\infty = 0.75$, incidence = 3°.

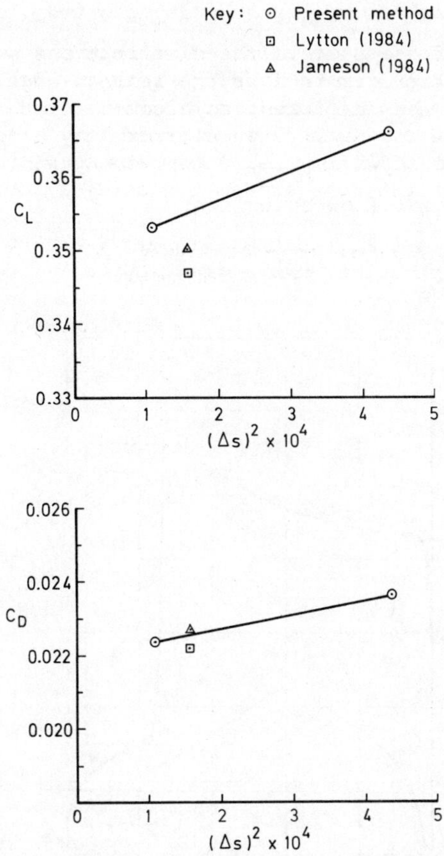

Fig. 12 Variation of C_L and C_D with $(\Delta s)^2$. NACA 0012, $M_\infty = 0.80$, incidence = 125°.

Figs. 13 and 14 help to explain the differences in lift and drag shown in Fig. 11 for the RAE 2822 aerofoil. The present method gives a significantly higher Mach number over the whole of the upper surface, with the difference being pronounced near the leading edge. This is sufficient to outweigh the additional lift, in Lytton's distribution, from having a shock wave that is both steeper and slightly further aft. The total pressure distributions in Fig. 14 indicate a possible cause of the difference in levels of Mach number. The total pressure p_0 is given by

$$p_0 = p\left(1 + \frac{\gamma - 1}{2} M^2\right)^{\frac{\gamma}{\gamma-1}}$$

and, in an exact solution, the quantity plotted $(1 - p_O)$ would be zero forward of the shock. The distributions show that there are much larger errors near the leading-edge in Lytton's result. These may be sufficient to account for the difference in Mach number. Incidentally, such errors may also account for the difference shown in Fig. 9 between Lytton's C_L and the values from potential flow solutions.

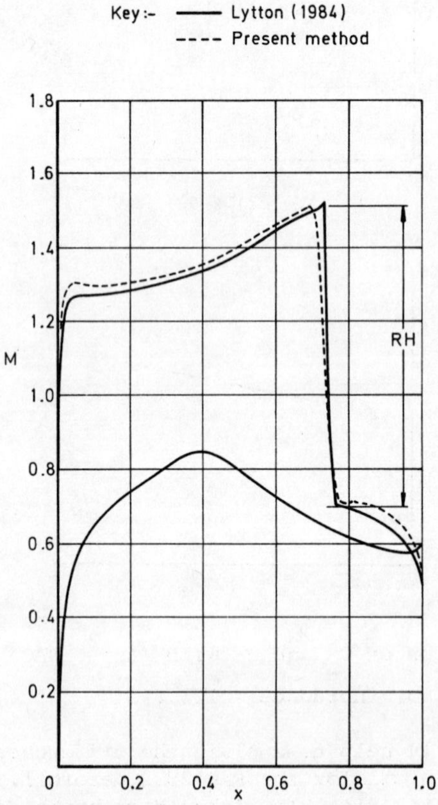

Fig. 13 Comparison of Mach number distributions. RAE 2822, $M_\infty = 0.75$, incidence = 3°.

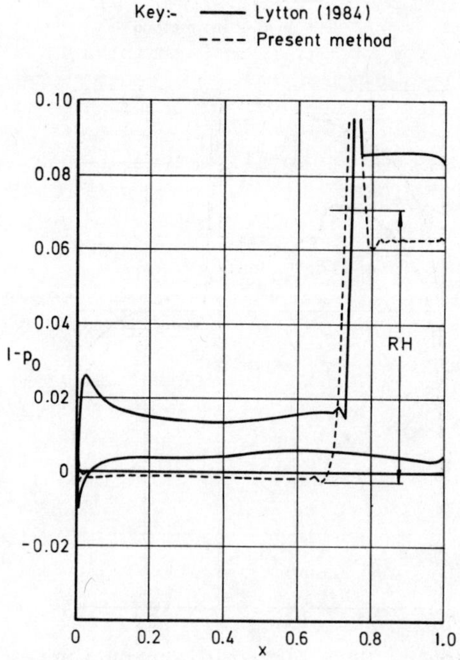

Fig. 14 Comparison of total pressure distributions. RAE 2822, $M_\infty = 0.75$, incidence = 3°.

Fig. 15 helps to explain the differences shown in Fig. 12 for the NACA 0012 aerofoil with $M_\infty = 0.80$. The differences in the Mach number distributions are small compared with those shown in Fig. 13, and yet they are sufficient to give a difference of nearly 2% in lift. Here, the differences are seen to be mainly at the shock, which is captured better by Lytton, and in the Mach number levels, which might again be attributable to differences in the magnitudes of the errors generated near the leading edge.

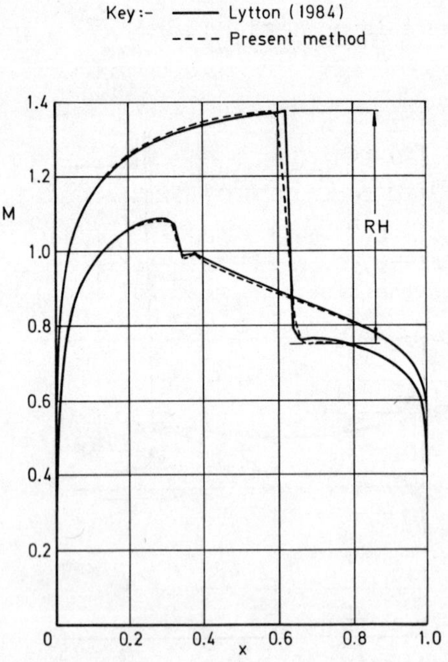

Fig. 15 Comparison of Mach number distributions. NACA 0012, $M_\infty = 0.80$, incidence = $1.25°$.

4.3 Convergence histories for present calculations, and practical implications

An extended convergence history for C_L and $\log_{10}\left|\frac{\delta\rho}{\Delta t}\right|_{mean}$ is shown in Fig. 16. The residual $\left|\frac{\delta\rho}{\Delta t}\right|_{mean}$ is derived from the change $\delta\rho$ at a vertex over the local time step Δt for the finest grid of each cycle, with the mean taken over the entire field. Variations with time, in CPU seconds (CRAY-1S), are shown*. Each arrow on the residual history curve points to a stage of the calculation over which the number of grid levels is kept constant (the stage begins with a sharp peak in the residual); the corresponding caption gives the number of cycles in that stage, the number of levels, and the cycle rate. It appears that in each stage the residual falls at a roughly uniform rate following an initial rapid fall; there is no sign of any levelling off. An examination of the convergence history for

---

* All calculations were performed on the CRAY-1S computer at the Royal Aircraft Establishment, using the vector-processing capability of the machine. The CPU times given in Fig. 16 to 20 were all obtained with the Lax-Wendroff multigrid scheme.

C_L shows that there is no need in practice to drive the residual down to 10^{-5} say. In this example the error in lift ΔC_L, where $\Delta C_L \equiv |C_L - C_L^*|$ and C_L^* is the converged result (to five significant figures), is within 0.5% of C_L^* after about 20 seconds, when the residual is just below 10^{-3}. The marks A, B and C on the C_L history are the times beyond which the error is less than 1%, 0.5% and 0.1% respectively. The same marks are used in the histories that follow.

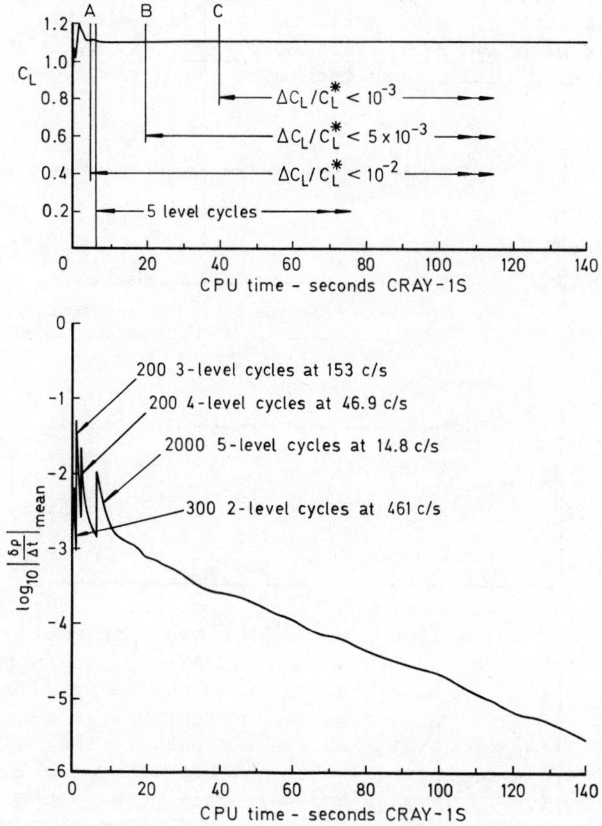

Fig. 16 Extended convergence history for C_L and $\log_{10}|\delta\rho/\Delta t|_{mean}$ RAE 2822, $M_\infty = 0.75$, incidence = 3°.

Convergence histories with an expanded time scale, for C_L, C_D and the residual, for the RAE 2822 at $M_\infty = 0.75$ and the NACA

0012 at $M_\infty = 0.80$, are shown in Figs. 17 and 18 respectively. In addition to histories for calculations that terminate with 5-level cycles and 256 x 32 cells on the finest grid, we show histories for calculations that terminate with 4-level cycles and 128 x 16 cells on the finest grid (in dotted line). For the latter histories, the 1%, 0.5% and 0.1% error marks are denoted by A', B' and C'. It can be seen that for both aerofoils there is for C_L and C_D a rapid approach, initially, towards the steady state and little further change thereafter; the 1% error mark in C_L is reached in about 5 seconds and 14 seconds respectively. The convergence histories for the calculations with 128 x 16 cells follow the histories for the calculations with 256 x 32 cells fairly closely but the various error marks in C_L and C_D are much more rapidly approached.

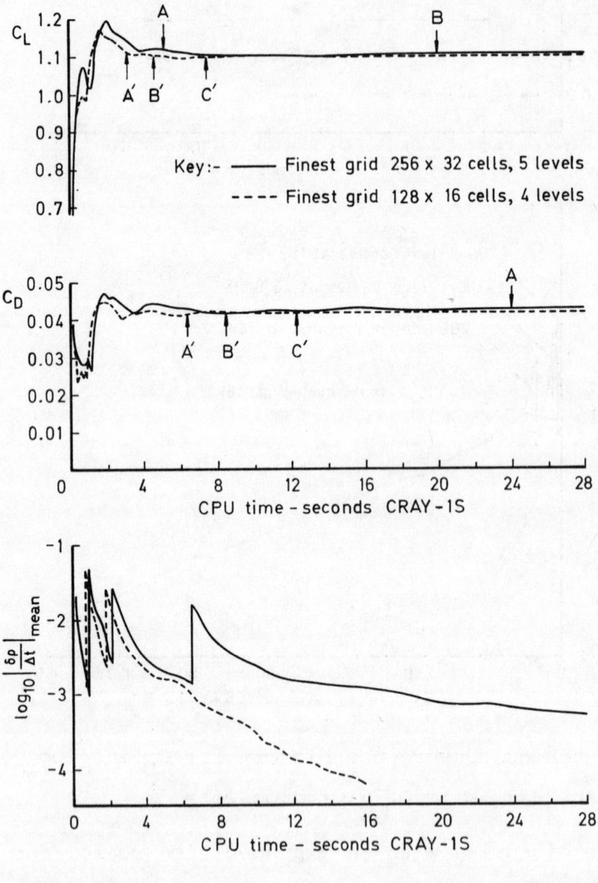

Fig. 17 Convergence history for C_L, C_D and $\log_{10}|\delta\rho/\Delta t|_{mean}$ RAE 2822, $M_\infty = 0.75$, incidence = 3°.

Since the converged values of C_L and C_D appear to differ, in taking 128 x 16 rather than 256 x 32 cells, by no more than the difference between corresponding results from different methods the practical advantage in confining calculations to the coarser grid seem obvious: not only would there be a appreciable savings in computer time, there would also be considerable savings in computer memory requirements. To confirm this assessment we consider the corresponding coarse-grid distributions of Mach number.

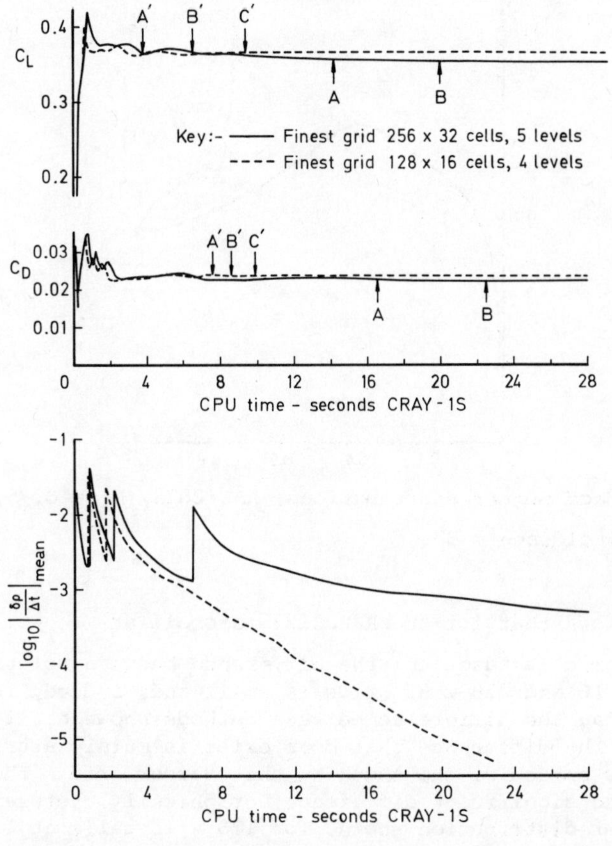

Fig. 18 Convergence history for C_L, C_D and $\log_{10}|\delta\rho/\Delta t|_{mean}$ NACA 0012, $M_\infty = 0.80$, incidence = 1.25°.

In Figs. 19 and 20 comparisons are shown of distributions from converged, 5-level, 256 X 32 cell solutions with distributions obtained after 100 4-level cycles where the finest grid consists of 128 x 16 cells. The time taken to reach the latter is 3.6 seconds CPU, starting from uniform flow.

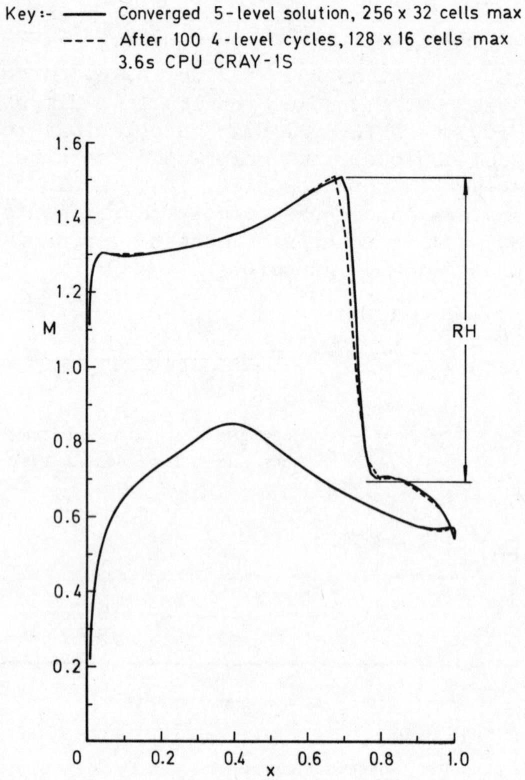

Fig. 19 Mach number distributions, RAE 2822, $M_{\infty} = 0.75$, incidence = 3°.

Fig. 19 shows that for the RAE 2822 aerofoil at $M_{\infty} = 0.75$ the 128 x 16 grid is adequate; the difference between solutions on the 128 x 16 and 256 x 32 grids is small and, indeed, is smaller than the difference between methods shown earlier in Fig. 13. The difference that does exist is mainly attributable to a wider spread of the shock on the coarser grid. There is, in fact, no significant difference, graphically, between the Mach number distribution shown, for 128 x 16 cells at 100 cycles, and the converged solution on the same grid. This might be expected from the position in this case of the 1% error mark A' - at about 3.2 seconds (Fig. 17). The same conclusion regarding the adequacy of the grid can be drawn for the other case, with the qualification that a computer time of 3.6 seconds is no longer enough. This would be expected from the convergence histories shown in Fig. 18. Fig. 20 shows the Mach number for the NACA 0012 aerofoil at $M_{\infty} = 0.80$. It can be seen that the weak shock on the lower surface is not well represented after 100 cycles on the coarse grid. Included on the Figure are points from the converged 128 x 16 solution, and these demonstrate that the 128 x 16 grid is adequate for the capture of such shocks. In fact the distribution after about 7 seconds is indistinguishable from the converged distribution, and it differs from the 256 x 32 distribution by no more than the differences shown in Fig. 15.

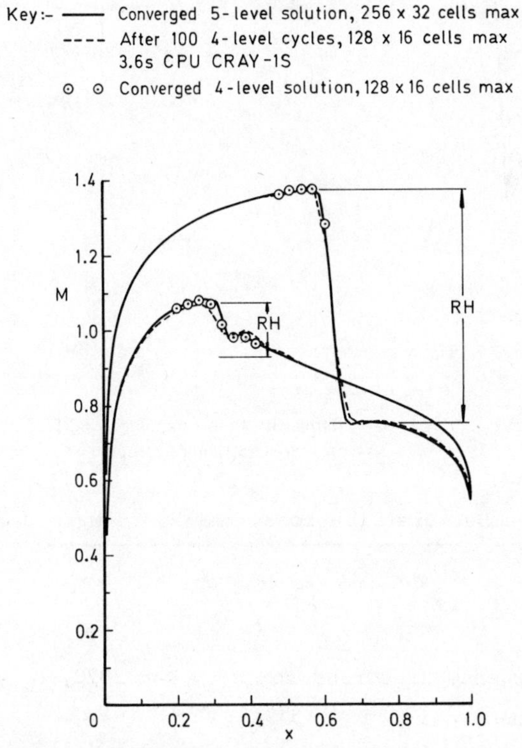

Fig. 20 Mach number distributions, NACA 0012, $M_\infty = 0.80$, incidence = 1.25°.

All the above results were obtained by use of the Lax-Wendroff multigrid scheme. For the cases marked by an asterisk in Table 1, use of the composite multigrid scheme yielded essentially identical converged results. However the convergence histories differ. Since the 128 x 16 grid seems fine enough for practical purposes we show in Figs. 21 and 22 comparisons of convergence histories, calculated using this grid, for the lift coefficient C_L of the RAE 2822 and NACA 0012 aerofoils respectively.

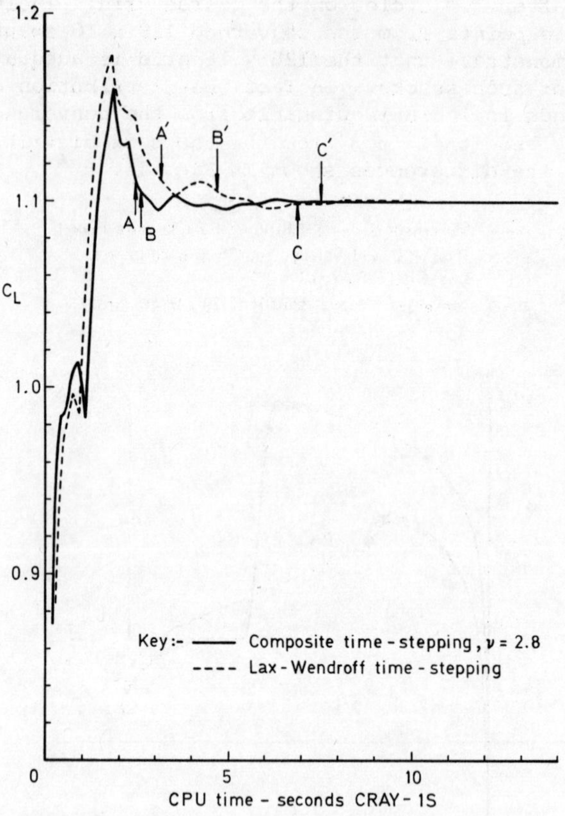

Fig. 21 Convergence histories for C_L, RAE 2822, M_∞ = 0.75, incidence = 3°

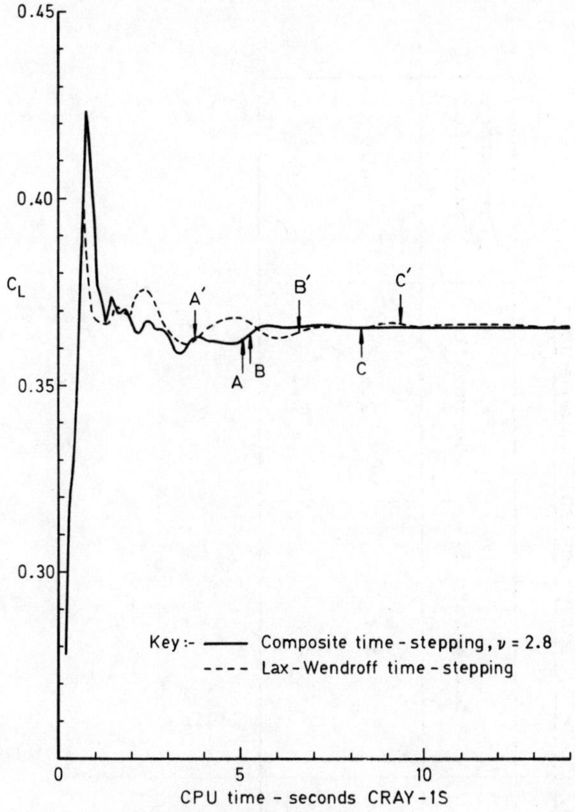

Fig. 22 Convergence histories for C_L, NACA 0012, $M_\infty = 0.80$, incidence = 1.25°.

The scales are not the same as in Figs. 17 and 18 but are expanded so as to show up more clearly the differences between the convergence histories (in broken and full lines). Also shown on Figs. 21 and 22 are the 1%, 0.5% and 0.1% error marks denoted, as in Figs. 17 and 18, by A, B and C. It can be seen that with the Courant number $\nu = 2.8$ the composite scheme yields significantly faster convergence. If we take a calculation to be sufficiently well converged for practical purposes at the 0.5% error mark, as suggested by the above discussion of Figs. 17 to 20, then the computing times for practical convergence are:

Table 2

CPU times (seconds, CRAY-1S) for practical convergence

Case	Lax-Wendroff	Composite	Percentage reduction
RAE 2822, $M_\infty = 0.75$, incidence $= 3°$	4.7	2.6	45%
NACA 0012, $M_\infty = 0.80$, incidence $= 1.25°$	6.6	5.3	20%

Thus, results of sufficient accuracy for practical purposes can be obtained with a relatively coarse 128 x 16 C-mesh in CPU times of 3-5 seconds. It seems worth noting, however, that the gains over the Lax-Wendroff scheme may not always be large, so that the Lax-Wendroff scheme may be preferred for its relative simplicity.

5. CONCLUDING REMARKS

The motivation in the present work has been the achievement of high accuracy at low cost in computational effort. The cell-vertex formulation of Ni (1981) and Denton (1981) has been adopted as a starting point because with it second-order accuracy may be easily obtained in principle, even for non-smooth grids. In achieving this accuracy, for highly stretched grids and at body boundaries, some extensions have been made, so that cell-vertex schemes now provide a firmer basis for practical calculations. To accelerate the convergence to the steady state, time-stepping multigrid schemes of Lax-Wendroff and multi-stage (Runge-Kutta) type have been developed. An alternative approach for hyperbolic problems has been suggested which has provided some guidance in the derivation of corresponding multigrid algorithms. Multi-stage time-stepping had been introduced by Jameson et al (1981) in a cell-average formulation of the Euler equations. Here a four-stage scheme has been coupled with the cell-vertex formulation and the scheme has been stabilized by the addition of a second-order time-change rather than a fourth-order dissipation. It has been demonstrated that use of the four-stage time-stepping for the calculations on the coarse grids in the multigrid cycle gives significant reductions in overall computing time.

While all the above improvements seem worthwhile, practical measures of accuracy and speed must include systematic comparisons with other methods and such comparisons have yet to be made. In the meantime there remains considerable scope for further development.

First, the accuracy of cell-vertex schemes does not appear to have been seriously tested for grids that are not smooth. Such grids are increasingly encountered as more complex configurations are treated.

Next, the capture of shock waves is open to criticism on more than one point. Empirical smoothing is applied, so that the strength and position of a captured shock, and the magnitide of the associated spatial oscillations that arise numerically, are sensitive to variations in the 'constants' adopted for the smoothing formula. This is, of course, a feature common to all shock-capturing schemes that are not of the upwind type. Here the constants have been chosen (partly by comparison with results from other methods) and then kept fixed for all the calculations performed, whatever the aerofoil, the grid, or the free-stream Mach-number. This is better than varying the constants to suit the case, but it would be better still if no empirical choice were involved. However, even if we accept the need for empirical smoothing there is still scope for improvement. The formula adopted here is essentially that proposed by Ni (1981), which has the merit of being simple but is, in fact, non-conservative. A new, conservative, formula may give results that are less sensitive to variations in the smoothing constants.

Finally, there seems to be considerable scope for improvement in the speed of multigrid algorithms for the Euler equations. Lax-Wendroff and multi-stage algorithms were respectively proposed by Ni (1981) and Jameson (1983). These were pioneering extensions of the multigrid concept to solution of the Euler equations, where the aim was to demonstrate feasibility rather than to find the fastest algorithm. It has been shown here that a combination of the Lax-Wendroff and multi-stage algorithms gives a significant reduction in computing time, but no attempt has been made to provide theoretical support, which might show how the composite scheme could be optimized, or to explore completely different alternatives.

REFERENCES

Baker, T.J., (1979). Unpublished results. Presented at GAMM Workshop on Numerical Methods for the Computation of Inviscid Transonic Flows with Shock Waves, Stockholm, 1979.

Bauer, F., Garabedian, P.R. and Korn, D.G. (1972). Supercritical Wing Sections. Lecture Notes in Economics and Mathematical Systems, No. 66, Springer-Verlag, Berlin, 1972.

Brandt, A., (1977). Multi-Level Adaptive Solutions to Boundary-Value Problems. *Math. Comp.*, Vol. **31**, No. 138, pp. 333-390.

Davis, R.L., (1983). The Prediction of Compressible, Laminar, Viscous Flows using a Time Marching Control Volume and Multigrid Technique. Paper 83-1896-CP, Proc. AIAA Sixth Computational Fluid Dynamics Conference, Danvers, Mass, 13-15 July, 1983.

Denton, J.D., (1981). An improved Time-Marching Method for Turbomachinery Flow Calculation. (Paper presented at Conference on Numerical Methods in Aeronautical Fluid Dynamics, University of Reading, UK, 30 March - 1 April, 1981) Numerical Methods in Aeronautical Fluid Dynamics (P.L. Roe, ed.) pp 189-210, Academic Press, 1982.

Flores J., Barton, J., Holst, T. and Pulliam, T., (1984). Comparisons of the Full Potential and Euler Formulations for Computing Transonic Airfoil Flow. Paper presented at 9th Int. Conf. Num. Methods in Fluid Dynamics, Cen-Saclay, France, 1984.

Jameson, A., Schmidt, W. and Turkel, E., (1981). Numerical Solution of the Euler Equations by Finite Volume Methods using Runge-Kutta Time-Stepping Schemes, AIAA Paper 81-1259.

Jameson, A., (1983). Solution of the Euler Equations by a Multigrid Method. Applied Mathematics and Computation, **13**, pp. 327-356.

Jameson, A., (1984). Numerical Solution of the Euler Equations for Compressible Inviscid Fluids. Princeton University, Report MAE 1643.

Lytton, C.C., (1984). Solution of the Euler Equations for Transonic Flow past a Lifting Aerofoil - the Bernoulli Formulation. RAE TR 84080.

Ni, R.-H., (1981). A Multiple Grid Scheme for Solving the Euler Equations. Paper 81-1025, Proc. AIAA Fifth Computational Fluid Dynamics Conference, Palo Alto, California, 22-23 June, 1981. (Also, with added appendices, *AIAA Journal,* Vol. **20**, No. 11, pp. 1565-1571, 1982).

Rizzi, A., (1981). Computational Mesh for Transonic Airfoils. In Numerical Methods for the Computation of Inviscid Transonic Flows with Shock Waves (A. Rizzi and H. Viviand, eds.) pp 222-254. Vieweg, Braunschweig/Wiesbaden, 1981.

MULTIGRID AND CONJUGATE GRADIENT ACCELERATION OF BASIC ITERATIVE METHODS

P. Sonneveld and P. Wesseling
*(Department of Mathematics and Informatics
Delft University of Technology, The Netherlands)*

and

P.M. de Zeeuw
*(Centre for Mathematics and Computer Science
Amsterdam, The Netherlands)*

1. INTRODUCTION

Let a discretisation of a partial differential equation be denoted by

$$Ay = b \tag{1.1}$$

This is a system of linear equations.

Stationary single-step iterative methods for solving (1.1) can be written as

$$y^{n+1} = y^n + B(b - Ay^n) \tag{1.2}$$

where the matrix B, which often is not formed explicitly, depends on the iterative method that is chosen.

If the discretisation is nonlinear, it is assumed that some linearisation leads to a linear system (1.1). The iterations (1.2) will then be inner iterations within an outer iteration on the nonlinearity.

In practical applications the number of unknowns in (1.1) can be quite large, and simple and straightforward iterative methods normally converge slowly. A significant recent development in numerical mathematics is the rise of preconditioned conjugate gradient methods (CG) and multigrid methods (MG). With respect to linear problems, both can be regarded as acceleration techniques for basic iterative methods. This provides us with a unifying point of view.

The purpose of this paper is to briefly review CG and MG from the viewpoint just mentioned, and to introduce the reader to the literature.

For completeness it should be mentioned that MG can also be applied directly to nonlinear problems, and this may lead to a more efficient solution method than linearisation followed by application of a fast linear solver. See (Hackbusch and Trottenberg, 1982) for a general introduction to MG. Up-to-date accounts of CG are given in (Hageman and Young, 1981; Golub and van Loan, 1983).

2. BASIC ITERATIVE METHODS

"Basic iterative method" (BIM for short) is a loose appellation for methods of type (1.2) that are simple and easy to implement. For a review of BIMs (including some that are not so simple) see (Young, 1971, or Varga, 1962). Examples are the Gauss-Seidel (GS) and SOR (successive over-relaxation) methods. The computational complexity of these methods for a typical problem such as the two-dimensional Poisson equation is $O(N^2)$ and $O(N^{3/2})$, respectively, with N the number of grid points. These numbers explain both the popularity of SOR in the sixties, and the large computer times needed for large N. Accelerated by CG or MG, the computational complexity becomes $O(N^{5/4})$ (conjecture) or $O(N)$, respectively, as will be discussed later.

The most popular BIMs used with CG and MG are GS methods and incomplete factorisation (IF) methods.

The reader is assumed to be familiar with GS methods. They come in various orderings (lexicographic, red-black etc.) and may be pointwise or blockwise (by lines or planes mostly). We give a short outline of IF methods.

An incomple LU-factorization (ILU) of the matrix A consists of a lower triangular matrix L and an upper triangular matrix U such that

$$LU = A + C \tag{2.1}$$

where C represents the error matrix. With C = O (complete factorization) L and U usually are much less sparse than A, which leads to large storage and computer time requirements. By allowing $C \neq O$ sparsity of L and U can be obtained. For example, if A represents a typical two-dimensional 7-point finite difference stencil of a second order elliptic partial difference operator on a rectangular grid, the sparsity patterns of L, U, A and C could be as in Fig. 2.1.

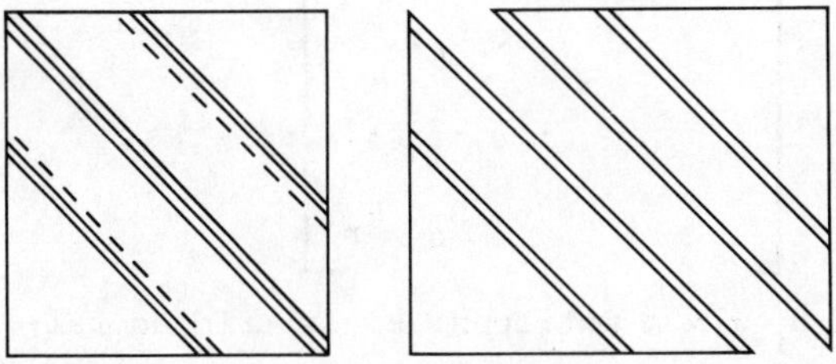

Fig. 2.1. Sparsity patterns. - - - : C

For simple formulae for the computation of L, U and C on a rectangular computational grid and theoretical results on the existence of L and U see (Meijerink and van der Vorst, 1977, 1981, Wesseling, 1982A, Sonneveld et al., 1985).

Equation (2.1) leads to the following iterative method for solving (1.1):

$$y^{n+1} = y^{n} + (LU)^{-1}(b - Ay^{n}) \tag{2.2}$$

Using (2.1) this can be rewritten as

$$y^{n+1} = (LU)^{-1}(b + Cy^{n}) \tag{2.3}$$

Hence, A is not needed, and L and U can be stored at the location of A. Usually C is not stored but computed. Hence, no extra storage is required. Because C is very sparse (cf. Fig. 2.1) the right hand side of (2.3) comes cheap. One iteration with (2.3) takes 18N operations, generation of L and U takes 21N operations, with N the number of unknowns: see for example (Wesseling, 1982A).

Incomplete block LU-factorization (IBLU or ILLU, incomplete line LU-factorization) of A goes as follows. On a rectangular grid with n vertical and n horizontal lines the matrix A has the following structure:

$$A = \begin{bmatrix} B_1 & U_1 & & & \\ L_2 & B_2 & U_2 & & \\ & L_3 & B_3 & U_3 & \\ & & \cdot & \cdot & \cdot \\ & & & \cdot & \cdot & \cdot \\ & & & & L_n & B_n \end{bmatrix} \tag{2.4}$$

with L_i, B_i and U_i m x m matrices; B_i is tridiagonal; L_i and U_i are lower and upper bidiagonal, with sparsity patterns $\{(j,j-1), (j,j)\}$ and $\{(j,j), (j,j+1)\}$ respectively. There exists a matrix D such that

$$A = (L + D)D^{-1}(D + U) \tag{2.5}$$

where

$$L = \begin{bmatrix} O & & & & \\ L_2 & O & & & \\ & L_3 & O & & \\ & & \cdot & \cdot & \\ & & & \cdot & \cdot \\ & & & L_n & O \end{bmatrix}, \quad U = \begin{bmatrix} O & U_1 & & & \\ & O & U_2 & & \\ & & \cdot & \cdot & \\ & & & \cdot & \cdot \\ & & & O & U_{n-1} \\ & & & & O \end{bmatrix},$$

$$D = \begin{bmatrix} D_1 & & & & \\ & D_2 & & & \\ & & \cdot & & \\ & & & \cdot & \\ & & & & D_n \end{bmatrix}$$

Equation (2.5) is sometimes called a line LU-factorization because the blocks in L, D and U correspond to (in this case horizontal) lines in the grid. Equation (2.5) can be rewritten as

$$A = L + D + U + LD^{-1}U \tag{2.6}$$

One finds that $LD^{-1}U$ is the following block-diagonal matrix:

$$LD^{-1}U = \begin{bmatrix} O & & & \\ & L_2D_1^{-1}U_1 & & \\ & & \cdot & \\ & & & \cdot \\ & & & & L_nD_{n-1}^{-1}U_{n-1} \end{bmatrix} \tag{2.7}$$

From (2.6) and (2.7) we deduce the following algorithm for the computation of D:

$$D_1 = B_1, \; D_i = B_i - L_iD_{i-1}^{-1}U_{i-1}, \; i = 2,3,\ldots n \tag{2.8}$$

The matrix D_i^{-1} is a full m x m matrix, which causes the cost of a line LU-factorization to be $O(nm^3)$, as for standard LU-factorization. An IBLU factorization is obtained if we replace $L_iD_{i-1}^{-1}U_{i-1}$ by its tridiagonal part. Thus, algorithm (2.8) is replaced by:

$$\tilde{D}_1 = B_1, \; \tilde{D}_i = B_i - \text{tridiag}\,(L_i\tilde{D}_{i-1}^{-1}U_{i-1}), \; i = 2,3,\ldots,n. \tag{2.9}$$

The IBLU factorization of A is now defined to be

$$A = (L + \tilde{D})\tilde{D}^{-1}(\tilde{D} + U) + R \tag{2.10}$$

with R the error matrix, and $\tilde{D}$ the block diagonal matrix with blocks $\tilde{D}_i$.

Detailed formulae for the computation of $\tilde{D}$ are given by (Concus et al., 1985; Sonneveld et al., 1985; Axelsson et al., 1983; Meijerink, 1983; Underwood, 1976).

From (2.10) the following iterative method is obtained:

$$(L + \tilde{D})\tilde{D}^{-1}(\tilde{D} + U)\delta y = b - Ay^n, \quad y^{n+1} = y^n + \delta y^n \qquad (2.11)$$

The cost of one iteration with (2.11) is 37N operations; the cost of $\tilde{D}$ and its triangular factorization is 36N operations. Storage of 6N reals is required for the triangular factorization of $\tilde{D}$ and for $\tilde{D}$. For details see (Sonneveld ed al., 1985).

Both ILU and IBLU can be implemented efficiently on vector computers: see (van der Vorst, 1982, 1985; Meurant, 1984; Hemker et al., 1983). Variants of ILU are discussed by Behie and Forsyth (1984), Gustafsson (1978) and Manteuffel (1979).

Extension of ILU to three dimensions is straightforward, and has been tested in combination with CG and MG by Behie and Forsyth (1983). A three-dimensional version of IBLU has been proposed by Meijerink (1983) and Kettler and Wesseling (1985).

3. CONJUGATE GRADIENT ACCELERATION OF BASIC ITERATIVE METHODS

For an introduction to conjugate gradients (CG) (and Tchebychev) methods, see (Hageman and Young, 1981, or Golub and van Loan, 1983). Here only a brief outline is given.

One way to look at CG is as follows. Assuming the matrix A to be large and sparse we allow ourselves to use A only for multiplication with vectors and nothing else. This means that polynomials in A can be built. A rather general algorithm then is

$$y^{n+1} = y^n + \alpha_n p^n, \quad p^n = \theta_n(A) r^0 \qquad (3.1)$$

Here α_n is a scalar to be determined, θ_n is a polynomial of degree n, and $r^0 = b - Ay^0$ is the initial residue. We find

$$r^{n+1} = b - Ay^{n+1} = \phi_{n+1}(A) r^0 \qquad (3.2)$$

with

$$\phi_{n+1}(A) = I - A\{\alpha_n \Theta_n(A) + \alpha_{n-1}\Theta_{n-1}(A) + \ldots + \alpha_n \Theta_0(A)\} r^0 \qquad (3.3)$$

Hence

$$\phi_n \in \Pi_n^1, \qquad (3.4)$$

$$\Pi_n^1 = \{\psi_n | \psi_n(0) = 1 \ , \ \psi_n \text{ is polynomial of degree} \leq n\} \qquad (3.5)$$

The distinguishing property of CG is that ϕ_n is constructed such that

$$||\phi_n(A)r^o|| \leq ||\psi_n(A)r^o|| \ , \ \forall \ \psi_n \in \Pi_n^1 \tag{3.6}$$

Depending on the choice of norm, different CG variants are obtained.

By itself CG does not achieve an impressive rate of convergence. It comes into its own when it is combined with preconditioning. Let B be the preconditioning matrix. Then a preconditioned CG method is given by:

$$\begin{aligned} &p^{-1} = 0, \ r^o = b - Ay^o, \\ &p^n = Br^n + \beta_n p^{n-1}, \ \beta_n = \rho_n/\rho_{n-1}, \ \rho_n = r^{n^T} Br^n, \\ &y^{n+1} = y^n + \alpha_n p^n, \ \alpha_n = \rho_n/\sigma_n, \ \sigma_n = p^{n^T} Ap^n, \\ &r^{n+1} = r^n - \alpha_n Ap^n \end{aligned} \tag{3.7}$$

see (Hestenes, 1956). With B = I one obtains an unpreconditioned CG method. In this case the norm in (3.6) has been chosen as follows:

$$||r||^2 = r^T A^{-1} r \tag{3.8}$$

Now we will explain why (3.7) can be regarded as an acceleration of the BIM (1.2), under the assumption that B is SPD, so that one may write $B = EE^T$. By substitution of $r = E^{-T}\tilde{r}$, $p = E\tilde{p}$, $y = E\tilde{y}$ one finds that (3.7) is CG applied to the preconditioned system $\tilde{A}\tilde{y} = \tilde{b}$ with $\tilde{b} = E^T b$, $\tilde{A} = E^T AE$. According to (3.2), (3.6) we have

$$E^T r^n = \phi_n(E^T AE)E^T r^o \tag{3.9}$$

with

$$||\phi_n(E^T AE)E^T r^o|| \leq ||\psi_n(E^T AE)E^T r^o|| \ , \ \forall \psi_n \in \Pi_n^1 \tag{3.10}$$

From (1.2) it follows that

$$E^T r^n = E^T \psi_n(AEE^T)E^{-T}E^T r_o, \psi_n(x) = (1 - x)^n \in \Pi_n^1 \tag{3.11}$$

Since $E^T(AEE^T)^k E^{-T} = (E^T AE)^k$, $\forall$ k eq. (3.11) can be rewritten as

$$E^T r^n = \psi_n(E^T AE)E^T r^o \tag{3.12}$$

Comparison of (3.9) - (3.12) shows that (3.7) will converge at least as fast as (1.2). The computation of Br^n in (3.7) is performed by means of one iteration with (1.2), since

$$Br^n = y^{n+1} - y^n \tag{3.13}$$

The rate of convergence of CG has been studied in a number of publications. From (3.6) one may deduce, as in (Daniel, 1967), that the required number of iterations n to reduce the residue by a factor ε satisfies

$$n \geq \frac{1}{2} \ln \frac{\varepsilon}{2} \, cond_2(BA)^{\frac{1}{2}} \tag{3.14}$$

with $cond_2$ the condition number in the Euclidean norm. For a detailed study of the rate of convergence see (van der Sluis and van der Vorst, 1985).

An effective and popular preconditioning is provided by incomplete Cholesky factorization, leading to the ICCG method, proposed by Meijerink and van der Vorst (1977). Its effectiveness is explained by taking a look at the spectrum of BA. Gustafsson (1978) proves that with a modified form of incomplete Cholesky factorization for the Poisson equation in two dimensions

$$cond_2(BA) = O(1/h) \tag{3.15}$$

so that because of (3.14) the computational cost of $O(N^{5/4})$, with N the number of gridpoints. Compared with older methods, very impressive rates of convergence are obtained with ICCG for a number of difficult problems by Kershaw (1978). An implementation leading to greater efficiency is given by Eisenstat (1981).

The restriction of CG to SPD problems is of course a severe one. Various CG type methods have been proposed for non-SPD systems. An example is the CGS (conjugate gradients squared) method, proposed by Sonneveld (1984). This method is defined as follows, in its preconditioned form:

$$\begin{aligned} &f^0 = b - Ay^0 , \quad g^{-1} = h^0 = 0, \\ &u = Bf^n + \beta_n h^n , \quad g^n = u + \beta_n(\beta_n g^{n-1} + h^n), \\ &h^{n+1} = u - \alpha_n BAg^n , \quad y^{n+1} = y^n + \alpha_n(u + h^{n+1}) , \\ &f^{n+1} = f^n - A\alpha_n (u + h^{n+1}) , \end{aligned} \tag{3.16}$$

with

$$\alpha_n = \rho_n/\sigma_n , \; \beta_0 = 0 , \; \beta_n = \rho_n/\rho_{n-1} ,$$

$$\sigma_n = \tilde{r}^{0^T} Bf^n , \; \sigma_n = \tilde{r}^{0^T} BAg^n$$

where $\tilde{r}^0$ is a vector to be chosen.
For a justification of this method see (Sonneveld, 1984; Sonneveld et al., 1985). It is related to the biconjugate gradient (bi-CG) method of Fletcher (1976). With bi-CG one has

$$r^n = \phi_n(BA)r^0 \tag{3.17}$$

With CGS one obtains at almost no extra cost

$$r^n = \phi_n^2(BA)r^0 \tag{3.18}$$

The polynomials in (3.17) and (3.18) are the same. It follows that if bi-CG converges, then CGS converges faster (the S in "CGS" is inspired by the squaring of ϕ_n). Furthermore, unlike bi-CG, CGS does not use A^T.

Bi-CG and CGS are but two examples of extensions of CG to non-SPD systems. We will not review other extensions that have been proposed, but restrict ourselves to mentioning the publications of Concus and Golub (1976), Vinsome (1976), Widlund (1978), Axelsson (1980), van der Vorst (1981) and Young and Jea (1980). A somewhat related class of methods is formed by the Tchebychev methods: see (Manteuffel, 1977, 1978; Hageman and Young, 1981). These methods can be very efficient, but their effectiveness depends on the choice of certain parameters.

Unlike the SPD case, in the non-SPD case the theory is far from complete. For CGS applied to a general system a rule of thumb is, that with ILU or IBLU preconditioning good convergence may be expected if A satisfies

$$a_{ii} \geq -\sum_{j \neq i} a_{ij} \ , \ a_{ij} \leq 0 \ , \ j \neq i \qquad (3.19)$$

Sonneveld (1984) and Sonneveld et al. (1985) estimate the cost of a CGS iteration to be 60 flops (ILU preconditioning) or 88 flops (IBLU preconditioning) per grid point.

4. MULTIGRID ACCELERATION OF BASIC ITERATIVE METHODS

The basic ideas of MG are quite general and have wide range of application, including other fields than partial differential equations. MG can be used not only to accelerate iterative methods, but also, for example, to formulate novel ways to solve nonlinear problems, including bifurcation problems and eigenvalue problems, or to devise algorithms that construct adaptive discretizations. The volume edited by Hackbusch and Trottenberg (1982) presents a useful survey of all aspects of MG. See also (Brandt, 1984).

We restrict ourselves here to the one aspect of MG mentioned in the title of this paper. This makes it possible to simplify MG, and to distinguish situations where its effectiveness is guaranteed. The significance of MG as an acceleration technique springs from the fact, that in principle a computational complexity of O(N) can be achieved. This has been proved rigorously under quite general circumstances; see (Hackbusch, 1982) for a survey of MG convergence theory. Work in this area is still going on.

If the BIM (1.2) converges, it usually has the property, exploited by MG, that the non-smooth part of error and residue is annihilated rapidly, whereas it takes many iterations (more and more as the mesh size decreases) to get rid of the smooth part. The property of smoothness will be defined precisely shortly. The fundamental MG idea is to approximate the smooth part of the error on coarser grids. In the MG context the BIM (1.2) is called a smoothing process.

A two-grid method can be defined and explained as follows. Let G be a fine grid and $\bar{G} \subset G$ a coarse grid. Let the sets of grid functions $G \to \mathbb{R}$ and $\bar{G} \to \mathbb{R}$ be denoted by Y and $\bar{Y}$, respectively. Let the coarse grid approximation of (1.1) be given by

$$\bar{A}\bar{y} = \bar{b} \qquad (4.1)$$

Furthermore, let there be given a prolongation operator P and a restriction operator R:

$$P : \bar{Y} \rightarrow Y \ , \ R : Y \rightarrow \bar{Y} \qquad (4.2)$$

A two-grid method for the acceleration of the BIM (1.2) can be formulated as follows. Let y^j be the current iterand, and let $y^{j+\frac{1}{2}}$ be the result of applying a coarse grid correction to y^j:

$$y^{j+\frac{1}{2}} = y^j + P\bar{A}^{-1}R(b - Ay^j) \qquad (4.3)$$

where we assume for the time being that the coarse grid problem is solved exactly. For the residue $r^j = b - Ay^j$ we find:

$$r^{j+\frac{1}{2}} = (I - AP\bar{A}^{-1}R)r^j \qquad (4.4)$$

We now make the following choice for $\bar{A}$, called Galerkin approximation:

$$\bar{A} = RAP \qquad (4.5)$$

A more obvious choice of $\bar{A}$ would be to discretise the differential equation on the coarse grid $\bar{G}$, but (4.5) leads to a more elegant exposition of MG principles, and has other advantages as well, to be mentioned later. It now follows from (4.4) that

$$r^{j+\frac{1}{2}} \in \text{Ker}(R) \qquad (4.6)$$

as noted by Hemker (1982) and McCormick (1982). In other words, $r^{j+\frac{1}{2}} \perp \text{Ker}^{\perp}(R)$, which justifies the appellation "Galerkin approximation" for (4.5). Following Hemker (1982) we relate the concept of smoothness to Ker(R):

Definition 4.1 The set of R-smooth grid functions is $\text{Ker}^{\perp}(R)$.

This definition makes sense intuitively, since in practice R consists of taking weighted averages with positive weights over the fine grid, so that a grid function in Ker (R) has many sign changes, and hence may be called non-smooth.

It remains to annihilate the non-smooth part of r, and this is done in the second part of two-grid iteration, called smoothing. This is done with the BIM (1.2):

$$y^{j+1} = y^{j+\frac{1}{2}} + B(b - Ay^{j+\frac{1}{2}}) \tag{4.7}$$

and we find:

$$r^{j+1} = (I - AB)r^{j+\frac{1}{2}} \tag{4.8}$$

The projection operator on Ker(R) is $I - R^T(RR^T)^{-1}R$, and we conclude from (4.8) and (4.6) that

$$r^{j+1} = (I - AB)(I - R^T(RR^T)^{-1}R)r^{j+\frac{1}{2}} \tag{4.9}$$

This leads us to the following definition:

Definition 4.2 The R-smoothing factor of the smoothing process (1.2) is

$$\rho_R = ||\ (I - AB)(I - R^T(RR^T)^{-1}R||$$

The concepts of smoothness and smoothing factor can also be related to the range of P, or be introduced by means of Fourier analysis as proposed by Brandt (1977). For definitions, see (Sonneveld et al., 1985). Fourier analysis leads to less precise exposition, but to smoothing factors that are easy to compute. See (Brandt, 1982, and Kettler, 1982) for more details.

If we postulate that

$$||I - AP\bar{A}^{-1}R|| \leq 1 \tag{4.10}$$

(which holds approximately if P and R are accurate enough) then we may conclude from (4.4) and (4.9) that

$$||r^{j+1}|| \leq ||(I - AB)(I - R^T(RR^T)^{-1}R)||\,||r^j|| \tag{4.11}$$

Without coarse grid correction, the residue reduction factor would be $||I - AB||$. Equation (4.11) shows how MG accelerates the convergence of the BIM (1.2), if the BIM reduced non-smooth residue components efficiently.

Choosing $\bar{A}$ according to (4.5) has the advantage that the user needs to specify the matrix and the right hand side on the finest grid only. Hence a multigrid code can be programmed such that it is perceived by the user just like any other subroutine for solving linear algebraic systems. Furthermore, (4.5) provides automatically a sound "homogenization" in cases where the coefficients or the right hand side vary rapidly. With difference approximation the implementation of the boundary conditions on the coarse grids is quite often not trivial. On the other hand, depending on the problem, computing $\bar{A}$ with (4.5) may be more expensive than constructing a finite difference approximation. Also $\bar{A}$ has to be stored (perhaps to be overwritten at a later stage by its incomplete factorization). With adaptive discretisation (4.5) is not feasible, since one does not know a priori what the finest grid will be. Results of numerical experiments comparing coarse grid Galerkin and finite difference approximation are given by Wesseling (1982A).

For second order equations prolongation may be defined by linear interpolation. If the "full multigrid" schedule is used, second order interpolation is also needed at certain stages in the schedule: see (Brandt, 1977). For restriction one may safely take $R = P^T$. For discussions of various other possibilities, and comparative experiments, see for example (Brandt, 1977, 1982; Stuben and Trottenberg, 1982; Wesseling, 1982A). However, when the coefficients in the given differential equation are strongly discontinuous one should use matrix-dependent prolongation instead of linear interpolation, in order to take into account the fact, that the solution is locally linear only in a piece-wise fashion. One may still use $R = P^T$. See (Alcouffe et al., 1981; Kettler, 1982).

The two grid method described above becomes a multigrid method if exact solution of the coarse grid is replaced by approximate solution, using MG with coarser grids. In the MGD-family of MG codes we do this with one two-grid iteration employing an additional coarser grid with doubled mesh-size, and so on recursively, until the coarsest grid (usually a 3 x 3 grid) is reached, where one iteration is performed according to (1.2). The resulting MG method is said to be of sawtooth type, because its schedule is represented in a natural way by the schematic of Fig. 4.1, which is a sawtooth curve.

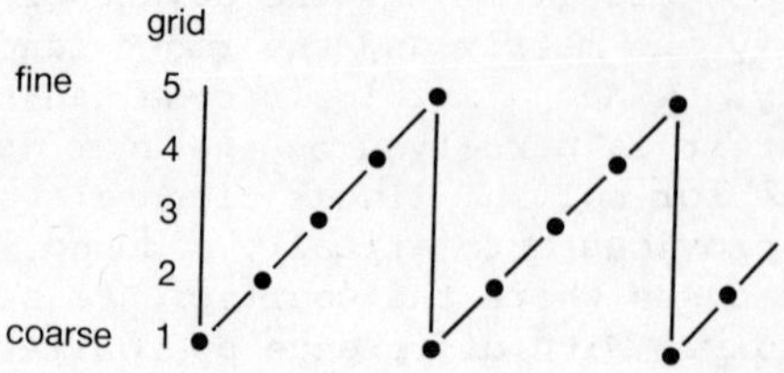

Fig. 4.1. Sawtooth multigrid schedule.
A dot represents a smoothing step.

Various more general MG schedules have been described, see for example (Brandt, 1977, 1982), (Stüben and Trottenberg, 1982). Some comparative experiments are described by Wesseling (1982A). The sawtooth schedule is the simplest possible MG schedule. One may wonder whether such a simple schedule can handle a sufficiently large variety of cases. Experience indicates that the answer is affirmative, see for example the experiments carried out by Wesseling and Sonneveld (1980), Kettler (1982), Wesseling (1982A,B), Hemker et al. (1983), McCarthy (1983), Sonneveld et al. (1985), and the application to transonic potential flow by Nowak and Wesseling (1984). We think that with an effective smoothing process and accurate coarse grid approximation, a simple MG schedule suffices for linear problems.

The MGD-family of MG codes consists at the moment of the subroutines MGD1 and MGD5. These are meant for solving finite difference discretisations of a general second order partial differential operator (possibly not self-adjoint, with a mixed derivative allowed) on a two dimensional rectangular uniform grid. The user provides only the matrix and the right hand side on the finest grid. Because the user is not allowed to influence these subroutines by setting parameters etc, they are called autonomous. MGD1 and MGD5 work like a "black box". MGD1 uses ILU smoothing, MGD5 employs IBLU. As a result, MGD5 is more robust. These programs are portable. In the near future, MGD1 will be available in the NAG library. Versions MGD1V and MGD5V have been designed for auto-vectorization on CRAY and CYBER-205, without sacrificing much on scalar machines. More details, and CPU-time measurements on CYBER-170, CYBER-205 and CRAY-1, and some design considerations related to these MG subroutines may be found in (Wesseling, 1982B, Hemker et al., 1983, 1984, 1985, Sonneveld et al., 1985). The MGD software may be obtained by sending a magnetic tape to the second author of the present paper.

MG software (the MGOO package) is also made available by the group at GMD, Bonn, see (Foerster and Witsch, 1982; Stüben et al., 1984). MGOO is for self-adjoint second order elliptic partial differential equations on a two dimensional rectangular domain. Unlike the MGD codes, the user is allowed some freedom in choosing the MG algorithm. There is a choice in MG schedules, smoothing processes and restriction operators. For the Poisson equation an especially fast version is available. A mixed derivative is not allowed.

Dendy (1982) describes a "black box multigrid" method that can handle self-adjoint equations with strongly discontinuous coefficients; the software mentioned before can handle this only when the discontinuities occur on grid lines that belong to at least a few coarse grids. The user has a choice of multigrid schedule and smoothing processes.

It is to be expected that results of comparative numerical experiments with the software mentioned above will become available in the near future.

5. DISCUSSION AND FINAL REMARKS

It will be clear from the foregoing that CG has a sound theoretical basis only in the SPD case. For MG the O(N) complexity is theoretically well-founded only if one regards the coefficients in the differential equation as fixed. However, in practice one often has a need for numerical methods the properties of which remain invariant as a coefficient tends to a limit. Examples are the convection-diffusion equation at high Péclet number:

$$u_i\phi_{,i} - \varepsilon\phi_{,ii} = f \qquad (5.1)$$

with $1/\varepsilon$ the Péclet number, or the Navier-Stokes equations at high Reynolds number, or the anisotropic diffusion equation:

$$-\varepsilon\phi_{,11} - \phi_{,22} = f \qquad (5.2)$$

Equation (5.2) also models the effect of having a computational grid with cells of very high aspect ratio. Testing in practice seems to be the only way to determine how a method works in cases like this. In fact, also in easier cases where $\varepsilon = O(1)$ practical tests are needed to find out what the efficiency of a method is; there may be large differences in computer time between two methods that both have complexity O(N).

Therefore there is a need for test problems of sufficient generality such that one may have a reasonable degree of confidence in those methods that have done well on these test problems. One important test equation is (5.1), with u_i in "all" possible directions, for example $u_1 = \cos\alpha$, $u_2 = \sin\alpha$, and testing with $\alpha = 0(15^{\circ})\ 345^{\circ}$. Having too few α values in one's test set can be quite misleading; see the results in (Hemker et al., 1983; Sonneveld et al., 1985). Another important test equation is (5.2), with the axes rotated over an angle α, resulting in

$$-(\varepsilon c^2 + s^2)\phi_{,11} - 2(\varepsilon - 1)sc\phi_{,12} - (\varepsilon s^2 + c^2)\phi_{,22} = f \tag{5.3}$$

with $c = \cos\alpha$, $s = \sin\alpha$. This introduces the additional complication of a mixed derivative. A mixed derivative occurs in practice when a non-orthogonal coordinate mapping is used.

The fact that the coefficients in (5.1), (5.3) are constant rather than variable does not make these problems easier, but more difficult. It may easily happen that the BIM that is accelerated with CG or MG is a bad preconditioner or smoother for certain combinations of ε and α. This has more serious consequences when the unfavourable ε,α values occur throughout the domain than when these values occur only locally.

Suitable test problems with strongly discontinuous coefficients are the problems of Stone and Kershaw: see (Kettler, 1982) for a description and further references.

Whether a particular MG method will work for a particular problem may be predicted by Fourier smoothing analysis. See Kettler (1982) for an extensive catalogue of Fourier smoothing analysis results.

Sonneveld et al. (1985) report on numerical experiments, solving (5.1) and (5.3) with six methods: (1): MGD1, (2): MGD5, (3) , (4): MGD1 with horizontal or alternating zebra line Gauss-Seidel smoothing, respectively, instead of ILU, (5): CGS with ILU preconditioning, (6): CGS with IBLU preconditioning. A zebra rather than a successive ordering was used because of considerations regarding possible use of vector computers. Equation (5.1) is discretised with upwind differences so that property (3.19) holds. It is found that (2) and (6) work efficiently in all cases. The performance of the other methods deteriorates in some cases. As far as MG is concerned, the observed behaviour corresponds with smoothing analysis results, in so far as these are available. There are special cases in which methods (2) or (6) are surpassed in efficiency by one or more of the other methods, as is to be expected. For example, if ε is large in eq. (5.2) a very effective and simple preconditioner/smoother is horizontal line Gauss-Seidel (zebra or successive). If one wants to handle as large a class of problems as possible with a single autonomous (black box) code without user-provided adaptions, IBLU should be used for smoothing or preconditioning.

The experiments just mentioned were performed on a 65 x 65 grid. Computing times with CG or MG were roughly the same. We have found that as the grid becomes larger, CG starts to lag behind, in accordance with theoretical computational complexity results. But it should be remembered that CG is easier to program than MG.

6. ACKNOWLEDGEMENT

The authors are indebted to their colleague Henk van der Vorst for helpful comments.

7. REFERENCES

Alcouffe, R.E., Brandt, A., Dendy, J.E. (Jr.), Painter, J.W., (1981). "The Multi-Grid Methods for the Diffusion Equation with Strongly Discontinuous Coefficients". *SIAM J. Scient. Stat. Comp.* **2**, 430-454.

Axelsson, O., (1980). "Conjugate Gradient Type Methods for Unsymmetric and Inconsistent Systems of Linear Equations". *Lin. Algebra and its Appl.* **29**, 1-16.

Axelsson, O., Brinkkemper, S. and Il'in, V.P., (1983). "On some Versions of Incomplete Block-Matrix Factorization Methods". *Lin. Algebra and its Appl.* **58**, 3-15.

Behie, A. and Forsyth, P., (1983). "Multigrid Solution of Three Dimensional Problems with Discontinuous Coefficients". *Appl. Math. and Comp.* **13**, 229-240.

Behie, A. and Forsyth, P. (1984). "Incomplete Factorization Methods for Fully Implicit Simulation of Enhanced Oil Recovery". *SIAM J. Sci. Stat. Comput.* **5**, 543-561.

Brandt, A.,(1977). "Multi-Level Adaptive Solutions to Boundary-Value Problems". *Math. Comp.* **31**, 333-390.

Brandt, A.,(1982). "Guide to Multigrid Development". In: Hackbusch and Trottenberg (1982), 220-312.

Brandt, A., (1984). "Multigrid Techniques: 1984 Guide, with Applications to Fluid Dynamics". Dept. of Applied Math., The Weizmann Institute of Science, Rehovot 76100, Israel.

Concus, P. and Golub, G.H., (1976). "A Generalized Conjugate Gradient Method for Nonsymmetric Systems of Linear Equations". In: R. Glowinski and J.L. Lions (eds.), Proc. of the Second Int. Symposium on Computer Methods in Applied Sciences and Engineering, Paris, 1975. *Lecture Notes in Economics and Math. Systems* **134**. *Springer-Verlag, Berlin.*

Concus, P., Golub, G.H. and Meurant, G., (1985). "Block Preconditioning for the Conjugate Gradient Method". *SIAM J. Sci. Stat. Comput.* **6**, 220-252.

Daniel, J.W., (1967). "The Conjugate Gradient Method for Linear and Nonlinear Operator Equations". *SIAM J. Numer. Anal.* **4**, 10-26.

Dendy, J.E. (Jr.), (1982). "Black Box Multigrid". *J. Comp. Phys.* **48**, 366-386.

Eisenstat, S.C., (1981). "Efficient Implementation of a Class of Preconditioned Conjugate Gradient Methods". *J. Sci. Stat. Comp.* **2**, 1-4.

Fletcher, R., (1976). "Conjugate Gradient Methods for Indefinite Systems". In: G.A. Watson (ed.), "Numerical Analysis". Proceedings, Dundee 1975. *Lecture Notes in Math.* **506**, 73-89, Springer-Verlag, Berlin.

Foerster, H. and Witsch, K. (1982). "Multigrid Software for the Solution of Elliptic Problems on Rectangular Domains: MG00 (Release 1)". In: Hackbusch and Trottenberg (1982), 427-460.

Golub, G.H. and van Loan, C.F., (1983). "Matrix Computation". *North Oxford Academic,* Oxford.

Gustafsson, I., (1978). "A Class of First Order Factorization Methods". *BIT* **18**, 142-156.

Hackbusch, W., (1982). "Multigrid Convergence Theory". In: Hackbusch and Trottenberg (1982), 177-219.

Hackbusch, W. and Trottenberg, U., (1982). "Multigrid Methods". Proceedings, Koln-Porz, 1981. *Lecture Notes in Math*. **960** Springer-Verlag, Berlin.

Hageman, L.A. and Young, D.M., (1981). "Applied Iterative Methods". Academic Press, New York.

Hemker, P.W., (1982). "A Note on Defect Correction Processes with an Approximate Inverse of Deficient Rank". *J. Comp. Appl. Math.* **8**, 137-139.

Hemker, P.W., Kettler, R., Wesseling, P. and de Zeeuw, P.M., (1983). "Multigrid Methods: Development of Fast Solvers". *Appl. Math. and Comp.* **13**, 311-326.

Hemker, P.W., Wesseling, P., de Zeeuw, P.M., (1984). "A Portable Vector Code for Autonomous Multigrid Modules". In: B. Engquist, T. Smedsaas (eds.), "PDE Software: Modules, Interfaces and Systems". *North Holland, Amsterdam*, 29-40.

Hemker, P.W. and de Zeeuw, P.M., (1985). "Some Implementations of Multigrid Linear Systems Solvers". In: Holstein, H. and Paddon, D.J. (eds.), "Multigrid Methods for Integral and Differential Equations". Clarendon Press, Oxford.

Hestenes, M.R., (1956). "The Conjugate-Gradient Method for Solving Linear Systems". Proc. Sympos. Appl. Math., vol. VI, Numerical Analysis, McGraw-Hill, New York.

Kershaw, D.S., (1978). "The Incomplete Choleski-Conjugate Gradient Method for the Iterative Solution of Systems of Linear Equations". *J. Comput. Phys.* **26**, 43-65.

Kettler, R., (1982). "Analysis and Comparison of Relaxation Schemes in Robust Multigrid and Preconditioned Conjugate Gradient Methods". In: Hackbusch and Trottenberg (1982), 502-534.

Kettler, R. and Wesseling, P., (1985). "Aspects of Multigrid Methods for Problems in Three Dimensions". Report 85-08, University of Technology, Dept. of Math. and Inf., P.O. Box 356, 2600 AJ Delft. To appear in *Appl. Math. and Comp.*

Manteuffel, T.A., (1977). "The Tchebychev Iteration for Nonsymmetric Linear Systems". *Numer. Math.* **28**, 307-327.

Manteuffel, T.A., (1978). "Adaptive Procedure for Estimating Parameters for the Nonsymmetric Tchebychev Iteration". *Numer. Math.* **31**, 183-208.

Manteuffel, T.A., (1979). "Shifted Incomplete Cholesky Factorization". In: I.S. Duff and G.W. Stewart (eds.), "Sparse Matrix Proceedings, 1978". *SIAM Publications, Philadelphia*.

McCarthy, G.J., (1983). "Investigations into the Multigrid Code MGD1". Report AERE R10889, Harwell, UK.

McCormick, S.F., (1982). "An Algebraic Interpretation of Multigrid Methods". *SIAM J. Numer. Anal.* **19**, 548-560.

Meijerink, J.A. and van der Vorst, H.A., (1977). "An Iterative Solution Method for Linear Systems of which the Coefficient Matrix is a Symmetric M-matrix". *Math. Comp.* **31**, 148-162.

Meijerink, J.A. and van der Vorst, H.A., (1981). "Guidelines for the Usage of Incomplete Decompositions in Solving Sets of Linear Equations as they occur in Practical Problems". *J. Comput. Phys.* **44**, 134-155.

Meijerink, J.A., (1983). "Iterative Methods for the Solution of Linear Equations based on Incomplete Factorization of the Matrix". Publication 643, Shell Research B.V., Kon. Shell Expl. and Prod. Lab., Rijswijk, The Netherlands, July 1983.

Meurant, G., (1984). "The Block Preconditioned Conjugate Gradient Method on Vector Computers". *BIT* **24**, 623-633.

Nowak, Z, and Wesseling P., (1984). "Multigrid Acceleration of an Iterative Method with Application to Transonic Potential Flow". In: R. Glowinski and J.L. Lions (eds.), "Computing Methods in Applied Sciences and Engineering, VI". *North Holland, Amsterdam.*

Sonneveld, P., (1984). "CGS, a Fast Lanczos-type Solver for Nonsymmetric Linear Systems". Report 84-16, University of Technology, Dept. of Math. and Inf., P.O. Box 356, 2600 AJ Delft. To appear in *SIAM J. Sci. Stat. Comp.*

Sonneveld, P., Wesseling, P. and de Zeeuw, P.M., (1985). "Multigrid and Conjugate Gradient Methods as Convergence Acceleration Techniques". In: Holstein, H. and Paddon, D.J. (eds), "Multigrid Methods for Integral and Differential Equations". Clarendon Press, Oxford.

Stüben, K. and Trottenberg, U., (1982). "Multigrid Methods: Fundamental Algorithms, Model Problem Analysis and Applications". In: Hackbusch and Trottenberg (1982), 1-176.

Stüben, K., Trottenberg, U, and Witsch, K., (1984). "Software Development Based on Multigrid Techniques". In: B. Engquist, T. Smedsaas (eds.): "PDE Software: Modules, Interfaces and Systems". *North Holland, Amsterdam.*

Underwood, R.R., (1976). "An Approximate Factorization Procedure Based on the Block Cholesky Decomposition and its Use with the Conjugate Gradient Method". Report NEDO-11386, General Electric Co., Nuclear Energy Div., San Jose, CA.

van der Sluis, A. and van der Vorst, H.A., (1985). "The Rate of Convergence of Conjugate Gradients". Dept. of Math. Utrecht University, Preprint Nr. 354. To appear in *Numer. Math.*

van der Vorst, H.A. (1981). "Iterative Solution Methods for Certain Sparse Linear Systems with a Non-Symmetric Matrix arising from PDE Problems". *J. Comput. Phys.* **44**, 1-19.

van der Vorst, H.A., (1982). "A Vectorizable Variant of some ICCG methods". *SIAM J. Sci. Stat. Comput.* **3**, 350-356.

van der Vorst, H.A., (1985). "The Performance of Fortran Implementations for Preconditioned Conjugate Gradients on Vector Computers". Report 85-09, University of Technology, Dept. of Math. and Inf., P.O. Box 356, 2600 AJ Delft.

Varga, R.S., (1962). "Matrix Iterative Analysis". *Prentice-Hall, Englewood Cliffs, New Jersey.*

Vinsome, P.K.W., (1976). "ORTHOMIN , an Iterative Method for Solving Sparse Sets of Simultaneous Linear Equations". Society of Petroleum Engineers, paper SPE 5729.

Wesseling, P. and Sonneveld, P., (1980). "Numerical Experiments with a Multiple Grid and a Preconditioned Lanczos type Method". In: R. Rautmann (ed.). "Approximation Methods for Navier-Stokes Problems". Proceedings, Paderborn 1979. *Lecture Notes in Math.* **771**, 543-562. Springer-Verlag, Berlin 1980.

Wesseling, P., (1982A). "Theoretical and Practical Aspects of a Multigrid Method". *SIAM J. Sci. Stat. Comp.* **3**, 387-407.

Wesseling, P., (1982B). "A Robust and Efficient Multigrid Method". In: Hackbusch and Trottenberg (1982), 614-630.

Widlund, O., (1978). "A Lanczos Method for a Class of Nonsymmetric Systems of Linear Equations". *SIAM J. Numer. Anal.* **15**, 801-812.

Young, D.M., (1971). "Iterative Solution of Large Linear Systems". *Academic Press, New York.*

Young, D.M. and Jea, K.C., (1980). "Generalized Conjugate-Gradient Acceleration of Nonsymmetrizable Iterative Methods". *Lin. Algebra and its Appl.* **34**, 159-194.

SELF ADAPTIVE MESH REFINEMENTS AND FINITE ELEMENT METHODS FOR SOLVING THE EULER EQUATIONS

B. Palmerio
(Universite de Nice, France)
V. Billey and J. Periaux
(AMD/BA, Saint Cloud, France)
A. Dervieux
(INRIA, Valbone, France)

INTRODUCTION

We are interested in transonic flow simulation around realistic aircraft geometries governed by the Euler equation. One of the most important problems is the grid generation around those complex geometries. In the particular case of Euler equations, due to its hyperbolic behaviour, it appears that the generation of adapted grids is a stiff problem. Our global strategy is to handle Finite Element-type triangular unstructured grids.

In this paper, we introduce for Euler solvers new improvements related to grid optimization. Indeed, in order to identify accurately the characteristics of the flow, the grid should satisfy the following important conditions:

(i) to take into account the geometry, body-fitted grids seem mandatory.
(ii) to be fine enough in the region where the equations to be solved need accuracy.

To reach objectives (i), (ii), we propose the following procedure:
- Step 1: An initial coarse grid Θ_o is generated, fulfilling only condition (i).
- Step 2: An adaptive procedure is applied, using intermediate simulations, in order to derive from the initial grid Θ_o a new grid Θ_k which will satisfy conditions (i) and (ii).

The goal of this paper is to describe such an adaptive procedure applied to Euler flow in a 2-D context, keeping in sight the future extension to 3-D.

Two families of adaptive procedures can be recognized:

In the *moving node* procedure, the nodes are moved while their number remains constant.

In the *enrichment* procedure, the location of the nodes is fixed but their number is increased.

For recent Euler computations with the first method we refer to Nakahashi and Deiwert (1985), Dwyer (1983), Lohner (1985).

We are interested here only in the second method. For solving Euler equations, this method has been recently illustrated by locally refined Finite Volume techniques combined or not with multigrids, see Usab and Murman (1983) and Berger and Jameson (1984) for the 2-D and also Baker, Jameson, Vermeland (1985) in a 3-D context. However Finite Element Methods have also been presented by Lohner (1985) and Palmerio (1984).

In this paper, the adaptive procedure has been applied to two different schemes:

(i) A Galerkin F.E.M. approximation using an artificial viscosity.

(2) A new second-order accurate upwind scheme with no viscosity parameter adjustment.

With scheme (1), we study a realistic industrial problem to illustrate the ability of the method for a better capture of the singularities (shocks, angular stagnation point,...).

The problem of spurious entropy generation at stagnation points is considered with Scheme (2).

1. ADAPTIVE PROCEDURE

1.1 Triangulation and degrees of freedom

We consider a 2-D F.E.M. type triangulation of the domain of computation Ω. This means that a polygonal region Ω_h is constructed, which is a "good" approximation of Ω. Then a triangulation Θ_h of Ω_h is constructed, satisfying the standard conditions (see Zienkiewicz, 1971).

For a given vertex i the neighbour vertices are those which belong to a triangle having i as a vertex. Let us denote by $N(i)$ the number of neighbours of a vertex i; a triangulation can be unstructured in the following sense: $N(i)$ can be different from $N(j)$ for two different vertices i and j (see Fig. 1).

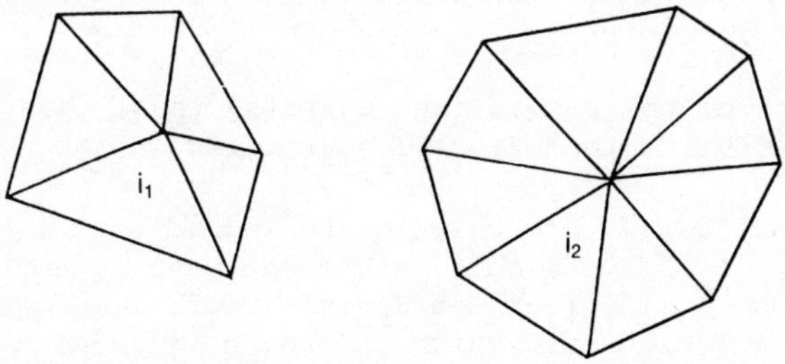

Fig. 1 Example of different number of neighbours of a vertex ($N(i_1) = 5$, $N(i_2) = 8$)

As far as mesh refinement is concerned, this means that we shall be able to increase locally the number of elements without changing the data structure describing the mesh. From this point of view, the unstructured grids provide more flexibility than F.D.M./F.V.M. ones, for which a locally refined mesh is described in a different manner from the initial (I,J,K)-mesh (extra loop and matching interfaces must exist in the computer program).

Representation of the unknowns

A piecewise linear continuous approximation is chosen: the degrees of freedom are values of the functions at nodes. The nodes are located at vertices of the triangulation (conformal element). This choice has been introduced previously in several papers (Angrand and Dervieux, 1984, Billey, 1984). Many arguments favouring this choice have been presented in the above papers and also in Dervieux (1985).

1.2 Enrichment algorithm

1. Firstly a criterion is used, to decide which triangles require additional effort on accuracy. For these triangles new points are introduced, located at the middle of each side.

2. The identification of the sides to be divided is then derived in a second step.

3. In the third step, new triangles are constructed by considering in the old triangles the number of sides containing

new points (see Fig. 2); three situations can appear:

- If there is only one point added, the former triangle is split into two new ones.

- If two points must be added three sub-triangles are created.

- With three new points (on each mid-side), then the triangle is divided into four sub-triangles.

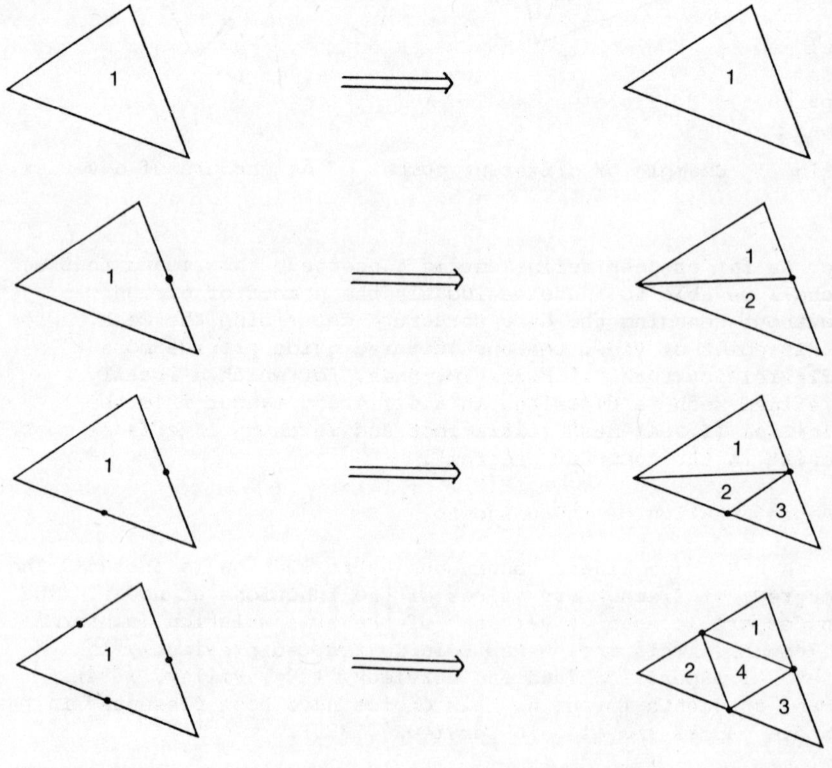

Fig. 2 Mesh enrichment procedure, formation (or not) of new triangles.

1.3 Criterion of enrichment

We consider a physical criterion highly advocated in steady Euler calculations. In these calculations, the entropy S should be purely convected, i.e. satisfying

$$\vec{U} \cdot \vec{\nabla} S = 0 \tag{1.1}$$

where $\vec{U}$ is the velocity of the flow, except in the vicinity of the shocks which act as sources of entropy.

This observation leads us to define a new criterion to determine the region to be refined, i.e. on each triangle where the scalar product $\vec{U}.\vec{\nabla}S$ is greater than a constant C times the maximum of $\vec{U}.\vec{\nabla}S$ (taken over all the triangles).

As a consequence, the regions of refinement will locate the elements where spurious numerical entropy production occurs and also the physical entropy production on shocks. C is a parameter controlling the level of refinement that we wish to perform.

2. APPLICATION TO EULER FLOWS

We present some experiments of the above methods using a currently used Galerkin scheme. In a second section we present experiments with a new upwind scheme much more suitable for adaptive enrichment.

2.1 A Richtmyer-Galerkin scheme

The Euler equations are denoted by:

$$W_t + (F(W))_x + (G(W))_y = 0 \tag{2.1}$$

$$F(W) = \begin{pmatrix} \rho u \\ \rho u^2 + p \\ \rho uv \\ (e+p)u \end{pmatrix}, \quad G(W) = \begin{pmatrix} \rho v \\ \rho uv \\ \rho v^2 + p \\ (e+p)v \end{pmatrix} \tag{2.2}$$

$$p = (\gamma - 1)(e - \frac{1}{2}\rho(u^2 + v^2)) \; ; \; \gamma = 1.4 \tag{2.3}$$

where W(x,y) are the usual conserved quantities (ρ,ρu,ρv,e), i.e. density, moments, and energy; u and v are the velocity components, p the pressure, e the total energy and γ the specific heat ratio.

An explicit Richtymer-Galerkin scheme is constructed; it consists of a predictor-corrector sequence (see Angrand and Dervieux, 1984, Angrand et al., 1983). We recall the two steps of this scheme in which V_h is the standard Galerkin space of approximation:

$$V_h = \{\phi \text{ continuous, and linear on each triangle of } \Theta_h\}$$

1. *Predictor step:* for each T_h triangle of Θ_h,

$$\hat{W}(T_h) = \frac{1}{area(T_h)} \iint_{T_h} W^n \, dx\, dy + \alpha\, \Delta t \int_{\partial T_h} [F(W^n) v_x + G(W^n) v_y] \, d\sigma \tag{2.4}$$

2. *Corrector step:* $W^{n+1} \in V_h$, $\forall \phi \in V_h$

$$\iint_{\Omega_h} \Sigma_o \frac{W^{n+1} - W^n}{\Delta t} \Sigma_o \phi \, dx\, dy = \beta_1 \iint_{\Omega_h} [F(W^n)\phi_x + G(W^n)\phi_y] \, dx\, dy$$

$$+ \beta_2 \iint_{\Omega_h} [F(\hat{W})\phi_x + G(\hat{W})\phi_y] \, dx\, dy + \int_{\partial\Omega_h} \phi \, [F(\overline{W}) \, v_x + G(\overline{W}) v_y] \, d\sigma \tag{2.5}$$

$$+ \chi \iint_{\Omega_h} f(W^n) < \vec{\nabla} W^n . \vec{\nabla} \phi > dx\, dy$$

with

$$\alpha = \frac{1 + \sqrt{5}}{2}, \quad \beta_1 = \frac{2\alpha - 1}{2\alpha}, \quad \beta_2 = \frac{1}{2\alpha}, \quad \chi = 0.8$$

α, β_1, and β_2 correspond to an optimal choice of the length of the predictor time step. The notation $v = (v_x, v_y)$ holds for the outward unit vector normal at $\partial\Omega_h$. The symbol Σ_o denotes a mass-lumping operator allowing the computation of W^{n+1} without solving a system. The boundary conditions are included in the $\overline{W}$ integral. The last term of the right hand side is an artificial viscosity one. For further details we refer to Angrand and Dervieux (Ibid.).

2.1.1. A first simplified numerical experiment: Transonic flow in a channel with a bump.

Our goal is to evaluate the behaviour of the scheme applied to meshes containing a stiff refinement. Thus we do not consider the adaptive procedure but a single refinement of the boundary triangles in the following problem, proposed in a GAMM Workshop (see Rizzi and Viviand, 1981): the geometry is a channel with a 4% thick circular bump; the Mach number at entrance is $M_\infty = .85$.

The comparison of the pressure outputs will be done with a simulation using a rather fine regular mesh of 1512 nodes (Fig. 3).

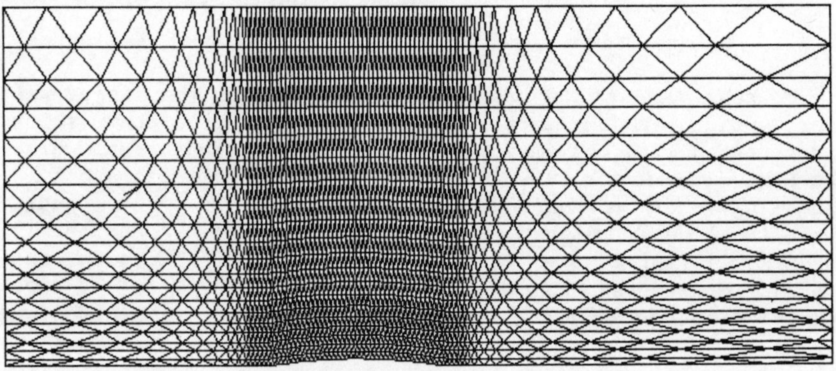

Fig. 3 Channel with a circular bump, fine mesh reference (1512 nodes).

The initial mesh is very coarse, with only 161 nodes (Fig. 4). The refinement is performed twice on the first row of triangles lying on the underneath bumped wall. The use of this procedure provides a mesh of 411 nodes (Fig. 5).

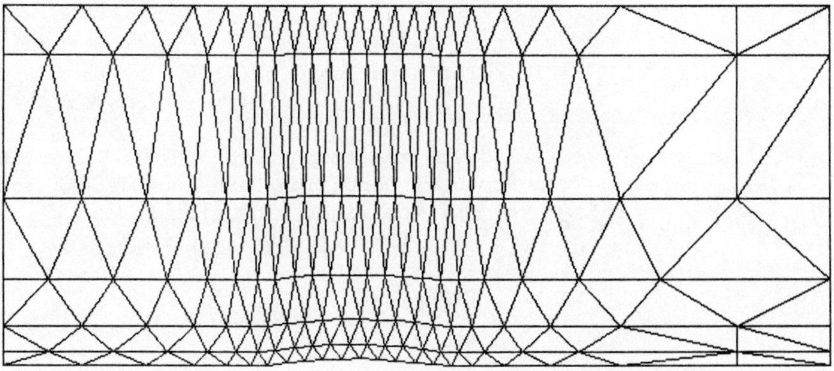

Fig. 4 Channel with a circular bump, coarse mesh (161 nodes)

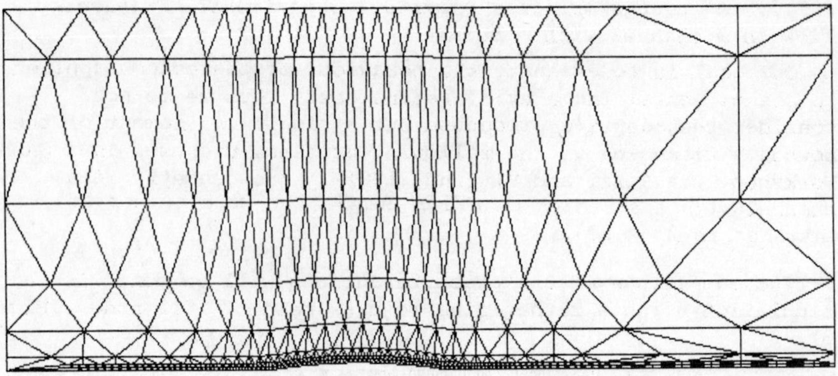

Fig. 5 Channel with a circular bump, refined mesh (411 nodes)

We consider now the pressure distribution on the bumped wall corresponding to the different grids: on Fig. 6, a sharp shock is obtained with the fine mesh by a costly computation, while in Fig. 7 the shock is spread also only over two strips but seems not sharp enough because of the coarseness of the mesh; with the locally refined mesh (Fig. 8), we obtain a sharper shock still spread over 2-3 strips of elements. The comparison on Fig. 9 with the fine mesh calculation proves that the shock location is correct.

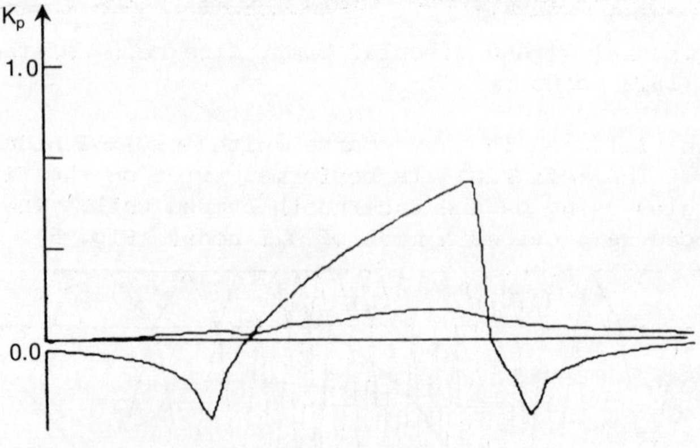

Fig. 6 Channel with a circular bump, pressure coefficient distribution with the fine mesh of 1512 nodes.

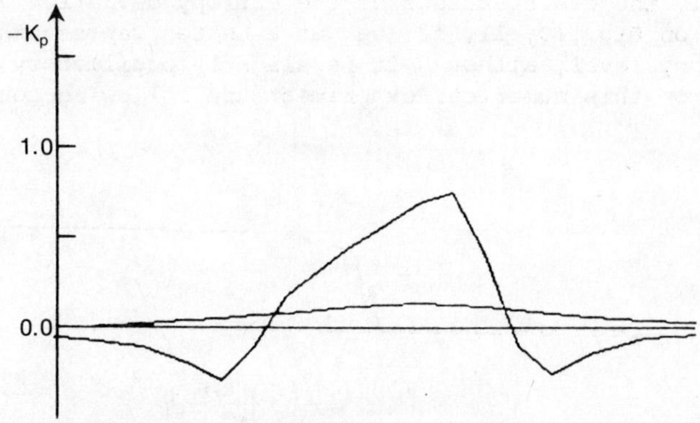

Fig. 7 Channel with a circular bump, pressure coefficient distribution with the coarse mesh of 161 nodes.

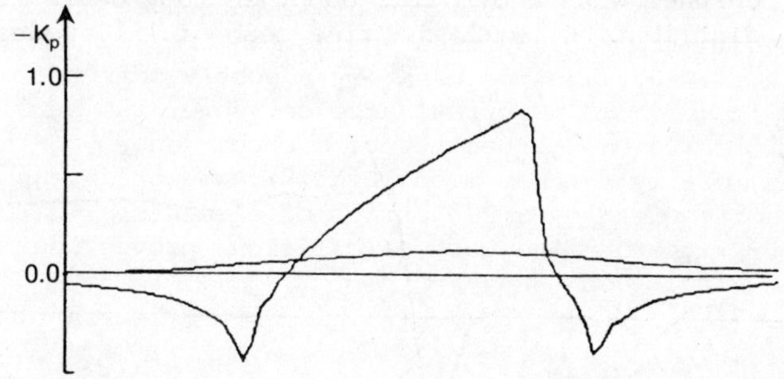

Fig. 8 Channel with a circular bump, pressure coefficient distribution with the refined mesh of 411 nodes.

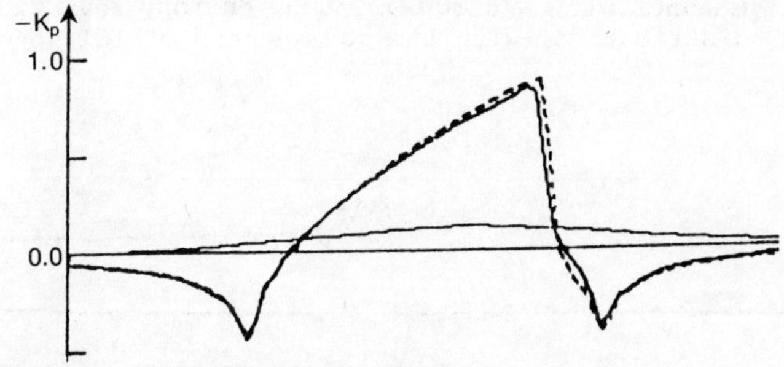

Fig. 9 Channel with a circular bump, comparison of the pressure coefficient distribution obtained with the fine mesh and the refined mesh.

Finally the distributions of the entropy deviation are also compared on Fig. 10, 11, 12; we get a better representation of the entropy level, although it is slightly oscillatory. We can derive from this numerical experiment the following conclusions.

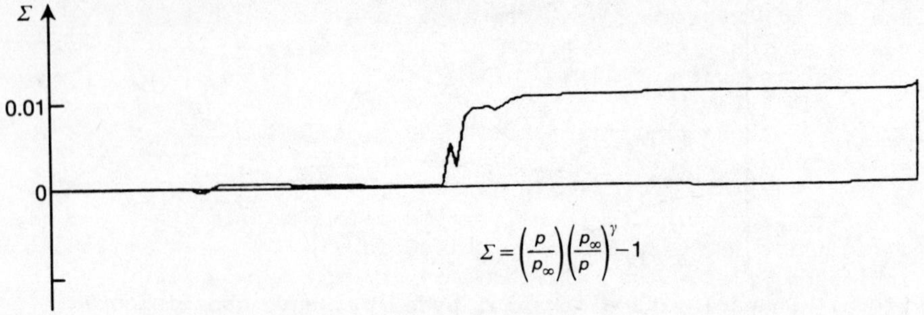

Fig. 10 Channel with a circular bump, entropy deviation distribution, with the fine mesh of 1512 nodes.

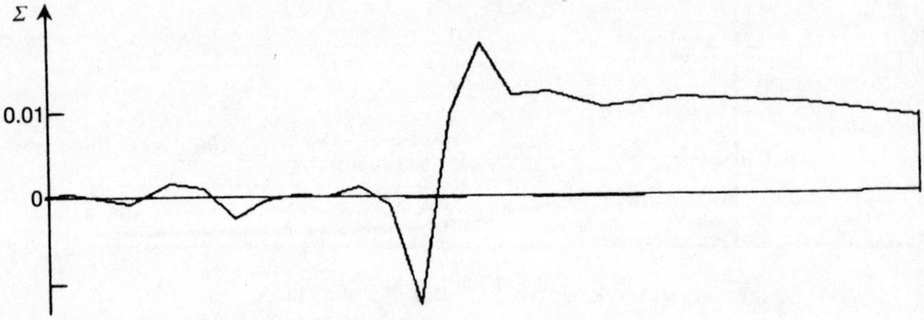

Fig. 11 Channel with a circular bump, entropy deviation distribution, with the coarse mesh of 161 nodes.

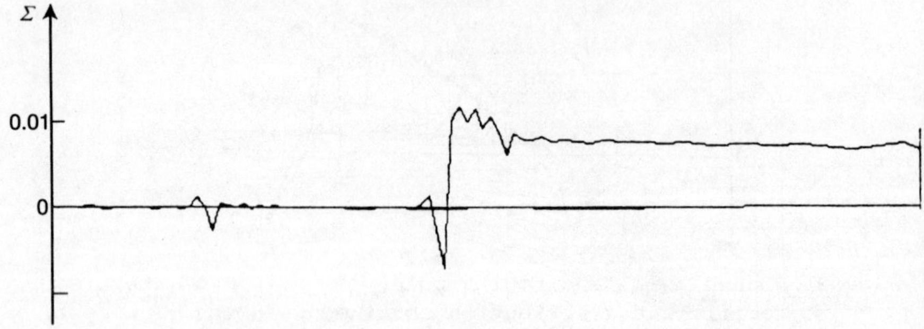

Fig. 12 Channel with a circular bump, entropy deviation distribution, with the refined mesh of 411 nodes.

1. The Finite Element Method behaves satisfactorily when we perform a rather stiff refinement.

2. In particular, the artificial viscosity model has also a good behaviour as far as shock profiles are concerned. Entropy distributions are also much improved but a sensitive tuning of the viscosity terms seems necessary to obtain oscillatory-free results.

2.1.2 A second numerical experiment of industrial interest: a supersonic flow around and inside an inlet.

In the following, we consider the complete adaptive algorithm. The problem to be solved is a 2-D flow around and inside an inlet. The half-geometry is considered, which reduces the domain to a channel ended by the engine (internal flow) and connected to the external flow. The Mach number at infinity is $M_{\infty} = 2.$ and the Mach number at the engine (outflow boundary) is $M_e = .27$.

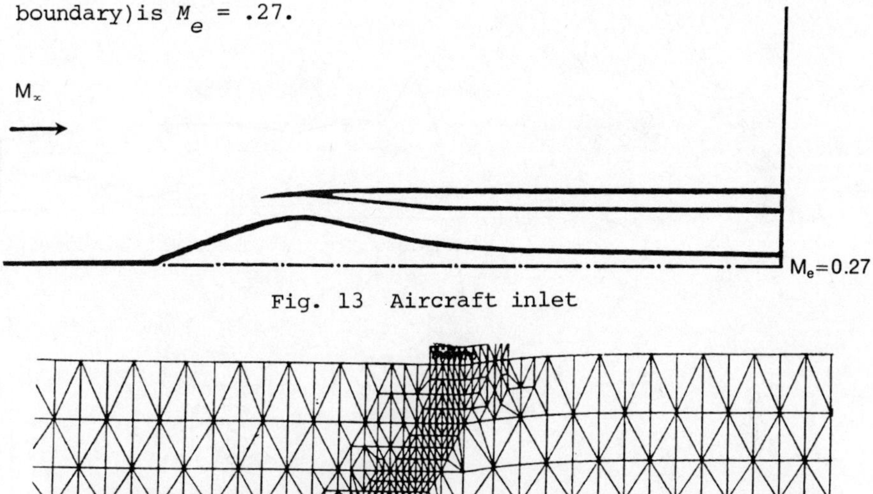

Fig. 13 Aircraft inlet

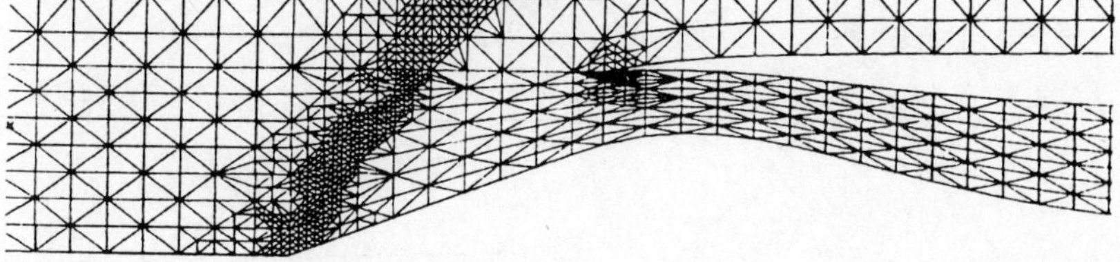

Fig. 14 Aircraft inlet, partial view of the final mesh (1925 nodes).

A double enrichment is performed via the entropy criterion. Starting from an initial regular mesh, the self adaptive refinement acts in the vicinity of the oblique shock and also of the sharp edge to provide local enrichment. Comparison of the K_p contours show a large improvement of the accuracy; the oblique shock is sharply captured, with only 3 strips of elements transition (Fig. 15, 16).

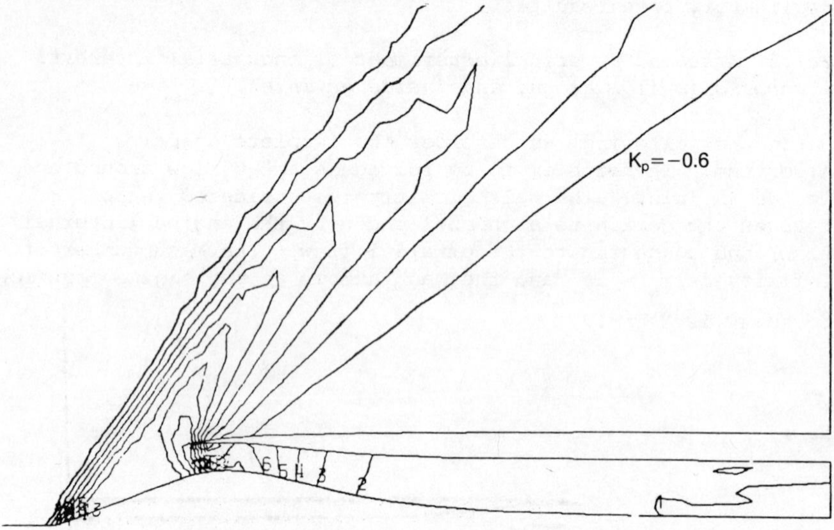

Fig. 15 Aircraft inlet, K_p contours, coarse mesh (1145 nodes)

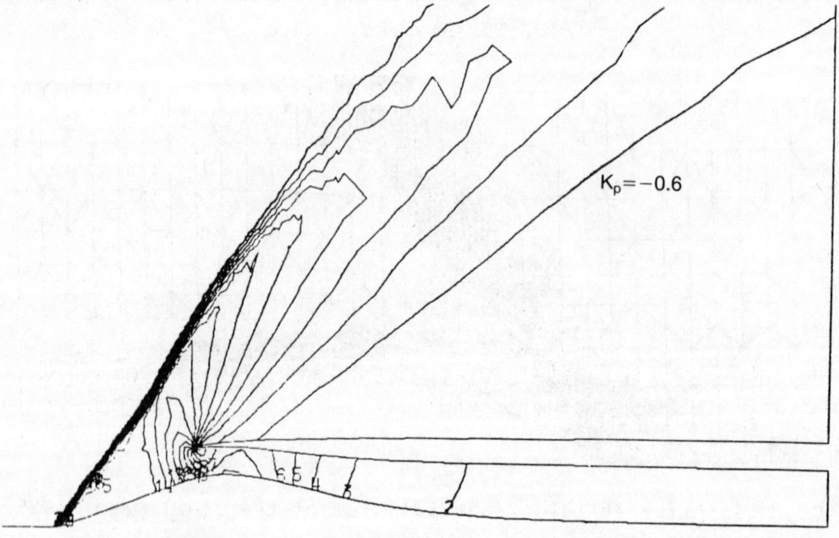

Fig. 16 Aircraft inlet, K_p contours, refined mesh (1925 nodes)

2.2 The upwind scheme

The combination of the above Galerkin Euler solver with grid refinement gives rather good results. But with this kind of centred scheme with added artificial viscosity, non-oscillatory entropy distributions are hardly obtained without adding extra viscosity. This problem is still more difficult when the mesh size varies rather rapidly.

Therefore we focus now on an upwind scheme where no parameter adjustment is needed.

This scheme, advocated previously by Van Leer (1983), has been derived for unstructured triangular meshes by Fezoui (1985).

We start from a Galerkin representation, using the above space V_h ; this enables us to construct approximate derivatives in V_h as follows:

$$\forall\, i \text{ vertex},\ \overline{W}_x(i) = \frac{\int_{supp(i)} \frac{\partial W}{\partial x}\, dx\, dy}{\int_{supp(i)} dx\, dy} \qquad (2.6)$$

where *supp*(i) is the support of the basis function related to vertex i. The resulting approximate gradient is used in the two phases of the time-stepping:

1. Prediction phase

$$\hat{W}_i = W_i^n - \frac{\Delta t}{2}\,[F'(W_i^n)\overline{W}_x(i) + G'(W_i^n)\,\overline{W}_y(i)] \qquad (2.7)$$

2. Correction phase

$$W_i^{n+1} = W_i^n - \frac{\Delta t}{area(\hat{i})} \int_{\partial\hat{i}} (F^*\ v_x + G^*\ v_y)\ \partial\sigma \qquad (2.8)$$

The second phase is a Finite Volume formulation with respect to cells $\hat{i}$ constructed around vertices i with the medians of neighbouring triangles; for the numerical integration (indicated by the superscript *) on the cell boundary $\partial\hat{i}$, we use integration values W_{ij} located on the mid-segments $[i,j]$ with vertex i and neighbouring vertices j as extremities:

$$W_{ij} = \hat{W}_i + \frac{1}{2}\begin{pmatrix}\overline{W}_x(i)\\ \overline{W}_y(i)\end{pmatrix}.\vec{ij}.$$

then a plus-minus flux splitting is applied to W_{ij} and W_{ji}; we refer to Fezoui (1985) for details.

Our purpose is to show that smooth (oscillation-free) entropy distributions can be obtained using this new scheme even in presence of strong local refinements: as a test case we consider a transonic flow around a NACA0012 airfoil with a Mach number at infinity equal to $M_\infty = .85$, without angle of attack. We start with an initial mesh containing 600 nodes shown on Fig. 17. The first value of the grid refinement parameter C is 3% and we obtain an enriched mesh with 1049 nodes presented on Fig. 18. Then, a second refinement is performed with the parameter C taken equal to 7%, but this time, the grid refinement has been performed only in the front part to avoid the addition of too many points near the shock. By that procedure, a final mesh with 1615 nodes is obtained (Fig. 19).

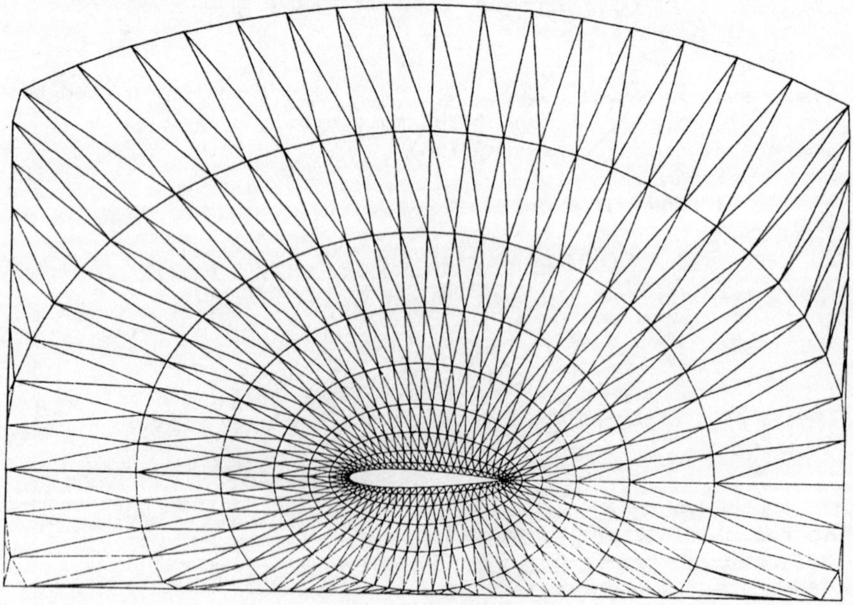

Fig. 17 NACA 0012 airfoil, blow-up of the coarse mesh (600 nodes).

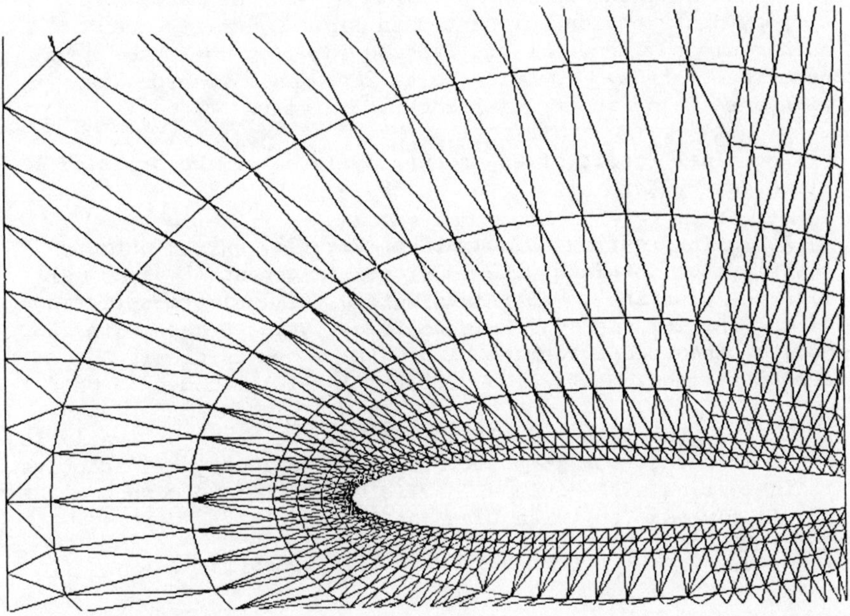

Fig. 18 NACA 0012 airfoil, blow-up of the medium mesh (1049 nodes).

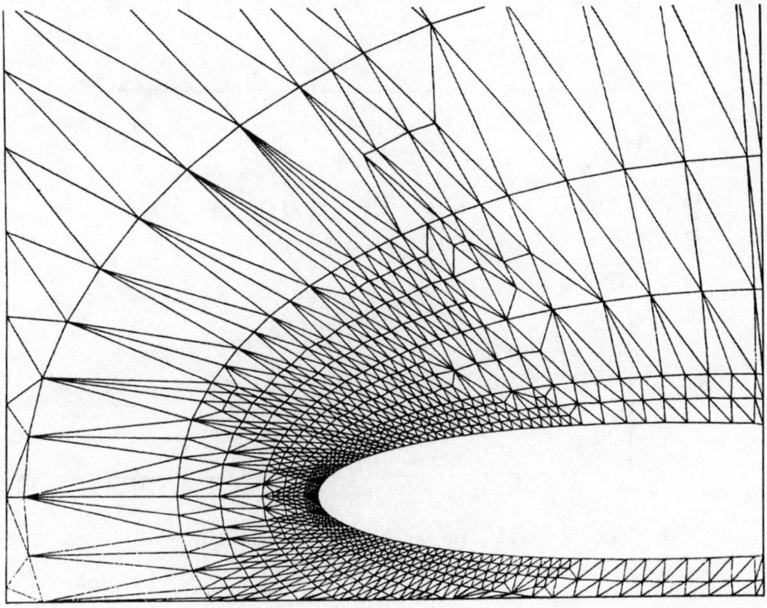

Fig. 19 NACA 0012 airfoil, blow-up of the final mesh (1615 nodes)

We can make the following comments: on one hand the entropy distributions presented on Fig. 20, 21, 22 keep a rather smooth aspect at the leading edge, on the other hand, numerical entropy reduction is achieved in a monotone manner, with a ratio of about 3 for each refinement.

The final result gives an entropy level of the order of 10^{-3} almost everywhere before the shock.

It is interesting to compare on Fig. 21, 22 the entropy distribution obtained with only one enrichment (1049 nodes) with the distribution obtained with a full uniform splitting of the initial mesh which leads to a regular O-mesh with 2280 nodes; the good agreement before the shock is clear, demonstrating the efficiency of the local refinement approach.

Clearly the final resulting mesh is not the best available NACA0012 mesh; our purpose is to promote local self-adaptive refinement as a flexible and efficient method to generate physically adapted triangulations from a priori *arbitrary* initial meshes.

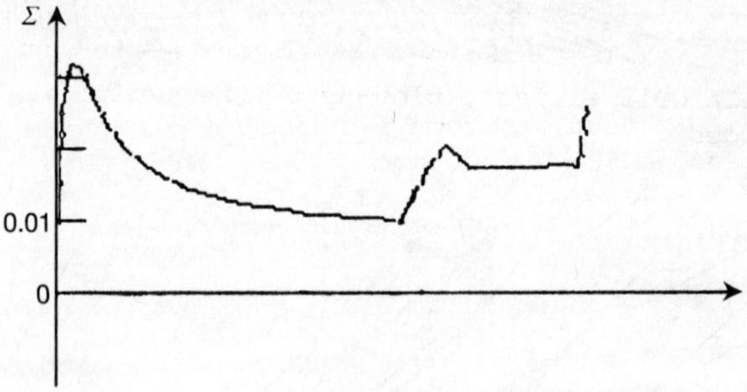

Fig. 20 NACA0012 test, entropy plot with the coarse mesh of 600 nodes.

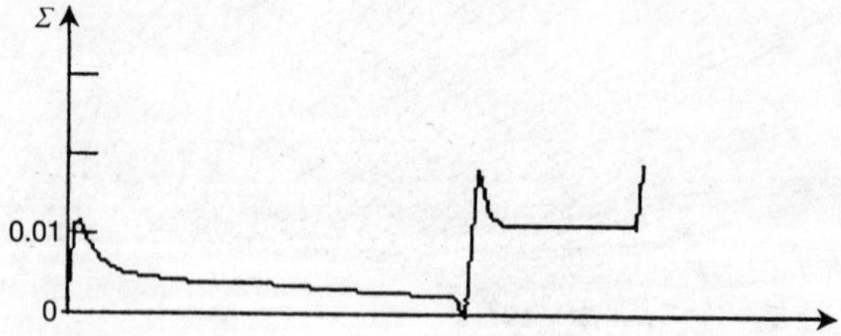

Fig. 21. NACA0012 test, entropy plot with the medium mesh of 1049 nodes.

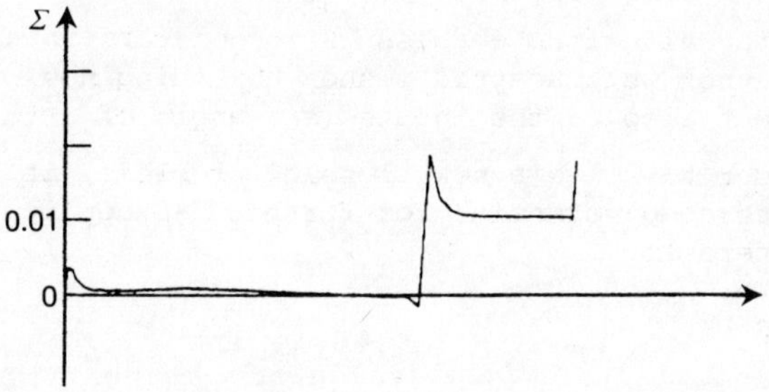

Fig. 22 NACA0012 test, entropy plot with the final mesh of 1615 nodes.

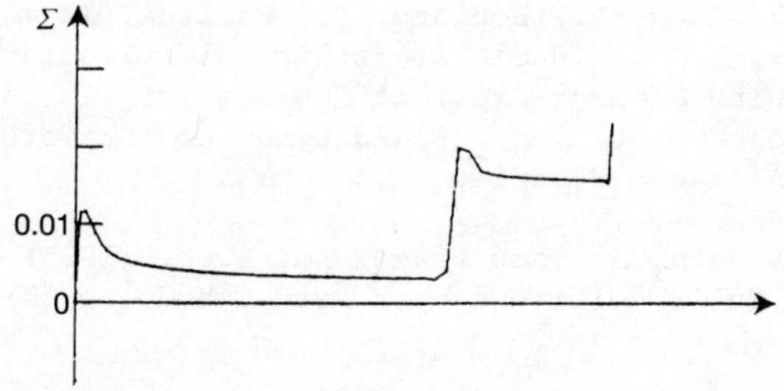

Fig. 23 NACA0012 test, entropy plot with the *uniformly* refined mesh of 2280 nodes

3. CONCLUSION

In this paper, we have described a self-adaptive mesh enrichment algorithm based on:

- Finite Element type unstructured meshes

- a physical criterion well adapted to a sensitive accuracy problem in Euler numerical simulations.

The corresponding enrichment procedure has been combined with two representative Euler solvers, to illustrate the accurate and efficient possibilities of these schemes when performed on severely unstructured meshes:

An oblique shock has been captured accurately with the Galerkin F.E.M. solver.

We have however pointed out the better behaviour of the upwind scheme to obtain smooth entropy distribution; this is likely to be due to the fact that the numerical viscosity of the upwind scheme is self adapted.

The resulting algorithm enables us to generate in a fast aerodynamic manner well adapted meshes depending not only on the geometry but also on the inputs (M_∞, angle of attack, etc...) in order to solve a new class of problems; it seems to have a quite good potential for further 3-D applications of industrial interest.

4. REFERENCES

Angrand, F., and Dervieux, A., (1984) Some explicit triangular Finite Element schemes for the Euler Equations, *Int. J. for Num. Methods in Fluids,* **4**, 749-764.

Angrand, F., Dervieux, A., Boulard, V., Periaux, J. and Vijayasundaram, G., (1983) Transonic Euler Simulations by means of Finite Element Explicit Schemes. In 6th AIAA Conference on Computational Fluid Dynamics (Danvers, Mass., USA) and AIAA paper 83-1924.

Baker, T.J., Jameson, A., and Vermeland, R.E., (1985) Three Dimensional Euler solutions with Grid Embedding, AIAA paper 85-0121.

Berger, M.J. and Jameson, A., (1984) An adaptive multigrid method for the Euler equations in 9th *Int. Conf. on Num. Methods in Fluid dynamics,* Lecture Notes on Physics, **218**, 92-97.

Billey, V., (1984). Resolution des Equations d'Euler par des Methodes d'Elements Finis ; Applications aux ecoulements 3-D de l'Aerodynamique, Thesis, University of Paris VI.

Dervieux, A., (1985). Steady Euler simulations using unstructured meshes, Von Karman Institute for Fluid Dynamics, Lecture Series 1985-04, Computational Fluid Dynamics, March 25-29, 1985.

Dwyer, M.A., (1983) A Discussion of some Criteria for the Use of Adaptive Gridding, AIAA paper 83-1932.

Fezoui, F., (1985) Resolution des equations d'Euler par un schema de Van Leer en elements finis, INRIA Report 358.

Lohner, R., Morgan, K., Peraire, J., Zienkiewicz, O.C., and Kong, L., (1985) Finite Element Methods for Compressible Flows, these proceedings.

Nakahashi, K. and Deiwert, G.S., (1985) A practical adaptive grid method for complex fluid flow problems in 9th *Int. Conf. on Num. Methods in Fluid Dynamics,* Lecture Notes in Physics, **218**, 422-426.

Palmerio, B., (1984) Self-adaptive F.E.M. algorithms for the Euler equations, *INRIA* Report 338.

Rizzi, A. and Viviand, H., (1981) Numerical methods for the computation of inviscid transonic flows with shock waves, *Notes on Numerical Fluid Dynamics,* **3** Vieweg and Sohn, Brauschweig/Wiesbaden.

Usab, W.J. Jr. and Murman, E.H., (1983) Embedded Mesh Solutions of the Euler Equations using a Multiple-Grid Method in 6th AIAA Conference on Computational Fluids Dynamics (Danvers, Mass. USA), July 13-15 1983.

Van Leer, B., (1983) Computational Methods for Ideal Compressible Flows, Von Karman Institute for Fluid Dynamics, Lectures Series 1983-04, Computational Fluid Dynamics, March 7-11, 1983.

Zienkiewicz, O., (1971) The Finite Element Method in Engineering Science, McGraw-Hill, London.

AN ADAPTIVE ORTHOGONAL GRID METHOD FOR 2D OR AXISYMMETRIC COMPRESSIBLE FLOW PROBLEMS

Luca Zannetti
(Dipartimento di Ingegneria Aeronautica e Spaziale, Politecnico di Torino, Italy)

1. INTRODUCTION

A finite difference time-dependent method has been proposed recently to solve inverse problems in ducts for compressible, inviscid, rotational flow fields (see Zannetti (1984)).

The method is extended in the present paper to deal with more general problems, including external flow direct problems. The method belongs to the category of adaptive grid techniques. The grid is body fitted and orthogonal and its metric factors depend on the flow properties. Once the steady solution is reached, the grid is formed by a set of streamlines and lines orthogonal to the streamlines.

The numerical computation is based on the lambda-formulation (see Moretti (1979), Zannetti et al. (1981), Moretti and Zannetti (1984)). The simple form of the boundary conditions and the orthogonality of the grid allow the qualities of the lambda-formulation to be fully exploited, particularly as regards consistency of the computation at the boundaries and accuracy in transonic flow fields.

2. MATHEMATICAL MODEL

Denote by s the arc length of the envelopes of the paths of the particles of a compressible fluid during a transient and by n the arc length of the lines orthogonal to the envelopes. Define a pair of curvilinear coordinates along the envelopes and their orthogonal trajectories ϕ and ψ , related to s and n by

$$ds = h_1 d\phi \ , \ dn = h_2 d\psi \qquad (2.1)$$

where h_1 and h_2 are metric factors.

Taking t and x as independent variables, the equations of motion are

$$\left.\begin{aligned} a_t - \kappa S_t + \frac{q}{h_1}(a_\phi - \kappa S_\phi) + \delta a \left(\frac{q_\phi}{h_1} + q\frac{\nu_\psi}{h_2}\right) &= 0 \\ q_t + q\frac{q_\phi}{h_1} + \frac{a}{h_1}\left(\frac{a_\phi}{\delta} - \frac{a}{\gamma}S_\phi\right) &= 0 \\ q\nu_t + q^2\frac{\nu_\phi}{h_1} + \frac{a}{h_2}\left(\frac{a_\psi}{\delta} - \frac{a}{\gamma}S_\psi\right) &= 0 \\ S_t + q\frac{S_\phi}{h_1} &= 0 \end{aligned}\right\} \qquad (2.2)$$

where a is the speed of sound, q the modulus of the flow velocity, ν is the flow angle, S is the entropy,

$$\delta = \frac{\gamma-1}{2} \quad \text{and} \quad \kappa = \delta\frac{a}{\gamma} \; ,$$

and all flow properties are normalised to suitable reference values.

Since we are not interested in a physical description of an actual transient, but are looking for the steady solution of (2.2), we assume h_1 and h_2 to be defined as the scale factors relative to the steady state. By so doing, we alter the physical meaning of equations (2.2) only so far as the time derivatives are different from zero, and the method we use reduces to a "pseudo" time-dependent method. When the flow is steady the particle paths coincide with the streamlines: thus it is natural to select the variable ψ as playing the role of stream function. Therefore the metric factor h_2 is given by

$$h_2 = \frac{1}{\rho q} A(\psi, t) \qquad (2.3)$$

where $A(\psi,t)$ is the derivative of an arbitrary stretching function. Similarly if the flow is irrotational, is is natural to select the variable ϕ as playing the role of velocity potential: therefore it is

$$h_1 = \frac{1}{q} B(\phi, t) \qquad (2.4)$$

If the flow is rotational, the variable ϕ loses such a physical meaning and the metric factor h_1 has to depend on the vorticity. By definition, in this set of orthogonal curvilinear coordinates, the vorticity is given by

$$\omega = -\frac{q_\psi}{h_2} - \frac{q\, h_{1\psi}}{h_1 h_2} . \tag{2.5}$$

By integrating (2.5), the metric factor h_1 can be expressed as:

$$h_1 = B(\phi)\,\frac{\exp\left[-\int_{\psi_0}^{\psi} \frac{\omega\, h_2}{q}\, d\psi\right]}{q} , \tag{2.6}$$

where the integral has to be computed along ψ = constant lines and the vorticity ω is evaluated by means of the relationship

$$\omega = \frac{TS_\psi - H^\circ_\psi}{h_2\, q} , \tag{2.7}$$

where H° is the total enthalpy. Equation (7) is again valid only at the steady state, in agreement with the general idea of disregarding the physical meaning of the transient and focusing interest on the steady solution.

We have already decided to sacrifice the physical description of the transient in favour of getting the steady solution on a grid that (i) is orthogonal, (ii) does not have to be defined a priori, (iii) adapts itself to the flow. In the same spirit we can simplify equations (2.2). In fact, when the flow is steady, the entropy is piecewise constant along the streamlines, holding a value that has to be prescribed at the inflow boundary and jumping across shock waves if shock waves occur. Assuming that we are able to evaluate the entropy jumps across shocks, as explained in (Zannetti, 1984) the last of (2.2) can be neglected, the derivatives S_ϕ can be eliminated as well, and the derivatives S_ψ can be regarded as known terms. With such assumptions the equations actually used for the computation are

$$\left.\begin{aligned}
&a_t + \frac{q}{h_1} a_\phi + \delta a\left(\frac{q_\phi}{h_1} + q\,\frac{\nu_\psi}{h_2}\right) = 0\\
&q_t + q\,\frac{q_\phi}{h_1} + \frac{a}{h_1}\,\frac{a_\phi}{\delta} = 0\\
&q\nu_t + q^2\,\frac{\nu_\phi}{h_1} + \frac{a}{h_2}\left(\frac{a_\psi}{\delta} - \frac{a}{\gamma} S_\psi\right) = 0
\end{aligned}\right\} \tag{2.8}$$

3. NUMERICAL TECHNIQUE

The same hyperbolic nature is shared by (2.8) and the original Euler equations. The same boundary conditions apply to both sets, as well. The numerical scheme we use is the lambda-scheme, as formulated in (Moretti and Zannetti, 1984).

Following this paper we rewrite (2.8) as follows:

$$\left.\begin{aligned} a_t &= \frac{\delta}{2}(f_1 + f_2 + f_4 + f_5) \\ q_t &= \frac{1}{2}(f_2 - f_1) \\ \nu_t &= \frac{1}{2q}(2f_3 + f_5 - f_4) \end{aligned}\right\} \qquad (3.1)$$

where

$$\left.\begin{aligned} \lambda_1 &= \frac{1}{h_1}(q - a) \\ \lambda_2 &= \frac{1}{h_1}(q + a) \\ \lambda_3 &= \frac{1}{h_1} q \\ \lambda_4 &= -\frac{a}{h_2} \\ \lambda_5 &= \frac{a}{h_2} \end{aligned}\right\} \qquad (3.2)$$

and

$$\left.\begin{aligned} f_1 &= -\lambda_1\left(\frac{a_\phi}{\delta} - q_\phi\right) \\ f_2 &= -\lambda_2\left(\frac{a_\phi}{\delta} + q_\phi\right) \\ f_3 &= -\lambda_3 q \nu_\phi \\ f_4 &= -\lambda_4\left(\frac{a_\psi}{\delta} - q\nu_\psi - \frac{a}{\gamma} S_\psi\right) \\ f_5 &= -\lambda_5\left(\frac{a_\psi}{\delta} + q\nu_\psi - \frac{a}{\gamma} S_\psi\right) \end{aligned}\right\} \qquad (3.3)$$

The form (3.1) of the equations of motion is the result of three linear combinations of four compatibility equations on four characteristic surfaces. Each compatibility equation contains derivatives taken along two lines on a characteristic surface, viz, the bicharacteristic, along which signals propagate, and another line. The linear combination above eliminates all the terms not related to bicharacteristics. Therefore the λ_j express the components of the velocities of propagation along the four selected bicharacteristics and the f_j terms express the convection of signals along the same bicharacteristics. The ϕ derivatives and the ψ derivatives are approximated by "upwind" differences, according to the direction of propagation of signals, providing consistency with the domain of dependence of each grid point.

4. THE FORMULATION OF THE PROBLEM

In Figs. 1a and 1b the channel and the computational grid are sketched in both the ϕ, ψ plane and the physical plane. The line AC is the inflow boundary, the line BD is the outflow boundary, and the lines AB and CD are the walls of the channel. At the inflow boundary all the flow properties are prescribed if the flow is supersonic, while the flow angle ν , the total temperature T_o and the total pressure P_o are prescribed if the flow is subsonic. At the outflow boundary no boundary conditions are required if the flow is supersonic, whereas the static pressure is given if the flow is subsonic.

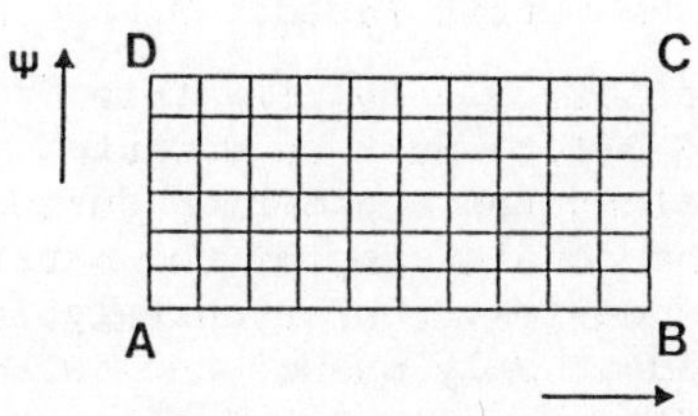

Fig. 1a.

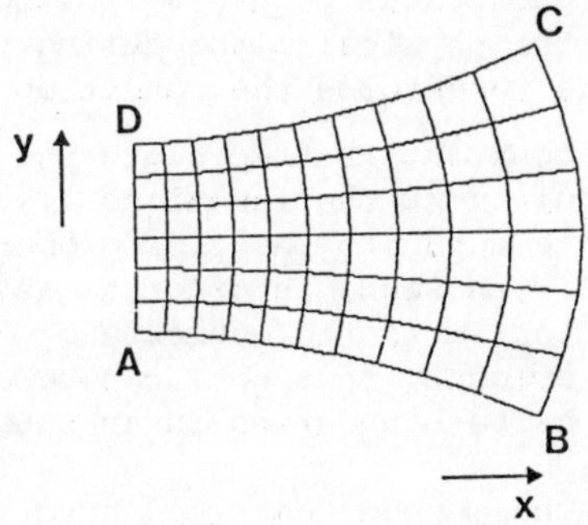

Fig. 1b.

When solving inverse problems in channels we look for the shape that the walls have to assume to satisfy the pressure distribution prescribed as boundary condition along them. The boundary conditions along the boundaries can be prescribed as function of the curvilinear coordinates ϕ and ψ: as a consequence the functions $A(\psi,t)$ and $B(\phi,t)$ of (2.3) and (2.6) can be assumed as constant, $A(\psi,t) = B(\phi,t) = 1$. The length and width of the channel are unknown a priori and are part of the solution, as is normal for an inverse problem. Once the steady solution is obtained, the shape of the walls, as well as the shape of any ϕ = constant or ψ = constant grid line, can be computed by numerical integration. The ϕ = constant grid lines are computed through

$$z = z_{in} + \int_{\psi_{in}}^{\psi} e^{i(\nu+\frac{\pi}{2})} h_2 d\psi \tag{4.1}$$

and the ψ = constant lines through

$$z = z_{in} + \int_{\phi_{in}}^{\phi} e^{i\nu} h_1 d\phi \tag{4.2}$$

where $z = (x+iy)$ is the complex coordinate on the physical plane.

When solving direct problems the geometry is known while the flow properties are unknown. It is convenient to prescribe the boundary conditions as function of the arc length s along the walls and the arc length n along the permeable boundaries. It follows that the values of the functions $A(\psi,t)$ and $B(\phi,t)$ can no longer be assumed a priori if we require that the computational points at the boundaries hold a fixed location on the physical plane during the transient. To this end we have to enforce the condition that the metric factors h_1 and h_2 be constant in time along the boundaries. To fulfil this requirement the functions $A(\psi,t)$ and $B(\phi,t)$ must be computed and adapted to the flow properties along the boundaries during the transient, in order to assure the consistency of the metric factors with the geometry of the boundaries. Unfortunately, as is evident, this goal can be achieved at only one of the two opposite boundaries in an internal flow.

The method described here is particularly suited to solve inverse problems, or to solve problems which are direct only for one solid wall, as in the case of mixed-direct-inverse problems where, for instance, the shape of one wall of a channel is prescribed while the other wall has to be designed in order to satisfy a prescribed pressure distribution along it, or for solving full direct problems with only one solid wall, as in external flows or in internal flows with a symmetry line.

5. TREATMENT OF BOUNDARIES

Since the grid is orthogonal, we can use a procedure at the boundaries which is both simple and consistent with the hyperbolic nature of the governing equations. The boundaries are computed following the ideas expressed in (Zannetti, 1981). A certain number of the f_j terms cannot be correctly approximated by finite difference at the boundaries since they express signals propagating towards the interior of the flow field. Such terms are considered as unknown in (3.1) and have to be evaluated using the boundary conditions.

At the inlet boundary, if the flow is subsonic, the terms f_2 and f_3 must be assumed as unknowns that have to be determined satisfying the boundary conditions. Here the flow angle ν and the total temperature are prescribed as a function of ψ and are constant in time. By considering their time derivatives, the following equations have to be satisfied:

$$\frac{a}{\delta} a_t + qq_t = 0 \qquad (5.1)$$

$$\nu_t = 0 .$$

The two unknown terms can be determined by substituting into (3.1), that is:

$$f_2 = \frac{qf_1 + a(f_1 + f_4 + f_5)}{a + q}$$
$$f_3 = \frac{f_4 - f_5}{2} \qquad (5.2)$$

The boundary points on the walls are treated in a similar way. When solving inverse problems, the static pressure distribution is prescribed along the walls as independent of time. The entropy being assumed constant in time along the walls, the speed of sound is constant in time, that is

$$a_t = 0 .$$

At the lower boundary, for instance the unknown term is f_5: by enforcing the boundary condition on the governing equations, the unknown term is computed using

$$f_5 = -(f_1 + f_2 + f_4) .$$

When solving direct problems, the flow angle ν is given along the solid walls. Therefore, at the lower wall the unknown term f_5 is computed using

$$f_5 = -(2f_3 - f_4) .$$

The outflow boundary is computed in the same way. If the flow is subsonic the static pressure is prescribed as the boundary condition. The unknown term f_1 is then given by

$$f_1 = -(f_2 + f_4 + f_5)$$

6. EXTENSION TO AXISYMMETRIC SWIRLING FLOW

The present computational technique can be extended directly to axisymmetric swirling flows. Keeping the same notation used so far, let us denote by v the tangential component of the flow velocity in an axisymmetric motion. The governing equations, with the simplifications introduced in (2.8) and (3.1) can be written;

$$a_t = \frac{\delta}{2}(f_1 + f_2 + f_4 + f_5) - \delta a \frac{q \sin\nu}{y}$$

$$q_t = \frac{1}{2}(f_2 - f_1) + \frac{v^2}{y} \sin\nu$$

$$\nu_t = \frac{1}{2q}(2f_3 + f_5 - f_4) + \frac{v^2}{y\,q} \cos\nu$$

$$v_t = f_6 - \frac{q\,v}{y} \sin\nu$$

with

$$f_6 = -\lambda_3\, v_\phi$$

The metric factors h_1 and h_2 are expressed by (2.7) and (2.4) respectively, as in the two-dimensional flow. In (2.7) ω now means the tangential component of the vorticity. For its computation (4.1) has to be replaced by

$$\omega = \frac{v^2}{y\,q}\sin\nu + \frac{TS_\psi - H^\circ_\psi + vv_\psi}{h_2\,q}$$

The computation can be carried on in the same way as in the two-dimensional case, except that during the integration in time the physical radial coordinate y has to be updated by taking the imaginary part of (4.2) or (5.1) at each computational point.

7. NUMERICAL EXAMPLES

Four numerical examples of the proposed method are described here. The first two refer to full inverse computations, the third to a mixed direct-inverse computation, the fourth to a full direct computation of a duct with a symmetry line.

The Ringleb flow has been selected as the test case. Two streamlines of the Ringleb flow, in the transonic region, have been chosen as the walls of a channel whose shape has to be determined by the present method. The original irrotational Ringleb flow has been modified to test the method when dealing with rotational flow fields. To this end a non-uniform total temperature is imposed on the flow field, while the total temperature is kept uniform. The resulting flow field is still analytically determined, the shape of the stream lines, the pressure and Mach number fields are unchanged, remaining the same as in the original irrotational flow field, while flow velocity and temperature are different.

The channel bounded by the streamlines $\phi = .85$ and $\psi = 1.1$ of the Ringleb flow has been selected. It has been made rotational by imposing a total temperature distribution with $T_o = 1$ at $\psi = .85$ and $T_o = 1.2$ at $\psi = 1.1$. The theoretical pressure acting along the walls has been taken as the design input. Figs. 2 and 3 show the theoretical channel, with the iso-velocity and iso-Mach patterns, respectively.

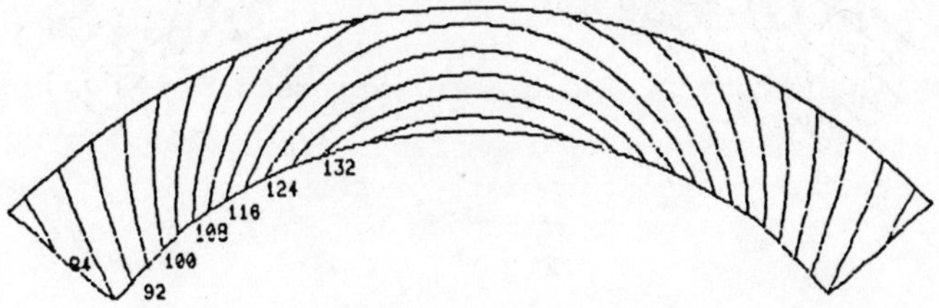

Fig. 2.

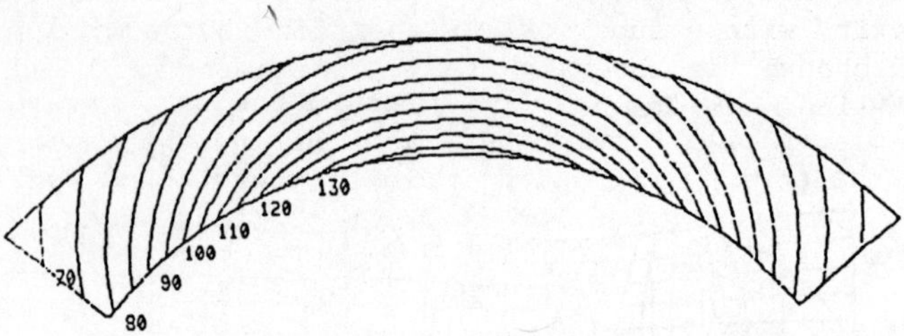

Fig. 3.

Figs. 4 and 5 show the computed shape, grid, iso-velocity and iso-Mach patterns.

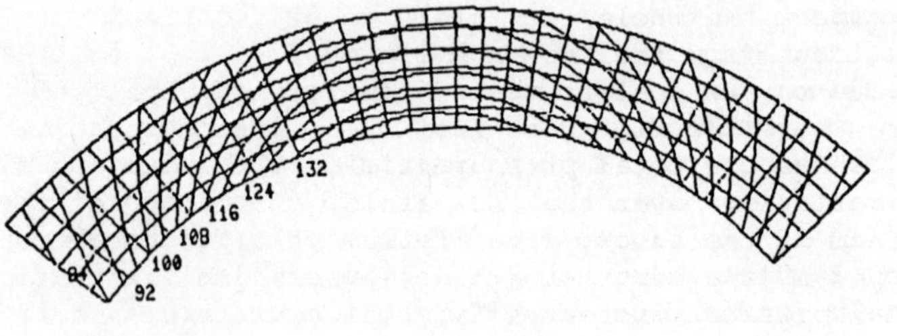

Fig. 4.

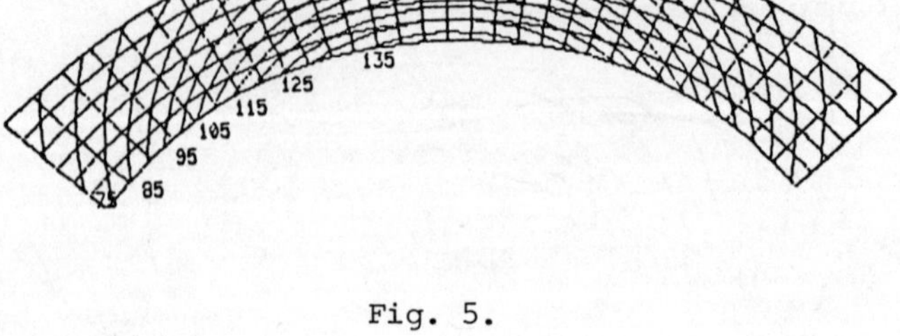

Fig. 5.

The second example refers to the design of a two-dimensional transonic channel. The pressure is prescribed along the walls to make the flow subsonic at the inlet and at the exit, with a supersonic transition in the middle. The computed channel is shown in Fig. 6, where the grid and the iso-Mach lines are drawn.

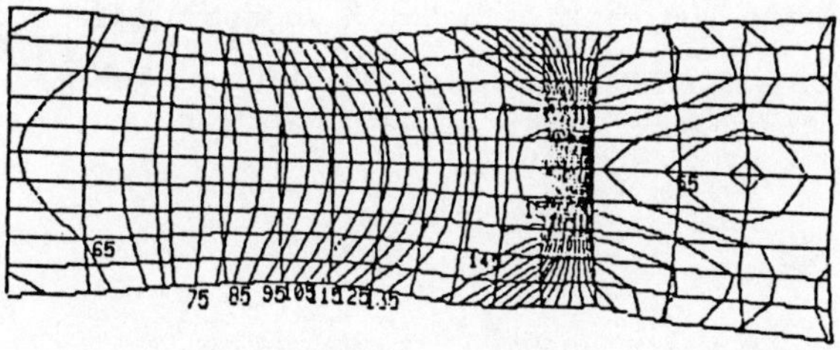

Fig. 6.

Since the pressure is prescribed as a smooth function of ϕ , the flow has to be shockless at the walls, but in general we cannot expect the whole flow field to be shockless. In fact, as is evident from the iso-Mach lines of Fig. 6, an imbedded shock, due to the coalescence of compression waves, appears in the core where the flow goes from supersonic to subsonic. Fig. 7. shows the total pressure field. It undergoes very little variation over the flow field, except on the region downstream of the shock. The minimum value of total pressure is reached at the centreline, where the shock is stronger. The total pressure increases from the centreline out along the shock and practically reaches the upstream unperturbed value where the shock vanishes.

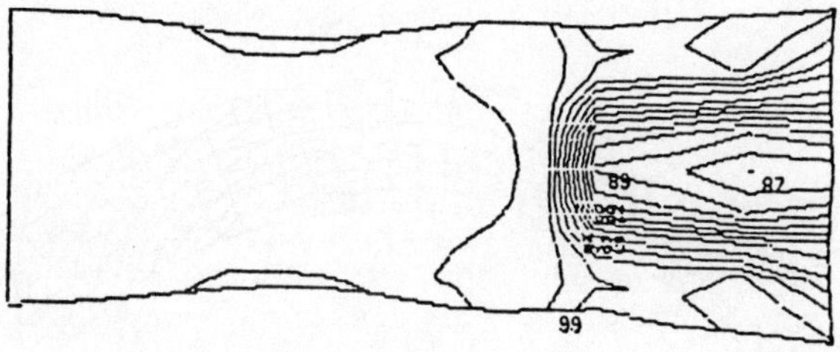

Fig. 7.

The third example is a two-dimensional mixed direct-inverse problem in a channel. The problem is direct for the lower boundary, where the shape of the wall is known and the pressure on it is unknown, and it is inverse for the upper wall whose shape is to be determined in order to satisfy a constant pressure distribution. As stated above when discussing the formulation of the problem, the locations of the nodes on the lower wall are kept fixed on the physical plane by taking the metric factor h_1 constant in time along the lower boundary during the computation. To this end, the function $B(\phi,t)$ is computed at the lower boundary, at each computational step, by

$$B(\phi,t) = h_1^* q$$

where h_1^* is the value assumed by the metric factor at the starting condition, at the time t=0. The inlet boundary has been treated in a similar way to assure that its length remains constant in time.

The computed channel is shown in Fig. 8 together with the grid and the iso-Mach lines. The shape of the lower wall, where the direct problem is solved, has been chosen to enhance the capability of the method to deal with awkward geometries.

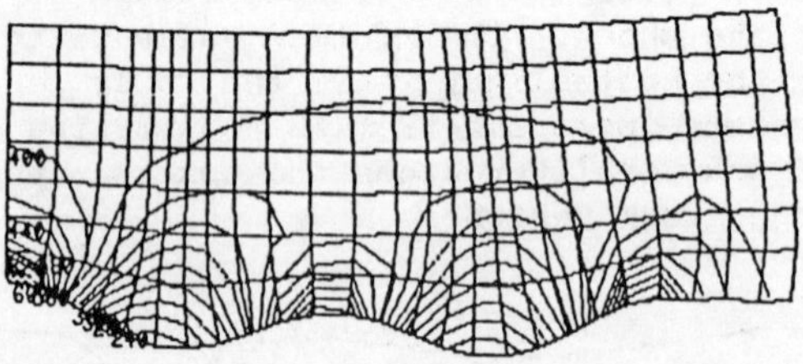

Fig. 8.

The last example refers to a full direct problem with a symmetry line. The computation was carried on as in the previous example except for the upper boundary which is now a symmetry line, treated as a solid straight wall. The nodes have a fixed location on the physical plane at the lower boundary during the computation, as explained above, while they move along the upper boundary in so far as the flow is not steady and the flow properties change in time, although the boundary conditions at this boundary do not change in time, being the slope of the wall constant. The result is shown in

Fig. 9. It shows a two-dimensional transonic nozzle with the grid and the iso-Mach lines.

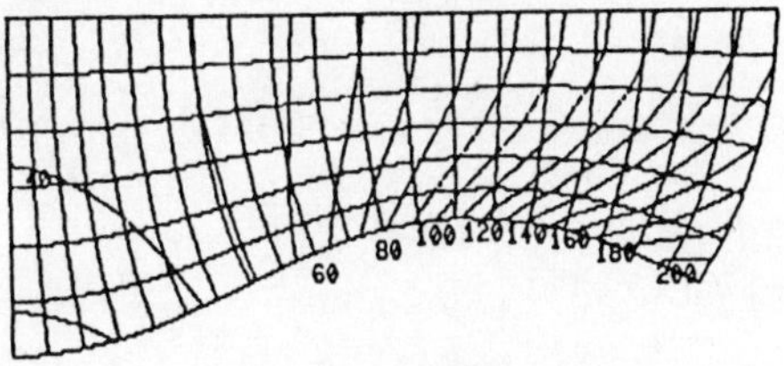

Fig. 9.

REFERENCES

Moretti G. (1979) "The λ-Scheme", Computers and Fluids, Vol. 7, No. 3, pp. 191-205.

Moretti. G. and Zannetti. L., (1984). "A New Improved Computational Technique for 2D Unsteady Compressible Flow." AIAA Journal, Vol. 22, No. 6, June 1984.

Zannetti. L. (1984). "A Natural Formulation for the Solution of 2D or Axisymmetric Inverse Problems." Int. Conf. on Inverse Design Concepts in Engineering Sciences." (ICIDES), Austin, Texas, Oct. 1984.

Zannetti. L. and Colasurdo. G. (1981). "Unsteady Compressible Flow: a Computational Method Consistent with the Physical Phenomena." AIAA Journal, Vol. 19, No. 7, pp. 852-856, July 1981.

Zannetti. L. (1981). "Numerical Treatment of Boundaries in Compressible Flow Problems." Notes on Numerical Fluid Mechanics, Vol. 5, pp. 337-343, Oct. 1981.

ARTIFICIAL VISCOSITY Q ERRORS FOR STRONG SHOCKS AND MORE ACCURATE SHOCK-FOLLOWING METHODS

W.F. Noh
(Lawrence Livermore National Laboratory, Livermore)

1. INTRODUCTION

The artificial viscosity (Q) method of von Neumann-Richtmyer (1950) has been (and is) a tremendously useful numerical technique for following shocks wherever and whenever they appear in the flow. As we shall see, it must be used with some caution, as serious Q-induced errors can occur in some strong shock calculations.

We investigate three types of Q errors:

1. Excess Q heating, of which there are two types: (i) excess Wall Heating on shock formation and (ii) Shockless Q Heating;
2. Q errors when shocks are propagated over a non-uniform mesh; and
3. Q errors in propagating shocks in spherical geometry.

We use as a basis of comparison, the Lagrangian formulation given in (Colella and Woodward, 1982) with $Q = C_O^2 \rho \ell^2 (u_x)^2$, and refer to this as our standard calculation. In Section 2, the Langrangian differential equations with Q [in plane ($\delta = 1$,), cylindrical, ($\delta = 2$), and spherical ($\delta = 3$) geometry] are given, and we include an artificial heat flux term $H = h_O^2 \rho \ell^2 |u_x| \varepsilon$ used in Noh's (Q&H) shock-following method (see Noh, 1983). For our comparisons, three Q's are defined: Q_L, $Q_L(v)$, and Q_E (where Q_L is our standard Q, above, and $Q_L(v)$ is the original definition given in (Colella and Woodward, 1982) (see also Richtmyer and Morton, 1967, p. 319) and depends on the geometry δ = 1, 2, or 3). Also two H's, H_L and H_E, are defined. Here, L refers to the normal Lagrange usage, in which the length ℓ is taken to be equal to the Lagrange mesh interval Δx in the difference formulations, and shocks are spread over a fixed number of mesh intervals ($\simeq 3$) regardless of their actual size. E refers to the Eulerian (or "fixed length") definitions in which ℓ is a constant ($\simeq \Delta x_{max}$), and shocks are spread over a fixed physical distance ($\simeq 3\Delta x_{max}$). Then, in the (Q&H) method, Q_L and H_L are used together, as are Q_E and H_E. Standard (staggered mesh) differencing is used (see Noh, 1985), and the nominal benchmark Lagrangian $Q_L = 2\rho/(\Delta u)^2$.

In Section 3, the Wall Heating Q error is investigated in test problem #1 (Figs. 8a,b). This is an infinite-strength, constant-velocity shock that is generated when a zero-temperature ($\gamma = 5/3$) gas is brought to rest at a rigid wall ($x = 0$). The excess Wall Heating error occurs in the first few zones near the wall (Fig. 1). Now, in nature, heat conduction prevents excess Wall Heating from developing, and thus the (Q&H) method (which includes an artificial heat flux term, H) also prevents the excess Wall Heating error. This is shown in Fig. 2.

The Shockless Q Heating error is investigated using a test problem called "Uniform Collapse" (see Noh, 1976). Here the fluid is shockless, although it is everywhere compressing, and, consequently, an energy error $\Delta\varepsilon$ is introduced for Q_L, Q_E, and $Q_L(v)$. In particular, the $Q_L(v)$ error is much larger (for $\delta = 3$) than for Q_L (i.e., it is shown that $\Delta\varepsilon_L(v) = \delta^2\Delta\varepsilon_L$), and this fact strongly favours our use of Q_L as the standard Lagrange formulation.

In Section 4, the second type of Q error is investigated by introducing a non-uniform mesh ($\Delta x_k = R\Delta x_k$ for constant R) into problem #1 (see Fig. 3). With $Q_L = 2\rho(\Delta u)^2$ as our standard Q, the Q errors for R = 1.05 and R = 1.25 (Figs. 4 and 5) are compared. The errors are seen to approach 100% for R = 1.25, and thus can be a serious concern. A calculation using Q_E (the fixed-length Q), taken together with H_E (in the (Q_E & H_E) method), eliminates both type #1 and #2 errors (Fig. 6). Unfortunately, using Q_E spreads shocks over a fixed physical distance of $\simeq 3\Delta x_{max}$, which is unacceptable in those regions where a smaller mesh interval occurs. A theoretical explanation for the non-uniform mesh error (see also Noh, 1985) shows that letting $\ell = \Delta x$ in the difference formulation implies

$\ell = \ell(x)$ in the differential equation formulation of Q_L, and it is this use of ℓ that generates this non-uniform mesh error. The (Q_L&H_L) method permits the use of a considerably smaller Q_L coefficient, C_O, and thus produces sharper shocks. This, in turn, produces smaller Q errors, as seen in Fig. 7. The (Q&H) method thus offers an acceptable procedure if the unequal zoning is not too severe.

In Section 5, the spherical geometry Q errors are investigated using Noh's spherical-shock test problem (see Noh, 1983 and Fig. 8d). This type #3 error is considerably more complicated than the previous Q errors, in that it depends both on the Q formulation [e.g., Q_L vs Q_L (v)] and whether or not Q is treated as a scalar viscosity as in (Von Neuman and Richtmyer, 1950) or as a tensor viscosity, as in (Shultz, 1964). The errors are seen to be truly enormous (Fig. 9), where the error for the standard Q_L formulation is nearly 600% (near the origin) and is nearly 1000% for the original Q_L(v) definition of Q given in (Von Neumann and Richtmyer, 1950). In Fig. 10, we see how serious this error can be as a function of mesh size and just how slow the solution converges to the exact value, $\rho^+ = 64$. Indeed, even for K = 800, there is still a considerable error near the origin. The explanation is given in Fig. 11, where it is shown that the error results from the finite shock thickness, and thus the inability of the Q method to select the correct preshock density. That is, the shock spreading picks $\rho^- < \rho^-_{exact} = 16$. We conclude that sharper shocks give smaller Q errors and indeed this is shown in Fig. 12, when the (non-Q) PPM method of Colella and Woodward (1982) produces very sharp shocks and has minimal error. This error in PPM is further reduced by using an adaptive mesh technique to capture the shocks, and the result using 400 zones (with an adaptive mesh) is equivalent to a normal 1200-zone (essentially converged) PPM problem.

In Section 6, Schultz's tensor Q formulation (T) of the hydrodynamic equations is given, along with his tensor Q_L(S) definition (see Schultz, 1964). Calculations using this tensor formulation (Fig. 13) are a significant improvement over the standard scalar (S) solutions (e.g., Fig. 9). However, there is only a slight improvement using Schultz's Q_L(S) over the standard Q_L. Consequently, we conclude that the tensor equations (T) are more important than which Q_L formulation is used. Again, as sharper shocks reduce the type =3 error, nearly exact results are obtained using a very small ($C_0^2 = \frac{1}{4}$) in the (Q_L&H_L) (T) method, where $Q_L = (\frac{1}{4})\rho(\Delta u)^2$ and $H_L = 10\rho|\Delta u|\Delta\varepsilon$ (Fig. 14). In Fig. 15, the various Qs, (Q&H), and the (non-Q) PPM method are compared. The best results are obtained from the tensor formulation using the (Q_L&H_L) (T) method and PPM with an adaptive mesh.

2. LAGRANGIAN FLUID EQUATIONS WITH ARTIFICIAL VISCOSITY (Q) AND HEAT FLUX (H)

Von Neumann and Richtymer (1950) introduced their artificial viscosity Q as a scalar quantity, and we take their formulation of the Lagrangian fluid equations as our standard. Also, the new (Q&H) shock-following method of Noh (1983) (which uses an artificial heat flux H in addition to the artificial viscosity Q to follow shocks) is included in the formulation.

2.1 *Differential Equations*

The independent Lagrange variables are r and t, where r is the initial position of the Eulerian (physical) coordinate (i.e., R(r,0) = r), and u, ρ, ε, and P are the velocity, density, internal energy, and pressure. The differential equations for plane (δ = 1), cylindrical (δ = 2), and spherical (δ = 3) geometries are then (letting $dm = \delta\rho^0 r^{\delta-1} dr = \rho^0 dr^\delta$):

$$
\begin{aligned}
u_t &= -\delta R^{\delta-1}(P+Q)_m && \text{momentum} \quad , \\
R_t &= u && \\
v &= 1/\rho = (R^\delta)_m && \text{mass} \quad , \qquad (2.1) \\
\varepsilon_t &= -(P+Q)v_t + \delta(R^{\delta-1}H)_m && \text{energy} \quad , \\
P &= P(\rho,\varepsilon) && \text{equation of state} \quad .
\end{aligned}
$$

2.2 *Definitions of Q and H*

In Q and H, we include linear terms (see Noh, 1976, 1983). These are used in some of the Q error comparisons to produce smoother shock profiles, but otherwise don't affect the Q errors. Also, the Qs and Hs are set to zero if the indicated tests fail to hold.

Standard Lagrange Q: if $u_r < 0$
$$Q_L(C_0^2, C_1) = C_0^2 \rho \ell^2 (u_r)^2 - C_1 \rho C_S \ell u_r \tag{2.2}$$

Standard Lagrange H: if $Q \neq 0$.
$$H_L(h_0^2, h_1) = h_0^2 \rho \ell^2 |u_r| \varepsilon_r + h_1 \rho C_S \ell \varepsilon_r \tag{2.3}$$

Eulerian fixed-length Q: if $u_R < 0$.
$$Q_E(C_0^2, C_1) = C_0^2 \rho \ell^2 (u_R)^2 - C_1 \rho C_S \ell u_R \tag{2.4}$$

Eulerian fixed-length H:
$$H_E(h_0^2, h_1) = h_0^2 \rho \ell^2 |u_R| \varepsilon_R + h_1 \rho C_S \ell \varepsilon_R \tag{2.5}$$

Original Lagrange formulation:
$$Q_L(v) = (C_0 \rho^0 \ell)^2 \rho \left(\frac{r}{R}\right)^{2\delta - 2} (v_t)^2 \tag{2.6}$$

if $v_t < 0$ (see Von Neumann and Richtmyer, 1950),

where C_0, C_1, h_0 and h_1 are constants, ℓ is a constant with the dimensions of length, and C_S is the local sound speed. The usage is (Q_L&H_L) and (Q_E&H_E). Equation (2.6) is the original Q formulation (see also Richtmyer and Morton (1967), p. 319) in terms of $v = (1/\rho)$, and we note it is the only Q here to depend on the geometry (δ). In the Lagrange formulation (L), the standard use is to take $\ell = \Delta r$ in the difference equations. This spreads shocks over a fixed number ($\simeq 3$) of mesh intervals (regardless of their size), while in the Eulerian formulation (E), ℓ is constant ($\simeq \Delta x_{max}$), and shocks are spread over a fixed physical length ($\simeq 3\ell$). Standard "staggered mesh" difference equations are used (see Noh (1985) for details), and the nominal Q_L benchmark difference formulation is $Q_L = 2\rho(\Delta u)^2$.

3. EXCESS Q HEATING

There are two excess Q heating errors: (1) excess wall (or piston) heating due to Q, which occurs on shock formation (e.g., at a rigid wall where a gas is brought to rest, and a shock is propagated away, or for the sudden startup of a piston); and (2) Q heating for shockless compressions (e.g., when $u_r < 0$, but no shock is present).

3.1 The Wall Heating Error Test Problem

Test problem #1 is that of a constant-state, constant-velocity shock of infinite strength (i.e., the pre-shock pressure $P^{-} = P^{O} = 0$). The shock is generated in a perfect ($\gamma = 5/3$) gas by bringing the cold ($\varepsilon^{O} = 0$) gas to rest at a rigid wall at $x = 0$. This is just the familiar constant-velocity (piston) shock, but in a frame of reference where the piston (here a rigid wall) is at rest (Figs. 8a, b). In Fig. 1, ρ^{+} is plotted for our "standard calculation" using $Q_L = 2\rho(\Delta u)^2$. The shaded area (the Wall Heating error) occurs typically in the first three zones next to the wall (or piston). That this error is unavoidable for any shock smearing method, is argued in (Noh, 1985).

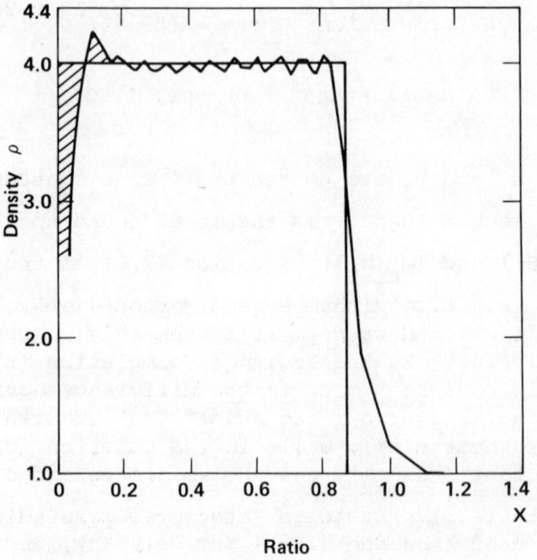

Fig. 1 $Q_L = 2\rho(\Delta u)^2$. Shaded area is "Wall Heating Error". Typically it is over three mesh intervals.

In real fluids, heat conduction is present, and Wall Heating does not occur, and this is the basis of Noh's (Q&H) method, Eqs. (2.1) and (2.3). In Fig. 2, $Q_L(0.67, 0.2)$ is used, plus the heat flux term $H_L(0,3/4) = 3/4\rho C_s\Delta\varepsilon$. As expected, the Wall Heating error is zero. We also note that the $(Q_L \& H_L)$ solution is considerably smoother, and this permits the use of much smaller Q constants C_0 and C_1 (with generally smaller Q errors).

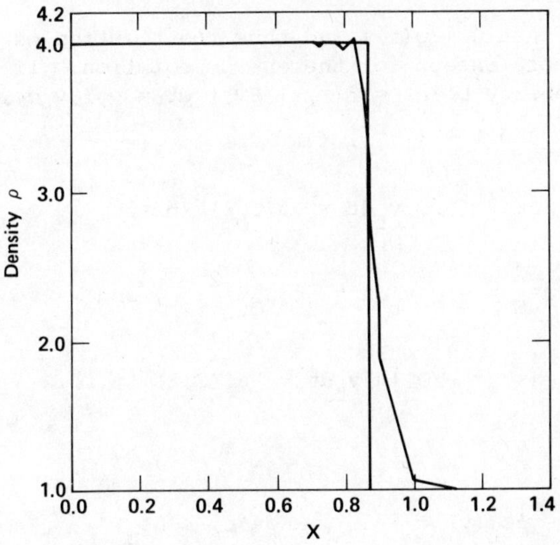

Fig. 2 The (Q&H) method, where $Q_L(\frac{2}{3}, \frac{1}{5}) = \frac{2}{3}\rho(\Delta u)^2 + \frac{1}{5}\rho C_s|\Delta u|$ and $H_L(0,\frac{3}{4}) = \frac{3}{4}\rho C_s \Delta\varepsilon$. The heat flux term, H, reduces the Wall Heating error to zero.

3.2 Shockless Q Heating Errors

This is the situation where a compression wave exists (i.e., $u_r<0$), and thus $Q \neq 0$; yet the exact solution is shockless. For this analysis, we consider the useful "Uniform Collapse Problem" (see Noh (1976), p. 60), in which a flow is everywhere undergoing a compression, but no shock develops. We consider a unit "sphere" ($0 \leq r \leq 1$), (for $\delta = 1$, 2, or 3), and to simplify the analysis of the energy errors due to Q, we assume pressure to be a function of density only, $P = P(\rho)$.

The initial values are $u(r,0) = -r$, $\rho = \rho^0$, $\varepsilon = \varepsilon^0$, and $P^0 = P(\rho^0)$, with boundary conditions $u(0,t) = 0$ and $u(1,t) = -1$. The exact solution is that the fluid simply coasts with its initial velocity ($u^0 = -r$) until all points uniformly collapse onto the origin $R = 0$, at time $t = 1$. It is easy to verify that the exact solution is given by

$$u(r,t) = -r, \quad R = r(1-t), \text{ and } v = (1/\rho) = (1-t)^{\delta}(1/\rho^0). \qquad (3.1)$$

Thus, $\rho = \rho(t)$, $P = P(\rho) = P(t)$, and $\varepsilon = \varepsilon(t)$ (since $\varepsilon_t = -Pv_t = \delta/\rho^0 \; P(t) \; (1-t)^{\delta-1}$).

It follows that $Q = Q(t)$, and thus the solution of Eq. (3.1) continues to hold except for the energy equation. If we let the error in energy be $\Delta\varepsilon = \int(\varepsilon_t + Pv_t)\,dt = -\int Qv_t dt$, then, from Eq. (3.1) and $\tau = (1-t)$,

$$\Delta\varepsilon_L = -\int Q_L v_t dt = -\delta(C_o \ell)^2 \ell n(\tau) \ , \tag{3.2}$$

$$\Delta\varepsilon_E = -\int Q_\varepsilon v_t dt = \frac{1}{2}\delta(C_o \ell)^2 \tau^{-2}, \text{ and} \tag{3.3}$$

$$\Delta\varepsilon_L(v) = -\int Q_L(v) v_t dt = -\delta^3 (C_o \ell)^2 \ell n(\tau). \tag{3.4}$$

We note that

$$\Delta\varepsilon_L(v) = \delta^2 \Delta\varepsilon_L, \text{ and } \Delta\varepsilon_L < \Delta\varepsilon_E. \tag{3.5}$$

Now, as $\tau \to 0$, the error $\Delta\varepsilon \to \infty$, and this Shockless Q Heating error can indeed be serious. From Eq. (3.5) and $(\delta = 3)$, we see that the error $\Delta\varepsilon_L(v)$ is nearly an order of magnitude greater than the error $\Delta\varepsilon_L$. It is due to arguments similar to these that Noh in 1956 suggested that the Q_L of Eq. (2.2) be taken as the standard Q formulation for all geometries $\delta = 1, 2$ or 3. We make the suggestion again (and form more reinforcement, see Figs. 9 and 10), since $Q_L(v)$ still seems to be in common use.

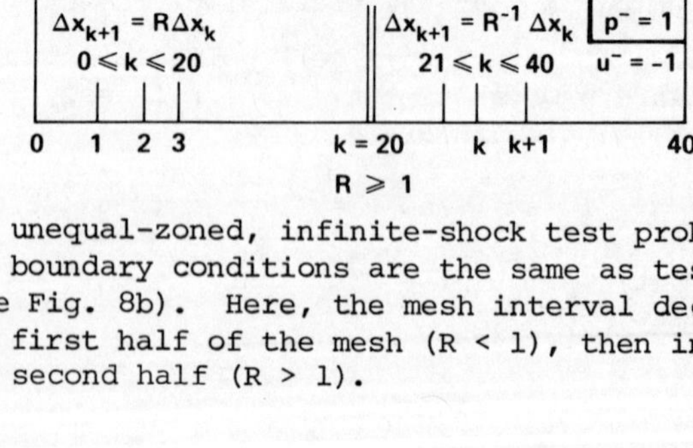

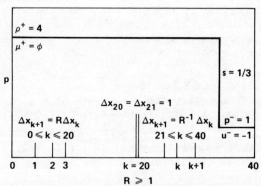

Fig. 3 The unequal-zoned, infinite-shock test problem. Initial and boundary conditions are the same as test problem #1 (see Fig. 8b). Here, the mesh interval decreases for the first half of the mesh (R < 1), then increases for the second half (R > 1).

4. Q ERRORS FOR A NON-UNIFORM MESH

The second type of Q error occurs when shocks are propagated over a mesh with unequal intervals. In problem #1, let

$$\Delta x_{k+1} = R\Delta x_k \ , \tag{4.1}$$

where R is a constant: $1 \lesssim R \lesssim 2$. We investigate the cases R = 1.05 and R = 1.25. In order to show the errors for both decreasing (R < 1) and increasing (R > 1) mesh intervals, we let the mesh decrease for the first half of the problem, then increase for the second half (Fig. 4).

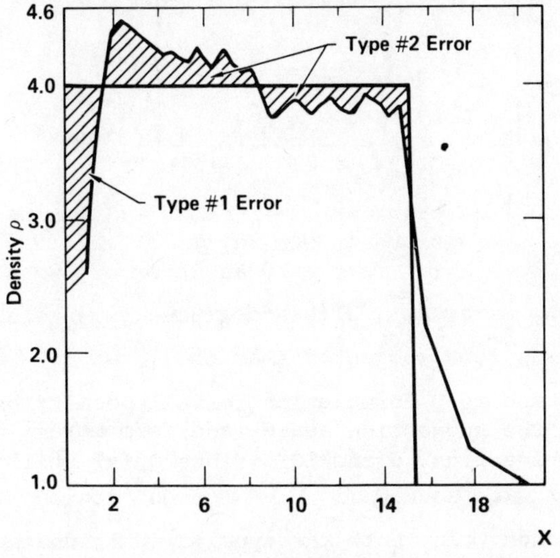

Fig. 4 Standard $Q_L(2,0) = 2\rho(\Delta u)^2$, and R = 1.25. The type #2 error ($\Delta\rho = \rho - 4$) is positive for the first half, where the mesh interval decreases (R < 1), and $\Delta\rho > 0$ for the second half (R > 1). The type #1 Wall Heating error is present in the first few zones next to the rigid wall on the left.

In Fig. 5, R = 1.05, and the density is plotted for our standard $Q_L = 2\rho(\Delta u)^2$. The total error is shaded, and again we see the familiar Wall Heating error in the first several zones. The new error, $\Delta\rho_L^+ = \rho_L^+ - \rho^+_{exact} = \rho_L^+ - 4$, is > 0 for the first half (decreasing mesh), and $\Delta\rho_L^+ < 0$ for the second half. This new (type #2) non-uniform mesh error grows with R and becomes very serious (≃100%) for R = 1.25. This is unfortunate, since it is not uncommon to use R = 2 in practice, and thus R = 1.25 might well be considered a modest zoning change.

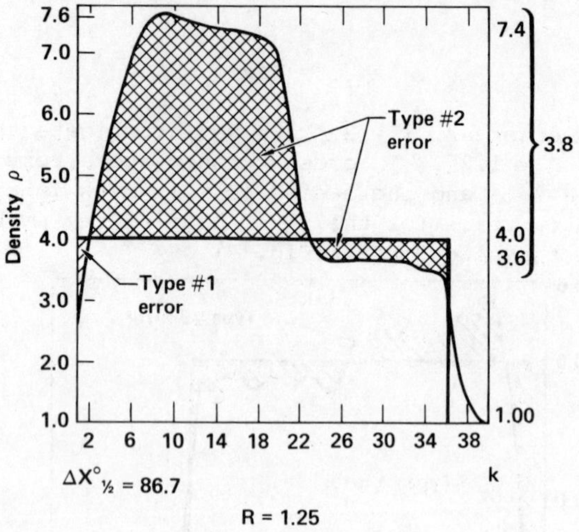

Fig. 5 Standard Q_L (2,0) = $2\rho(\Delta u)^2$, and R = 1.25. Here the error is very serious (≃100%).

The good news is shown in Fig. 6, where R = 1.05, and the Eulerian "fixed length" Q_E is seen to eliminate the non-uniform mesh error. When Q_E is used in conjunction with H_E in the (Q_E&H_E) method, then both the type #1 (Wall Heating) error and the type #2 (non-uniform mesh) error are completely eliminated. The bad news is that very large Q_E constants are necessary [where $(C_O \ell)^2 = (C_O \Delta x_{max})^2 = 6$ (for $C_O = 1$) and $C_1 \Delta x_{max} = 0.8$ (for $C_O = 0.33$)], and this is seen to spread the shock over a large number of smaller zones. In this regard, the use of Q_E is not satisfactory.

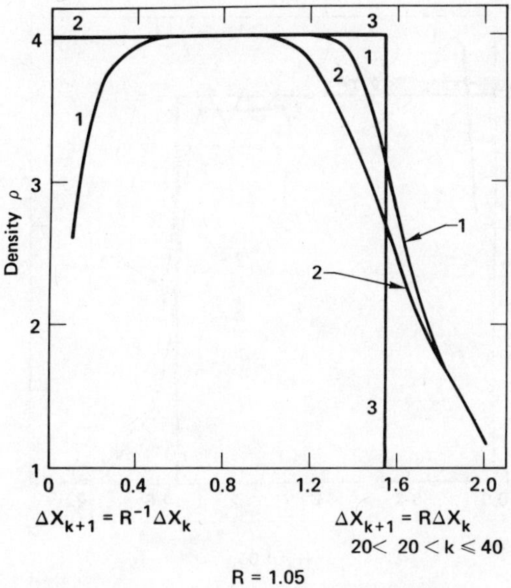

Fig. 6 R = 1.05, and we compare the fixed-length Q_E with the $(Q_E \& H_E)$ method.

(1) $Q_E(6,0.8) = 6\rho(\frac{\Delta u}{\Delta X})^2 + 0.8\rho C_s|\frac{\Delta u}{\Delta x}|$; this eliminates the non-uniform mesh error.

(2) $Q_E(6,0.8)$ & $H_E(0,6) = 6\rho C_s(\frac{\Delta\varepsilon}{\Delta X})$, which eliminates both the type #1 Wall Heating error and the non-uniform mesh type #2 error. However, there is too much shock spreading over the finely zoned region using Q_E for this to be a practical way to minimize these errors.

(3) Exact solution.

A more practical solution is to use the $(Q_L \& H_L)$ method. This eliminates the Wall Heating error, and (as mentioned earlier), the use of the heat flux H permits the use of a much smaller Q_L constant C_O (and C_1), which, in turn, reduces the type #2 (non-uniform) mesh error. In Fig. 7, R = 1.05 and $Q_L = \rho(\Delta u)^2$ (i.e., $C_O^2 = 1$) is compared with the $(Q_L \& H_L)$ method, with a corresponding reduction in the non-uniform zoning error to around 3%.

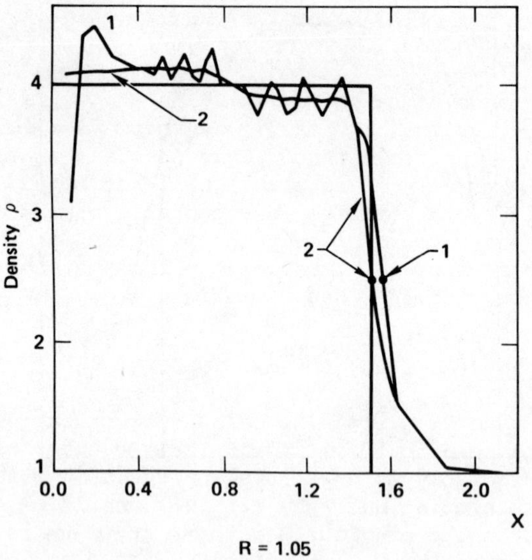

Fig. 7 R = 1.05, and we compare Q_L (with reduced coefficient $C_O = 1$) with the (Q_L&H_L) method.

(1) $Q_L(1,0) = \rho(\Delta u)^2$. The solution is noisy, but type #1 and #2 errors are reduced.

(2) $Q_L(1,0) = \rho(\Delta u)^2$ & $H_L(0, \frac{2}{3}) = \frac{2}{3}\rho C_s \Delta\varepsilon$. Using both Q_L and H_L (with small Q coefficients) eliminates the Wall Heating type #1 error altogether and reduces the non-uniform mesh type #2 errror to ≈ 3%. The (Q_L&H_L) method is much smoother than Q_L alone and may be a practical compromise for mesh-interval changes that are not too large.

4.1 Theoretical Discussion

In the standard Lagrange difference formulation of Q_L, the length ℓ in Eq. (2.2) is taken to be $\ell = \Delta x$, and thus $(\ell u_x) \to (\Delta u)$. Now, when an unequal mesh interval is used, Δx is no longer a constant, and this implies that $\ell = \ell(x)$ in the differential formulation of Q_L in Eq. (2.2). In fact, where Δx_k is given by Eq. (4.1), $\ell(x)$ is found to be

$$\ell(x) = 2[(R-1)x + \Delta x_o]/(R+1). \tag{4.2}$$

See Noh (1985) for details and further explanation of the non-uniform mesh error.

One consequence of using Eq. (4.2) in Eq. (2.2) in the differential equations, Eq. (2.1), is that steady travelling shocks are no longer solutions. This is clear since the shock width (instead of being a constant) will now be proportional to $\ell(x)$. This would still be an acceptable numerical approximation for shocks if only the proper shock jump conditions held, but our numerical experiments show that this is unfortunately not the case.

5. Q ERRORS IN SPHERICAL ($\delta = 3$) GEOMETRY

The third type of Q error is related to strong shock propagation in spherical (or cylindrical) geometries. This error is considerably more serious (up to 1000% error in (excess) shock heating near the origin), and is also more complicated than the previous Q errors. This third type of Q error depends on the Q formulation [i.e., Q_L of Eq. (2.2) vs. $Q_L(V)$ of Eq. (2.6)] and also seems to depend on whether Q is treated as a scalar or a tensor viscosity in the formulation of the hydrodynamic equations. In particular, a tensor formulation (Section 6) due to Schultx (1964) gives sharper shocks than our standard use of Q_L in Eq. (2.1), and this is instrumental in reducing this third type of Q error.

Test problem #3 is just the spherical ($\delta = 3$) generalization of test problem #1 where the post (infinite) shock solutions (u^+, ρ^+, ε^+, and P^+) are, again, constant step-value functions (Fig. 8d).

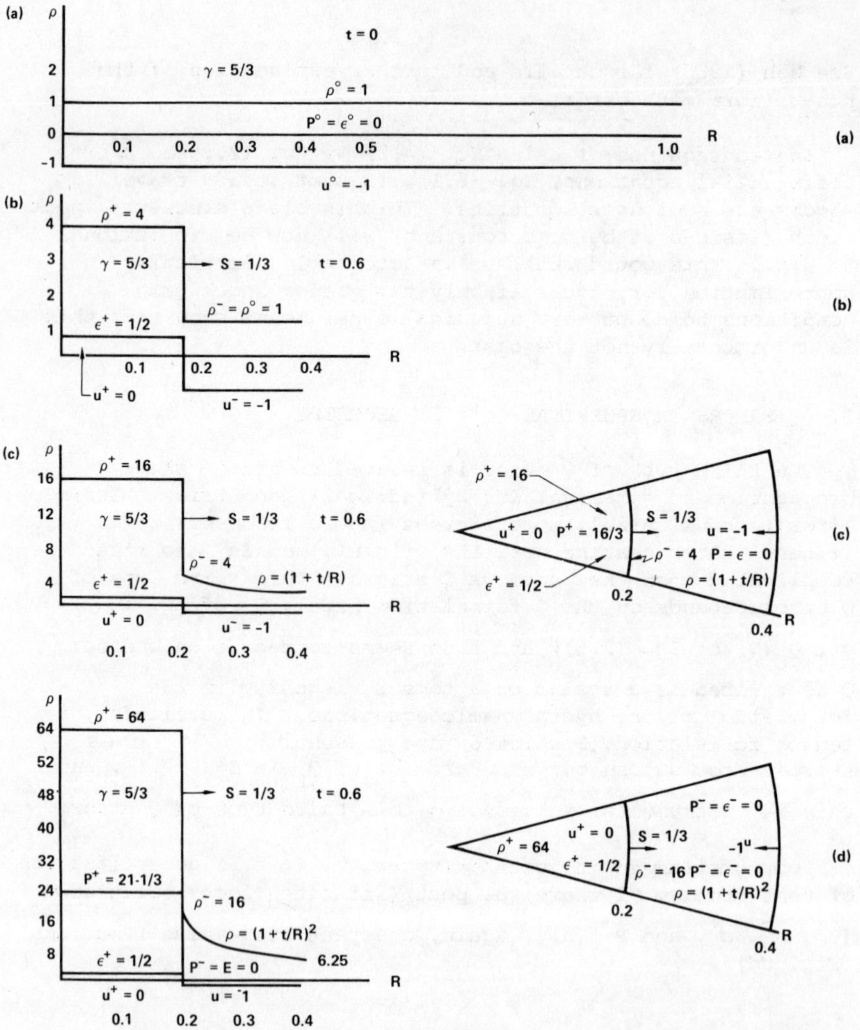

Fig. 8 The exact solution at t = 0.6 to Noh's (1983) Generic Constant-Velocity Shock problems: (a) initial conditions; (b) plane geometry ($\delta = 1$) with a shock generated at a rigid wall; (c) a shock reflection from the axis of symmetry for a cylinder ($\delta = 2$); and (d) shock reflection from the centre of a sphere ($\delta = 3$). All solutions have constant post-shock states, and all have the same constant shock speed, $S = \frac{1}{3}$ (for initial conditions (a) and $\gamma = 5/3$). The essential difference is the preshocked density $\rho^- = 1$ for δ; $\rho^- = 4$ for $\delta = 2$; $\rho^- = 16$ for $\delta = 3$; and $\rho = \rho^O (1+t/R)^{\delta-1}$ in front of the shock.

Our standard test problem has 100 mesh intervals ($\Delta r = 0.01$), and the results are compared at time $t = 0.6$. Since the shock speed is $S = 1/3$, then 80% of the mesh (i.e., 80 mesh points) have been traversed by the shock, and one would expect accurate results. Unfortunately, this is not the case, as is seen in Fig. 9, where the standard $Q_L(2) = 2\rho(\Delta u)^2$ is compared with the original $Q_L(V) = 2(\Delta r)^2 \rho(\frac{r}{R})^4 (\frac{\Delta v}{\Delta t})^2 = 2\rho[\Delta u + \frac{2u\Delta R}{R}]^2$.

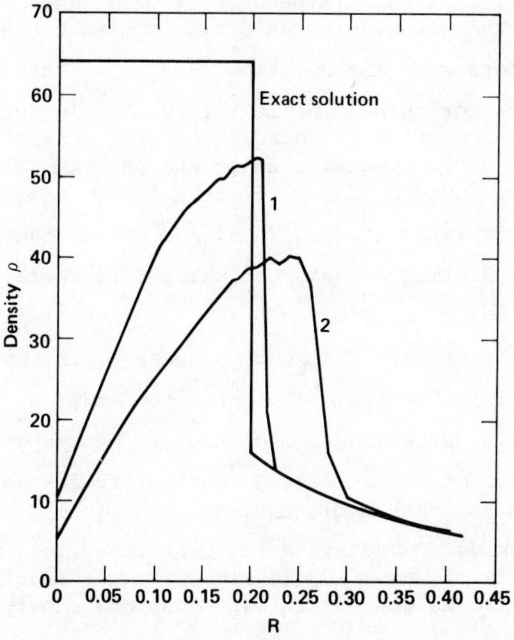

Fig. 9 (1) Standard $Q_L = 2\rho(\Delta u)^2$.

(2) $Q_L(v) = 2(\Delta r)^2 \rho(\frac{r}{R})^4 (\frac{\Delta v}{\Delta t})^2$ [the original von Neumann-Richtmyer Q formulation, Eq. (2.6)]. Here, Q_L is superior to $Q_L(V)$, but both Qs are in serious error. The correct solution is $\rho^+ = 64$.

Here, both are compared with the exact solution, $\rho^+ = 64$. The numerical results are strikingly poor, and in fact, hardly bear any resemblance to the exact solution. The error for the standard $Q_L = 2\rho(\Delta u)^2$ is seen to be on the order of 600% near the origin and 20% behind the shock, while the error for the original von Neumann-Richtmyer Q [here $Q_L(V)$ of Eq. (2.6)] is roughly 1000% in the central region and nearly 40% behind the shock. Clearly this third type of Q error depends on the Q formulation. The solution using $Q_L(V)$ is seen to be definitely inferior to the solution using Q_L. There are several reasons for this. One is related to the Shockless Q Heating Error of Section 3. Here, the first zone of test problem #3 is just a special case of the Uniform Collapse test problem, and we found in Eq. (3.5) that the Q_L energy errors went as $\Delta\varepsilon_L < \Delta\varepsilon_L(V) = \delta^2\Delta\varepsilon_L$. Thus, for $\delta = 3$, the error using $Q_L(V)$ is 9 times as large as using Q_L. There is an even more disquieting error in using $Q_L(V)$, in that it preheats the gas ahead of the shock. This occurs because, in the preshocked region (Fig. 9d), $V = 1/\rho = (1 + \frac{t}{R})^{-2}$, and thus $V_t \leq 0$, and $Q_L(V) \neq 0$. This preheating is, of course, not physical (and is another instance of a shockless Q Heating error - note that Q_L does vanish as it should), and this error combines with the the large Shockless Q Heating error near the origin to produce the poor results of curve (2) in Fig. 9. It's amazing that the shock solution is as good as it is. Just how slowly the $Q_L(V)$ solution converges is shown in the comparisons of Fig. 10, where the results are plotted for various mesh intervals: $K = 50$ ($\Delta r = 0.02$); $K = 100$ ($\Delta r = 0.01$), up to $K = 800$, where $\Delta r = 0.00125$. Even at $K = 800$, the numerical solution still has an unacceptable error. These results show that $Q_L(v)$ of Eq. (2.6) is a poor formulation and is essentially the reason that our definition of Q_L given by Eq. (2.2) is taken to be the standard Q (for $\delta = 1$, 2, and 3). We stress this point since $Q_L(V)$ still seems to be in common use.

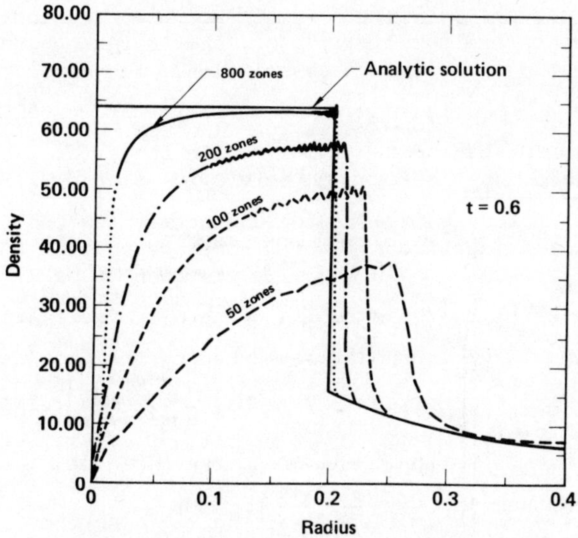

Fig. 10 This example shows the truly large errors (in density) using the original $Q_L(V) = 2(\Delta r)^2 (r/R)^4 (\Delta V/\Delta t)^2$ for various mesh intervals (Δr). The comparisons are $t = 0.6$ and $\Delta r = 0.01$, 0.01, 0.005, and 0.00125, or $K = 50$, 100, 200, and 800. This shows that the convergence of the density to the correct value $\rho^+ = 64$ is very slow indeed, and even for $K = 800$, the error is unacceptable.

Now, of course, these are still serious errors in the use of the standard $Q_L = 2\rho(\Delta u)^2$. This difficulty is analyzed in Fig. 11. The problem is associated with the shock smearing due to Q. Because of the finite shock thickness, the calculation "senses" an incorrect (too small) jump-off value of the preshocked density (ρ^-). That is, the shock smearing selects a $\rho^- < \rho^-_{exact} = 16$. This error is a maximum at early times and becomes less serious as time advances as the (similarity) solution spreads out the preshocked region over more and more mesh points. Thus, a given shock thickness produces less and less error as time increases. The key, then, to more accuracy is to sharpen shocks as much as possible.

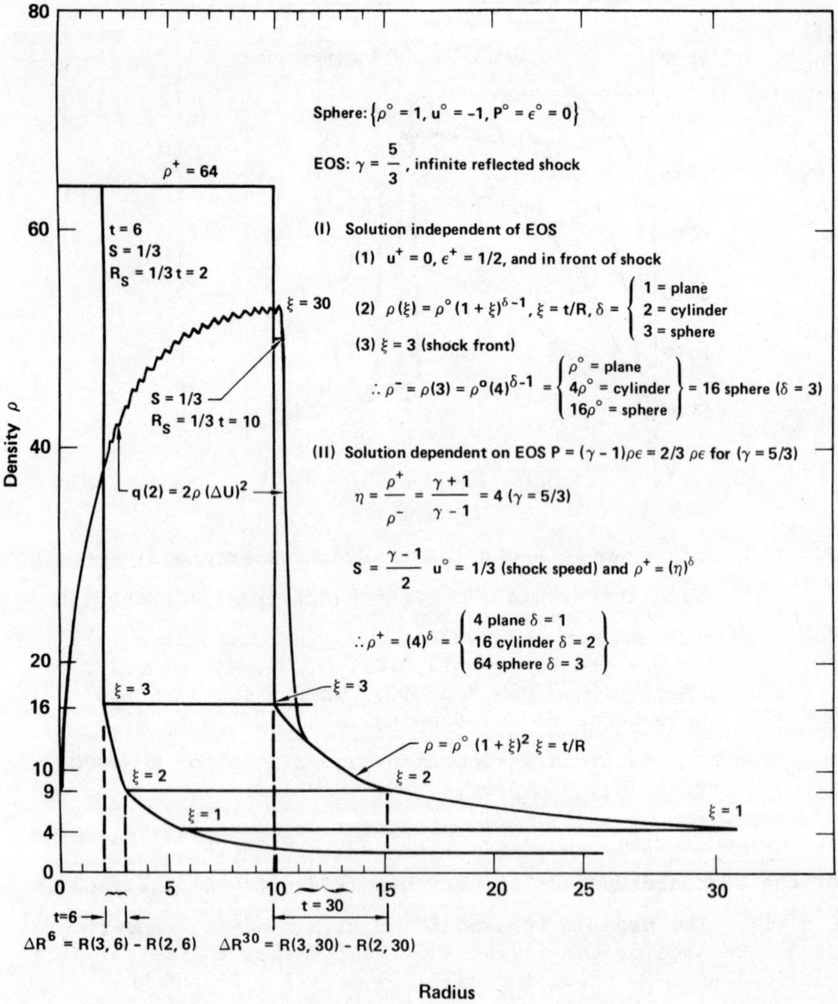

Fig. 11 The solution for Noh's spherical test problem (Fig. 8d) is given at two different times ($t = 6$ and $t = 30$) for the scale variable, $\xi = t/R$. As t increases, the preshock density profile is spread over a physically greater and greater distance. Hence, the preshock value $\rho^- = 16$ should be progressively easier to resolve numerically as time advances. The wiggly line is the numerical solution using the standard $Q_L = 2\rho(\Delta u)^2$. The numerical error is so large ($20\% \leq \varepsilon \leq 600\%$) that it hardly resembles the exact solution, $\rho^+ = 64$.

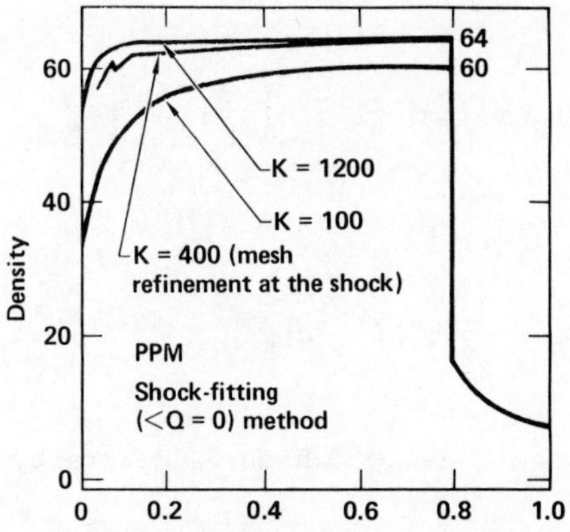

Fig. 12 The (non-Q) PPM method of Colella and Woodward (1982) has very narrow shocks (1 or 2 mesh intervals), and, for the standard test problem, (K = 100) gives superior results. PPM using mesh refinement (i.e., a shock-capturing adaptive mesh with K = 400) is equivalent to the standard PPM using K = 1200. It is clear that using an adaptive mesh is a very important procedure for accurately tracking shocks.

The non-Q PPM method of Colella and Woodward (1982) produces very sharp shocks (of the order of one-to-two mesh widths), and their results are considerably more accurate than the use of the standard Q_L. This is shown in Fig. 12. Also, the PPM results on the standard K = 100 problem are compared with their very accurate "adaptive mesh shock-following procedure" using K = 400. The K = 400 results are also shown to be nearly as accurate (converged) as the standard PPM with K = 1200. The effect of using an adaptive mesh is to minimize the actual (i.e., physical) shock thickness, and, as we have argued in Fig. 11, is all-important in determining the correct preshocked value $\rho^- = 16$. Clearly, using an adaptive mesh for resolving shocks is an important procedure, and such a method would be equally effective for any Q [or (Q&H)] shock-following procedure.

6. SCALAR VERSUS TENSOR Q FORMULATIONS

Schultz (1964) proposed that Q should be treated as a tensor viscosity and gave the following (T) formulation of the hydrodynamic equations (for δ = 1, 2, and 3). We include the von Neumann-Richtmyer scalar (S) formulation, Eq. (2.1), again for comparison. Also, the use of the artificial heat flux (H) remains the same:

$$\rho u_t + P_R = -Q_R \;, \qquad\qquad \text{Scalar (S)}$$

$$\rho(\varepsilon_t + PV_t) = -Q[u_R + (\delta-1)\frac{u}{R}] - \frac{1}{R^{\delta-1}}[R^{\delta-1}H]_R \;, \tag{6.1}$$

$$\rho u_t + P_R = -[Q_R + (\delta-1)\frac{Q}{R}] \;, \qquad\qquad \text{Tensor (T)}$$

$$\rho(\varepsilon_t + PV_t) = -Qu_R - \frac{1}{R^{\delta-1}}[R^{\delta-1}H]_R \;. \tag{6.2}$$

Schultz also defined a new Q, which we denote by

$$Q_L(S) = C_0^2 \rho \ell^2 |u_{rr}|^{3/2} |u_r|^{1/2} , \tag{6.3}$$

if $u_r < 0$, and 0 otherwise (and where $\ell = \Delta r$ in the difference formulation as usual).

Now, Schultz's $Q_L(S)$ doesn't produce a Shockless Q Heating error (since $u_{rr} = 0$ for the Uniform Collapse Problem of Section 3), and we thus might expect superior results for our spherical test problem #3. Indeed, the results (Fig. 13) using $Q_L(S)(T)$ and $Q_L(T)$ [i.e., using Eq. (6.2)] are significantly better than using the scalar (S) equations, Eq. (6.1), but there is essentially no improvement using $Q_L(S)$ over Q_L. We conclude, then, that the major improvement occurs because shocks are narrower (for any Q) in the tensor (T) formulation. The reasons why shocks are sharper is not entirely clear, but it follows, in part, from the formulation Eq. (6.2), where we note that there is less QdV shock heating than for the scalar equations, Eq. (6.1). That is, QdV $\to Qu_R$ independently of geometry (δ = 1, 2, or 3) in Eq. (6.2). The (Q_L&H_L) (T) method [i.e., Eq. (6.2)] with a small Q constant $C_0^2 = 1/4$ is particularly accurate. This is shown in Fig. 14, where we compare $Q_L(1/4)(T)$ (which indeed has a narrower shock, but is extremely noisy) with [$Q_L(1/4)$&$H_L(10)$](T). These (Q_L&H_L)(T) results are reasonably smooth behind the shock and are essentially exact. Thus, we find the best all-round results for the 100-zone test problem are given by the (Q_L&H_L)(T) shock-following method using Schultz's tensor formulations, Eq. (6.2) or by using the (non-Q) PPM method with an adaptive shock-following mesh.

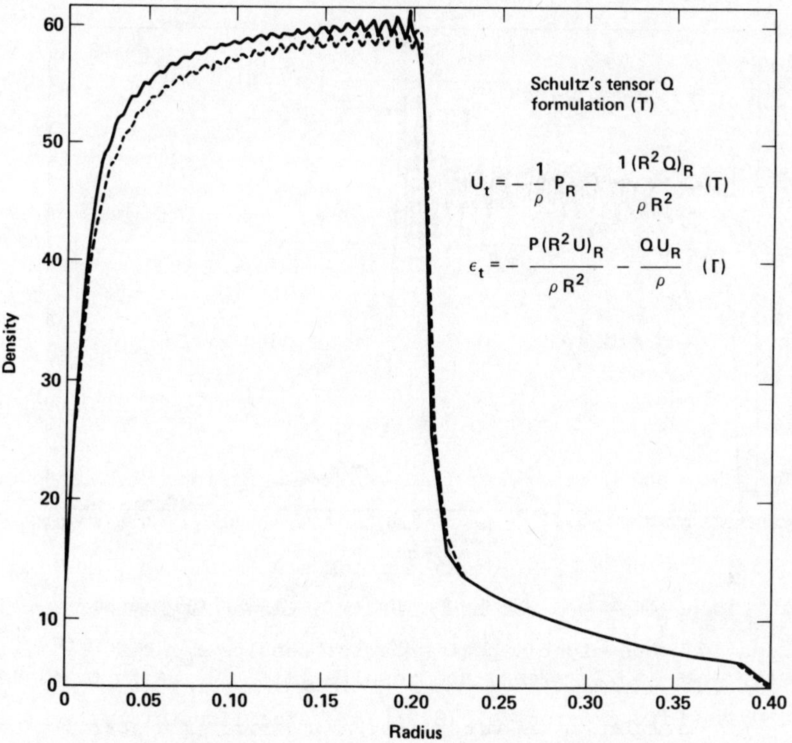

Fig. 13 The lower curve uses our standard $Q_L(2,0) = 2\rho(\Delta u)^2$, but is cast in Schultz's tensor formulation, Eq. (6.2). The upper curve also uses Schultz's tensor for simulation, Eq. (6.2), along with his $Q_L(S) = \rho|\Delta^2 u|^{3/2}|\Delta u|^{1/2}$ of Eq. (6.3). We see that the results are essentially the same for either Q_L formulation. We conclude, then, that it is the tensor use of Q that is important, rather than the particular choice of Q_L. Consequently, we stay with the standard $Q_L = 2\rho(\Delta u)^2$ usage.

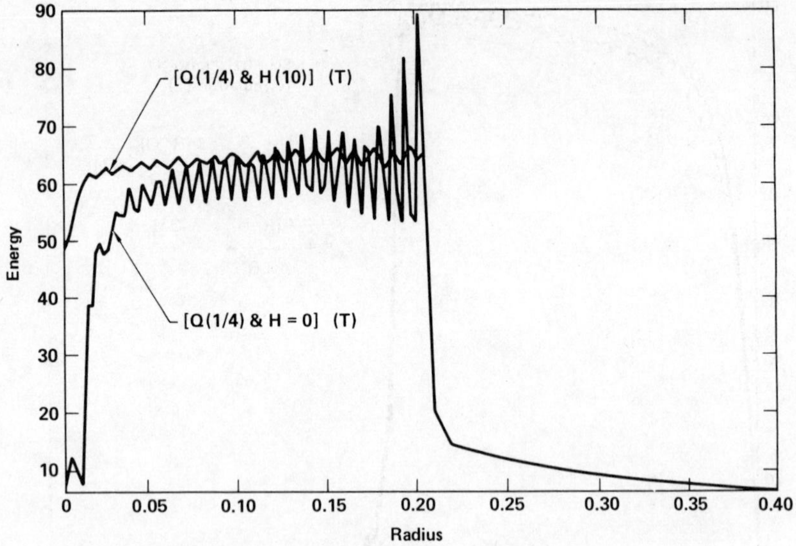

Fig. 14 A comparison of Q_L(T) and the $(Q_L \& H_L)$(T) method [i.e., Eq. (6.2)] using a very small Q_L constant $C^2 = 1/4$. Here, the $Q_L(1/4) = 0.25\rho(\Delta u)^2$ (T) formulation is very noisy, but produces a narrow shock. The sharp shock remains in the [Q_L(1/4 & H_L(10)] (T) method [where $H_L(10) = 10\rho|\Delta u|\Delta\varepsilon)$], and we see that most of the post-shock noise is damped, and the density and energy errors are nearly zero. Thus, the $(Q_L \& H_L)$ (T) method is a preferred shock-following procedure.

The results are summarized in Fig. 15, where we compare the various Q_Ls, $(Q_L \& H_L)$, and the PPM method.

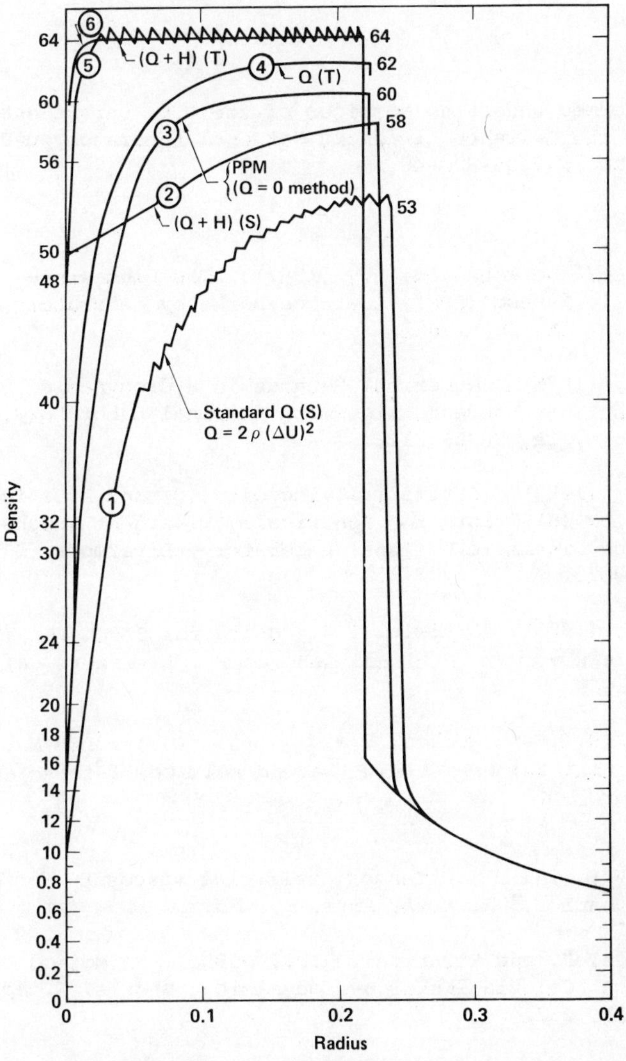

Fig. 15 We compare all of the methods for the standard K = 100 test problem and for both the scalar (S) and tensor (T) formulations [Eqs. (6.1) and (6.2)]. The (non-Q) PPM Method lies between the scalar Q_L (S) and (Q_L & H_L) (S) results, but is not as accurate as the Q_L (T) and the (Q_L & H_L) (T) results. PPM with mesh refinement, curve (5), and the (Q_L & H_L) (T) method, curve (6), give essentially the converged (exact) solution and are the preferred methods.

Work performed under the auspices of the U.S. Department of Energy by the Lawrence Livermore National Laboratory under contract No. W-7405-ENG-48.

REFERENCES

Collela, P., and Woodward, P., (1982). The Piecewise-Parabolic Method (PPM), Lawrence Berkeley Laboratory, Berkeley, CA, LBL-14661.

Noh, W.F., (1976). Numerical Methods in Hydrodynamic Calculations, Lawrence Livermore National Laboratory, Livermore, CA, UCRL-52112.

Noh, W.F., (1983). Artificial Viscosity (Q) and Artificial Heat Flux (H) Errors for Spherically Divergent Shocks, Lawrence Livermore National Laboratory, Livermore, CA, UCRL-89623.

Noh, W.F., (1985). Artificial Viscosity for Strong Shocks, Lawrence Livermore National Laboratory, Livermore CA, UCRL-53669.

Richtmyer, R.D. and Morton, K.W., (1967) Difference Methods for Initial-Value Problems, Second Edition, *Interscience*, (New York).

Schultz, W.D., (1964). Tensor Artificial Viscosity for Numerical Hydrodynamics," *J. Math. Phys.* 5, 133.

Von Neumann, J. and Richtmyer, R.D., (1950). A Method for the Numerical Calculation of Hydrodynamical Shocks, *J. Appl. Phys.* 21, 232.

A SHOCK-FITTING SCHEME FOR THE EULER EQUATIONS USING DYNAMICALLY OVERLYING MESHES

C.M. Albone
(RAE, Farnborough)

1. INTRODUCTION

The problem of representing, accurately, embedded shock waves that occur in transonic flow with a subsonic free stream is considered. In most existing schemes (see Lytton, 1984, Jameson et al., 1981, and Hall, 1985) for solving the Euler equations, embedded shocks are captured as regions of high flow gradient spread over a few mesh intervals. Shock fitting, as in (Moretti, 1973), provides the mechanism by which shocks may be represented by surfaces of discontinuity; such a mechanism eliminates problems of shock resolution and removes the arbitrary degree of smoothing which is present in the currently popular non-upwind Euler algorithms such as those of Hall (1985) and Jameson et al. (1981). The problem is to devise a practical shock-fitting scheme, capable of dealing with flows in which little is known about the shock configuration.

In the present method (described more fully in (Albone, 1985)), shock fitting is approached using overlying meshes whose relative position changes in the course of the calculation. For a single embedded shock wave, this involves a shock-orientated mesh that floats through the main mesh as the fitted shock finds its natural position within the field. This approach has the advantage of aligned shock fitting without placing extra constraints on the main mesh. On both meshes, an upwind, non-conservative, time-split algorithm is employed. Shocks are first captured on the main mesh, and then, after detection, are fitted as part of the calculation on the floating mesh. At each mesh point on the shock the Rankine-Hugoniot equations for a moving shock are applied and simultaneously the shock speed is evaluated. These speeds are then used to define a new position of the fitted shock, and the floating mesh moves with the shock to a new location within the main mesh. The use of an upwind algorithm on the main mesh is not essential for the fitting scheme reported here, but it is preferred. Its use on the floating shock-orientated mesh considerably simplifies the logic of the fitting scheme.

The method has been tested for the simple two-dimensional model problem of supercritical flow past a circular-arc aerofoil. The main mesh is rectangular and the surface boundary condition is formulated in the thin-aerofoil approximation. Results from capturing and fitting have been obtained over a range of supercritical Mach numbers.

2. THE SHOCK-CAPTURING ALGORITHM

The Euler equations are solved here in non-conservative form and are referred to rectangular Cartesian axes, (x, z). They are

$$\frac{\partial \underline{q}}{\partial t} + \underline{B}\,\frac{\partial \underline{q}}{\partial x} + \underline{D}\,\frac{\partial \underline{q}}{\partial z} = 0\ , \qquad (2.1)$$

where $\underline{q} = \begin{pmatrix} A \\ u \\ w \\ s \end{pmatrix}$, $\underline{B} = \begin{vmatrix} u & c & 0 & 0 \\ c & u & 0 & \frac{-c^2}{\gamma(\gamma-1)} \\ 0 & 0 & u & 0 \\ 0 & 0 & 0 & u \end{vmatrix}$, $\underline{D} = \begin{vmatrix} w & 0 & c & 0 \\ 0 & w & 0 & 0 \\ c & 0 & w & \frac{-c^2}{\gamma(\gamma-1)} \\ 0 & 0 & 0 & w \end{vmatrix}$

and $A = \dfrac{2c}{\gamma - 1}$.

The velocity components in the x and z directions are u and w, respectively; c is the speed of sound, t is time and s is entropy $\left[= \ln\,(p/\rho^{\gamma})\right]$, where p is pressure and ρ is density.

A first-order-accurate algorithm, for the solution of equation (2.1) is constructed from two one-dimensional upwind operators. It has been found convenient to use as dependent variables the interdependent set P, Q, E, F and S, where $P = u + A$, $Q = u - A$, $E = w + A$ and $F = w - A$; their dependency is given by the relation $P - Q = E - F$. The set P, Q, E and F are often referred to as Riemann variables, since in one-dimensional homentropic flow each is a Riemann invariant. The resulting discretization scheme for equation (2.1) is

$$\vec{\delta}_t P = -(u+c)\delta_x P - (w)\delta_z u - \left(\frac{w+c}{2}\right)\delta_z E + \left(\frac{w-c}{2}\right)\delta_z F + \frac{c^2}{\gamma(\gamma-1)}\delta_x S, \tag{2.2}$$

$$\vec{\delta}_t Q = -(u-c)\delta_x Q - (w)\delta_z u + \left(\frac{w+c}{2}\right)\delta_z E - \left(\frac{w-c}{2}\right)\delta_z F + \frac{c^2}{\gamma(\gamma-1)}\delta_x S, \tag{2.3}$$

$$\vec{\delta}_t E = -(w+c)\delta_z E - (u)\delta_x w - \left(\frac{u+c}{2}\right)\delta_x P + \left(\frac{u-c}{2}\right)\delta_x Q + \frac{c^2}{\gamma(\gamma-1)}\delta_z S, \tag{2.4}$$

$$\vec{\delta}_t F = -(w-c)\delta_z F - (u)\delta_x w + \left(\frac{u+c}{2}\right)\delta_x P - \left(\frac{u-c}{2}\right)\delta_x Q + \frac{c^2}{\gamma(\gamma-1)}\delta_z S, \tag{2.5}$$

$$\vec{\delta}_t S = -(u)\delta_x S - (w)\delta_z S\ , \tag{2.6}$$

where $\vec{\delta}$ is a forward difference operator and δ is a difference operator which may be forward or backward ($\overleftarrow{\delta}$). For all space differences in equations (2.2) to (2.6), except those for S in equations (2.2) to (2.5), δ is interpreted as $\overleftarrow{\delta}$ if the bracketed coefficient (propagation speed) is positive and as $\vec{\delta}$ if the bracketed coefficient is negative. Where δ operates on S in equations (2.2) to (2.5), it has the same interpretation as for P in equation (2.2), Q in equation (2.3), E in equation (2.4) and F in equation (2.5).

Since the scheme is explicit, the allowable time step is governed by the CFL condition. The propagation speeds, u, u + c, u - c, w, w + c and w - c are evaluated as an upwind average. Given a uniform distribution of entropy far upstream, the solution obtained from equations (2.2) to (2.6) (in the absence of shock fitting) can result only in homentropic flow.

Solid surface boundary conditions in the present formulation are linearised. Flow tangency for the circular-arc aerofoil is therefore satisfied on the chord line z = o. Flow variables on a line of dummy points below z = o are set by symmetrical reflection of A, S and the tangential velocity component, and by antisymmetrical reflection of the normal velocity component. Far-field boundary conditions are those corresponding to a uniform stream. Where the boundary is homentropic, there is no need to test for inflow or outflow or to test for subsonic or supersonic flow; the upwind algorithm will select precisely what it wants from the complete set. For non-homentropic boundaries (subsonic outflow in the presence of shock fitting), boundary conditions for Q and S, required by the difference scheme, are necessarily evolutionary. That for S is obtained by solving equation (2.6) on the downstream boundary. That for Q is obtained by solving equation (2.2) for P on the downstream boundary and by using the specified downstream pressure.

3. SHOCK FITTING

3.1 Detection and sharpening

The shock-fitting procedure is intended to correct the errors of non-conservative capturing by representing the shock as a discontinuity across which the Rankine-Hugoniot equations are imposed. The first step is to detect any captured shock waves and determine their location and shape. For the present simple model problem, captured shocks are nearly normal to the x direction, and so detection can be based upon the appropriate sign change of $u - c$.

At this stage, the distribution of pressure etc, close to the shock is that for a smeared, captured shock. It is desirable to sharpen this distribution before beginning the fitting process. In order to carry this out, dependent variables adjacent to the shock location are adjusted by using linear extrapolation from neighbouring points on the same side of the shock wave. This adjustment is never repeated once the fitting and moving schemes are operating.

3.2 Interaction between fixed and floating meshes

The two interacting meshes are shown in Fig. 1. Points on the fixed mesh are indicated by either squares or triangles, and those on the floating mesh by circles or diamonds. Squares denote mesh points at which the Euler algorithm can be applied without any risk of differencing across the shock wave. Triangles denote those points where there is a risk (not a certainty) of differencing across the shock. Boundary values at the diamond points, required for the calculation on the floating mesh, are obtained using bilinear interpolation from values on the fixed mesh. Flow variables are then updated at circle points on the floating mesh, and simultaneously shock speed is evaluated at each mesh point on the shock wave as described in section 3.3. Next, flow variables at the square points on the fixed mesh and at the fixed mesh boundaries are updated. Finally, interpolation is again used, but now to update flow variables at triangle points on the fixed mesh from the updated circle point values on the floating mesh. It should be noted that interpolation involves mesh points all of which lie on only one side of the shock, and so interpolation across the shock cannot take place. Further, the floating mesh has been made large enough so that all interpolated values are obtained wholly from points that are updated using the Euler algorithm. The calculated shock speed is used to change the location and shape of the shock and so move and distort the floating mesh as described in section 3.4.

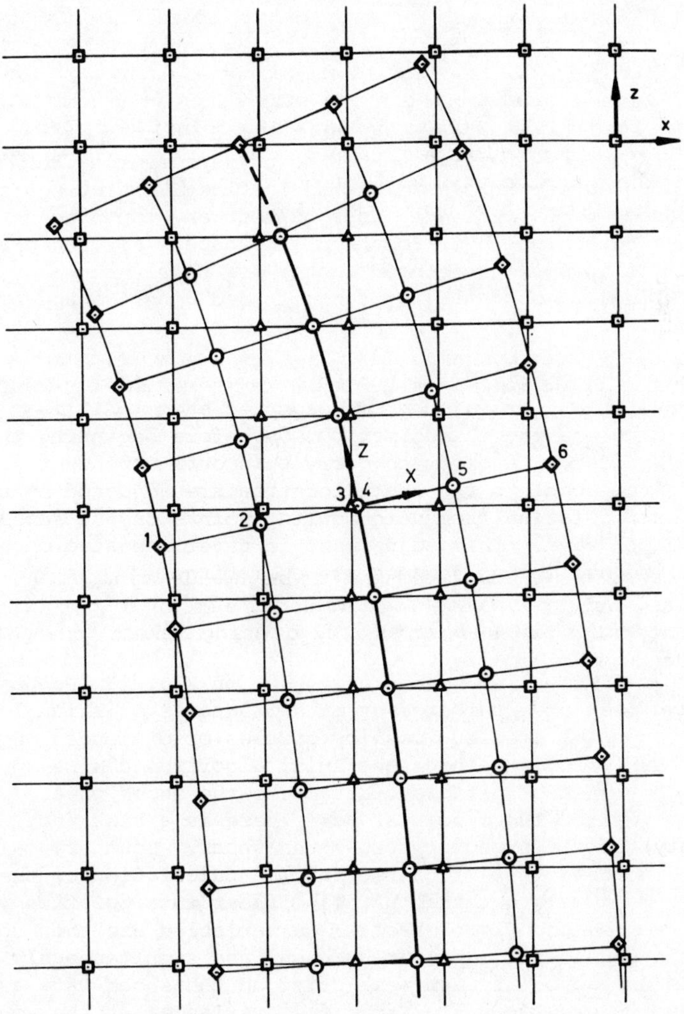

Fig. 1. Shock-orientated mesh overlying the fixed mesh.

The floating mesh is defined with the coordinate direction normal to the shock as X and that along the shock as Z, the shock itself being X = 0. Mesh intervals in the Z direction are formed by the intercepts of the shock with lines z = constant; those in the X direction are chosen just large enough for variables at diamond points in Fig. 1, when obtained by interpolation from points on the fixed mesh, to involve only those points marked as squares. On z = o, the shock is normal to the body surface tangent. The mesh is extended in the -Z direction to form a set of dummy points, which comprise its lower boundary and in the +Z direction beyond the last point at which fitting takes place to form its upper boundary. On every line Z = constant there are six points numbered 1 to 6 as shown in Fig. 1. Points 3 and 4 are coincident, but have distinct values of flow variables; point 3 lies on the upstream side of the shock and point 4 on the downstream side. Points 1 and 6 constitute the upstream and downstream boundaries respectively of the mesh.

3.3 Calculation on the floating mesh

We use as dependent variables the set equivalent to those used on the fixed mesh, but with velocity components defined in the X and Z directions and denoted by U and W respectively. The set is P_f, Q_f, E_f, F_f and S defined by

$$P_f = U + A, \quad Q_f = U - A, \quad E_f = W + A \quad \text{and} \quad F_f = W - A \ ,$$

where the suffix f denotes floating mesh. Initial values are set at all points on the floating mesh, immediately following shock detection and sharpening, by using bilinear interpolation from the fixed mesh.

Updating of flow variables takes place at all circle points using the Euler algorithm of section 2 recast in the curvilinear system (X, Z). The difference equations take the form

$$\vec{\delta}_t P_f = -(U+c)\,\delta_x P_f - \frac{(W)}{R}\,\delta_z U - \frac{1}{R}\left(\frac{W+c}{2}\right)\delta_z E_f + \frac{1}{R}\left(\frac{W-c}{2}\right)\delta_z F_f$$

$$+ \frac{c^2}{\gamma(\gamma - 1)}\,\delta_x S - \frac{K}{R}\,(cU - W^2)\ , \qquad (3.3.1)$$

$$\vec{\delta}_t Q_f = -(U-c)\delta_x Q_f - \frac{(W)}{R}\delta_z U + \frac{1}{R}\left(\frac{W+c}{2}\right)\delta_z E_f - \frac{1}{R}\left(\frac{W-c}{2}\right)\delta_z F_f$$

$$+ \frac{c^2}{\gamma(\gamma-1)}\delta_x S + \frac{K}{R}(cU + W^2), \qquad (3.3.2)$$

$$\vec{\delta}_t E_f = -\frac{1}{R}(W+c)\delta_z E_f - (U)\delta_x W - \left(\frac{U+c}{2}\right)\delta_x P_f + \left(\frac{U-c}{2}\right)\delta_x Q_f$$

$$+ \frac{c^2}{R\gamma(\gamma-1)}\delta_z S - \frac{KU}{R}(W+c), \qquad (3.3.3)$$

$$\vec{\delta}_t F_f = -\frac{1}{R}(W-c)\delta_z F_f - (U)\delta_x W + \left(\frac{U+c}{2}\right)\delta_x P_f - \left(\frac{U-c}{2}\right)\delta_x Q_f$$

$$+ \frac{c^2}{R\gamma(\gamma-1)}\delta_z S - \frac{KU}{R}(W-c), \qquad (3.3.4)$$

$$\vec{\delta}_t S = -(U)\delta_x S - \frac{(W)}{R}\delta_z S, \qquad (3.3.5)$$

where $R = 1 + KX$, and K is the shock curvature. The interpretation of the difference operator δ is the same as that for the fixed mesh algorithm described in section 2. Updating at points 2, 3 and 5 (Fig. 1) is performed using equations (3.3.1) to (3.3.5). At point 4, only equation (3.3.2) is solved; the remaining equations are replaced by the Rankine-Hugoniot equations for a moving shock, with upstream conditions taken as those at point 3. In this way due account is taken of the manner in which upstream propagation on the $(U - c)$ characteristic affects point 4, and the number of equations enables us to evaluate the shock speed as well as the flow variables at point 4. The shock speed, V, is given by the implicit equation

$$U_3 + \frac{2c_3\left(1-M_3^2\right)}{(\gamma+1)M_3} - \frac{2c_3}{(\gamma-1)}\left\{1 + \frac{2(\gamma-1)\left(\gamma M_3^2+1\right)\left(M_3^2-1\right)}{(\gamma+1)^2 M_3^2}\right\}^{\frac{1}{2}} = Q_f\big|_4, \qquad (3.3.6)$$

where $Q_f|_4$ is the value of Q_f at point 4 obtained from equation (3.3.2), the suffix 3 denotes evaluation at mesh point 3 and $M_3 = (U_3 - V)/c_3$. The solution for V is obtained iteratively and is then used (together with flow conditions at point 3) to evaluate P_f, E_f, F_f and S at point 4 from the remaining Rankine-Hugoniot equations.

A study of the terms in equations (3.3.1) to (3.3.5) reveals the advantage of combining an upwind scheme with a shock-orientated mesh. No differencing can take place across a shock as long as $U > c$ at point 3 and $U < c$ at point 4. For instance, we see that at point 3, U, $U + c$ and $U - c$ are positive and so only backward differences are employed. Similarly at point 4 $U - c$ is negative in equation (3.3.2) and so differences of Q_f and S in that equation are forward.

Boundary conditions for the floating mesh calculation are supplied at points 1 and 6 on each line Z = constant, and at all points on the top and bottom boundaries. As with far-field boundary conditions on the fixed mesh, the algorithm is capable of sorting out what information it requires from the complete set available at the boundaries.

3.4 Determination of a new shock shape, position and extent

It appears at first sight that this should be straightforward. Each mesh point (Z_k) on the shock should be moved a distance $\Delta t V(Z_k)$ at each time step in a direction normal to the shock wave at that point (where Δt is the local time increment and $V(Z_z)$ is the shock speed) to define a new shock shape and position. A problem with this approach is that unsmooth shock shapes can be generated, and this may lead to numerical instability as shown by (Moretti, 1974) for bow shocks. Here, we have a further difficulty in that shocks terminate in the field with vanishingly small strength. The scheme adopted for an embedded shock with a subsonic free stream involves representing its shape by a constrained polynomial. This gives the shock an effective stiffness and so constrains the location of its weak free end. Although constraining the natural shape of a shock is in principle undesirable, its effect upon shock shape is very small for embedded shocks, which usually have low curvature. The unconstrained new shape is evaluated straightforwardly and then the constrained shape is obtained here by approximating the unconstrained shape by a single, weighted, least-squares-fit cubic which is normal to the aerofoil surface tangent at z = 0. The weighting is

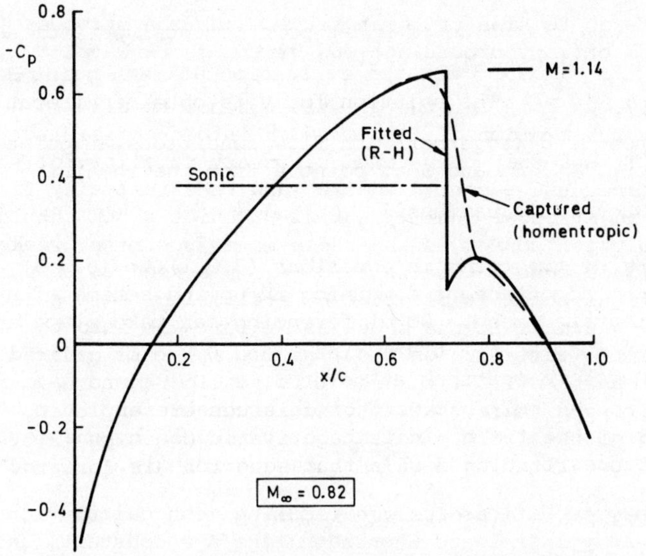

Fig. 2. Fitted and captured shock waves: 10% thick circular arc

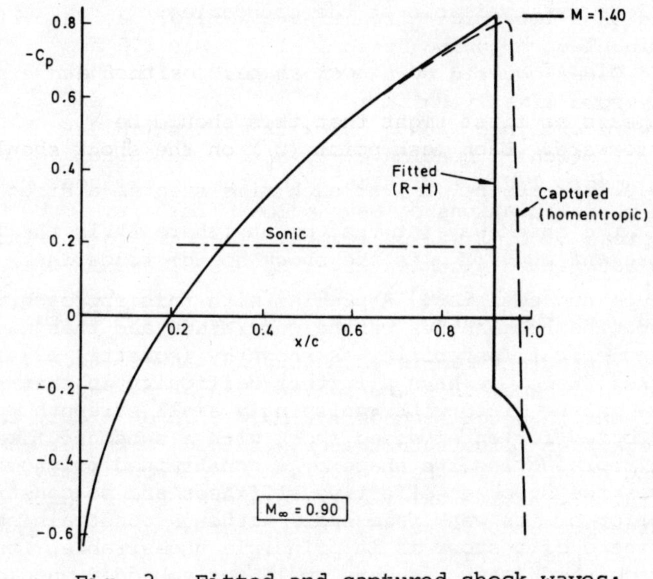

Fig. 3. Fitted and captured shock waves: 10% thick circular arc

applied so as to bias the best fit towards the stronger (and hence more entropy producing) end of the shock wave.

Finally, we consider the question of how to allow the extent of penetration of fitted shock into the field to change. It has only been found necessary to allow for shock shrinkage because captured shocks (and thus initially fitted shocks) are stronger than a Rankine-Hugoniot shock should be. As the solution converges the shock moves upstream, weakens and its field penetration is reduced. The need to reduce the extent of the shock is detected by evaluating the upstream normal Mach number at the shock point furthest into the field each time the variables on the floating mesh are updated. If this is less than 1.01, the shock is considered to have so little strength there that fitting is unnecessary at that point and so the field penetration is reduced by one mesh point and the floating mesh shrinks accordingly.

4. RESULTS FOR A CIRCULAR-ARC AEROFOIL

The choice of test problem was dictated by the need to provide a simple geometric environment in which to test the overlying mesh shock-fitting scheme. Calculations were performed on a 10% thick circular-arc aerofoil at zero lift. The rectangular computing region, $-1.5 \leqslant x \leqslant 2.5$ and $0 \leqslant z \leqslant 1.0$ was covered by a uniform mesh, with the aerofoil chord line lying on $z = 0$ between $x = 0$ and $x = 1.0$. Results were obtained with a 128 x 32 mesh which gave a uniform 3.125% chord spacing everywhere. Calculations showed that the aerofoil has a critical Mach number of about 0.79. A series of calculations at Mach numbers (M_∞) from 0.80 to 0.90 in steps of 0.02 was performed. The results showing captured and fitted shock waves at $M_\infty = 0.82$ and 0.90 are contained in Figures 2 and 3. These were considered to be converged when the average normalised residual fell to 5×10^{-7}. Using a three-level multigrid convergence acceleration scheme, this was achieved in 5 cpu seconds for $M_\infty = 0.82$ and in 8 cpu seconds for $M_\infty = 0,90$ on a Cray 1s. The two examples illustrate that the method works for weak and strong shocks. In the steady state, entropy jumps discontinuously from a value of zero ahead of the shock to its Rankine-Hugoniot value and thereafter it remains constant on streamlines.

5. CONCLUDING REMARKS

A new shock-fitting scheme for the Euler equations is presented which involves overlying meshes, one of which is shock-orientated. The scheme is embedded in an explicit non-conservative upwind shock-capturing algorithm, which in turn has been linked with a multigrid acceleration technique, resulting in a rapidly converging method. The use of an overlying, shock-orientated mesh simplifies the logic of shock fitting and avoids overconstraining the main mesh at the cost of a small amount of interpolation. Further, shock-fitting appears not to conflict with multigrid. The scheme is robust, and there is no apparent reason why the analysis should not be extended to more general two-dimensional problems or to three dimensions.

6. REFERENCES

Albone, C.M., (1985). "An upwind, multigrid, shock-fitting scheme for the Euler equations". *RAE Technical Report* 85004.

Hall, M.G., (1985). "Cell-vertex multigrid schemes for solution of the Euler equations". Published in these Conference Proceedings.

Jameson, A., Schmidt, W. and Turkel, E., (1981). "Numerical solution of the Euler equations by finite volume methods using Runge-Kutta time-stepping schemes". *AIAA Paper* 81-1259.

Lytton, C.C., (1984). "Solution of the Euler equations for transonic flow past a lifting aerofoil - the Bernoulli formulation". *RAE Technical Report* 84080.

Moretti, G., (1973). "Experiments in multidimensional floating shock fitting". Polytechnic Inst. of New York, *PIBAL* Report 73-18.

Moretti, G., (1974). "A pragmatic analysis of discretization procedures for initial - and boundary-value problems in gas dynamics and the influence on accuracy - or - Look Ma, No Wiggles!" Polytechnic Inst. of New York, *POLY-AE/AM* Report 74-15.

LAGRANGE-GALERKIN MIXED FINITE ELEMENT APPROXIMATION OF THE NAVIER-STOKES EQUATIONS

Endre E. Süli
*(Department of Mathematics, University of Belgrade, Yugoslavia)**

1. INTRODUCTION

Recently a new method has been proposed for the numerical treatment of convection-diffusion equations (Benque et al., 1980; Russell, 1980). It is based on combining the method of characteristics with finite element or finite difference procedures. For convection dominated problems it has smaller time truncation errors than those of standard methods, as shown by Douglas and Russell (1982).

Pironneau (1982) gave a detailed analysis of this method for the Navier-Stokes equations. He proved an error bound $h + \Delta t + h^2/\Delta t$, which is suboptimal in space, but he allowed the kinematic viscosity to go down to zero.

In this paper we present optimal error estimates for the method when the nonlinear convection term and the viscous term are roughly equivalent in importance. Using the Lagrangian representation of the flow, we discretize the total derivative of the velocity along the trajectories and we combine this special discretization with the mixed finite element method.

2. FORMULATION OF THE PROBLEM

2.1 *The differential system*

We consider the nonstationary Navier-Stokes problem:

$$\begin{aligned} &u' + (u.\nabla)u - \nu\Delta u + \nabla p = f \\ &\qquad\qquad\qquad \nabla.u = 0 \text{ in } \Omega \times (0,T], \qquad (2.1.1) \\ &u|_{\partial\Omega} = 0, \quad u|_{t=0} = u_0, \end{aligned}$$

*Now at the Oxford University Computing Laboratory.

where Ω is a bounded domain in R^2, T is a fixed positive number, u is the velocity field in R^2, p is the kinematic pressure, ν the kinematic viscosity, f a density of body forces per unit mass and u_o the initial velocity field. For simplicity, we assume homogeneous boundary data and the domain Ω to be convex polygonal.

For $1 \leq p \leq \infty$, let $L^p(\Omega)$ denote the usual Lebesgue space with the norm $||.||_{L^p}$. For $m \geq 0$, $W^{m,p}(\Omega)$ will denote the Sobolev space of index m with the norm $||.||_{W^{m,p}}$ (Adams, 1975). For p=2 we write $H^m(\Omega)$ for $W^{m,2}(\Omega)$. $H^1_o(\Omega)$ will denote the closure of the set of infinitely differentiable functions with compact support in Ω for the norm $||.||_{H^1}$. Let X be any of the spaces just defined. Then X^2 will denote the topological product X x X.

We need the following special spaces, well suited to incompressible flow:

$$H = \{ u \in L^2(\Omega)^2 : \nabla.u = 0, u.n|_{\partial\Omega} = 0\},$$

$$V = \{u \in H^1_o(\Omega)^2 : \nabla.u = 0\},$$

$$L^2_o(\Omega) = \{ p \in L^2(\Omega) : \int_\Omega p \, dx = 0\}.$$

For $1 \leq p \leq \infty$, $L^p(0,T;X)$ will denote the set of strongly measurable functions u: $(0,T) \rightarrow X$ such that

$$||u||_{L^p(0,T;X)} = \left(\int_0^T ||u(t)||^p_X \, dt \right)^{1/p} < \infty, \quad 1 \leq p < \infty,$$

or

$$||u||_{L^\infty(0,T;X)} = \operatorname*{ess\,sup}_{t\in(0,T)} ||u(t)||_X < \infty, \quad p = \infty.$$

Finally, $C(0,T;X)$ will denote the set of continuous functions u: $[0,T] \rightarrow X$ normed by

$$||u||_{C(0,T;X)} = \max_{t\in[0,T]} ||u(t)||_X .$$

The mixed finite element approximation of problem (2.1.1) is based on the following saddle-point weak formulation.

For f and u_o given,

$$f \in L^2(O,T;L^2(\Omega)^2),\ u_o \in V$$

find (u,p) such that $(u,p) \in L^2(O,T;H_o^1(\Omega)^2 \times L_o^2(\Omega))$,

$$\frac{d}{dt}(u,v) + ((u.\nabla)u,v) + \nu(\nabla u,\nabla v) - (\nabla.v,p) = (f,v) \quad \forall\, v \in H_o^1(\Omega)^2,$$

$$(\nabla.u,q) = O \quad \forall\, q \in L_o^2(\Omega), \tag{2.1.2}$$

$$u(O) = u_o\,,$$

where the inner products are to be interpreted to be in $L^2(\Omega)^2$.

It is well known that problem (2.1.2) has a unique solution (Temam, 1983). Moreover, $(u,p) \in L^2(O,T;H^2(\Omega)^2 \times H^1(\Omega))$, $u \in C(O,T;V)$, $u' \in L^2(O,T;H)$. Hence the material derivative of u

$$D_t u = u' + (u.\nabla)u$$

belongs to $L^2(O,T;L^2(\Omega)^2)$ and (2.1.2) may be rewritten as

$$(D_t u,v) + \nu(\nabla u,\nabla v) - (\nabla.v,p) = (f,v) \quad \forall\, v \in H_o^1(\Omega)^2,$$

$$(\nabla.u,q) = O \quad \forall\, q \in L_o^2(\Omega), \tag{2.1.3}$$

$$u(O) = u_o.$$

The crucial aspect of the method is the approximation of the material derivative.

2.2 *Lagrangian representation of the flow*

The Lagrangian representation of the flow is based on the function

$$X : (x,s;t) \in \Omega \times (O,T)^2 \rightarrow X(x,s;t) \in \Omega,$$

where $X(x,s;t)$ represents the position at time t of the particle of fluid which is at point x at time $t=s$. This means that $t \to X(x,s;t)$ is the parametric representation of the trajectory of this particle. If we are given the velocity field u, then the trajectories may be determined from the initial value problem:

$$\frac{d}{dt} X(x,s;t) = u(X(x,s;t),t)$$
$$X(x,s;s) = x. \tag{2.2.1}$$

If u belongs to $L^\infty(0,T;W^{1,\infty}(\Omega)^2)$, then $X(.,s;t)$ is a Lipschitz continuous bijection of Ω onto Ω and its Jacobian is equal to one almost everywhere on $\Omega \times (0,T)^2$.

Let M denote a positive integer, $\Delta t = T/M$ and $t_m = m.\Delta t$. For $x \in \Omega$, $s=t_{m+1}$, $t_m \le t < t_{m+1}$, $0 \le m \le M-1$, (2.2.1) implies that

$$x - X(x,t_{m+1};t_m) = \int_{t_m}^{t_{m+1}} u(X(x,t_{m+1};t),t)\,dt$$
$$\tilde{=} \Delta t.u(X(x,t_{m+1};t_{m+1}),t_{m+1}) = \Delta t.u(x,t_{m+1}). \tag{2.2.2}$$

Let $\psi = (1 + |u|^2)^{1/2}$ and $\tau = (1,u_1,u_2)/\psi$. Then,

$$D_t u(x,t_{m+1}) = \psi(x,t_{m+1}) \frac{\partial u(x,t_{m+1})}{\partial \tau(x,t_{m+1})} .$$

The backward difference approximation

$$\frac{\partial u(x,t_{m+1})}{\partial \tau(x,t_{m+1})} \tilde{=} \frac{u(x,t_{m+1}) - u(X(x,t_{m+1};t_m),t_m)}{\{|x-X(x,t_{m+1};t_m)|^2 + |t_{m+1}-t_m|^2\}^{1/2}}$$

combined with (2.2.2) gives:

$$D_t u(x,t_{m+1}) \tilde{=} \frac{u(x,t_{m+1}) - u(X(x,t_{m+1};t_m),t_m)}{\Delta t} .$$

Now we turn to the problem of spatial discretization of (2.1.3).

2.3 Formulation of the finite element procedure

Let h denote a discretization parameter tending to zero and let X_h and M_h be two finite element spaces such that $X_h \subset H^1_o(\Omega)$ and $M_h \subset L^2_o(\Omega)$. We define $W_h = (X_h)^2$ and $V_h = \{v_h \in W_h: (\nabla . v_h, q_h) = 0 \ \ \forall q_h \in M_h\}$.

Assume that W_h (resp. M_h) is associated with a uniformly regular polygonalization C_h of Ω and piecewise polynomial functions of some fixed degree ℓ (resp. ℓ-1), $\ell \geq 1$.

We relate the continuous and discrete spaces by the following hypotheses:

H1) There exists a positive constant C_1 such that

$$\inf_{v_h \in W_h} ||v-v_h||_{H^1} \leq C_1 h^r ||v||_{H^{r+1}}$$

$$\forall v \in H^{r+1}(\Omega)^2 \cap H^1_o(\Omega)^2 \quad \forall r: 1 \leq r \leq \ell.$$

H2) There exists a positive constant C_2 such that

$$\inf_{q_h \in M_h} ||q-q_h||_{L^2} \leq C_2 h^r ||q||_{H^r}$$

$$\forall q \in H^r(\Omega) \cap L^2_o(\Omega) \quad \forall r: 0 \leq r \leq \ell.$$

H3) (The Babuška-Brezzi condition) There exists a positive constant C_3 such that

$$\inf_{q_h \in M_h} \sup_{v_h \in W_h} \frac{(\nabla . v_h, q_h)}{||v_h||_{H^1} ||q_h||_{L^2}} \geq C_3.$$

The approximations of the velocity and the pressure at $t=t_m$ will be denoted by u^m_h and p^m_h. Now, $(u^{m+1}_h, p^{m+1}_h) \in W_h \times M_h$ is defined to be the solution of the discrete problem:

$$(d_t u_h^{m+1}, v_h) + \nu(\nabla u_h^{m+1}, \nabla v_h) - (\nabla . v_h, p_h^{m+1}) = (f^{m+1}, v_h) \quad \forall v_h \in W_h,$$

(2.3.1)

$$(\nabla . u_h^{m+1}, q_h) = 0 \quad \forall q_h \in M_h,$$

where

$$d_t u_h^{m+1}(x) = \frac{u_h^{m+1}(x) - u_h^m(X_h^m(x,t_{m+1};t_m))}{\Delta t}$$

and $X_h^m(x,t_{m+1};.)$ is the solution of the initial value problem:

$$\frac{d}{dt} X_h^m(x,t_{m+1};t) = u_h^m(X_h^m(x,t_m;t)), \quad t_m \leq t < t_{m+1},$$

(2.3.2)

$$X_h^m(x,t_{m+1};t_{m+1}) = x.$$

The initial approximate velocity $u_h^o \in V_h$ is determined as the H_o^1 - elliptic projection of u_o onto V_h:

$$(\nabla(u_h^o - u_o), \nabla v_h) = 0 \quad \forall v_h \in V_h .$$

The theory of mixed finite element methods shows that for u, u_o and p sufficiently smooth, the error $u-u_h$ is optimal in space (order h^ℓ in H^1) and time (first order). To be more specific, for a normed space X with the norm $|| . ||_X$ let

$$\ell^p(0,T;X) = \{ v: \{t_1,\dots,t_M\} \to X : ||v||_{\ell^p(0,T;X)} =$$

$$= (\Delta t \sum_{i=1}^{M} || v(t_i)||_X^{\,p})^{1/p} < \infty\}, \; 1 \leq p < \infty,$$

$$\ell^\infty(0,T;X) = \{ v: \{t_1,\dots,t_M\} \to X : ||v||_{\ell^\infty(0,T;X)} =$$

$$= \max_{1 \leq i \leq M} ||v(t_i)||_X < \infty\}, \; p = \infty.$$

Assume that

$$f \in C(0,T;L^2(\Omega)^2),$$
$$u_o \in H^{s+1}(\Omega)^2 \cap W^{1,\infty}(\Omega)^2 \cap V, \qquad (2.3.3)$$

where $s \geq 1$. Let us organize our regularity hypotheses on the corresponding solution (u,p) of (2.1.2) according to the results in which they are used. We assume

$$\begin{aligned} &u \in L^\infty(0,T;H^{s+1}(\Omega)^2 \cap W^{1,\infty}(\Omega)^2 \cap V)\\ &u' \in L^2(0,T;H^s(\Omega)^2 \cap H)\\ &u'' \in L^2(0,T;H)\\ &p \in L^\infty(0,T;H^s(\Omega) \cap L_o^2(\Omega))\\ &p' \in L^2(0,T;H^{s-1}(\Omega)) \end{aligned} \qquad (2.3.4)$$

$$\begin{aligned} &u \in L^\infty(0,T;H^{s+1}(\Omega)^2 \cap W^{1,\infty}(\Omega)^2 \cap V)\\ &u' \in L^2(0,T;H^{s+1}(\Omega)^2 \cap H)\\ &u'' \in L^2(0,T;H)\\ &p \in L^\infty(0,T;H^s(\Omega) \cap L_o^2(\Omega))\\ &p' \in L^2(0,T;H^s(\Omega)). \end{aligned} \qquad (2.3.5)$$

Theorem 1 (Süli, 1985)

Suppose that the hypotheses H1, H2, H3, (2.3.3) and (2.3.4) hold and assume that the discretization parameters obey the relation

$$\Delta t = o(h). \qquad (2.3.6)$$

Then

$$||u-u_h||_{\ell^2(0,T;H_o^1(\Omega)^2)} + ||p-p_h||_{\ell^2(0,T;L^2(\Omega))} \leq C_4(h^r + \Delta t)$$

$$\forall\, r: 1 \leq r \leq \min(\ell,s),$$

$$||u-u_h||_{\ell^\infty(0,T;H_o^1(\Omega)^2)} \leq C_5(h^r + \Delta t) \quad \forall\, r: 1 \leq r \leq \min(\ell,s),$$

for h sufficiently small. The sizes of the Δt terms depend principally on $||D_t^2 u||_{L^2(0,T;L^2(\Omega)^2)}$, where $D_t^2 u = D_t(D_t u)$.

Remark 1

For reasons of stability we require that the time step vanishes faster than the spatial mesh ($\Delta t = o(h)$). Since $O(h^r) = o(h)$ for $r > 1$, this constraint is important only in the case min $(\ell,s) = 1$.

Theorem 2 (Süli, 1985)

Suppose that the hypotheses H1, H2, H3, (2.3.3) and (2.3.5) hold and assume that the discretization parameters obey the relation (2.3.6). Then,

$$||u-u_h||_{\ell^\infty(0,T;L^2(\Omega)^2)} \leq C_6(h^{r+1} + \Delta t) \quad \forall r: 1 \leq r \leq \min(\ell,s),$$

for h sufficiently small. The size of the Δt term depends principally on $||D_t^2 u||_{L^2(0,T;L^2(\Omega)^2)}$.

Remark 2

Since $O(h^{r+1}) = o(h)$ for $r \geq 1$, the mesh restriction (2.3.6) is insignificant asymptotically.

Example

In the Hood-Taylor mixed finite element method the elements $K \in C_h$ are quadrilaterals and the approximating spaces are defined as

$$W_h = \{v \in H_o^1(\Omega)^2 \ : \ v_{i|K} \in Q_2(K),\ i = 1,2;\ \forall K \in C_h\},$$

$$M_h = \{q \in L_o^2(\Omega) \cap C(\bar{\Omega}) \ : \ q_{|K} \in Q_1(K);\ \forall K \in C_h\}.$$

Theorems 1 and 2 for h sufficiently small and $\ell = 2$ give the following error estimates:

$$||u-u_h||_{\ell^2(0,T;H_o^1(\Omega)^2)} + ||p-p_h||_{\ell^2(0,T;L^2(\Omega))} \leq C_4(h^2 + \Delta t),$$

$$||u - u_h||_{\ell^\infty(0,T;L^2(\Omega)^2)} \leq C_6(h^3 + \Delta t).$$

For m=0,...,M-1, the discretization procedure (2.3.1) leads to algebraic systems of the form:

$$\begin{bmatrix} A & B^T \\ B & 0 \end{bmatrix} \begin{bmatrix} U^{m+1} \\ P^{m+1} \end{bmatrix} = \begin{bmatrix} G^{m+1} \\ 0 \end{bmatrix}$$

with a symmetric matrix - the same matrix on each time level. In practice, the vector G^{m+1} is calculated by numerical quadrature rules (e.g. Gaussian rules). We fix the quadrature points x_k, k=1,...,K at time level m+1 and we ask where they came from at level m. In other words, (2.3.2) has to be solved only for $x=x_k$, k=1,...,K. Since the spatial mesh is fixed for all time the implementation of the method is simple.

3. ACKNOWLEDGEMENT

This paper is based on the research performed while the author was visiting the Institute for Computational Fluid Dyanmics at Oxford and Reading.

The author is grateful to Professor K.W. Morton for suggesting this study and for numerous fruitful discussions during the course of the work.

4. REFERENCES

Adams, R.A., (1975). Sobolev spaces. Academic Press, New York, San Francisco, London.

Benque, J.P., Ibler, B., Keramsi, A., Labadie, G., (1980). A Finite Element Method for Navier-Stokes Equations. Proceedings of the 3rd International Conference on Finite Elements in Flow Problems, Banff, Alberta, Canada, pp. 295-301.

Douglas, J.Jr., (1982). Numerical Methods for Convection Dominated Diffusion Problems Based on Combining the Methods of Characteristics with Finite Element or Finite Difference Procedures. *SIAM J. Numer. Anal.*, V. **19**, No5, pp. 871-885.

Pironneau, O., (1982). On the Transport-Diffusion Algorithm and its Applications to the Navier-Stokes Equations. *Numer. Math.*, **38**, pp. 309-332.

Russell, T.F., (1980). An Incompletely Iterated Characteristic Finite Element Method for a Miscible Displacement Problem. *Ph. D. Thesis, University of Chicago.*

Süli, E.E., (1985). Lagrange-Galerkin Mixed Finite Element Approximation of the Navier-Stokes Equations. *Ph. D. Thesis*, University of Belgrade.

Temam, R., (1983). Navier-Stokes Equations and Nonlinear Functional Analysis. SIAM, Philadelphia, Pennsylvania.

OPTIMAL ERROR ESTIMATION FOR PETROV-GALERKIN APPROXIMATIONS IN TWO DIMENSIONS

B.W. Scotney
(Department of Mathematics, University of Ulster)

1. INTRODUCTION

We examine the optimality of Petrov-Galerkin approximations for the linear diffusion-convection equation in two dimensions. The Riesz Representation Theorem provides a relationship between the trial space S^h - test space T^h pairing and the optimality of the associated Petrov-Galerkin method. We present an analysis which leads to optimal error estimates for the Petrov-Galerkin approximation $U \in S^h$ of the form $|u-U|_1 \leq C \inf_{V \in S^h} |u-V|_1$ involving the smallest possible constant C, where $|\cdot|_1$ denotes the norm $(\int |\underline{\nabla}.|^2 d\Omega)^{\frac{1}{2}}$.

The diffusion-convection problem takes the form

$$\begin{aligned} -a\nabla^2 u + \underline{\nabla}.(\underline{b}u) &= f \qquad \text{in } \Omega , \\ u = g \text{ on } \partial\Omega_1 , \qquad \partial u/\partial n &= \Phi \text{ on } \partial\Omega_2 \end{aligned} \tag{1.1}$$

for a quantity u. Here, a is a constant scalar diffusion coefficient and $\underline{b} = \underline{b}(\underline{x})$, $\underline{x} \in \Omega$, is a vector convective velocity. We suppose that Ω is a bounded open region in R^2 with a polygonal boundary $\partial\Omega = \partial\Omega_1 \cup \partial\Omega_2$ and $\partial\Omega_1 \cap \partial\Omega_2 = \phi$; $\partial u/\partial n$ denotes differentiation in the direction of the outward normal on $\partial\Omega_2$.

We adopt the notation

$$W^{m,p} = \{w : D^a w \in L^p \ \forall\ a \text{ such that } |a| \leq m\}, \tag{1.2}$$

and we denote by H^1 the space $W^{1,2}$ - see Oden and Reddy (1976). There exists a continuous linear mapping $\gamma_o : H^1 \to L^2(\partial\Omega_1)$, and we suppose that g is such that there exists $G \in H^1$ with $\gamma_o(G) = g$. We define the space of functions

$$H^1_{E_o} = \{v \in H^1 \,|\, \gamma_o v = 0 \text{ on } \partial\Omega_1\} . \tag{1.3}$$

We may then define

$$H^1_E = \{v \in H^1 \mid v-G \in H^1_{E_o}\} . \tag{1.4}$$

We shall also use the space H^1/R. As general references, see Ciarlet (1978), Girault and Raviart (1979).

The weak formulation of problem (1.1) is to find $u \in H^1_E$ such that

$$B(u,v) = \langle f,v\rangle \qquad \forall\, v \in H^1_{E_o} . \tag{1.5}$$

We make the following assumptions :

$$f \in L^2,\ 0 < a,\ \underline{b} \in [W^{1,\infty}]^2,\ \underline{n}.\underline{b} \geq 0 \text{ on } \partial\Omega_2,\ \underline{\nabla}.\underline{b} = 0 . \tag{1.6}$$

For $v \in H^1_E$ we may write $v = v_o + G$, where $v_o \in H^1_{E_o}$. We may thus reformulate (1.5) as: find $u_o \in H^1_{E_o}$ such that

$$B(u_o,w) = F(w) \qquad \forall\, w \in H^1_{E_o} , \tag{1.7}$$

where the bilinear form $B(.,.) : H^1/R \times H^1_{E_o} \to R$ is given by

$$B(w_1,w_2) \equiv \langle a\underline{\nabla}w_1,\underline{\nabla}w_2\rangle + \langle \underline{\nabla}.(\underline{b}w_1),w_2\rangle \quad \forall\, w_1 \in H^1/R,\ w_2 \in H^1_{E_o} , \tag{1.8}$$

and the linear functional $F(.) : H^1_{E_o} \to R$ is given by

$$F(w) = \langle f-\underline{\nabla}.(\underline{b}G),w\rangle - \langle a\underline{\nabla}G,\underline{\nabla}w\rangle \quad \forall\, w \in H^1_{E_o} . \tag{1.9}$$

We may establish the existence and uniqueness of weak solutions $u_o \in H^1_{E_o}$ to (1.7) and $u \in H^1_E$ to (1.5) via the Lax-Milgram Theorem -see Ciarlet (1978).

We shall assume that the trial space $S^h \subset H^1$, and that it is of dimension N, spanned by a basis $\{\phi_i, i = 1,\dots,N\}$. The space S^h_o is defined as

$$S^h_o = S^h \cap H^1_{E_o} . \quad (1.10)$$

We will consider only g for which there exists $G^h \in S^h$ such that $\gamma_o(G^h) = g$. We may then define

$$S^h_E = \{V = G^h + W \mid W \in S^h_o\} . \quad (1.11)$$

The Galerkin approximation to (1.5) is to find $U \in S^h_E$ such that

$$B(U,V) = \langle f,V\rangle \qquad \forall V \in S^h_o . \quad (1.12)$$

Since $S^h_o \subset H^1_{E_o}$ the existence and uniqueness of $U \in S^h_E$ are guaranteed by the Lax-Milgram Theorem, and the following error estimate due to Babuška and Aziz (1972) holds:

$$|u-U|_T \leq [1 + C_1/C_2] \inf_{V \in S^h_E} |u-V|_T . \quad (1.13)$$

Here, $|\,.\,|_T$ denotes the norm $\langle T.,T.\rangle^{\frac{1}{2}}$ where T is the operator $T. = a^{\frac{1}{2}}\underline{\nabla}.$ derived from $T^*T = \frac{1}{2}(L + L^*)$ where L^* is the formal adjoint of the differential operator $L. = -a\nabla^2. + \underline{\nabla}.(\underline{b}.)$. If the trial space is characterised by the estimate

$$|u-U_I|_{H^1} \leq C\, h^{k-1}|u|_k , \quad (1.14)$$

where $U_I(\underline{x}) \in S^h$ is the finite element interpolate of $u(\underline{x})$, and h represents an element size, (see Strang and Fix (1973)), the ratio C_1/C_2 in (1.13) may be replaced by a mesh Péclet number $|\underline{b}|h/a$. The Galerkin method may therefore become "near-optimal", but if the Péclet number is large, this will occur only for extremely small values of h. This is precisely the situation when the flow is convection-dominated.

Many authors have proposed Petrov-Galerkin methods for problem (1.1). A test space $T^h \subset H^1$ other than S^h is employed. Setting $T_o^h = T^h \cap H_{E_o}^1$, the Galerkin system (1.12) is replaced by the problem of finding $U \in S_E^h$ such that

$$B(U,V) = \langle f,V\rangle \qquad \forall\, V \in T_o^h. \tag{1.15}$$

T_o^h has the same dimension as S_o^h and is spanned by a basis $\{\psi, i = 1,\ldots,N\}$. The problem of choosing S^h to adequately represent the solution to (1.5) remains, but the principal problem now is the selection of the test space for a given trial space.

2. OPTIMAL ERROR ESTIMATION

It is convenient to introduce the operators

$$T_1 \equiv a^{\frac{1}{2}}\underline{\nabla}\,, \quad T_2 \equiv a^{\frac{1}{2}}\underline{\nabla} - \underline{b}/a^{\frac{1}{2}}. \tag{2.1}$$

We may then write the Petrov-Galerkin formulation (1.15) as: find $U \in S_o^h$ such that

$$\langle T_2U, T_1V\rangle + \int_{\partial\Omega_2} \underline{b}.\underline{n}UVds = F(V) \quad \forall\, V \in T_o^h\,. \tag{2.2}$$

Let H_1 denote the Hilbert space $H_{E_o}^1$ equipped with the inner product

$$\langle v,w\rangle_1 = \langle a^{\frac{1}{2}}\underline{\nabla}v, a^{\frac{1}{2}}\underline{\nabla}w\rangle = \langle T_1v, T_1w\rangle\,. \tag{2.3}$$

We note that $\langle .,.\rangle_1$ provides a norm for H^1/R, and we denote by H the Hilbert space H^1/R equipped with the inner product (2.3). Then using the Riesz Representation Theorem, there exists a map $R_1 : H_1 \to H_1$ such that

$$\langle T_2w_1, T_1w_2\rangle + \int_{\partial\Omega_2} \underline{b}.\underline{n}w_1w_2ds = \langle T_1w_1, T_1(R_1w_2)\rangle \;\forall\, w_1 \in H,\; w_2 \in H_1. \tag{2.4}$$

Hence we may consider generating an approximation $U \in S_o^h$ to the problem defined by (1.7), (1.8) and (1.9) using a test space which is related to the trial space through the transformation

$$R_1 T_o^h = S_o^h. \tag{2.5}$$

The existence of an inverse $R_1^{-1} : H_1 \to H_1$ is guaranteed - see Oden and Reddy (1976).

Using (2.4), the Petrov-Galerkin formualtion (1.15) may be written as : find $U \in S_o^h$ such that

$$\langle T_1 U, T_1(R_1 V)\rangle = F(V) \quad \forall\, V \in T_o^h. \tag{2.6}$$

Hence if the choice of test space in (2.5) could be achieved exactly, the approximation obtained from (2.6) is optimal in the norm $|a^{\frac{1}{2}}\underline{\nabla}.|$ defined by T_1. The extent to which a given Petrov-Galerkin formulation fails to satisfy (2.5) may be used to identify the extent to which optimality of the associated approximation is lost:

Theorem 2.1 (Morton (1982)).

For each $v \in H$ let $R_1 : H_1 \to H_1$ be such that

$$B(v,w) = \langle v, R_1 w\rangle_1 \quad \forall\, w \in H_1. \tag{2.7}$$

Then if the constant Δ_1 is such that

$$\inf_{W \in T_o^h} |V - R_1 W|_1 \le \Delta_1 |V|_1 \quad \forall\, V \in S_o^h \tag{2.8}$$

and U is the Petrov-Galerkin solution to problem (1.15), and u is the solution to (1.5), the following estimate holds:

$$|u-U|_1 \le (1-\Delta_1^2)^{-\frac{1}{2}} \inf_{V \in S_E^h} |u-V|_1 . \tag{2.9}$$

Proof

See Morton(1982). (For a proof of the Riesz Representation Theorem, see Adams (1975).)

The smallest constant $(1-\Delta_1^2)^{-\frac{1}{2}}$ in the approximation property (2.8) measures the degree of success in approximating the trial space by the test space transformed using the Riesz representation operator R_1. It is this smallest constant which controls the optimality of the associated Petrov-Galerkin solution.

We present a general technique for evaluating Δ_1 as

$$\Delta_1 = \sup_{V \in S_o^h} \inf_{W \in T_o^h} \frac{|V - R_1 W|_1}{|V|_1} . \qquad (2.10)$$

Any element $V \in S_o^h$ may be written as

$$V = \sum_{i=1}^{N} V_i \phi_i \qquad (2.11)$$

and any element $W \in T_o^h$ as

$$W = \sum_{i=1}^{N} W_i \psi_i . \qquad (2.12)$$

We denote by $\underline{V}$ the N-vector with components V_j, $j = 1,\ldots,N$, etc. By introducing the three N x N matrices A, B and C whose (i,j) entries are

$$\left.\begin{aligned} A_{ij} &= \langle R_1\psi_i, R_1\psi_j\rangle_1 , \\ B_{ij} &= \langle R_1\psi_i, \phi_j\rangle_1 , \\ C_{ij} &= \langle \phi_i, \phi_j\rangle_1 , \end{aligned}\right\} \begin{aligned} i &= 1,\ldots,N, \\ j &= 1,\ldots,N, \end{aligned} \qquad (2.13)$$

respectively, we may obtain

$$\Delta_1^2 = \sup_{\underline{X} \in R^N} \left(1 - \frac{\underline{X}^T Q^T Q \underline{X}}{\underline{X}^T \underline{X}}\right) , \qquad (2.14)$$

where $Q = A^{-\frac{1}{2}} B C^{-\frac{1}{2}}$.

The constant Δ_1^2 may thus be obtained by computing the largest eigenvalue of the matrix $I - Q^T Q$. Alternatively we may compute the smallest eigenvalue λ_1 satisfying the generalised eigenvalue problem

$$B^T A^{-1} B \underline{V} = \lambda_1 C \underline{V} , \qquad (2.15)$$

(see, for example, Wilkinson (1965)), and thence obtain the error constant $(1-\Delta_1^2)^{-\frac{1}{2}}$ via the relationship $1-\Delta_1^2 = \lambda_1$.

For the general diffusion-convection problem in one dimension, it is possible to construct explicit analytic expressions for the Riesz representation operator R_1 and for its inverse R_1^{-1}. The matrices, A, B and C in (2.13) may thus be constructed, allowing an analytic study of the generalised eigenvalue problem (2.15). This is performed by Scotney (1985), resulting in optimal error estimates of the form (2.9) in which the optimal constant $(1-\Delta_1^2)^{-\frac{1}{2}}$ is evaluated in terms of the mesh Peclet number and simple functions of the test space. The analysis is used to generate "near-optimal" test spaces by the selection of upwinding parameters which minimise the constant in the optimal error estimate. Griffiths and Lorenz (1978) have also investigated the optimality of Petrov-Galerkin methods, though their error estimates are based on the non-optimal bound (1.13).

3. DECOMPOSITION IN TWO DIMENSIONS

In two dimensions the Riesz representation operators relating the optimal test space to the trial space are no longer explicitly available. We consider a decomposition of $[L^2]^2$ into the direct sum of curl-free and divergence-free subspaces. This enables us to perform an analysis in which the Riesz representation operator is not explicitly required, at the expense of approximating a family of self-adjoint boundary value problems, each associated with an element in a basis for the test space.

We remove the dependence on the explicit form for the Riesz representer R_1 from the matrices in (2.13). The matrix C does not involve R_1, and we may use (2.4) to immediately write

$$B_{ij} = \langle R_1\psi_i, \phi_j\rangle_1 = B(\phi_j, \psi_i) \quad . \tag{3.1}$$

The removal of R_1 form the matrix A is described below.

We restrict consideration to the diffusion-convection problem (1.1) with Dirichlet boundary conditions. We shall further assume that Ω is a bounded simply connected Lipschitz domain, (see Oden and Reddy (1976)). The map $R_1 : H_1 \to H_1$ is defined by (2.4) with T_1 and T_2 given by (2.1) so that, using Green's Theorem and $\underline{\nabla}.\underline{b} = 0$,

$$\langle a\underline{\nabla}v, \underline{\nabla}w + (\underline{b}/a)w - \underline{\nabla}(R_1 w)\rangle = 0 \quad \forall\, v \in H,\ w \in H_1 \ . \tag{3.2}$$

Equation (3.2) is used to write the (i,j) entry in A in the form

$$A_{ij} = \langle R_1\psi_j, R_1\psi_i\rangle_1 = \langle a\underline{\nabla}(R_1\psi_j), \underline{\nabla}(R_1\psi_i)\rangle$$

$$= \langle a\underline{\nabla}(R_1\psi_j), \underline{\nabla}\psi_i + (\underline{b}/a)\psi_i\rangle. \qquad (3.3)$$

Since $(\underline{b}/a) \in W^{1,\infty}$, $(\underline{b}/a)\psi_i \in [H_o^1]^2$, and we consider the following decomposition theorem:

Theorem 3.1

Let $\Omega \subset R^2$ be a bounded simply connected Lipschitz domain. Then since $(\underline{b}/a)\psi_i \in [L^2]^2$ there exists a function $q_i \in H^1$ and a function $z_i \in H^1$ such that

$$(\underline{b}/a)\psi_i = \underline{\nabla}q_i + \underline{\nabla}xz_i \ , \qquad (3.4)$$

$$\partial q_i/\partial n = (\underline{b}/a)\psi_i . \underline{n} = 0 \quad \text{on } \partial\Omega.$$

Proof

See Girault and Raviart (1979), pp. 29-30. Moreover, $q_i \in H^1$ is determined uniquely up to an additive constant, giving a unique $q_i \in H$. Further, $z_i \in H_o^1$ may be determined uniquely as the solution of

$$\langle\underline{\nabla}xz_i, \underline{\nabla}xw\rangle = \langle(\underline{b}/a)\psi_i, \underline{\nabla}xw\rangle \quad \forall\, w \in H_o^1 \ , \qquad (3.5)$$

$$z_i = 0 \quad \text{on } \partial\Omega$$

- see Girault and Raviart (1979). We note that in two dimensions the vector curl, $\underline{\nabla}xa$, denotes $\underline{\nabla}xa = (\partial a/\partial y, -\partial a/\partial x)^T$, whilst the scalar curl, $\nabla x\underline{a}$, denotes $\nabla x\underline{a} = \partial a_2/\partial x - \partial a_1/\partial y$, where $\underline{a} = (a_1, a_2)^T$. Using Green's Theorem, $z_i \in H_o^1$ is the unique solution of

$$\langle\underline{\nabla}z_i, \underline{\nabla}w\rangle = \langle\nabla x((\underline{b}/a)\psi_i), w\rangle \quad \forall\, w \in H_o^1 \ , \qquad (3.6)$$

$$z_i = 0 \quad \text{on } \partial\Omega.$$

The solution $z_i \in H_o^1 \cap H^2$, (Grisvard (1976)).

We now use (3.4) to rewrite equation (3.3) as

$$A_{ij} = \langle a\underline{\nabla}(R_1\psi_j), \underline{\nabla}(\psi_i + q_i) + \underline{\nabla}xz_i)\rangle . \qquad (3.\ 7)$$

By use of (3.2) and (3.4) again, we may remove q_i from (3.7) to obtain

$$A_{ij} = \langle a\underline{\nabla}\psi_j + \underline{b}\psi_j, \underline{\nabla}\psi_i + (\underline{b}/a)\psi_i\rangle - \langle a\underline{\nabla}xz_j, \underline{\nabla}xz_i\rangle . \quad (3.8)$$

This is the representation introduced in Morton and Scotney (1985). The operator R_1 has thus been removed from the matrix A at the expense of introducing the functions z_i and z_j.

4. APPLICATION TO ERROR ESTIMATION

We apply the analysis introduced in Section 2 and Section 3 to evaluate the optimal constant $(1-\Delta_1^2)^{-\frac{1}{2}}$ in the error estimate (2.9) for the "upwind" Petrov-Galerkin approximation of Heinrich, Huyakorn, Mitchell and Zienkiewicz (1977) and for the Galerkin approximation to the diffusion-convection problem (1.1) in two dimensions. For ease of presentation we consider a rectangular domain $\Omega = (x_a, x_b) \times (y_a, y_b)$ discretised using a uniform grid characterised by a parameter h with nodes at positions $\{(x_k, y_l) = (x_a + kh, y_a + lh),\ k=0,\ldots,N;\ l=0,\ldots,M\}$, and $x_N = x_b$, $y_M = y_b$. S^h is the space of continuous functions which are bilinear on each element. The test functions $\psi_i(x,y)$ are tensor products of the one-dimensional test functions displayed in Fig. 4.1 - see Heinrich et al (1977) for details.

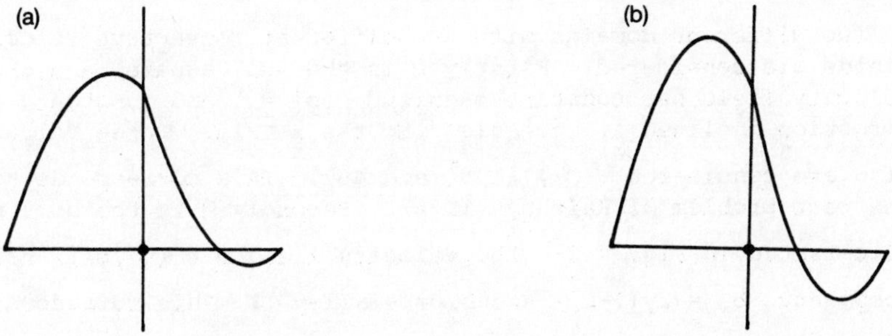

Fig. 4.1 Test functions in one dimension given by Heinrich et al (1977) at mesh Péclet number (a) bh/a = 5, (b) bh/a = 50.

For each test function $\psi_i \in T_o^h$, we approximate the self-adjoint boundary value problem (3.6) by a Galerkin formulation to find $Z_i \in \tilde{S}_o^{\hat{h}}$ such that

$$<\underline{\nabla}Z_i, \underline{\nabla}\hat{\phi}_k> = <\nabla x((\underline{b}/a)\psi_i), \hat{\phi}_k> \forall\, \hat{\phi}_k \in \tilde{S}_o^{\hat{h}}\,, \qquad (4.1)$$

where $\tilde{S}_o^{\hat{h}}$ is a piecewise bilinear finite element trial space on the support of ψ_i characterised by an element size $\hat{h}$. In our computations we have taken $\hat{h} = h/10$. An approximation Z_i to z_i is thus constructed for each $\psi_i \in T_o^h$. Each entry in the matrix A is thence constructed using (3.8). The matrices B and C are constructed using (3.1) and (2.13) respectively.

The optimal constant $(1-\Delta_1^2)^{-\frac{1}{2}}$ in (2.9) is obtained numerically by application of a conjugate gradient algorithm to determine

$$(1-\Delta_1^2)^{-\frac{1}{2}} = \sup_{V \in R^N} \left(\frac{\underline{V}^T C \underline{V}}{\underline{V}^T B^T A^{-1} B \underline{V}}\right)^{\frac{1}{2}} . \qquad (4.2)$$

To the extent that (4.1) is an approximation to (3.6), our analysis is, of course, not exact.

The Galerkin formulation is analysed in the same way simply by replacing ψ_1 by ϕ_i in (4.1).

Two different domains with two different convective velocity fields are considered. Firstly Ω is the unit square and the velocity field has constant magnitude $|\underline{b}| = 1$ and constant direction inclined at an angle θ to the x-axis. Three values of θ are considered : 0°, 21.8° and 45°. This corresponds to the test problem of Raithby (1976). Secondly Ω is the domain illustrated in Fig. 4.2; the velocity field $\underline{b} = (b_1, b_2)^T$ has components $b_1 = 2y(1-x^2)$ and $b_2 = -2x(1-y^2)$. This corresponds to the test problem proposed by Hutton (1981).

	$(1-\Delta_1^2)^{-\frac{1}{2}}$					
	Heinrich et al (1977)			Galerkin		
bh/a	$\Theta = 0^\circ$	$\Theta = 21.8^\circ$	$\Theta = 45^\circ$	$\Theta = 0^\circ$	$\Theta = 21.8^\circ$	$\Theta = 45^\circ$
0	1.0000	1.0000	1.0000	1.0000	1.0000	1.0000
2	2.6567	2.7161	2.7861	2.9562	2.9762	2.9997
10	4.4689	4.4152	4.4859	7.3729	7.3824	7.5488
50	8.1608	6.2446	6.0324	28.731	21.560	20.277
100	9.7248	6.6202	6.3690	56.828	38.602	37.473
10^3	12.257	6.9963	6.7132	566.16	366.30	359.26
10^4	12.649	7.0360	6.7500	4837.9	3660.9	3277.5

Table 4.1 Optimal error constants for the problem of Raithby (1976)

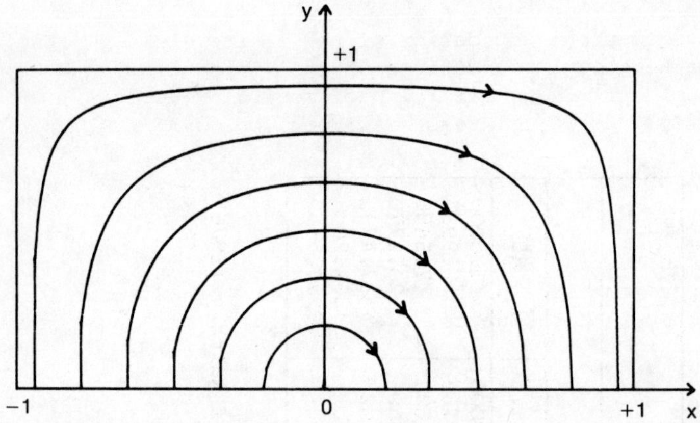

Fig. 4.2 Domain and streamlines in test problem of Hutton (1981).

Our results are displayed in Table 4.1 and in Table 4.2.

	$(1-\Delta_1^2)^{-\frac{1}{2}}$	
bh/a	Heinrich et al (1977)	Galerkin
0	1.0000	1.0000
2	2.8522	2.8833
10	7.7980	8.3276
50	15.023	24.710
100	17.117	45.848
10^3	19.631	405.96
10^4	19.928	3183.6

Table 4.2 Optimal error constants for the problem of Hutton (1981).

ACKNOWLEDGEMENTS

I wish to thank Professor K.W. Morton and Dr. E.E. Süli for their invaluable help with this work.

REFERENCES

Adams, R.A., (1975) Sobolev Spaces. Academic Press, New York.

Babuška, I. and Aziz, A.K., (1972) Survey lectures on the mathematical foundation of the finite element method. The Mathematical Foundations of the Finite Element Method with Applications to Partial Differential Equations. (Ed. A.K. Aziz), Academic Press, New York, pp. 3-363.

Ciarlet, P.G., (1978) The Finite Element Method for Elliptic Problems. North Holland Publ. Comp., Amsterdam.

Girault, V. and Raviart, P.-A., (1979) Finite element approximations of the Navier-Stokes equations. Lecture Notes in Mathematics, 749, Springer-Verlag, Berlin.

Griffiths, D.F. and Lorenz, J., (1978) An analysis of the Petrov-Galerkin finite element method. *Comp. Meth. Appl. Mech. and Eng.,* **14**, pp. 39-64.

Grisvard, P., (1976) Behaviour of the solutions of an elliptic boundary value problem in a polygonal or polyhedral domain. Numerical Solution of Partial Differential Equations III (SYNSPADE 1975). (Ed. B. Hubbard), Academic Press, New York, pp. 207-274.

Heinrich, J.C., Huyakorn, P.S., Mitchell, A.R. and Zienkiewicz, O.C., (1977) An upwind finite element scheme for two-dimensional convective transport equations. *Int. J. Num. Meth. Eng.,* **11**, pp. 131-143.

Hutton, A.G., (1981) The numerical representation of convection. IAHR Working Group Meeting, May 1981.

Morton, K.W., (1982) Finite element methods for non-self-adjoint problems. Lecture Notes in Mathematics, 965. (Ed. P.R. Turner), Springer-Verlag, Berlin, pp. 113-148.

Morton, K.W. and Scotney, B.W., (1985) Petrov-Galerkin methods and diffusion-convection problems in 2D. The Mathematics of Finite Elements and Applications V. Proc. MAFELAP 1984. (Ed. J.R. Whiteman), Academic Press, London, pp. 343-366.

Oden, J.T. and Reddy, J.N., (1976) An Introduction to the Mathematical Theory of Finite Elements. Wiley-Interscience, New York.

Raithby, G.D., (1976) Skew upstream differencing schemes for problems involving fluid flow. *Comp. Meth. Appl. Mech. and Eng.,* **9**, pp. 153-164.

Scotney, B.W., (1985). Ph.D. Thesis, Dept. of Mathematics, University of Reading.

Strang, G. and Fix, G.J., (1973) An Analysis of the Finite Element Method. Prentice-Hall, New York.

Wilkinson, J.H., (1965). The Algebraic Eigenvalue Problem. O.U.P.

A PSEUDO-SPECTRAL SOLUTION OF VORTICITY-STREAM FUNCTION EQUATIONS USING THE INFLUENCE MATRIX TECHNIQUE

J.M. Vanel and R. Peyret
(Département de Mathématiques, Université de Nice, France)

and

P. Bontoux
(I.M.F.M., Université d'Aix-Marseille 11, France)

1. INTRODUCTION

In this paper we present a fast and accurate method for computing unsteady two-dimensional incompressible flows. The method makes use of the vorticity and the stream function as dependent variables approximated by means of Chebyshev polynomial expansions. When using these variables the delicate question is the derivation of proper boundary conditions for the vorticity. The treatment of these conditions must preserve accuracy, stability and should avoid the use of an iterative solution procedure. The method developed here makes use of the influence matrix technique. This technique which has been found successful for the velocity-pressure equations (Kleiser and Schumann, 1980; Le Quéré and Alziary de Roquefort, 1982) has been applied until now to vorticity-stream function equations only in cases which reduce to one-dimensional problems (Tuckerman, 1983; Dennis and Quartapelle, 1983). Our purpose is to give a two-dimensional formulation of the method, to point out the theoretical and numerical difficulties associated with it, and finally to present numerical results illustrating the properties of the method.

2. EQUATIONS AND NUMERICAL APPROXIMATION

The Navier-Stokes equations within the vorticity-stream function formulation are

$$\frac{\partial\omega}{\partial t} - \frac{1}{Re}\nabla^2\omega = -\vec{V}.\nabla\omega + f \equiv F(\psi,\omega;f) \qquad (2.1)$$

$$\nabla^2\psi + \omega = 0 \qquad (2.2)$$

where the stream function ψ and the vorticity ω are connected to the velocity $\vec{V}$ by

$$\vec{V} = (u,v) = \left(\frac{\partial\psi}{\partial y}, -\frac{\partial\psi}{\partial x}\right), \quad \omega = \frac{\partial v}{\partial x} - \frac{\partial u}{\partial y}. \qquad (2.3)$$

These equations are solved in D ($-1< x,y<1$) with the boundary conditions on $\Gamma = \partial D$.

$$\psi=g, \quad \frac{\partial\psi}{\partial\nu} = h \quad (\vec{\nu} = \text{unit normal to } \Gamma). \qquad (2.4a,b)$$

The initial condition at $t = 0$ is

$$\vec{V} = \vec{V}^{o}, \text{ then } \omega = \omega^{o}. \qquad (2.5)$$

In (2.1), Re is the Reynolds number and f is a given forcing term.

The time-discretization is a combination of the 2nd-order Backward Euler scheme for the viscous term with an Adams-Bashforth type evaluation of the convective term, i.e:

$$\frac{1}{2\Delta t}(3\omega^{n+1} - 4\omega^{n} + \omega^{n-1}) - \frac{1}{Re}\nabla^2\omega^{n+1} = 2F^{n} - F^{n-1} \qquad (2.6)$$

$$\nabla^2\psi^{n+1} + \omega^{n+1} = 0 \qquad (2.7)$$

where n refers to the time $t_n = n\Delta t$. This second order scheme is preferred to the usual Crank-Nicolson/Adams-Bashforth scheme for its better stability properties:

(1) the higher Fourier or Chebyshev modes are better damped,

(2) for moderate Re the critical time-step is larger, and for small Re the scheme is unconditionally stable, as found by Zakaria (1985) for the advection-diffusion equation solved with a Collocation-Chebyshev method.

So, at each time-step the following Stokes-type problem has to be solved:

$$\left.\begin{aligned} &\nabla^2\omega - \lambda\omega = G \qquad (2.8) \\ &\nabla^2\psi + \omega = 0 \qquad (2.9) \end{aligned}\right\} \text{ in D}$$

$$\psi = g, \quad \frac{\partial \psi}{\partial \nu} = h \text{ on } \Gamma \tag{2.10a,b}$$

where ω and ψ are written for ω^{n+1} and ψ^{n+1}, respectively.

The approximation of (2.8) - (2.10) makes use of the Chebyshev polynomial expansion:

$$\begin{pmatrix} \omega \\ \psi \end{pmatrix} = \sum_{n=o}^{N} \sum_{m=o}^{M} \begin{pmatrix} \hat{\omega}_{n,m} \\ \hat{\psi}_{n,m} \end{pmatrix} T_n(x) T_m(y) \tag{2.11}$$

The Chebyshev coefficients of G in (2.8) are calculated by the pseudo-spectral technique which consists in performing the products in the physical space and derivations in the spectral space, these two spaces being connected through a FFT algorithm.

The unknowns $\hat{\omega}_{n,m}$, $\hat{\psi}_{n,m}$ are obtained through the Tau-Method (see Gottlieb-Orszag, 1977) associated with the matrix diagonalization technique of Haidvogel and Zang (1979) for the solution of the systems.

The classical difficulty associated with the formulation lies in the determination of boundary conditions for the vorticity. The technique which consists in using $\omega = -\nabla^2\psi$ on the boundary has been employed in association with a Tau-Chebyshev method by Elie et al. (1983). We present here another solution to this problem which makes use of the influence matrix method. The advantage of this method lies in the strong coupling at the same time-level (n + 1) between ω and ψ on the boundary, leading to good accuracy in time and stability without requiring any iterative procedure.

3. INFLUENCE MATRIX METHOD

3.1 Description of the Method

Thanks to the linearity of the problem, the solution of (2.8)-(2.10) can be written as

$$\begin{pmatrix} \omega \\ \psi \end{pmatrix} = \begin{pmatrix} \tilde{\omega} \\ \tilde{\psi} \end{pmatrix} + \sum_{i=1}^{J} \gamma_i \begin{pmatrix} \omega_i \\ \psi_i \end{pmatrix} \tag{3.1.1}$$

In this formula, $(\tilde{\omega},\tilde{\psi})$ is the solution of Problem A:

$$\begin{cases} \nabla^2\tilde{\omega}-\lambda\tilde{\omega} = G \text{ in } D \\ \tilde{\omega} = \tilde{\omega}_o \text{ (arbitrary) on } \Gamma \end{cases} \tag{3.1.2}$$

$$\begin{cases} \nabla^2\tilde{\psi} = - \tilde{\omega} \text{ in } D \\ \tilde{\psi} = g \quad \text{on } \Gamma \quad . \end{cases} \tag{3.1.3}$$

In the summation in the equation (3.1.1) the number J is connected with the number of collocation points (ξ_j, η_j) on Γ. These (ω_i, ψ_i) are solutions of Problem B:

$$\begin{cases} \nabla^2\omega_i - \lambda\,\omega_i = 0 \quad \text{in } D \\ \omega_{ij} = \delta_{ij} \quad \text{on } \Gamma \end{cases} \qquad \begin{cases} \nabla^2\psi_i = - \omega_i \quad \text{in } D \\ \psi_i = 0 \quad \text{on } \Gamma \end{cases} \tag{3.1.4}$$

where j refers to the points (ξ_j,η_j) and δ_{ij} is the Kronecker symbol . Then the J coefficients γ_i are determined such that ψ satisfies the boundary conditions (2.10b), i.e.

$$\left(\frac{\partial\psi}{\partial\nu}\right)_j = h_j = \left(\frac{\partial\tilde{\psi}}{\partial\nu}\right)_j + \sum_{i=1}^{J} \gamma_i \left(\frac{\partial\psi_i}{\partial\nu}\right)_j \tag{3.1.5}$$

This equation, written for all j, yields an algebraic system for the γ_i's. The matrix $A = [\,(\partial\psi_i/\partial\nu)_j]$ of this system is the "influence matrix".

Note that: (1) Problem B is independent of the time-level considered. Consequently, the calculation of ω_i, ψ_i, A and its inverse A^{-1} can be done once and for all before starting the time-integration. (2) It is more economical to define the boundary values of ω

$$\omega_j = \tilde{\omega}_{oj} + \gamma_j \quad \text{on } \Gamma \tag{3.1.6}$$

and to solve a problem A [with (3.1.6) as boundary condition for ω] rather than to store the J fields (ω_i, ψ_i) and to perform the linear combination in (3.1.1).

3.2 Problem of corners and determination of J

The presence of corners in the domain D leads to difficulties. Two methods have been considered to surmount these.

Firstly, we assume that f,g and h are such that $\psi \in C^2$ so that ω at a corner C can be expressed by:

$$\omega|_C = - \nabla^2\psi|_C \qquad (3.2.1)$$

which is known from the condition (2.10a).

So, in Method 1, we prescribe

$$\tilde{\omega}|_C = \tilde{\omega}_{o|C} = - \nabla^2\psi|_C \qquad (3.2.2)$$

$$\omega_{i|C} = 0 \qquad (3.2.3)$$

and the equation (3.1.5) is written at each collocation point on Γ except the corners; hence $J = 2(N + M -2)$.

In Method 2, f,g and h are assumed such that $\nabla^4\psi \in C^o$ so that the equation (2.8) is satisfied up to the corner C. Now, together with (3.2.2) and (3.2.3), we can prescribe

$$\nabla^2\tilde{\omega}|_C = - \lambda\nabla^2\psi|_C + G|_C \qquad (3.2.4)$$

$$\nabla^2\omega\ |_C = 0 \qquad (3.2.5)$$

Because of these additional conditions, the equation (3.1.5) must be written at $J = 2(N + M - 4)$ collocation points only. At the four remaining points, the values of $\tilde{\omega} = \tilde{\omega}_o$ and ω_i are determined so that (3.2.4) and (3.2.5) are satisfied. Here, we have chosen these special points S adjacent to the corners (Fig. 1), that is to say in the regions where the density of collocation points is high.

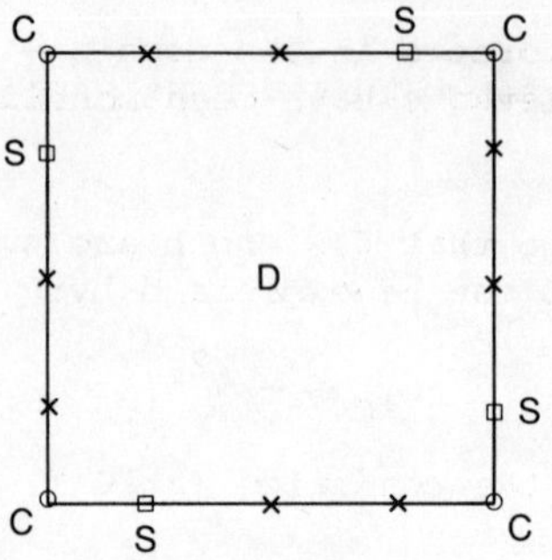

Fig. 1

3.3 Difficulty associated with the Tau-Method

The use of the Tau-Method to solve problem B for ψ_i leads to a difficulty arising because the last weighted residuals, corresponding to the higher modes, are not considered (they are replaced by the boundary conditions). So it appears that there exists a linear combination $\overline{\omega}$ of the ω_i such that the only non-zero coefficients have indices $n = N - 1$ or N, or $m = M - 1$ or M, precisely those coefficients which are not taken into account in the calculation of the corresponding stream function. Consequently we have a numerical solution $\overline{\psi} = 0$ and the matrix A is not invertible. This difficulty can be surmounted if the whole spectrum of the vorticity ω_i is considered when solving the equation for ψ_i. This is done by increasing the number of polynomials in the expansion (2.11) for ψ, which goes up to $N + 2$, $M + 2$ now.

We are thus led to the problem of expressing $\partial\psi/\partial x$ and $\partial\psi/\partial y$, defined by $(N + 3)(M + 3)$ coefficients, at the usual $(N + 1)(M + 1)$ collocation points $(\cos n\pi/N, \cos m\pi/M)$. The evaluation of G is done by simply neglecting the last two coefficients in either direction. On the boundary, two variants have been considered:

<u>Method 1, 2:</u> all the coefficients of $\partial\psi/\partial\nu$ are taken into account.
<u>Method 1', 2'</u>: the two last modes of $\partial\psi/\partial\nu$ are neglected.

The numerical results have shown that 1 is much more accurate than 1', and that 2' is slightly better than 2.

4. NUMERICAL RESULTS

All the calculations reported here have been made with N = M.

4.1 Exact unsteady solution

We consider the solution

$$\psi_{ex} = \frac{2\pi - 1 + \sin 2\pi t}{2\pi}\ \psi_o,\quad \psi_o = (1-x^2)^2(1-y^2)^2\ e^{x(y-1)} \tag{4.1.1}$$

which defines f, g and h. Here we compare the accuracy of various time-discretization schemes:

(1) 1st-order Backward Euler/Adams-Bashforth

(2) 2nd-order Backward Euler/Adams-Bashforth type [Eqs. (2.6) - (2.7)].

(3) Crank-Nicolson/Adams-Bashforth.

The calculations have been made with Re = 1, N = 20 and $0.1 \leq \Delta t \leq 0.005$. With these values of N and Δt, the leading part of the error comes from the time-discretization. Fig. 2 shows the normalized mean quadratic error:

$$\overline{E}_\omega = \underset{t}{\mathrm{Max}}\{\|\omega - \omega_{ex}\|_{L2}/\|\omega_{ex}\|_{L2}\}\ . \tag{4.1.2}$$

The presence of three levels in time in the schemes necessitates starting-up procedures with modified schemes which, in the present calculations, are first order only. Hence, the maximum in (4.1.2) is taken when the effect of the initialization has disappeared: the error $\overline{E}_\omega$ is then periodic in time. From Fig. 2, it can be seen that scheme (2) is slightly more accurate than scheme (3), at least for the special solution computed here.

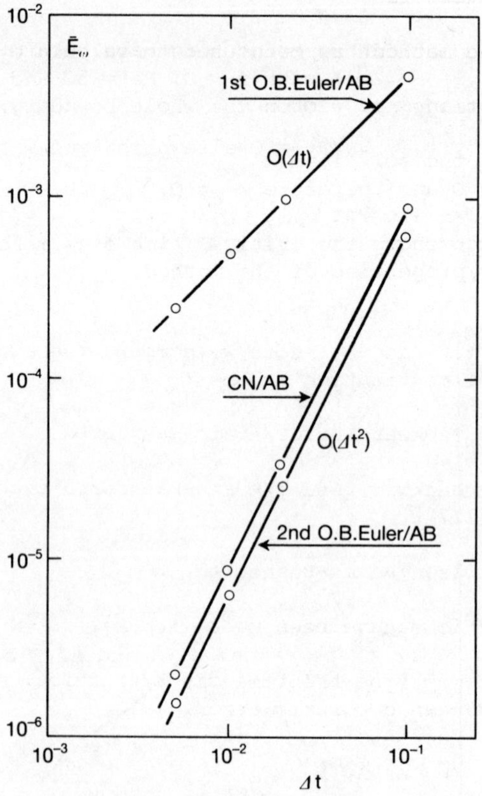

Fig. 2 Error $\overline{E}_\omega$

4.2 Exact steady solution

In order to compare the accuracy of methods 1 and 2' we have considered the steady solution $\psi_{ex} = \psi_o$, using the scheme (2.6) - (2.7) with Re = 1. From Table 1, it is seen that method 2' is more accurate than method 1 for small N. The accuracy is comparable with some advantage for Method 1 when N is larger.

N	10	12	16	20
Method 1	1.223×10^{-4}	5.828×10^{-6}	9.179×10^{-11}	4.045×10^{-11}
Method 2'	2.928×10^{-5}	3.507×10^{-7}	1.723×10^{-10}	3.573×10^{-10}

Table 1: Error $\overline{E}_\omega$

4.3 Steady Cavity flow

Finally, the method has been used to calculate the steady solution of (2.1) - (2.2) with f = 0, Re = 50 and with the boundary conditions: $\psi = 0$ on the whole boundary; $\partial\psi/\partial y = -(1 - x^2)^2$ at y = 1, $\partial\psi/\partial\nu = 0$ elsewhere; and with the initial condition $\vec{V}^o = 0$ and therefore $\omega^o = 0$.

Fig. 3, which shows the critical time-step, makes clear the good stability properties of the method.

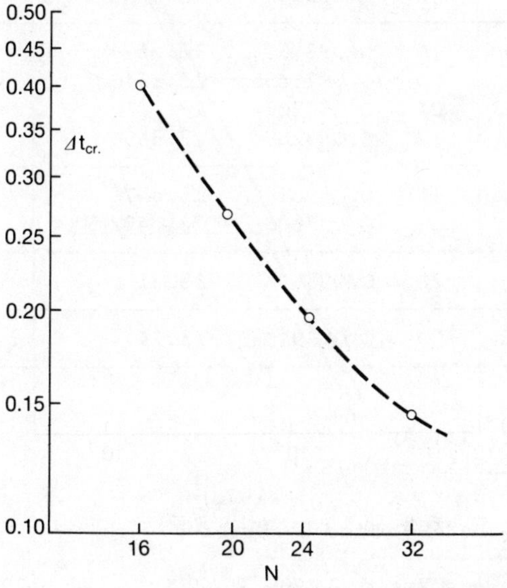

Fig. 3 Critical time-step

Table 2 gives some results. Here, the variables have been redefined by using the side L of the square as reference length, so that $0 \leq X, Y \leq 1, \Psi = \psi/2$. $\Omega = 2\omega$ and $Re_L = 100$. refer to (Peyret and Taylor, 1983) for results given by other methods. Because of the strong variation of Ω on the upper side Y = 1, the consideration of the maximum on collocation points only gives a bad idea of the convergence of the method with respect to N. Since the spectral approximation gives the solution at every $X \in [0,1]$, we have also listed in the last column of Table 2 the maximum value of $\Omega(X,1)$ on 201 equally spaced points.

Table 2. Results for cavity flow, $Re_L = 100$

Method	N	$Max\|\psi\|^{(e)}$	$Max\|\Omega(X,1)\|^{(e)}$	$Max\|\Omega(X,1)\|^{(f)}$
pseudo-spectral 1	16	0.083139	13.4235	13.5108
	20	0.082692	13.1519	13.4357
	24	0.083315	13.4149	13.4357
	32	0.083403	13.3418	13.4462
pseudo-spectral 2'	16	0.083158	13.3475	13.4527
	20	0.082695	13.1790	13.4506
	24	0.083315	13.4260	13.4490
	32	0.083403	13.3441	13.4441
pseudo-spectral (a)	16	0.083057	13.3574	13.4485
	20	0.082545	13.1810	13.4453
	24	0.083201	13.4315	13.4522
	32	0.083265	13.3496	13.4517
Collocation (V,p) (b)	16	0.083686	13.3574	13.4428
	32	0.083685	13.3422	13.4445
Hermitian (c)	20	0.0835	13.31	-
Finite-Diff. (d)	20	0.0829	13.14	-

(a) Elie et al. (1983),
(b) Ouazzani and Peyret (1983),
(c) 4th-order, Bontoux et al. (1978),
(d) 2nd-order, Peyret and Taylor (1983), p. 205, run 16,
(e) taken on collocation points,
(f) taken on 201 equally spaced points and obtained at X=0.62.

Fig. 4 shows the evolution of $\Omega(X,1)$. On the scale of the figure the difference between N = 16 and N = 32 cannot be distinguished. Fig. 5 is a blow-up of a very small region near X = 0. This provides an illustration of the high accuracy of Chebyshev approximation near the boundaries. Also, it is interesting to note the good agreement of the present results with those obtained by a Collocation - Chebyshev - Artificial Compressibility method (Ouazzani and Peyret, 1983) for the velocity-pressure equations, the latter method needing no special treatment at a corner.

The CPU time (CRAY 1S) is 0.03s/time-step for N = 20 and 0.078s/time-step for N = 32. The preprocessing is roughly equivalent to 4N time-steps.

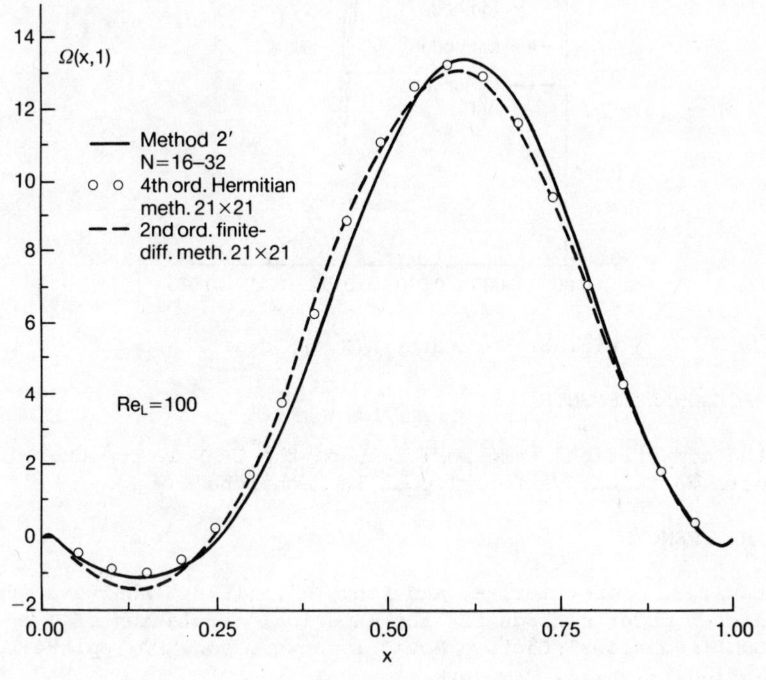

Fig. 4 Vorticity $\Omega(X, 1)$

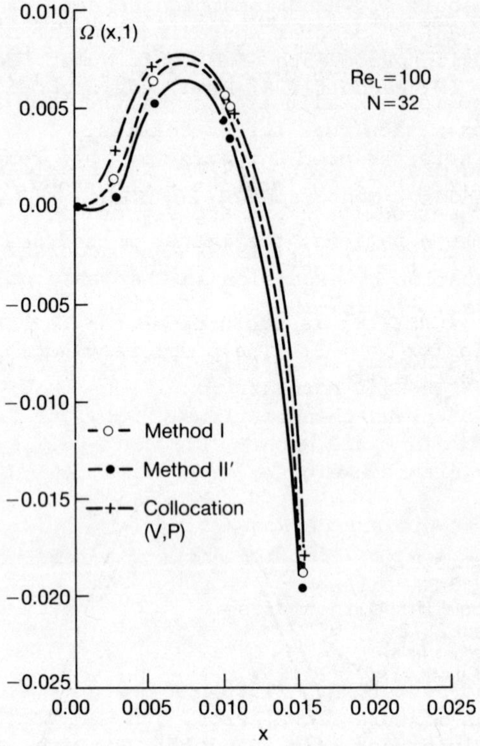

Fig. 5 Vorticity $\Omega(X, 1)$ near $X = 0$

5. ACKNOWLEDGEMENTS

The computations were performed on the CRAY 1S computer of "Centre de Calcul Vectoriel pour la Recherche".

6. REFERENCES

Bontoux, P., Forestier, B. and Roux, B., (1978) "Analysis of higher order methods for the numerical simulation of confined flows",Lecture Notes in Phys., vol. 90, pp. 94-102. Springer-Verlag, New York.

Dennis, S.C.R. and Quartapelle, L., (1983) "Direction solution of the vorticity-stream function ordinary differential equations by a Chebyshev approximation", *J. Comp. Phys.*, **22**, 448-468.

Elie, F., Chikhaoui, A., Randriamampianina, A., Bontoux, P. and Roux, B., (1983). "Spectral approximation for Boussinesq double diffusion", Proc. 5th GAMM Conf. Numer. Methods in Fluid Mech., (M. Pandolfi, R. Piva, eds). pp. 57-64, Vieweg, Braunschweig.

Gottlieb, D. and Orszag, S.A., (1977) Numerical Analysis of Spectral Methods. Monograph No. 26, SIAM, Philadelphia.

Haidvogel, D.B. and Zang, T., (1979) "The accurate solution of Poisson's equation by expansion in Chebyshev polynomials". *J. Comp. Phys.*, **30**, 167-180.

Kleiser, L. and Schumann, U., (1980) "Treatment of incompressibility and boundary conditions in 3-D numerical simulations of plane channel flows", Proc. 3th GAMM Conf. Numer. Methods in Fluid Mech., (E.H. Hirschel ed.), pp. 165-173, Vieweg, Braunschweig.

Le Quéré, P., and Alziary de Roquefort, T., (1982) "Sur une méthode spectrale semi-implicite pour la résolution des équations de Navier-Stokes d'un écoulement bidimensionnel visqueux d'un fluide incompressible". C.R. Acad. Sci. Paris, **294**, 11. pp. 941-943.

Ouazzani, J. and Peyret, R., (1983) "A pseudo-spectral solution of binary gas-mixture flows", Proc. 5th GAMM Conf. Numer. Methods in Fluid Mech., (M. Pandolfi, R. Piva, eds.), pp. 275-282, Vieweg, Braunschweig.

Peyret, R. and Taylor, T.D., (1983) Computational Methods for Fluid Flow. Springer-Verlag, New York.

Tuckerman, L., (1983) "Formation of Taylor vortices in spherical Couette flow", Ph.D. Thesis, M.I.T., Cambridge, Mass.

Zakaria, A., (1985) "Etude de divers schémas pseudo-spectraux de type collocation pour la résolution des équations aux dérivées partielles. Application aux équations de Navier-Stokes". Thèse 3ième cycle, Départ. Math., Université de Nice.

THE USE OF CHEBYCHEV POLYNOMIALS IN SPECTRAL FLUID DYNAMIC SIMULATIONS

P.R. Voke and H.M. Tsai
(The Turbulence Unit, Queen Mary College)
and
M.W. Collins
(Thermofluids Engineering Research Centre, The City University)

1. BACKGROUND

This paper reports certain results of a mathematical nature arising from the use of two spectral (Fourier-Fourier-Chebychev) simulation codes for plane Poiseuille flow, both of which have been used for full simulation (no modelling) and large-eddy simulation (incorporating a subgrid scale model).

Fluid dynamic simulations take advantage of the fact that we know the equations of motions governing turbulence, and that we have large computing machines at our disposal. By choosing a computational box significantly larger than the expected correlation length of the turbulence under study, we can reproduce convincingly the properties of homogeneous, isentropic turbulence at low Reynolds numbers. Periodic boundary conditions are imposed on the box, and some random disturbance is set up, which is allowed to evolve in time according to the Navier-Stokes equations. The equations are treated in "k-space", that is, in terms of the Fourier coefficients. The nonlinear terms are computed by Fourier transformation into "x-space", giving values of velocity fields at a sequence of discrete mesh points at which the nonlinear product of fields are evaluated. Both rigorous and approximate methods are known for dealing with the aliasing errors that arise from this process when the product is retransformed to k-space.

There are severe limitations on the use of such full simulations of the Navier Stokes equations. The finite number of Fourier modes that can be used in a real digital simulation means that the Reynolds number must be very low. (The truncated Fourier expansion must provide a basis spanning the attractor of the flow in function space, which therefore needs to have a rather small dimension.) As the Reynolds number increases it rapidly becomes impossible simultaneously to resolve all the important scales present. The only hope of performing simulations at higher Reynolds numbers appears to be some form of large eddy simulation, defined as a simulation in which only a limited range of scales is resolved, and eddies below the resolution of the grid are modelled. For a review of the field, see Voke and Collins (1983).

The introduction of one or more boundary walls changes the flows fundamentally. Even at Reynolds numbers just above transition, the rates of change of many important dynamical quantities become large as the wall is approached. In the context of a finite-difference (or mixed spectral finite-difference) computation, it is usual to employ mesh stretching in the direction perpendicular to the wall in order to achieve adequate numerical resolution of the dynamics, (Moin and Kim 1982). At higher Reynolds numbers this becomes unwieldy; certain large-eddy simulations attempt to model the wall layer dynamics as well as subgrid turbulence, (Schumann 1975).

2. CHEBYCHEV POLYNOMIALS

Chebychev polynomial expansions represent an alternative to finite differences with mesh stretching. Orszag and Kells (1980) have used a fully spectral simulation to perform full simulations of plane Poiseuille flow using Chebychev polynomials in one direction and Fourier expansions in the other two directions.

The attraction of Chebychev expansions is that they provide a spectral representation consistent with arbitrary nonperiodic boundary conditions, avoiding the Gibbs phenomenon. In addition they are closely related to cosine expansions, and Chebychev transformation may be performed as efficiently as cosine transformation, using fast Fourier transform (FFT) algorithms. The natural collocation points for the computation of nonlinear terms lie on a cosine distribution between the walls, reflecting the underlying weighting used to define the polynomials.

3. THE FDS SIMULATION CODE

The first of our codes, FDS, solves the fully coupled incompressible Navier-Stokes equations for the primitive variables u_i (velocity) and p (pressure/density), which may be written in the form:

$$\partial_t u_i = -\partial_i p - u_j \partial_j u_i + \nu\, \partial^2 u_i + F_i \qquad (3.1)$$

$$\partial_i u_i = 0 \qquad (3.2)$$

F_i is some forcing term such as a unit pressure gradient. The equations take a similar form if nondimensionalised using the wall shear velocity and half channel width H, with ν then representing the reciprocal of the Reynolds number based on these quantities. We Fourier transform these equations in the x_1 and x_2 directions, (the Fourier wavenumbers are k_1 and k_2) and expand in a finite Chebychev series in the x_3 direction perpendicular to the walls. Simple order finite differencing is used in time, explicit for the inertial term and implicit for the others, giving a set of equations for quantities u_i^+ and p^+ at the advanced time in terms of explicitly calculated quantities R_i:

$$\frac{2}{\Delta t} u_i^+ + ik_i p^+ - \nu\left(\frac{\partial^2}{\partial x_3^2} - k_1^2 - k_2^2\right) u_i^+ = R_i, \ i=1,2 \tag{3.3}$$

$$\frac{2}{\Delta t} u_3^+ + \frac{\partial p}{\partial x_3} - \nu\left(\frac{\partial^2}{\partial x_3^2} - k_1^2 - k_2^2\right) u_3^+ = R_3 \tag{3.4}$$

$$ik_1 u_1^+ + ik_2 u_2^+ + \frac{\partial u_3^+}{\partial x_3} = 0 \tag{3.5}$$

The derivatives with respect to x_3 have been left in the x-space form here, but are computed by the well known recursion relation for the Chebychev coefficients of a derivative function (Fox and Parker, 1968)

$$u'(k_3) = u'(k_3+2) + 2(k_3+1)u(k_3+1) \tag{3.6}$$

This recursive algorithm is inconvenient to computes explicitly on vector processors (it is best done on a large number of coefficient sequences in parallel). The equations are integrated twice with respect to x_3 to obtain the standard form of the left hand side. The matrices are given explicitly in Voke and Collins (1983 and 1985). The numerical integrations should be the precise algorithmic inverses of the corresponding differentiation algorithms, if the left hand side matrices are to be fully consistent with the right hand side terms.

A graphical technique has been used to assist in the derivation of the matrices for the complex set of equations (3.4-3.6). The derivative algorithm for a sequence of Chebychev coefficients $u(k_3)$ $(k_3=0,N_3)$ is represented by figure 1, in which each horizontal arrow represents simple addition, and each diagonal arrow addition of the term multiplied by the appropriate $2(k_3+1)$ value.

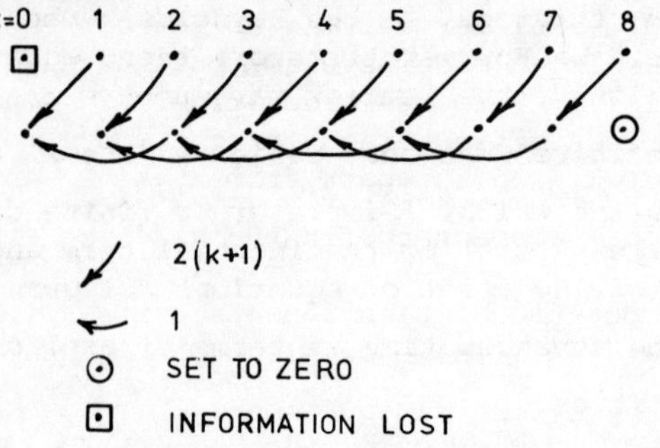

Fig. 1 Information flow graph; differentiation of Chebychev coefficient sequence.

The reverse algorithm, used for setting up the matrix for a first order differential equation, is shown in figure 2a. It shows that one degree of freedom is unaccounted for; this is fixed by the boundary conditions of the problem. Double integration is shown in figure 2b, and requires two conditions.

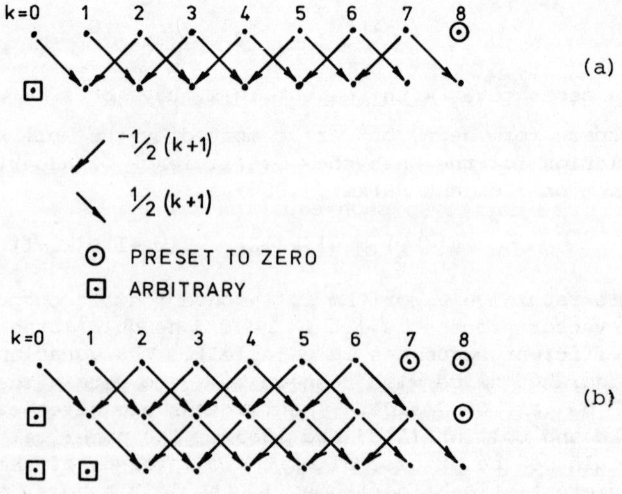

Fig. 2 Information flow graphs; integration of Chebychev coefficient sequence; (a) single integration (b) double integration.

Each distinct route from a point on the top line to one on the bottom line represents a contribution to the appropriate matrix element, calculated from the product of the factors associated with the arrows in the path. This graphical technique proved useful for setting up the correct matrix elements for the second order system, but has also been used to set up the matrix for a fourth order problem. Consistent, nonsingular matrices are obtained for the second order system for all values of k_1 and k_2 apart from $k_1=k_2=0$, which requires a single additional element to fix the mean pressure.

A further computational technique is worthy of mention. The equations (3.3) to (3.5) have complex coefficents, but by multiplying 3.3 by i ($i^2=-1$), equations for the unknowns iu_1, iu_2, u_3 and p with purely real coefficents result. The matrices are banded, with a bandwidth of 20, but not all the diagonals are present. They are roughly equivalent to hexadiagonal matrices. Since the pattern of nonzero elements is the same for various values of k_1 and k_2 (unless either is zero), sparse matrix techniques are used to factorise and solve the system.

4. THE CHANEL SIMULATION CODE

This code was written independently by Antonopoulos-Domis (1983), and uses time stepping and Fourier-Chebychev expansion methods similar to those described above. However, it is based on a derived equation for u_3, obtained by taking the Laplacian (∂^2) of the u_3 momentum equation, and substituting for $\partial^2 p$ from the normal Poisson equation for pressure. This gives an equation for $\psi=\partial^2 u_3$ of the form

$$\partial_t \psi = S + \nu\partial^2\psi; \; \psi = \partial^2 u_3 \tag{4.1}$$

where S is a nonlinear term computable entirely in terms of velocity components. Equation (4.1) is solved three times for combinations of zero or unit boundary conditions on ψ. Three different u_3 solutions result, (the Poisson equation is solved with $u_3=0$ on the boundary), and a linear combination found for which $\partial_3 u_3=0$ at the boundary.

The CHANEL code then finds the pressure at the advanced time algebraically, and solves for u_1 and u_2 from (3.1), incorporating the pressure gradients into the right hand sides.

In all cases the viscous terms are treated implicitly in time. Unfortunately Antonopoulos-Domis overlooked a still more efficient way to solve for u_1 and u_2, using the equation for the vorticity component ω_3, with boundary condition $\omega_3=0$:

$$\partial_t\omega_3 = \Omega_3 + \nu\partial^2\omega_3 \; ; \quad \omega_3=(\partial_1 u_2-\partial_2 u_1)/2 \qquad (4.2)$$

This equation is simply one component of the curl of equations (3.1), and Ω_3 is the third component of the curl of the inertial term in (3.1). Once ω_3 and u_3 are known at the advanced time, it is trivial algebra to find u_1 and u_2 in k-space from (4.2b) and (3.3). This method is to be incorporated into the CHANEL code shortly. No pressure solution is needed.

In spite of the necessity for multiple solutions, the method proves to be much more efficient than that used by FDS, because of the decoupling of the unknown fields.

5. RESULTS

Both codes have been used for full simulation of very low Reynolds number flows, around transition (FDS) or buoyancy-driven (a code derived from CHANEL), but have also been used for large-eddy simulations of low Reynolds number turbulence. The Reynolds numbers studied in the simulations are in the range Re=1000 to 10000, based on maximum velocity and channel half width, or H^+=30 to 300 based on shear velocity.

The full simulations around Re=1000 show transition to turbulence in the presence of finite disturbances between Re=1000 and 1100 (Voke and Collins 1985). 16 Fourier modes were used in each of the x_1 and x_2 directions, with 17 Chebychev polynomials used in the x_3 direction. Sustained turbulence occurred at H^+=40 (Re=1600). In spite of the low spectral resolution, there is no evidence of a pile-up of energy in the top Fourier modes for $H^+ \lesssim 40$.

At higher Reynolds numbers, a subgrid scale model is needed, and a relatively crude eddy viscosity model was used. For simulations at H^+=300, an apparently sustainable turbulent state was found even using 16 x 16 x 17 modes. However, an unphysical oscillation with respect to x_3 is seen in many important parameters, figures 3 and 4.

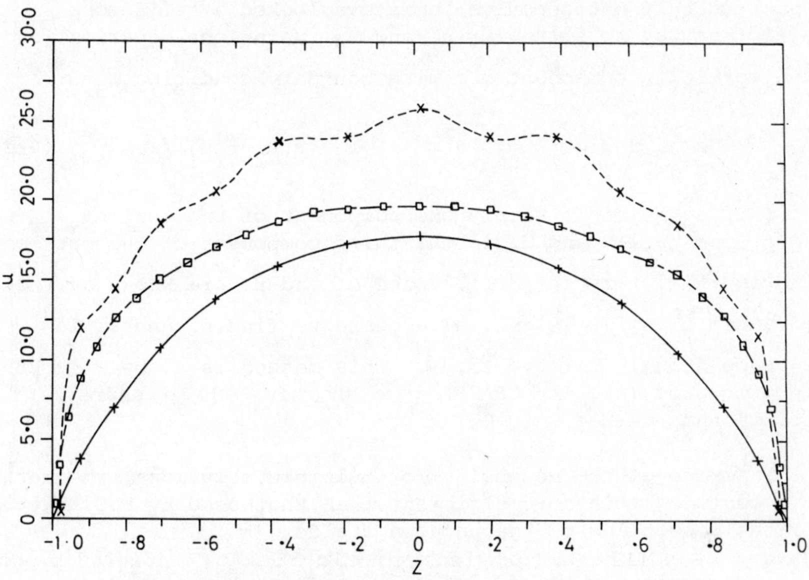

Fig. 3 Profiles of streamwise velocity u, averaged over planes parallel to the walls, predicted by code FDS. 16 x 16 Fourier modes.

______ Full simulation, $H^+=40$, 17 Chebychev modes. $t^+=tu_\tau/H=5.0$.

— — Large-eddy simulation, $H^+=194$, 33 Chebychev modes, $t^+=3.0$.

-x---- Large-eddy simulation, $H^+=300$, 17 Chebychev modes, $t^+=2.75$.

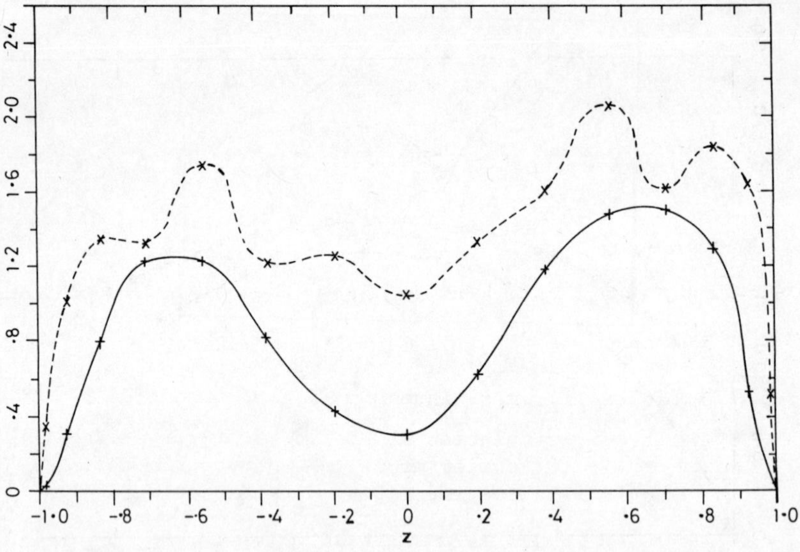

Fig. 4 Profiles of turbulent kinetic energy, predicted by code FDS. 16 x 16 Fourier modes.

_____ Full simulation, H^+=40, 17 Chebychev modes, t^+=5.0.

---x- Large-eddy simulation, H^+=300, 17 Chebychev modes, t^+=2.6.

This type of lateral wiggle is also found by CHANEL in several important quantities when using the same number of modes, figures 5, 6 and 7.

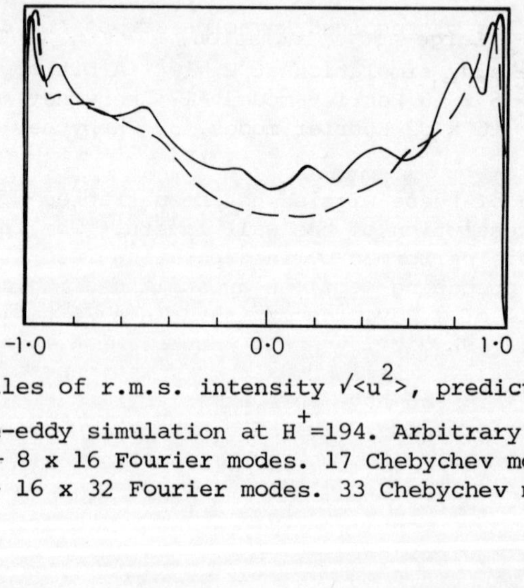

Fig. 5 Profiles of r.m.s. intensity $\sqrt{\langle u^2 \rangle}$, predicted by CHANEL. Large-eddy simulation at H^+=194. Arbitrary normalisation

——— 8 x 16 Fourier modes. 17 Chebychev modes.

----- 16 x 32 Fourier modes. 33 Chebychev modes.

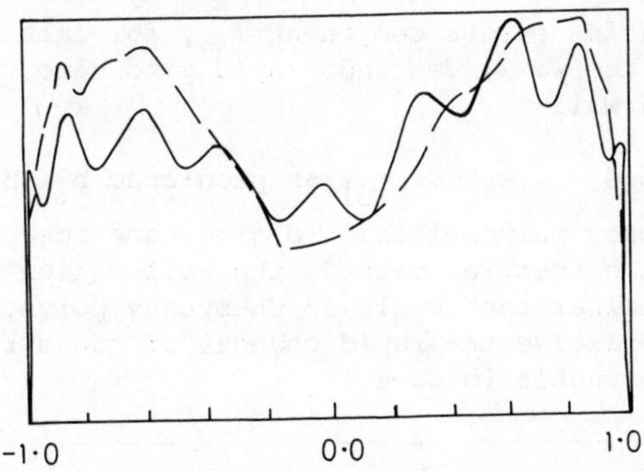

Fig. 6 Profiles of r.m.s. intensity $\sqrt{\langle v^2 \rangle}$, predicted by CHANEL. Large-eddy simulation at $H^+=194$. Arbitrary normalisation
——— 8 x 16 Fourier modes, 17 Chebychev modes.
----- 16 x 32 Fourier modes. 33 Chebychev modes.

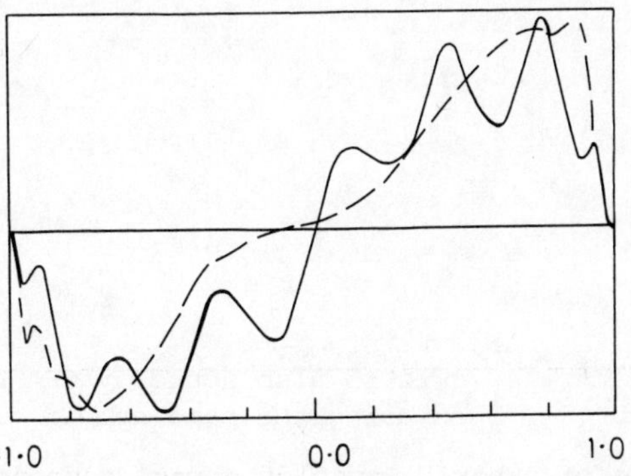

Fig. 7 Profiles of resolved Reynolds stress $\langle uw \rangle$; CHANEL. Large-eddy simulation at $H^+=194$. Arbitrary normalisation
——— 8 x 16 Fourier modes, 17 Chebychev modes.
----- 16 x 32 Fourier modes. 33 Chebychev modes.

The origin of these wiggles has been tracked down to a lack of adequate resolution of the wall layers. The large-eddy simulations are performed by incorporating into the equations given in the preceding sections an additional term that is the divergence of a Reynolds stress. This we term the "subgrid acceleration". Physically, this acceleration must be zero at the walls, and it is well known that components of Reynolds stresses are zero and have zero normal derivatives at walls. To ensure this, our eddy viscosity models, computed at the collocation points in x-space, incorporate an exponential damping factor of the Van Driest (1956) type. This results in a spanwise Reynolds stress component τ_{31}, for instance, that peaks close to the wall, descends rapidly to zero, and has zero gradient at the wall.

Figure 8 shows a graph of τ_{31} as predicted by the FDS code, using 17 Chebychev polynomials. To give some idea of the resolution of the spectral method, the collocation points are marked. It is clear that while 33 Chebychev polynomials might be adequate to resolve the rapid changes of the stress, 17 polynomials are unable to do so.

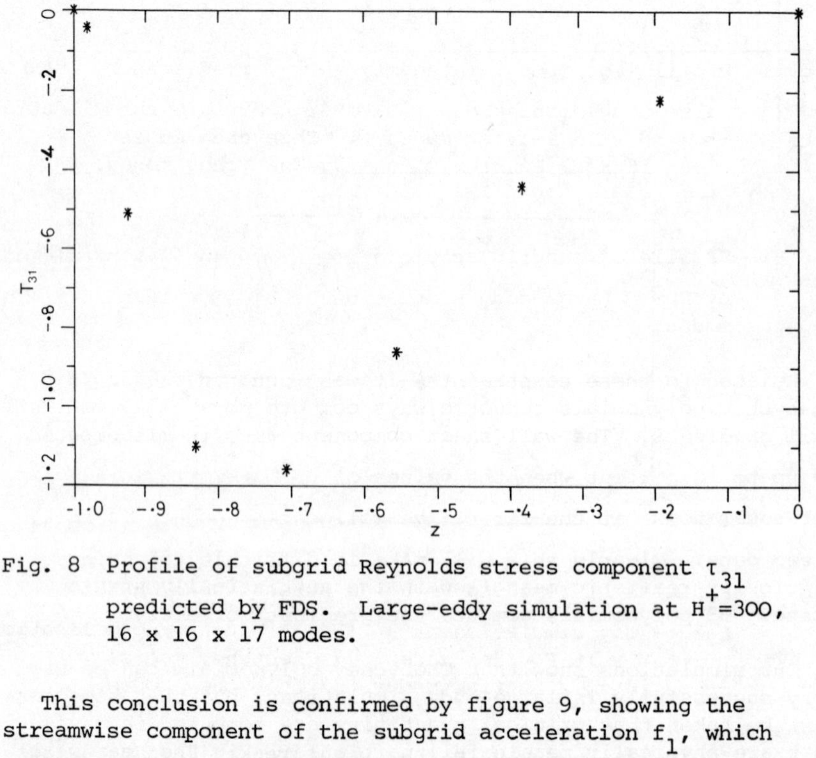

Fig. 8 Profile of subgrid Reynolds stress component τ_{31} predicted by FDS. Large-eddy simulation at $H^+=300$, 16 x 16 x 17 modes.

This conclusion is confirmed by figure 9, showing the streamwise component of the subgrid acceleration f_1, which arises predominantly from $\partial_3\tau_{13}$. This is not equal to zero at the wall, using 17 polynomials. In the code FDS the value at the wall is fixed to zero artificially, resulting in smooth predictions for turbulent flow using 17 or 33 polynomials. The code CHANEL does not fix f_1 to zero at the wall, but obtains good agreement with the experimental results of Kreplin and Eckelmann (1979) using 33 polynomials.

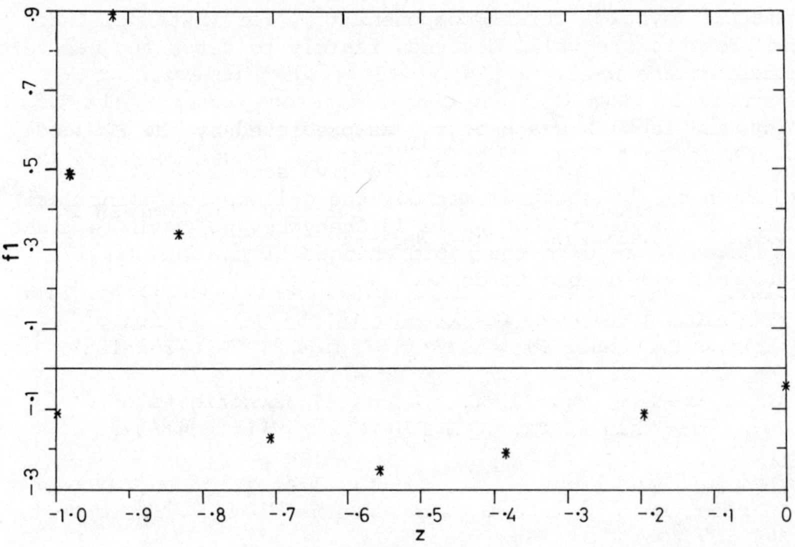

Fig. 9 Profile of subgrid acceleration component f_1, predicted by FDS. Large-eddy simulation at $H^+=300$. 16 x 16 x 17 modes.

Related to these constraints, it was found that even 33 Chebychev polynomials cannot always compute physically realistic wall gradients. The wall shear component $\partial_3 u_1$ is miscomputed by up to 10 percent when the values of u_1^+ are artificially set equal to x_3^+ at the first five collocation points in initial field data. Clearly this will make it difficult to achieve a physically realistic mean flow in the statistically stationary state. 65 polynomials compute the gradient correctly.

Our simulations show that Chebychev polynomials can be used very successfully in large-eddy simulations, but that some care must be taken that critical quantities are computed in ways that are physically meaningful, particularly in the near wall region.

ACKNOWLEDGEMENTS

The authors wish to record the vital contributions of M. Antonopoulos-Domis and B.A. Splawski who wrote the CHANEL code, and the work of Professor D.C. Leslie who directed the creation and use of this code.

REFERENCES

Antonopoulos-Domis, M., (1983). Numerical Simulation of Turbulent Flows in Plane Channels, Proceedings of the Third International Conference on Numerical Methods in Laminar and Turbulent Flow, 113-123. Pineridge Press, Swansea, U.K.

Fox, L. and Parker, I.B., (1968). Chebychev Polynomials in Numerical Analysis. Oxford University Press.

Kreplin, H. and Eckelmann, H., (1979). Behaviour of the Three Fluctuating Velocity Components in the Wall Region of a Turbulent Channel Flow". *Phys. Fluids* **22** (7), 1233-1239.

Moin, P. and Kim, J., (1982). Numerical Investigation of Turbulent Channel Flow. *J. Fluid Mech,* **118**, 341-377.

Orszag, S.A. and Kells, L.C., (1980). Transition to Turbulence in Plane Poiseuille and Plane Couette Flow. *J. Fluid Mech.* **96**, 159-205.

Shumann, U., (1975). Subgrid Scale Model for Finite Difference Simulations of Turbulent Flows in Plane Channels and Annuli. *J. Comput. Phys.* **18**, 376-404.

Van Driest, E.R., (1956). On Turbulent Flow near a Wall. *J. Aero. Soc.* **23**, 1007-1011.

Voke, P.R. and Collins, M.W., (1983). Large-Eddy Simulation: Retrospect and Prospect. *AERE-R* 10716, AERE Harwell, Oxfordshire, England.

Voke, P.R. and Collins, M.W., (1985). Direct and Large-Eddy Simulations of Plane Poiseuille Flow Using the FDS Fluid Dynamic Simulation Codes. AERE report, To be published.

STEADY VISCOUS FLOW PAST A CIRCULAR CYLINDER

B. Fornberg
(Exxon Research and Engineering Company, Annandale)

1. INTRODUCTION

Viscous flow past a circular cylinder becomes unstable around Reynolds number Re = 40. Although this first instability can be suppressed by inclusion of symmetry, other instabilities and limitations (wall effects etc.) make experimental results for the steady flow unreliable past Re = 100 (e.g. Grove et al. 1964). The only earlier numerical calculation to achieve both convergence and accuracy past Re = 120 was by the present author (Fornberg, 1980). A change in trends was discovered near Re = 300, but a simultaneous growth of numerical errors prevented us then from resolving its nature.

With a numerical technique based on Newton's method and made possible by the use of a supercomputer, we have now been able to obtain steady (but unstable) solutions up to Re = 600. It is found that the wake bubble (region with recirculating flow) grows in length approximately linearly with Re throughout this whole range. The width is found to increase like $Re^{\frac{1}{2}}$ up to Re = 300 at which point a transition to linear increase with Re begins. At the highest Reynolds numbers we reached, the wake resembles a pair of translating, uniform vortices, both touching the centre line. This observation forces a reassesment of earlier models for the structure of high Reynolds number wakes (e.g. Smith 1979).

2. MATHEMATICAL FORMULATION

With a cylinder of radius 1 and the Reynolds number based on the diameter, the steady-state Navier-Stokes equations (expressed in streamfunction Ψ and vorticity ω) take the form:

$$\Delta\Psi + \omega = 0 \qquad (2.1)$$

$$\Delta\omega + \frac{1}{2}\,\mathrm{Re}\,\{\frac{\partial\Psi}{\partial x}\cdot\frac{\partial\omega}{\partial y} - \frac{\partial\Psi}{\partial y}\cdot\frac{\partial\omega}{\partial x}\} = 0, \qquad (2.2)$$

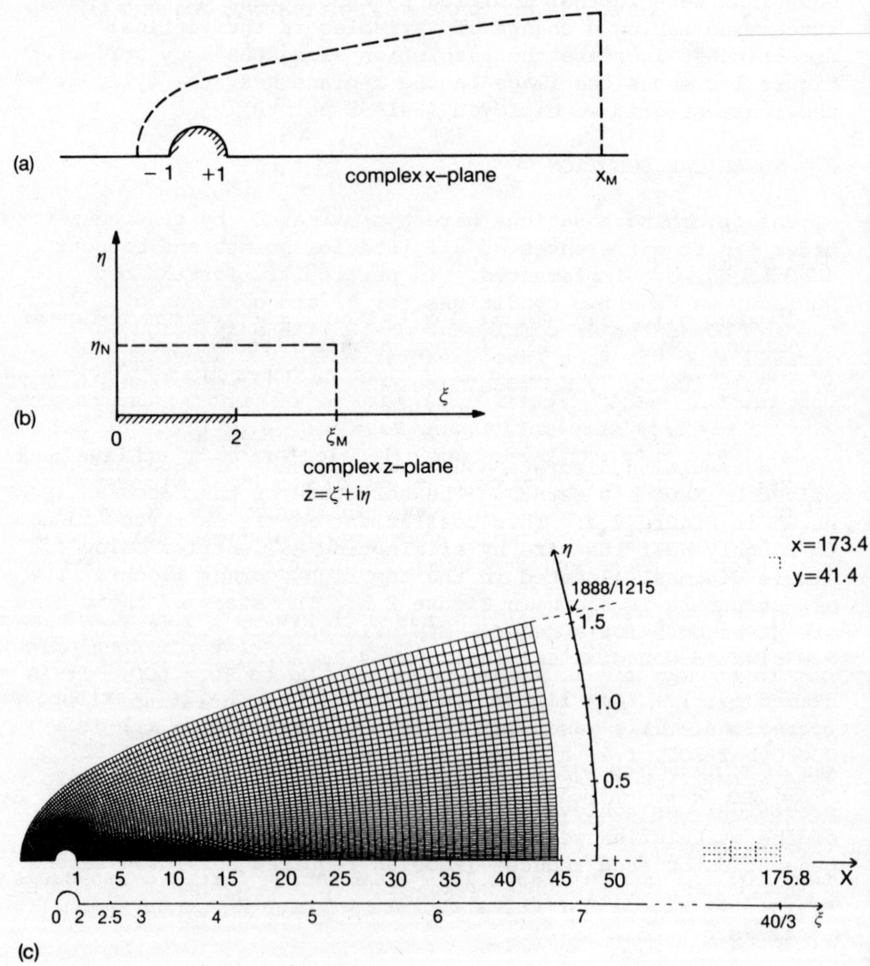

Fig. 1 a. Region in X-plane to be mapped to a rectangle.
b. Corresponding region in the Z-plane.
c. Image near the cylinder of a 241*49 grid.

Near the cylinder, all the vorticity is concentrated within a region like the one sketched in Figure 1 a. The conformal mapping $Z = X^{\frac{1}{2}} + X^{-\frac{1}{2}}$ maps such regions in the physical $X = x + iy$ plane to rectangles in the Z-plane (Figure 1 b). In the new coordinates, the governing equations remain unchanged apart from ω being replaced by $\omega/|dZ/dX|^2$ in (2.1). These equations were further modified by subtracting out the free stream and making a change of variables in the vertical direction to increase the resolution along the body surface. Figure 1 c shows the image in the X-plane near the cylinder of the coarsest grid we employed (241*49 points).

3. NUMERICAL SOLUTION

The governing equations were approximated by centred second order finite differences at all interior points and boundary conditions were implemented. In particular, for the body surface, we have two conditions for Ψ (and none for ω). Along the top boundary, we complement $\omega = 0$ with a relation on Ψ derived from the fact that Ψ satisfies $\Delta\Psi = 0$ above the boundary with only the decaying modes present (for details, see Fornberg, 1985).

The resulting discrete non-linear systems of equations were solved by Newton's method. The structure of the Jacobian is shown in Figure 2 a. This coefficient matrix is first reduced to roughly half its size by eliminating all entries below the single diagonal (located in the top right corner block). The new structure is shown in Figure 2 b. The sizes of the blocks are given both for a general M*N grid and for our 541*109 grid. Row pivoted Gaussian elimination was employed (first for the banded part). Each linear system requires 5.1×10^{10} arithmetic operations. This was found to take 403 seconds on a 2-pipe CDC Cyber 205, i.e. a sustained average performance of 127 Mflops (Million floating point operations per second, 64 bit precision) was achieved. The times required for other parts of the calculation were negligible (for example the generation of each reduced Jacobian required only 0.04 second).

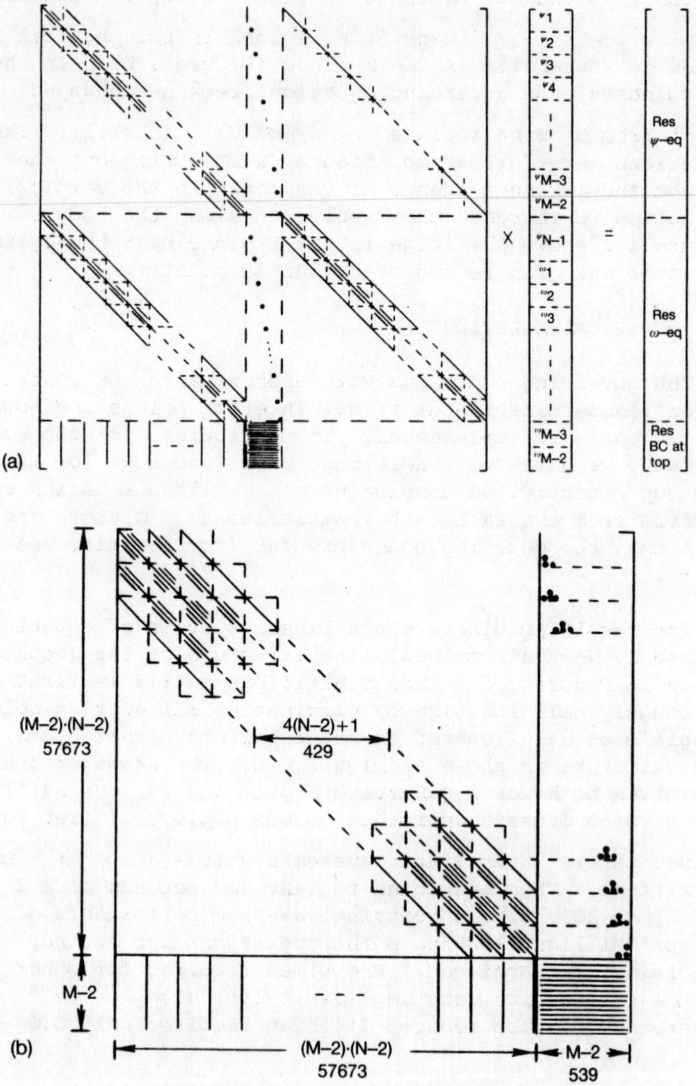

Fig. 2 a. Linear system in Newton's method.
b. Structure of the reduced Jacobian.

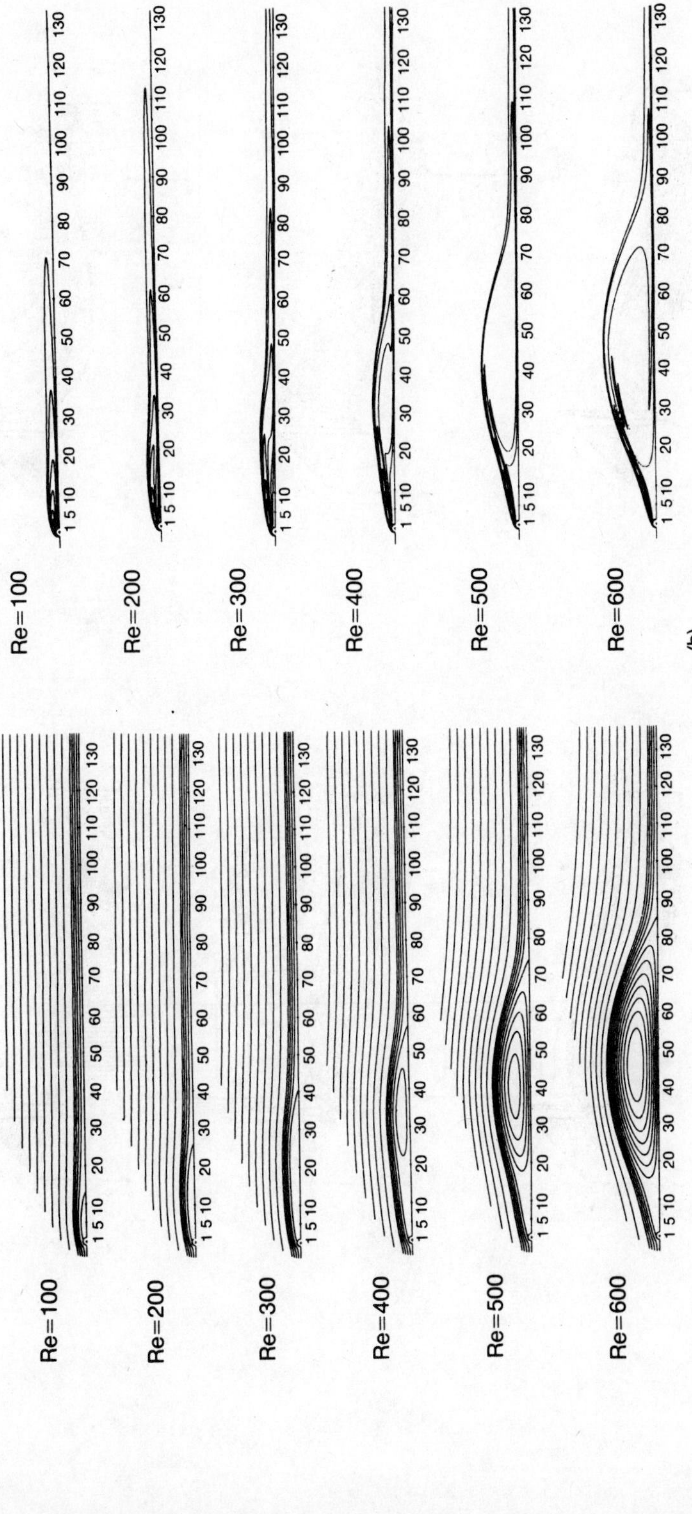

Fig. 3 a. Contour lines for the streamfunction.
b. Contour lines for the vorticity.

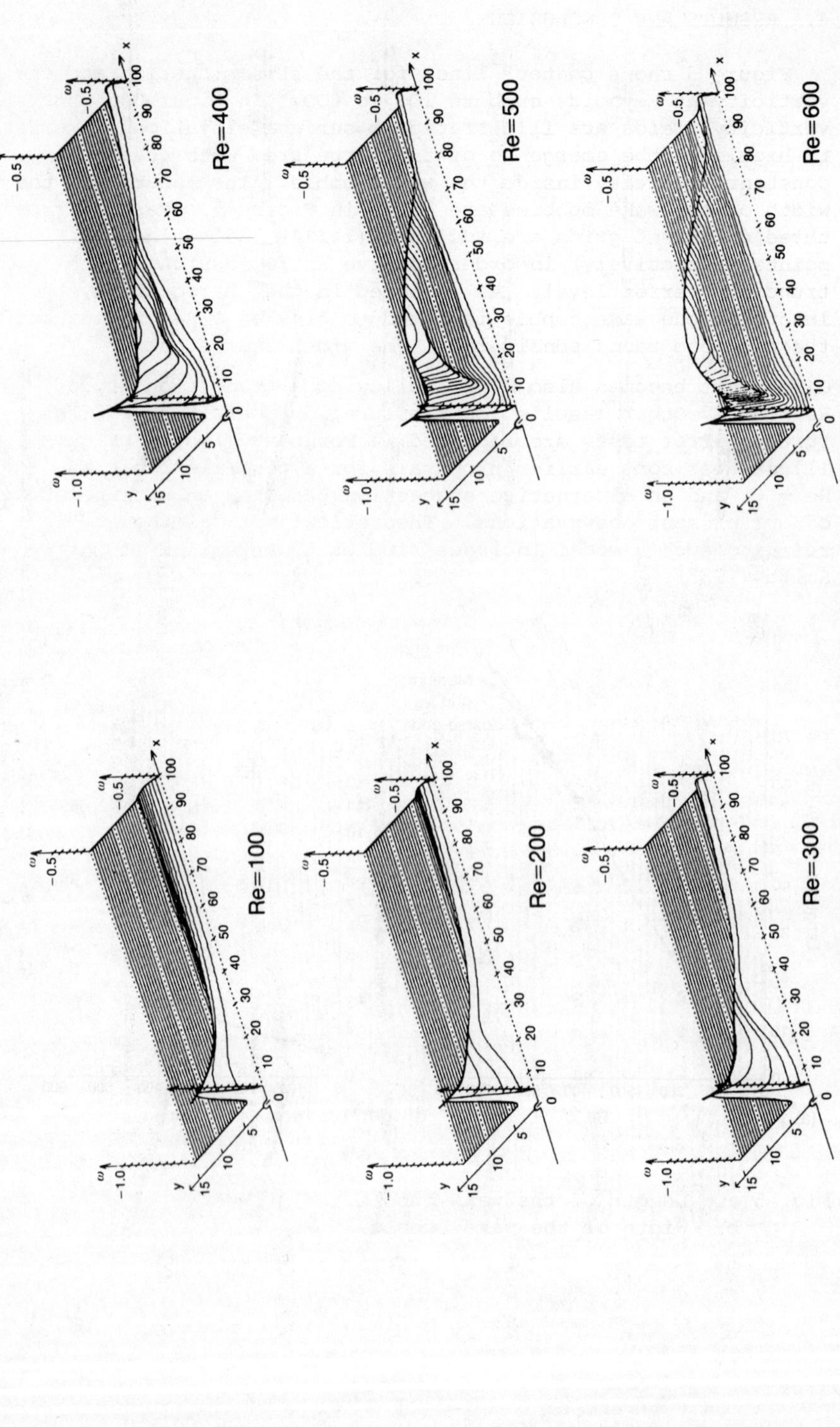

Fig. 4. The vorticity fields downstream from the cylinder.

4. RESULTS AND CONCLUSIONS

Figure 3 shows contour lines for the streamfunction and the vorticity at Reynolds numbers 100 to 600. In Figure 4, the vorticity fields are illustrated as surface elevations in order to highlight the emergence of the large area with nearly constant vorticity inside the wake bubble. The length and the width of the wake bubbles are shown in Figure 5. Results from three different grids are included (241*49, 361*73 and 541*109 points respectively) in order to give an impression of the truncation error level. As we noted in the introduction, the length of the wake bubble is found to grow as O(Re) throughout the complete range considered. The width starts out as $O(Re^{\frac{1}{2}})$ but becomes also O(Re) following a transition around Re = 300. Other results (on pressure, drag etc.) as well as further error tests are reported in Fornberg (1985). Figure 6 illustrates some earlier proposals for a possible limit as $Re \to \infty$ and an alternative suggestion based on an extrapolation of our present observations. Theoretical work aiming at refining such a model includes studies by Peregrine (1985) and Smith (1985).

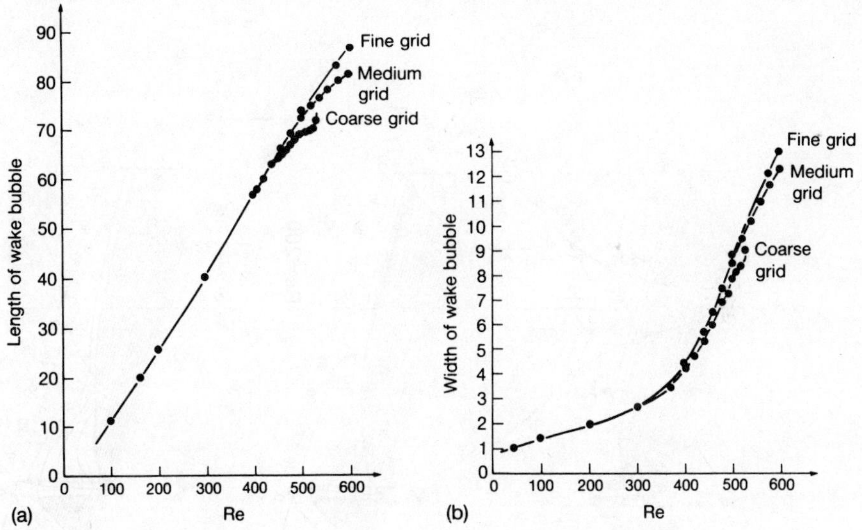

Fig. 5 a. Length of the wake bubble.
b. Width of the wake bubble.

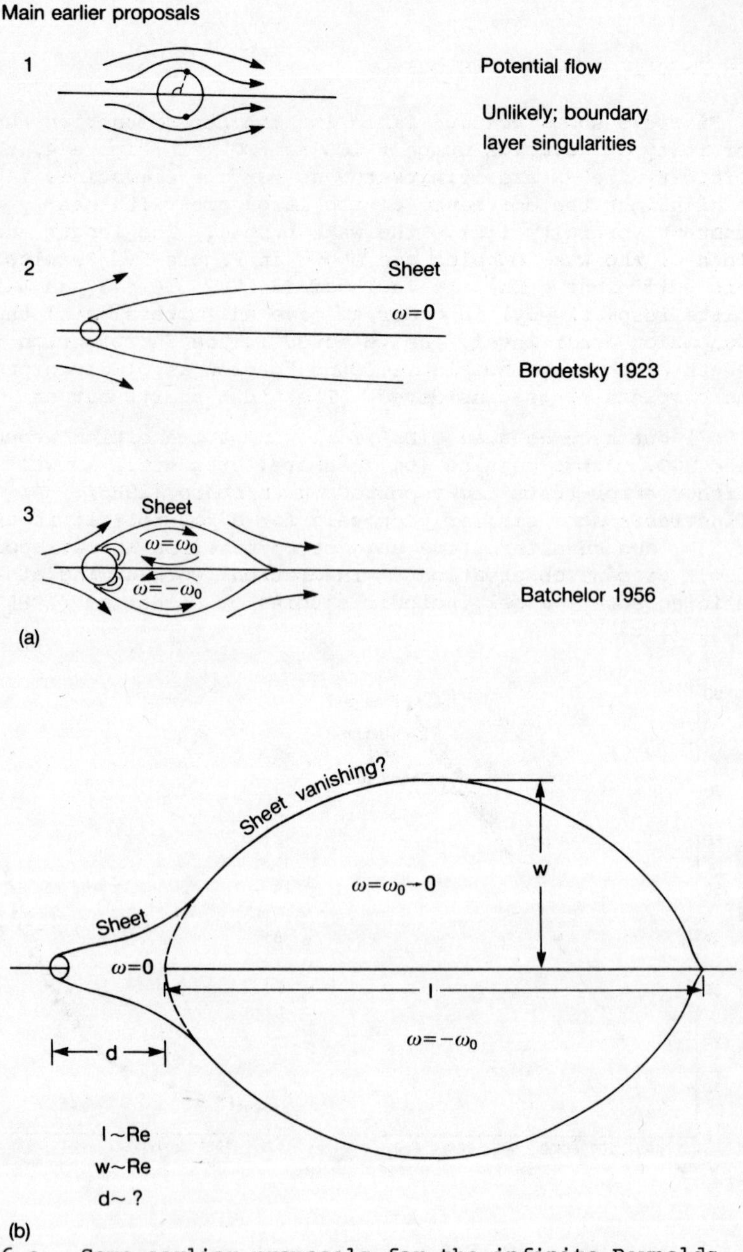

Fig. 6 a. Some earlier proposals for the infinite Reynolds number limit.

b. Possible structure at high Reynolds numbers based on trends observed in present work.

5. ACKNOWLEDGEMENTS

Part of this study was performed while the author was working at the California Institute of Technology. It was then supported by Control Data Corporation, Department of Energy (Office of Basic Energy Sciences) and the John Simon Guggenheim Memorial Foundation. Figures 1-5 are reproduced from Fornberg (1985) with permission from Journal of Computational Physics (Academic Press Inc.).

6. REFERENCES

Batchelor, G.K., (1956). A proposal concerning laminar wakes behind bluff bodies at large Reynolds number. *J. Fluid Mech.* **1**, 388-398.

Brodetsky, S., (1923). Discontinuous fluid motion past circular and elliptic cylinders. *Proc. Roy. Soc.* **A102**, 542-553.

Fornberg, B., (1980). A numerical study of steady viscous flow past a circular cylinder. *J. Fluid Mech.* **98**, 819-855.

Fornberg, B., (1985). Steady viscous flow past a circular cylinder up to Reynolds number 600'. To appear in *J. Comp. Phys.*

Grove, A.S., Shair, F.H., Petersen, E.E. and Acrivos, A., (1964). An experimental investigation of the steady separated flow past a circular cylinder. *J. Fluid Mech.* **19**, 60-80 (with 5 plates).

Peregrine, D.H., (1985). A note on the steady high-Reynolds-number flow about a circular cylinder. *J. Fluid Mech.* **157**, 493-500.

Smith, F.T., (1979). Laminar flow of an incompressible fluid past a bluff body: the separation, reattachment, eddy properties and drag. *J. Fluid Mech.* **92**, 171-205.

Smith, F.T., (1985). A structure for laminar flow past a bluff body at high Reynolds number. *J. Fluid Mech.* **155**, 175-191.

ON THE NUMERICAL SOLUTION OF 3-D INCOMPRESSIBLE FLOW PROBLEMS

N.S. Wilkes, C.P. Thompson, J.R. Kightley, I.P. Jones and A.D. Burns

(AERE, Harwell)

1. INTRODUCTION

This paper discusses some aspects of the numerical solution of three-dimensional incompressible flow problems using finite difference methods. These problems have traditionally been very demanding in computer resources, but developments in hardware, algorithms and software mean that they can now be solved on a routine basis using realistic grids. It is important, however, to take full advantage of these developments where possible. For example, in order to obtain significant benefits from a vector processor such as a CRAY-1 it is important to structure the software to enable vectorisation to occur and to minimise recursive work. New algorithms, based upon preconditioned conjugate gradients, for the efficient solution of systems of linear equations have been adopted in many fields, particularly in oil reservoir simulation. These methods offer the potential for being reasonably efficient on both scalar and vector processors. They have not, however, been widely used for the solution of turbulent flow problems. Other algorithmic developments have occurred in the treatment of velocity-pressure coupling using the primitive variable formulation of the Navier-Stokes equations. These points are discussed in this paper and illustrated by examples taken from the prediction of a realistic three-dimensional flow.

2. BASIC ALGORITHM

Most finite difference algorithms used for solving the incompressible Navier-Stokes equations in primitive variable formulation are based upon the SIMPLE algorithm, see Patankar (1980). The differential equations are first linearised and then discretised by integration over mass control volumes in the usual way. The algorithm then proceeds as follows:-

1) Guess the pressure field.

2) Solve the momentum equations for the velocity field.

3) Correct the velocity so that it satisfies the mass conservation equation by solving a pressure-correction equation, and then update the pressure field.

4) Solve the transport equations for any additional scalars, for example turbulence quantities and temperature.

(2.1)

A number of different treatments of step 3 exist, for example the SIMPLE algorithm, and its variants SIMPLER (Patankar (1980)) and PISO (Issa (1983)).

In a steady state calculation, one returns to step 2 with the latest estimate of the pressure field and cycles through steps 2 to 4 until convergence is reached. Because of the non-linearity of the system, coupling between dependent variables may be more important than a highly accurate solution for each individual transport equation. It is important, therefore, to do the minimum amount of work in the inner iterations which will not impair the convergence of the outer iterations. The robustness of this method, however, will be increased by having converged inner iterations, although this may be more expensive in computer time. Experience has shown that the pressure-correction equation should be solved more accurately at every stage in order to ensure that mass continuity is preserved in the fluid. This is important from the point of view of the robustness of the code, as well as providing the best rate of convergence of the outer iterations.

Iteration is also used at each time step of a transient problem when implicit methods are used. The PISO algorithm for the velocity-pressure coupling, when used in transient problems, is a semi-implicit non-iterative method. For this algorithm, the correction for mass continuity is such that one cycle through (2.1) yields errors in the variables which are proportional to the same power of the time step as time truncation errors; see Issa (1983). Therefore, given adequate convergence of the individual transport equations, it is no longer necessary to iterate around (2.1), but more work must be done in the linear algebra of that one cycle.

The calculation may be divided into two phases:-

a) the calculation of matrix coefficients, and

b) the solution of individual discretised transport equations.

If the domain of computation has a regular structure, which can be achieved by padding out and including dummy equations at the location of internal solid regions, a high degree of vectorisation may be obtained for the calculation of the matrix coefficients. Boundary conditions can be vectorised with care, including those for logarithmic boundary layers. Hybrid differencing may also be coded in a vectorisable form.

Hence 100% of the coefficient calculation has been written as vectorised code, resulting in a ratio of overall execution times in scalar and vector modes of between 3:1 and 4:1 on a CRAY-1. Traditionally, somewhere between 30 and 80% of the CPU time spent in solving these problems has been spent in doing the linear algebra. Because of the high degree of vectorisation of the coefficient calculation, the linear algebra will take a disproportionate part of the computations unless very efficient techniques are used.

In the next sections, therefore, we examine efficient methods for solving the linear equations which may exploit their particular form and will vectorise effectively. Many of these algorithmic developments are highly effective on scalar computers as well. Since, as has already been seen, a large proportion of the work is devoted to solving the pressure-correction equation, its efficient solution is of particular importance and so will be considered separately.

3. PRESSURE-CORRECTION EQUATION

It is usual practice to ensure that discretisation of each of the transport equations results in a coefficient matrix which is diagonally dominant and hence positive definite. This is not quite true of the pressure-correction equation in an incompressible calculation, since the matrix is singular, corresponding to the freedom to add an arbitrary constant to the pressure. Another unusual feature of this particular equation is that its coefficient matrix is symmetric, which arises because the differential equation of which it is a discrete analogue contains no first order derivative terms. Such a symmetric positive semi-definite system is ideal for solution by the conjugate gradient (CG) method. This section describes the experience gained with the implementation of this method in some three-dimensional flow codes.

Since discretisations of partial differential equations typically yield matrices with a poor eigenvalue structure (evenly spread over a wide range), the matrix must be pre-conditioned. This is equivalent to solving the original system by means of the Preconditioned CG algorithm (PCG), which resembles classical CG apart from the additional calculation of $M^{-1}\underline{\phi}$, (see Concus, Golub and O'Leary (1976)). M is the preconditioning matrix, which is in some sense 'close' to the coefficient matrix, A. The PCG algorithm will vectorise straight-forwardly, except perhaps for this preconditioning phase. The correct choice of preconditioning is therefore important. A balance is sought in the amount of recursive work entailed by this phase of the calculation. A complicated but accurate preconditioning will make each iteration more costly, yet may give a better convergence rate so fewer iterations are required.

Those preconditionings which have been employed include diagonal scaling, incomplete Cholesky factorisation (the ICCG algorithm), truncated incomplete Cholesky introduced by van der Vorst (1982), and the Nested Factorisation form of block pre-conditioning of Appleyard et al. (1981, 1983).

Detailed testing of these algorithms on realistic flow problems has shown that there is no single best algorithm for all problems on all computers, which is one good reason for maintaining a range of preconditionings. Indeed, a powerful feature of the PCG method is its flexibility. ICCG is found to be most useful, in that it is efficient on both scalar and vector machines, reliable on the test problems so far studied, and can handle the singularity of the system provided a 'column sum constraint' is not enforced.

One consideration in selecting the right preconditioning to use is the amount of storage available. On a computer such as a CRAY-1 which does not have virtual memory, this is a real concern when solving problems on meshes with over 40000 grid points (see section 5). The symmetry of the coefficient matrix can be exploited by storing, say, only the lower triangle of it. Enough workspace is then freed for the CG part of the algorithm. Each of the above preconditionings is economical on storage provided no fill-in is allowed in the incomplete factorisations.

An important feature of fluid flow calculations is that often we have no a priori knowledge of the relative sizes of terms in the matrix. Of the above-mentioned preconditionings, truncated incomplete Cholesky and Nested Factorisation are dependent on the order in which dimensions are scanned, which determines which terms lie close to the diagonal of the matrix and which are far away. Examples are reported by Kightley and Jones (1985) of problems which will converge using such anisotropic preconditionings with one ordering of the unknowns but diverge for another ordering. For this reason, an isotropic preconditioning such as incomplete Cholesky is preferred.

Details of all these algorithms and a comparison of their relative performance on model problems and on matrices arising from realistic flow problems may be found in Kightley and Jones (1985).

These PCG solvers have been implemented for the pressure-correction equation in an existing 3-D turbulent flow code. The cost of transient flow predictions has been reduced substantially - by up to 25% of the cpu time - on both a CRAY-1 and scalar computers, as compared with Stone's Strongly Implicit Procedure (SIP) with a fixed parameter, which was the method previously used. Table 1 shows the number of iterations of SIP and ICCG required to reduce the residual norm in the pressure-correction equation to a preset level, for each of the two pressure-correction steps of the PISO algorithm. The first few time steps of a typical turbulent flow in cylindrical geometry are illustrated. The work of ICCG is only very slightly greater than that of SIP, so relative cpu times can easily be deduced from the figures. Greater reliability was also obtained, as demonstrated by smaller deviations in the numbers of PCG iterations required. The numbers of iterations required for SIP could be reduced by using a cyclically varying parameter. The amount of work per iteration would increase significantly however, since the matrix has to be re-factorised. The best choice of parameter to use is also not known.

4. OTHER TRANSPORT EQUATIONS

Efficient solution of the equations for transport of all the other dependent variables is often less critical to the overall performance of a flow prediction code, because less accuracy is required in the solution due to the semi-implicit treatment of non-linear terms. Greater diagonal dominance, enhanced by under-relaxation, also makes the solution correspondingly easier. Standard solution methods are point- or line-relaxation schemes, and incomplete factorisation techniques such as Stone's method and ADI. Latterly, some of these have been coupled with multigrid acceleration. Line relaxation is widely used to good effect in 2-D calculations and can be applied in 3-D also. Assembler-coding of the tridiagonal solver on the CRAY-1 can halve the cpu time in this essentially

Table 1

Iterations on first and second pressure-correction steps of PISO algorithm. Grid size is 16×16×16.

Time step	SIP 1st step	SIP 2nd step	PCG 1st step	PCG 2nd step
1	55	43	18	15
2	66	45	24	15
3	24	8	14	12
4	16	7	17	7
5	33	7	15	6
6	12	7	16	11
7	23	10	16	12
8	34	18	17	14
9	32	19	17	15
10	36	25	17	15

scalar part of the computation, and has been found to be quite effective on many problems.

A number of adaptations of CG techniques to the unsymmetric case have been developed. These include solving the Normal equations $(A^T A)\underline{x} = A^T\underline{b}$, the bi-CG algorithm, and the Minimal Residual algorithm, see Appleyard et al. (1981 , 1983). A recent innovation is the CG-squared algorithm of Sonneveld, Wesseling and de Zeeuw (1983), communicated by Wesseling (1985). This has approximately the same cost per iteration as bi-CG and up to twice the rate of convergence. In view of the diagonal dominance of the systems to be solved by this method, diagonal scaling of the matrix may be sufficient pre-conditioning in many cases. Vectorisation is then very good, with vector lengths of the order of the number of unknowns which enhances efficiency and is necessary on some vector computers. It also is economical in storage, taking the same amount as bi-CG. This method has been implemented and some promising results have been obtained. Further investigations are in progress and will be reported in detail later by Kightley (1985).

5. ILLUSTRATIVE EXAMPLE

The methods described in the foregoing sections have been implemented into the code FLOW3D (Jones et al. (1985)) and tested in predictions of the flow in the NTH experiment (Hjertager and Magnussen (1981a)) depicted in Fig. 1. This consists of a narrow free air jet, issuing from a square duct into a rectangular enclosure, and leaving through a symmetrically placed square outlet of twice the linear dimensions of the inlet. Two symmetry planes were exploited, so that it was only necessary to perform computations in a quarter of the flow domain. This is an easy operation to specify with FLOW3D. The experiment was simulated on several different grids ranging from 9×9×9 to 31×31×42 in order to test the sensitivity of the results to the number of grid points.

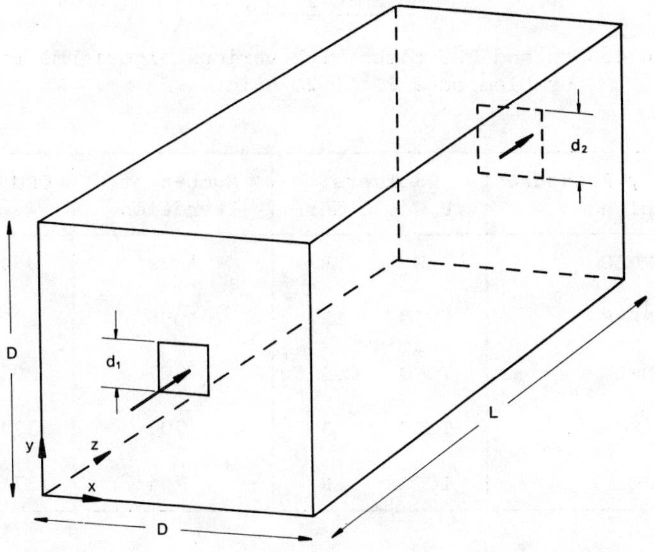

$D = 100\text{mm}$ $\frac{L}{D} = 3.0$ Two symmetry planes

$\frac{d_1}{D} = 0.1$ $\frac{d_2}{D} = 0.2$ Inlet velocity = 18.5ms

Fig. 1 Geometrical configuration of NTH Experiment

Details of a number of predictions of the flow in this experiment on a 20×20×20 grid are given in Table 2. The numbers of iterations and corresponding CPU times to achieve convergence are shown for different combinations of the solver methods and the velocity-pressure coupling. The convergence criterion used for these results is for the mass source residual to fall below a fixed preset level, which roughly corresponds to the point at which all variables agree in their first three significant figures with the answer which would be obtained by continuing the calculation to a much lower level of the residual. As this is a steady state calculation, a tight tolerance on the residual of each linear equation is not needed, so instead a fixed algorithm was used. The pressure-correction equation was solved using 4 iterations of ICCG or 6 sweeps of line relaxation with the orientation of the lines alternating between the two directions perpendicular to the main flow direction. On the unsymmetric equations, 1 sweep of line relaxation (LR) or 1 iteration of CG-squared (CGS) was performed. The work is roughly equivalent in each case.

Table 2

Iteration counts and CPU times with various algorithms for NTH problem on a 20×20×20 grid.

Velocity-Pressure coupling	Solvers P-C	Solvers Other	Number of Iterations	CPU Time (sec)
SIMPLE	LR	LR	1189	334.6
SIMPLE	ICCG	LR	1095	329.8
SIMPLE	ICCG	CGS	947	285.6
PISO	ICCG	LR	796	324.0
PISO(r)	ICCG	LR	794	371.3

A comparison is also made between the SIMPLE and PISO algorithms for solving the velocity-pressure coupling. The PISO algorithm requires the coefficients from all the momentum equations for the second of its two pressure-correction steps. This is extremely demanding on storage. For large problems it is not feasible to hold these coefficients in memory, so they must be held out of core. As I/O may incur a cost over and above that of CPU usage, an alternative strategy is to re-calculate the coefficients using the latest available estimate to the velocities at every stage. The results with out of core working are labelled PISO, and with re-calculated coefficients PISO(r). This re-calculation is strictly a modification of the true PISO algorithm, but will be a small change if the coefficients are not rapidly varying during the course of the iteration, or for transient problems, in time.

The differences in CPU times are not large because a comparatively small amount of work is performed in the linear algebra on this steady state calculation. On other problems that are not reported here, the difference between velocity-pressure coupling algorithms has been more marked. The improved mass conservation obtained by use of the PISO algorithm permits an increase in under-relaxation factors to be made, which may result in a large reduction in the number of cycles through the dependent variables. As well as a possible reduction in CPU time, this has the effect of increasing robustness. On this particular problem the choice of velocity-pressure coupling algorithm is not so important, since the rate of convergence of the turbulence quantities restricts the overall convergence rate of the coupled system. A two-stage correction of k and ε recommended by Issa (1983) has not been used here, but may relax this restriction.

In Fig. 2, predictions are plotted against experiment for the centre-line values of the mean axial velocity W, and the isotropic turbulence velocity $w' = \sqrt{(2k/3)}$. The grids were of dimensions 20×20×27 and 31×31×42, the inlet being resolved by 16 and 64 mass control volumes respectively. The grids were non-uniform along the z-axis, with fine resolution near the inlet, and a fully developed profile was inserted a short distance upstream of the inlet, as depicted in Fig. 3. There are noticeable discrepancies between the predicted and the experimental results; W decreases too rapidly (i.e. the rate of spread of the jet is over-predicted), and the predicted rise in k occurs further upstream than is observed experimentally. Whilst it cannot be completely discounted, numerical diffusion is unlikely to be responsible for the disagreement between the predictions and the experiment because of the agreement between the numerical results with different grid sizes. It does appear therefore that the standard k-ε model is over-predicting the rate of spread of the jet.

In order to compensate for this over-expansion, Hjertager and Magnussen (1981b) extended the potential core of the jet downsteam; inlet conditions were moved forward a distance consistent with experimental findings. They claimed that it

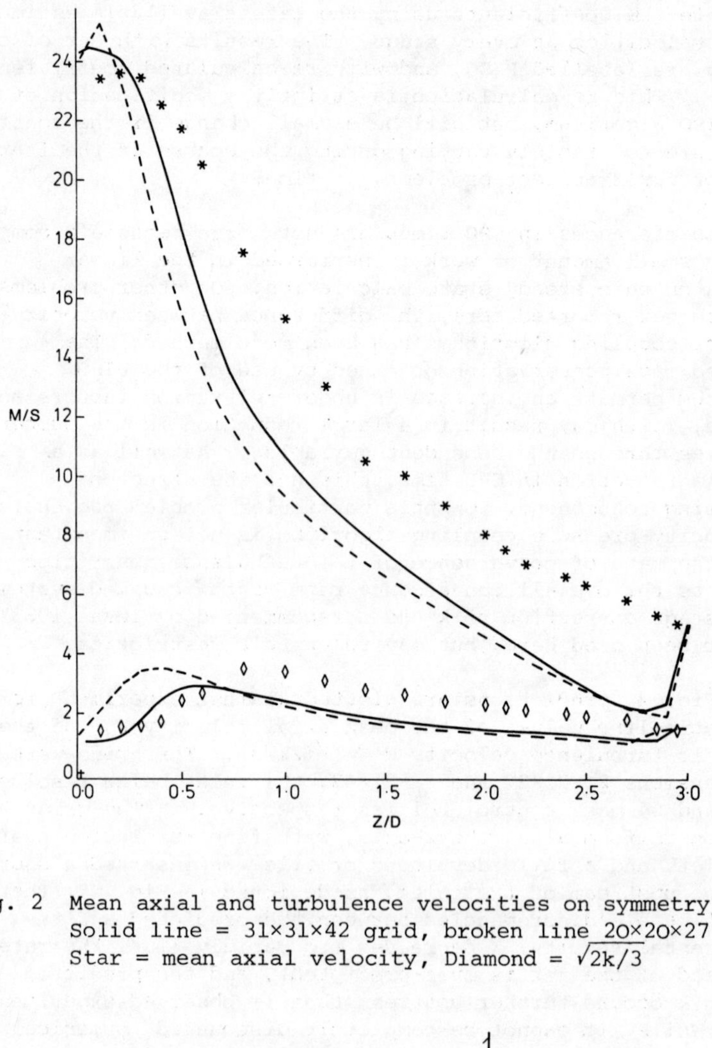

Fig. 2 Mean axial and turbulence velocities on symmetry axis. Solid line = 31×31×42 grid, broken line 20×20×27 grid. Star = mean axial velocity, Diamond = $\sqrt{2k/3}$

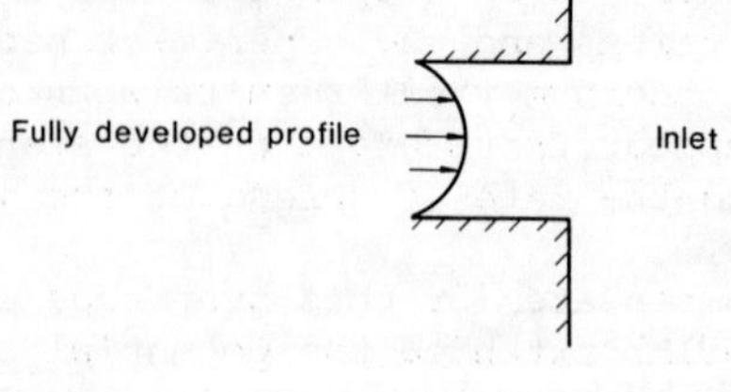

Fig. 3 Inlet boundary conditions without lip

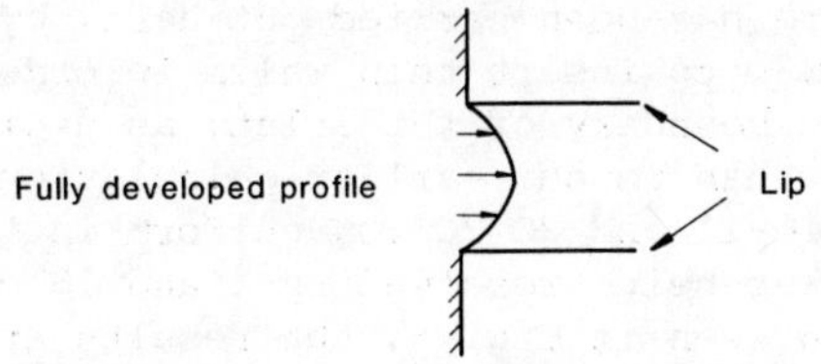

Fig. 4 Inlet boundary conditions with lip

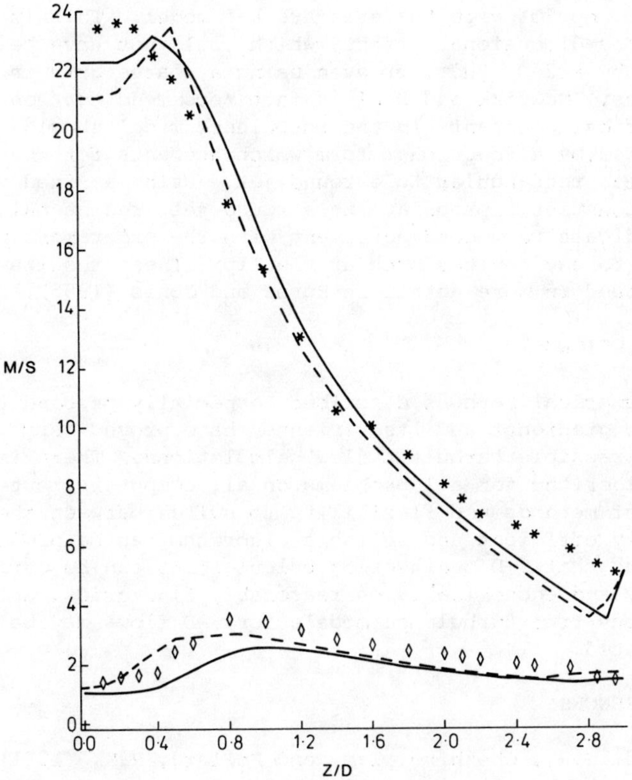

Fig. 5 Mean axial and turbulence velocities on symmetry axis, with modified inlet.
Solid line = 20×20×20 grid. Broken line = 16×16×11 grid. Star = mean axial velocity, Diamond = $\sqrt{2k/3}$.

was necessary because of the coarseness of their 9×9×9 grid. A similar procedure has been carried out here, by using the facilities of FLOW3D to insert thin walls in order to attach a small lip to the boundary of the inlet, as depicted in Fig. 4. Coarser grids than in our earlier calculations were used in this case (16×16×11 and 20×20×20, uniform in the main flow direction, the inlet being resolved by 4 and 16 control volumes respectively). As seen in Fig. 5, the results are closer to the experiment than the previous results on finer grids. Since difficulties have been encountered here on finer grids without the lip, the problem is more likely to be due to inadequacies in the turbulence model. Indeed, as noted by McGuirk and Rodi (1979), a three-dimensional square jet very quickly becomes round, and the rate of spread of a round jet is over-predicted by 40% with the standard k-ε model. This is a truly three-dimensional effect, which could not have been detected by a 2-D study, or even using a coarse grid in three dimensions. McGuirk and Rodi in fact recommend that one of the empirical constants in the turbulence model should be represented by a functional form which accounts for the transition from a rectangular to a round jet. Using a fixed value of this constant appropriate to a round jet, recent calculations indicate very good agreement with the experiment without recourse to any devices such as the lip. These results will be discussed in more detail in Burns and Jones (1985).

6. CONCLUSIONS

The numerical methods described, especially preconditioned conjugate gradients and its variants, have proved very useful in incompressible turbulent flow calculations. There is no 'best' algorithm for all problems on all computers, but with a range of methods and flexibility to switch between them a reasonably efficient and reliable algorithm can be produced. This means that 3-D engineering calculations can be performed routinely and economically on reasonably fine grids, and predictions from turbulence models for 3-D flows may be investigated.

7. REFERENCES

Appleyard, J.R., Cheshire, I.M. and Pollard, R.K. (1981) Special techniques for fully-implicit simulators. AERE-CSS 112.

Appleyard, J.R. and Cheshire, I.M. (1983) Nested factorisation. SPE 12264, Proceedings of Seventh SPE Symposium on Reservoir Simulation, 315-324.

Burns, A.D. and Jones, I.P. (1985) In preparation.

Concus, P., Golub, G.H. and O'Leary, D.P. (1976) A generalised conjugate gradient method for the numerical solution of elliptic partial differential equations. In "Sparse Matrix Communications", (J.R. Bunch and D.J. Rose, eds.), Academic Press, New York and London, 309-332.

Hjertager, B.H. and Magnussen, B.F. (1981a) Laser Doppler velocity measurement in the three dimensional flow of a jet in a square enclosure. Letters in *Heat Mass Transfer*, **8**, 171-186.

Hjertager, B.H. and Magnussen, B.F. (1981b) Calculation of turbulent three dimensional jet induced flow in rectangular enclosures. *Computers and Fluids*, **9**, 395-407.

Issa, R.I. (1983) Solution of the implicitly discretised fluid flow equations by operator splitting. Report FS/82/15 Imperial College, London.

Jones, I.P., Kightley, J.R., Thompson, C.P. and Wilkes, N.S. (1985) FLOW3D, A computer code for the prediction of laminar and turbulent flow and heat transfer: Release 1. To be presented at HTFS Research Symposium, Sept. 1985.

Kightley, J.R. (1985) In preparation.

Kightley, J.R. and Jones, I.P. (1985) A comparison of conjugate gradient preconditionings for three-dimensional problems on a CRAY-1. To appear in *Comp. Phys. Comm.* (AERE pre-print CSS 162).

McGuirk, J.J. and Rodi, W. (1979) The calculation of three dimensional turbulent free jets. Turbulent Shear Flows I, (F. Durst et al., eds), Springer.

Patankar, S.V. (1980) Numerical heat transfer and fluid flow. Hemisphere.

Sonneveld, P., Wesseling, P. and de Zeeuw, P.M. (1983) Multigrid and conjugate gradient methods as convergence acceleration techniques. Proc. Multigrid Conference, Bristol, to appear.

van der Vorst, H.A. (1982) A vectorisable variant of some ICCG methods. *SIAM J. Sci. Stat. Comput.*, **3**, 350-356.

Wessling, P. (1985) Multigrid and conjugate gradient methods as convergence acceleration techniques. These proceedings.

HIGH RAYLEIGH NUMBER SOLUTIONS OF THE BUOYANCY DRIVEN CAVITY PROBLEM

J.S. Rollett and S. Sivaloganathan
(Oxford University Computing Laboratory)

1. INTRODUCTION

The practical importance of fluids to transport or remove heat needs little illustration. It ranges from the circulation of coolant through a nuclear reactor to the mounting of a power transistor on a block with cooling fins. Equally apparent is the importance of thermal processes in meteorology and various other branches of geophysics.

We shall be concerned with flows in which temperature variations are introduced by some mechanism independent of the flow dynamics; in particular we shall not consider flow induced temperature variations arising from adiabatic expansion or compression or from viscous dissipation.

Temperature variations within a convective flow give rise to variations in fluid properties (e.g. density and viscosity). An analysis including the full effects is still too complicated and some approximation is essential. We use the quations in a form known as the Boussinesq approximations, where all density variations are neglected except in so far as they give rise to a buoyancy force term (see Tritton (1977)).

In section 2.1, we detail the governing equations and the finite element discretisation used. In section 2.2, we consider the buoyancy driven cavity test problem (see Jones and de Vahl Davis (1983)), we describe the general features of the flow, and note that as the Rayleigh number is increased, the numerical solution of the problem becomes increasingly difficult. Many of the numerical problems encountered are associated with the existence of singular solutions and in section 2.3, we present an efficient continuation procedure for solution at high Rayleigh numbers and present some brief results for the velocity vector fields at various Rayleigh numbers. Finally, we conclude in section 3 with some comments and appraisal.

2. THE BUOYANCY DRIVEN CAVITY TEST PROBLEM

2.1 In the Boussinesq approximations, the equations expressing conservation of momentum, mass and energy are as follows - here and throughout the summation convention for repeated indices has been used ($1 \leq n, m \leq 2$):-

$$u_m \frac{\partial u_n}{\partial x_m} - \frac{\partial \sigma_{nm}}{\partial x_m} - (Ra)(Pr)T\delta_{n2} = 0 \qquad \underline{x} \in \Omega \tag{2.1}$$

$$\frac{\partial u_m}{\partial x_m} = 0 \qquad \underline{x} \in \Omega \tag{2.2}$$

and $$u_m \frac{\partial T}{\partial x_m} - \frac{\partial \tau_m}{\partial x_m} = 0 \qquad \underline{x} \in \Omega, \tag{2.3}$$

where σ_{nm} is the modified viscous stress tensor given by:

$$\sigma_{nm} = -p\delta_{nm} + (Pr)\left(\frac{\partial u_n}{\partial x_m} + \frac{\partial u_m}{\partial x_n}\right) \tag{2.4}$$

$$\tau_m = \partial T/_{\partial x_m} \tag{2.5}$$

and δ_{nm} is the Kronecker delta. The non-dimensionalisation of Mallinson and de Vahl Davis (1977) has been adopted. Hence u_n, T and P are dimensionless velocities, temperature and pressure, $Pr(= \nu/k)$ is the Prandtl number and $Ra(= g\alpha D^3(T_h^* - T_c^*)/k\nu)$ is the Rayleigh number, where ν is the kinematic viscosity, k the diffusity, α the coefficient of volumetric expansion and g the gravitational acceleration. D is a characteristic length and T_h^*, T_c^* are characteristic hot and cold temperatures.

Buoyancy flows are difficult to treat theoretically because of the coupling between the momentum and energy equations. The velocity field is governed by the temperature field but the temperature field depends through the advection of heat on the velocity field. It is impossible to determine one independently of the other as can be done in forced convection.

The above equation system is now completed by a set of boundary conditions. These can be fairly general in nature, but for our purposes they are of the form:

$$u_n = 0 \qquad \text{on } \partial\Omega = \partial\Omega_1 \cup \partial\Omega_2 \tag{2.6}$$

$$T = \hat{T} \qquad \text{on } \partial\Omega_1 \tag{2.7}$$

$$\partial T/\partial n = 0 \qquad \text{on } \partial\Omega_2. \tag{2.8}$$

Clearly a wide range of dynamical behaviours can be expected depending on the relative importance of the buoyancy force relative to the other terms. We consider one extreme case, where the fluid would be at rest in the absence of temperature variations, the buoyancy force being the sole cause of motion. This is known as free convection or natural convection.

To construct finite element approximations, we require a weak formulation. The procedure is described briefly and can be found in more detail in Hutton (1974), Mitchell and Wait (1977) and Oden (1972).

Let $S_u(\Omega)$ be the space of smooth vector valued functions vanishing on the boundary $\partial\Omega$, $S_p(\Omega)$ the space of smooth scalar functions defined on Ω and let $S_T(\Omega)$ be the space of the smooth scalar functions vanishing on part of the boundary $\partial\Omega_1$. Then we have the following equivalent integral formulation with solutions $\underline{u}$, p, T s.t $\underline{u} \in S_u(\Omega)$, $p \in S_p(\Omega)$ and $(T - \hat{T}) \in S_T(\Omega)$ and

$$\int_\Omega u_m \frac{\partial u_n}{\partial x_m} v_n \, \underline{dx} + \int_\Omega \sigma_{nm} \frac{\partial v_n}{\partial x_m} \underline{dx} + (Ra)(Pr)\int_\Omega T v_n \, \underline{dx}\, \delta_{n2} = 0$$

$$\forall \, \underline{v} \in S_u(\Omega) \tag{2.9}$$

$$\int_\Omega \frac{\partial u_m}{\partial x_m} \cdot r \, \underline{dx} = 0 \qquad \forall\, r \in S_p(\Omega) \tag{2.10}$$

$$\int_\Omega u_m \frac{\partial T}{\partial x_m} \cdot w \, \underline{dx} - \int_\Omega \tau_m \frac{\partial w}{\partial x_m} \underline{dx} \quad \forall\, w \in S_T(\Omega). \tag{2.11}$$

Since no derivatives of pressure appear and at most first order derivatives of velocities and temperature, the restriction to smooth functions can be relaxed and $S_u(\Omega)$ completed to include vector functions vanishing on $\partial\Omega$ with square integrable first derivatives, likewise $S_T(\Omega)$ is completed to include scalar functions vanishing on part of the boundary $\partial\Omega_1$ with square integrable first derivatives and $S_p(\Omega)$ extended to include all square integrable scalar functions. The integral formulation is now the weak formulation of the differential problem.

Finite element approximations are now constructed in the standard manner (see Hutton (1974), Mitchell & Wait (1977)). The region Ω is covered with N finite elements Ω_e $(1 \le e \le N)$ with a total of M nodes, M_1 of which do not lie on the boundary $\partial\Omega$ and M_2 of which do not lie on the boundary $\partial\Omega_1$. At each node, a global velocity/temperature basis function $W_i(\underline{x})$ is constructed. For our purposes, the $W_i(\underline{x})$ are modified biquadratics of the serendipity class (see Zienkiewicz (1971)) which take the value 1 at the ith node and zero at all other nodes.

Each function $W_i(\underline{x})$, $1 \le i \le M_1$ is continuous in Ω and since $i > M_1$ for nodes on $\partial\Omega$, these vanish on $\partial\Omega$. Therefore these functions can be used to generate an M_1- dimensional subspace of $S_u(\Omega)$

$$S_u(\Omega) = \left\{ \underline{\tilde{u}} : \underline{\tilde{u}} = \sum_{i=1}^{M_1} \underline{\tilde{u}}_i \, W_i(\underline{x}) \right\} :$$

putting $\tilde{u}_{i,m} = 0$ for $M_1 < i \le M$, we can write

$$\underline{\tilde{u}} = \sum_{i=1}^{M} \underline{\tilde{u}}_i \, W_i \, (X). \tag{2.12}$$

Likewise, using the above argument, we can use $W_i(\underline{x})$, $1 \leq i \leq M_2$, to generate an M_2- dimensional subspace of $S_T(\Omega)$

$\tilde{S}_T(\Omega) = \left\{ \tilde{T} \ : \ \tilde{T} = \sum_{i=1}^{M_2} \tilde{T}_i \ W_i \ (\underline{x}) \right\}$ and the boundary condition of (2.8) is approximated by

$$\hat{T} = \sum_{i=M_2+1}^{M} \tilde{T}_i \ W_i(\underline{x})$$

where $\tilde{T}_i = \hat{T}_i$ for $M_2 < i \leq M$, and $\hat{T}_i$ is the value of $\hat{T}$ at node i) then:

$$\tilde{T} = \sum_{i=1}^{M} \tilde{T}_i \ W_i(\underline{x}) \ . \tag{2.13}$$

Due to certain theoretical considerations (see Girault and Raviart (1981)), the pressure approximations should be generated by polynomial basis functions of lower degree than those defining velocities. Thus a bilinear function $X_s(\underline{x})$ is constructed interpolating 1 at the s-th corner node and zero at all other corner nodes. The approximating space $\tilde{S}_p(\Omega)$ is then given by:

$$\tilde{S}_p(\Omega) = \{ \tilde{p} : \tilde{p} = \sum_{s=1}^{M} \tilde{p}_s X_s(\underline{x}) \} \ ,$$

where the suffix S ranges over only corners nodes

$$\tilde{p} = \sum_{s=1}^{M} \tilde{p}_s \ X_s(\underline{x}) . \tag{2.14}$$

Substituting (2.12), (2.13) and (2.14) into (2.9), (2.10) and (2.11) the following discrete systems of equations emerges, non-linear in the nodal values of velocity and temperature

$$f_{i,n} \equiv B_{ijk,m} \tilde{u}_{k,m} \tilde{u}_{j,n} + (Pr) A_{ij,mm} \tilde{u}_{j,n} + (Pr) A_{ij,mn} \tilde{u}_{j,m} - C_{si,n} \tilde{p}_s$$

$$- (Ra)(Pr) D_{ij,n} T_j = 0 \tag{2.15}$$

$$f_{i,3} \equiv \delta_{is} C_{si,m} \tilde{u}_{i,m} = 0 \tag{2.16}$$

$$f_{i,4} \equiv Bi_{kj,m}\tilde{u}_{j,m}\tilde{T}_k - A_{ij,mm}\tilde{T}_j = O \qquad (2.17)$$

where

$$A_{ij,mm} = \int_\Omega \frac{\partial w_i}{\partial x_n} \cdot \frac{\partial w_j}{\partial x_m} \underline{dx} \quad B_{ijk,m} = \int_\Omega w_i \frac{\partial w_j}{\partial x_m} w_k \underline{dx}$$

$$C_{si,m} = \int_\Omega X_s \frac{\partial w_i}{\partial x_m} \underline{dx} \ \& \ D_{ij,n} = \left\{\int_\Omega w_i w_j \underline{dx}\right\} \delta_{n_2} .$$

Applying Newton-Raphson linearisation, on the rth step we obtain:

$$J^{(r)}_{ij,nm} \Delta\tilde{u}^{(r)}_{j,m} - \delta_{sj} C_{si,n} \Delta\tilde{p}^{(r)} - (Ra)(Pr) D_{ij,n} \Delta\tilde{T}^{(r)}_j = - f^{(r)}_{i,n}$$

$$\delta_{is} C_{sj,m} \Delta\tilde{u}^{(r)}_{j,m} = - f^{(r)}_{i,3}$$

$$B_{ijk,m} \tilde{T}^{(r)}_k \Delta\tilde{u}^{(r)}_{j,m} + (B_{ikj,m}\tilde{u}^{(r)} - A_{ij,mm}) \Delta\tilde{T}^{(r)}_j = - f^{(r)}_{i,4}$$

$$\text{where } J_{ij,nm} = B_{ikj,m}\tilde{U}_{k,n} + (B_{ijk}\tilde{U}_{k,p} + (Pr)A_{ij,pp})\delta_{nm}$$

$$+ (Pr)A_{ij,mn} .$$

The (r+l) -th iterates for pressure, temperature and velocities are then obtained by:

$$\tilde{P}^{(r+l)}_s = \tilde{P}^{(r)}_s + \Delta\tilde{P}^{(r)}_s , \quad \tilde{T}^{(r+l)}_i = \tilde{T}^{(r)}_i + \Delta\tilde{T}^{(r)}_i ,$$

$$\tilde{u}^{(r+l)}_{i,n} = \tilde{u}^{(r)}_{i,n} + \Delta\tilde{u}^{(r)}_{i,n} .$$

The iterative sequence is started by initial approximations which can either be a creeping flow approxmation or a solution from a lower Rayleigh number.

The governing matrix of the system can be considered to be composed of an M x M set of sub-matrices of the following form

$$J^* = \begin{bmatrix} J_{ij}^{(r)} & -C_{ij} & F_{ij} \\ C_{ij}^{T} & 0 & 0 \\ G_{ij}^{T} & 0 & K_{ij} \end{bmatrix} \qquad (2.18)$$

where the zero block on the diagonal corresponds to the pressure variables. When using a Gaussian elimination strategy, pivoting is avoided by eliminating all velocity variables prior to pressure elimination. Fill-in-thus ensures a non-zero pressure pivot.

2.2 The buoyancy driven cavity problem is summarised in Fig. 1 together with a typical finite element grid used. The square cavity has differentially heated side walls (hot and cold) and both horizontal walls are adiabatic.

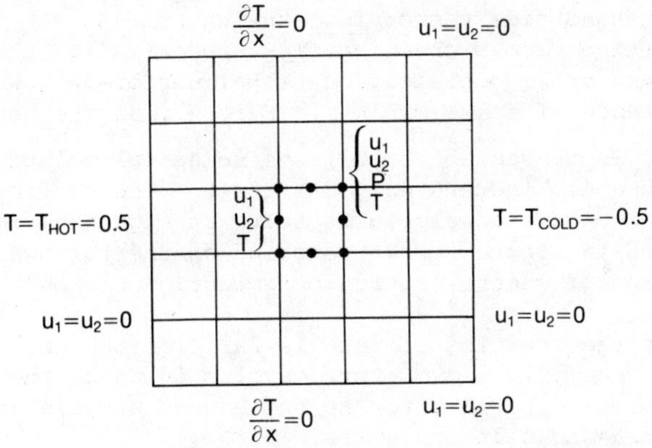

Fig. 1

The buoyancy driven problem is a square region at a fixed Prandtl number of .71 was proposed as a test problem by Jones et al. (1979) for validating solution procedures for the Navier-Stokes equations. The principal features of the flow are well documented (see Winters (1980), Marakatos and Pericleous (1984), Marshall et al. (1978)). For Ra = 10^3, the flow is virtually symmetric about both axes. At Ra = 10^4, we lose some of the symmetry, the flow concentrating towards the vertical walls, but by Ra = 10^5, the distortion is considerable with a weak two eddy structure in the core of the fluid. At Ra = 10^6, the flow is concentrated in the boundary layers with a weak re-circulating flow in the centre. At the highest Rayleigh numbers, waves are generated in the lower region of the hot up-going boundary layer and the upper region of the cold down-going boundary layer breaking into a turbulent core and the cellular structure of the flow is lost. At these Rayleigh numbers the steady state equations cease to be valid, and we should switch over to turbulence calculations.

2.3 When solving at a fixed Rayleigh number, we are at each stage solving a linear system of the form:

$J^* \underline{\delta x} = - \underline{f}$ where J^* is given by (2.18) and $(\underline{\delta x})^T = (\Delta\underline{u}^T, \Delta p^T, \Delta T^T)$

which is equivalent to

$$(J^*)^T (J^*) \underline{\delta x} = - (J^*)^T \underline{f} \; ; \qquad (2.19)$$

so the Newton Raphson technique is equivalent (for n equations in n unknowns) to Gauss-Newton minimization of the two-norm of the right hand side vector $\underline{f}$ of the equations. It is evident when starting from a creeping flow approximation that Newton-Raphson and other optimisation techniques break down (due to the existence of false minima). Further, as the Rayleigh number is increased above 10^6, the numerical solution of the problem becomes increasingly difficult. The continuation procedure we have developed to solve at high Rayleigh numbers is summarised in Figure 2. It is based on the following simple observation of general Newton behaviour:

(a) When the starting iterate is far from the true solution, there are commonly two or three cycles in which the shifts are of similar size. The error at the end of a cycle is hardly smaller than that at its start.

(b) There is then a cycle in which the shifts are smaller than those that went before, by a factor between, say, 0.5 and 0.1.

(c) Finally quadratic convergence sets in, with shifts which are more than one order of magnitude smaller than preceding shifts.

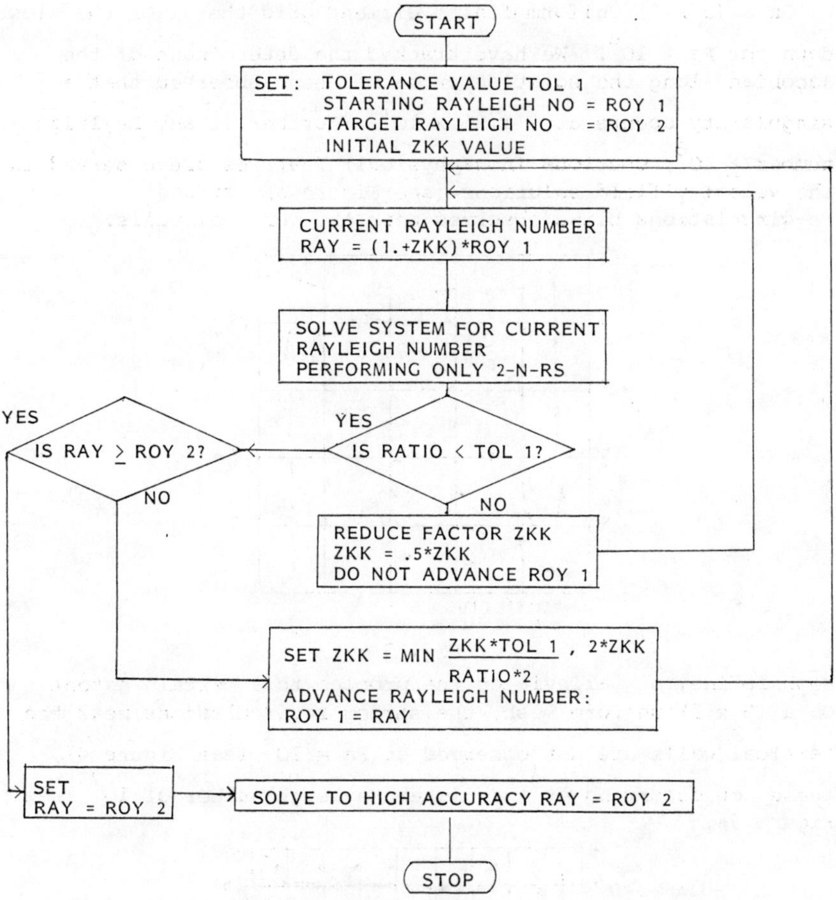

Fig. 2 Flow Diagram of Continuation Strategy

At stage (a) we are close to the size of error at which the process may not converge. We must avoid this. At stage (c) we are changing the parameters so little that we are not making progress efficiently. We wish to operate, so far as possible, at or near stage (b). We can use the ratio of second Newton shift to first Newton shift as a good indicator of whether we are in the region of "safe convergence". In essence, this means that we are tracking the solution in a tubular neighbourhood of the true solution curve.

In practice, we do not find that we have to reject many steps, since the algorithm tends to shorten its step length well before any problem is encountered.

On a (5 x 5) uniform finite element grid the algorithm slows down for Ra $\geq 10^6$. We have tracked the determinant of the Jacobian along the solution curve and have observed that a singularity occurs at ~ 1.26×10^7. Further at any Rayleigh number $\geq 10^6$, spurious (non-physical) features are observed in the velocity field solutions (see Figure 3), strong re-circulations being observed near the vertical walls.

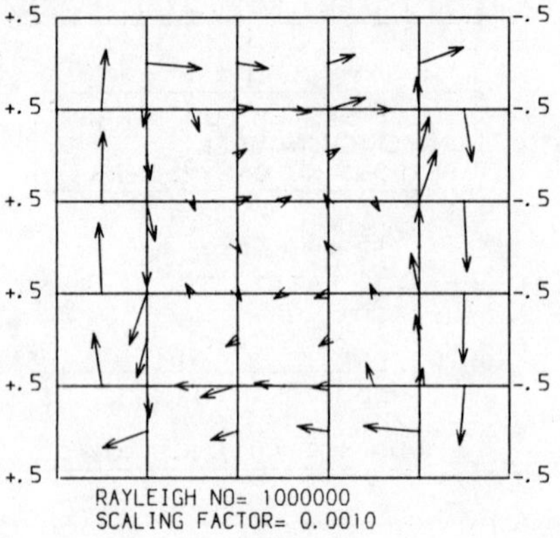

Fig. 3

Mesh refinement alleviates the problem to a certain extent. On a (9 x 9) uniform mesh, the strong recirculations near the vertical walls are not observed at Ra = 10^6 (see Figure 4), these now occurring at the higher Rayleigh number of 10^7 (see Figure 5).

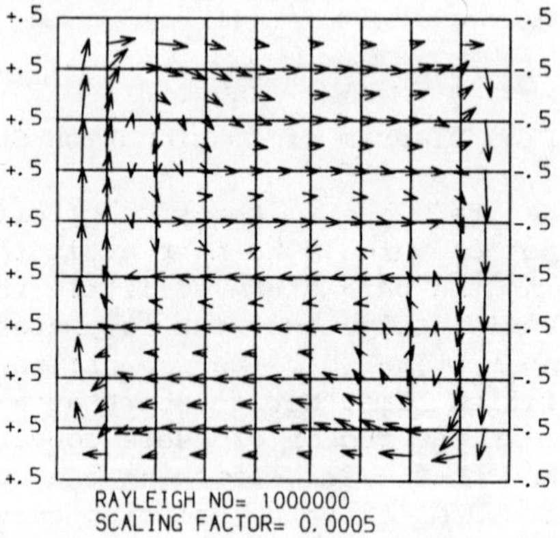

Fig. 4

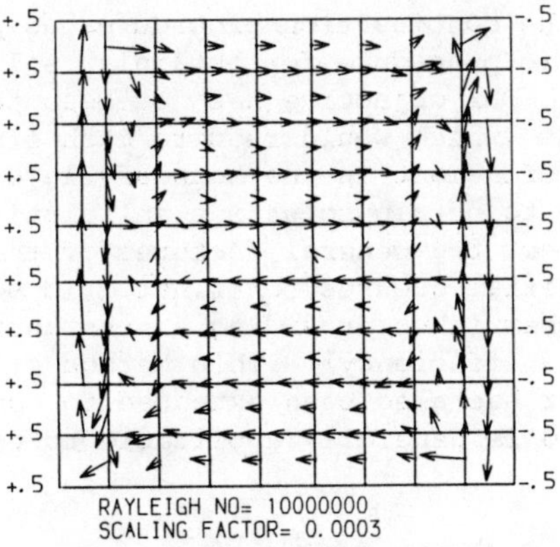

Fig. 5

A series of tests on successively finer meshes shifts the singularity from a Rayleigh number of ~ 1.2×10^7 (on a 5 x 5 grid) to a Rayleigh number of ~ 5.5×10^7 (on a 9 x 9 grid). A non-uniform 5 x 5 grid (with narrow elements near the walls) shifts the singularity to ~ 10^8, and a non-uniform 9 x 9 grid shows no signs of the singularity up to ~ 8×10^8. Figure 6 is a solution obtained at a Rayleigh number of 10^8 on a non-uniform 9 x 9 grid. We need much more mesh refinement (especially on the corners) to pick out the general flow features, and in any case we are now well into the turbulent regime and should switch to solving the turbulent equations.

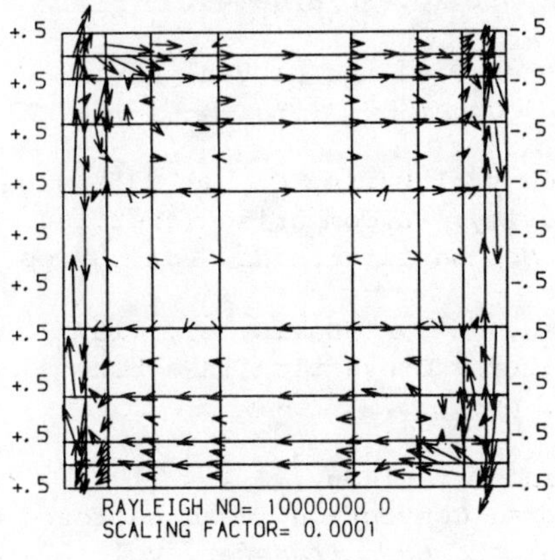

Fig. 6

In practice, the continuation procedure has proved an efficient automatic procedure for obtaining solutions at high Rayleigh numbers. Our objective has not been to obtain highly accurate solutions (which would require much finer grids and very much finer refinement in the boundary layers than we have used), but rather to obtain solutions on relatively crude meshes which reflect the general features of the flow, and that can be used as initial guesses for finer grid solutions at the same Rayleigh number (thus providing a substantial saving in computing time and efficiency). This method of continuation in Rayleigh number has also been extended to continuation in Prandtl number and is generally applicable to continuation in any parameter.

ACKNOWLEDGEMENT

One of us (S.S.) wishes to thank the Science and Engineering Research Council for a CASE Studentship, and the Central Electricity Generating Board for their support of this Studentship, including the computing facilities used for the calculations.

REFERENCES

Girault, V. and Raviart, P.A., (1981) Finite Element Approximations of the Navier-Stokes Equations. Springer-Verlag, Lecture Notes in Mathematics, No. 749.

Hutton, A.G., (1974) A Survey of the Theory and Application of f.e.m. in the analysis of viscous incompressible Newtonian flow. CEGB Report No. RD/B/N3049.

Jones, I.P., Roache, P.J. and de Vahl Davis, G., (1979) Back Cover of *Journal of Fluid Mechanics*, **95**.

Jones, I.P. and de Vahl Davis, G., (1983) Natural Convection in a square cavity: a comparison exercise. *Int. Journal for Numerical Methods in Fluids,* Vol. 3, pp. 227-248.

Mallinson, G.D. and de Vahl Davis, G., (1977) Three dimensional Natural Convection in a box. *Journal of Fluid Mechanics,* **83**, pp 1-31.

Marakatos, N.C. and Pericleous, K.A., (1984) Laminar and Turbulent Natural Convection in an enclosed Cavity. *Int. Journal of Heat and Mass Transfer,* Vol. 27 (No.65) pp. 755-772.

Marshall, R.S., Heinrich, J.C. and Zienkiewicz, O.C., (1978), Natural Convection in a square enclosure by f.e.m., Numerical Heat Transfer, Vol. 1 pp 315-330.

Mitchell, A.R. and Wait, R., (1977) The finite element method in p.d.e's. Wiley.

Oden, J.T., (1972) Finite Elements of Non-linear Continua. McGraw-Hill.

Tritton, D.J., (1977) Physical Fluid Dynamics, Van Nostrand Reinhold Co.

A TECHNIQUE TO SOLVE INCOMPRESSIBLE FLOW EQUATIONS COUPLED WITH A STRESS EQUATION

M.F. Webster
(ICFD, Department of Mathematics, University of Reading)

1. SCOPE OF PAPER

A numerical scheme is presented which solves equations of Navier-Stokes type for steady planar two-dimensional incompressible flows. The flow variables considered are stream function ψ, vorticity ω and stress ζ. The equation system to be solved is then stated in non-dimensional form as

$$\nabla^2\psi = -\omega, \qquad (1.1a)$$

$$\nabla^2\omega = R\underset{\sim}{u}.\underset{\sim}{\nabla}\omega + W\underset{\sim}{u}.\underset{\sim}{\nabla}\zeta, \qquad (1.1b)$$

with the definition of velocity $\underset{\sim}{u} = \underset{\sim}{\nabla} \times \psi = (\frac{\partial\psi}{\partial y}, -\frac{\partial\psi}{\partial x})$ and where R is a Reynolds number and W a stress parameter (W is a Weissenberg number for the Non-Newtonian fluid model). Equation system (1.1) is completed by a stress description equation. Attention has initially been directed towards a stress term that arises in a simple Non-Newtonian fluid model where

$$\zeta = \nabla^2\omega. \qquad (1.2)$$

This description is derived from the so-called second-order model. It is an overall aim that the techniques proposed should be sufficiently general to find application in any flow situation that may be described by a coupled system of the incompressible flow equations and a stress equation system, eg. some models for turbulent flows and buoyancy driven flows.

The full system of equations (1.1) and (1.2) is considered in a form with (1.1b) replaced by

$$W\underset{\sim}{u}.\underset{\sim}{\nabla}\zeta - \zeta = -R\underset{\sim}{u}.\underset{\sim}{\nabla}\omega, \qquad (1.3)$$

representing two elliptic Poisson equations and a hyperbolic equation.

It is discretised by finite difference methods on a uniform grid: the usual five-point operator is used for (1.1a) and (1.2) with (1.3) replaced by a Crank-Nicolson type scheme. The resulting system of (nonlinear) equations is solved iteratively in a sequence of Picard-type iterations at the outer or nonlinear level. At the linear or inner level a combination of inner iterations for the elliptic equations (SOR) and direct marching scheme for the hyperbolic equation is used. The convergence of the iteration and the stability of the difference schemes are analysed for a forward-facing step problem. Results are given for various model problems covering a range of values for the two-parameter family (R,W).

2. HISTORICAL BACKGROUND

The full system of nonlinear equations is solved iteratively. This paper concentrates on a traditional approach of linearisation by decoupling, introducing a set of linear equations and an outer Picard-type iteration. The present iterative scheme is compared to two similar iterative schemes proposed by Crochet and Pilate (1976) and Davies (1983). With integer $n \geqq 0$ indicating an outer iteration number these schemes may be summarised as follows:

$$W\underset{\sim}{u}^{n}.\underset{\sim}{\nabla}(\nabla^{2}\omega^{r_1}) + R\underset{\sim}{u}^{n}.\underset{\sim}{\nabla}\omega^{r_2} - \nabla^{2}\omega^{r_3} = 0, \qquad (2.1a)$$

$$\nabla^{2}\psi^{n+1} + \omega^{n+1} = 0\ ,\ \underset{\sim}{u}^{n+1} = \underset{\sim}{\nabla} \times \psi^{n+1}, \qquad (2.1b)$$

where for Scheme 1 (Crochet) $r_1 = r_2 = r_3 = n+1$; Scheme 2 (Davies) $r_1 = n$, $r_3 = n+1$, $R \equiv 0$; Scheme 3 (Present) $r_1 = r_3 = n+1$, $r_2 = n$.

Scheme 1 yields a third-order differential equation for ω leading to convergence difficulties in the corresponding inner iteration. Converged solutions of the outer iteration were reported for the (R,W) values of {(1,.1), (10,.2), (100,.4), (500,.8), (1000,1.4)}. A critical upper limit on W was observed for each selected R value, though this limit increased with increase in R. Scheme 2, with R = 0, gives an efficient inner ω iteration but W is effectively limited by the numerical smoothing of the source term $W\underset{\sim}{u}^{n}.\underset{\sim}{\nabla}(\nabla^{2}\omega^{n})$. Converged solutions were reported for (R,W) of {(0,.1), without filtering; (0,W), $W \leqq 10$ with filtering}. The third and present scheme differs from schemes 1 and 2 by the inclusion of a three-step outer iteration, and the solution for $\nabla^2\omega$ at the linear level by a direct marching scheme. No upper limit on W is found for converged solutions for $R \leqq 10$, though scheme 3 is dependent upon restriction of R.

The dependence of the outer iteration on R and W may be investigated through a linearised perturbation analysis using a single Fourier mode (see Morton et al., 1985). It may be deduced that low frequencies, (i.e. long wave-lengths) $|\underset{\sim}{\kappa}|$, give convergence difficulties for all three schemes. Scheme 2 also suffers at high frequencies leading to a bound on $|\underset{\sim}{\kappa}| \propto W^{-1}$ (see Tanner 1982) and convergence difficulties with increasing W. For scheme 1 at fixed R, as W increases from zero a limiting value may be predicted $W \propto R\,|\underset{\sim}{\kappa}|^{-2}_{min}$ which will increase with R. The present scheme presents much less severe restrictions on W for small R, but the situation is expected to deteriorate more rapidly with increasing R.

Acceptability of converged solutions for small R is based upon the Tanner/Geisekus theorem, (see Tanner 1966), for creeping flow (R = O). At W = O the solution for ζ is trivial, $\zeta = \nabla^2\omega = 0$, and from Tanner's theorem the velocity field $\{\underset{\sim}{u}\}_{W=0}$ also satisfies the problem $\forall\, W > 0$.

3. FULL STATEMENT OF MODEL PROBLEM

The solution of the forward-facing step or contraction geometry is considered for the model equations discussed earlier and used in the following form:

$$W\underset{\sim}{u}^{n}.\underset{\sim}{\nabla}\zeta^{n+1} - \zeta^{n+1} = -R\underset{\sim}{u}^{n}.\underset{\sim}{\nabla}\omega^{n}, \tag{3.1a}$$

$$\nabla^2\omega^{n+1} = \zeta^{n+1}, \tag{3.1b}$$

$$\nabla^2\psi^{n+1} = -\,\omega^{n+1}, \quad \underset{\sim}{u}^{n+1} = \underset{\sim}{\nabla} \times \psi^{n+1} = (u,v)^{n+1}. \tag{3.1c}$$

The region of solution is shown in Fig. 1: ABCD is a fixed boundary, FE a symmetry boundary, AF the inlet boundary and DE the outlet. The flow is assumed to be fully-developed at inlet and outlet. With, firstly, a statement of the minimum boundary assumptions required, followed by the implied conditions actually used, the appropriate boundary conditions are

Fixed boundary ABCD

$$u = v = 0 \quad \Rightarrow \psi = 0, \; \omega = -\frac{\partial^2 \psi}{\partial n^2}, \; \zeta = 0; \qquad (3.2a)$$

Inflow boundary AF

$$u \text{ given}, \; v = \frac{\partial v}{\partial x} = 0 \quad \Rightarrow \psi, \; \omega \text{ given} ; \qquad (3.2b)$$

Outflow Boundary DE

$$v = 0, \; \frac{\partial \omega}{\partial x} = \frac{\partial \zeta}{\partial x} = 0 \quad \Rightarrow \frac{\partial \psi}{\partial x} = \frac{\partial \omega}{\partial x} = 0, \; \underline{\zeta = 0}; \qquad (3.2c)$$

Symmetry boundary EF

$$\frac{\partial u}{\partial y} = 0, \; v = 0 \quad \Rightarrow \psi = \text{constant}, \; \omega = 0, \; \zeta = 0. \qquad (3.2d)$$

For given values of R and W, the last remaining parameter in the problem is the step ratio AF:DE taken as 4:1.

Commencing from W = 0 for (3.1) and introducing small W > 0 involves consideration of a singular perturbation problem. One may expect boundary layers for ζ to result dependent upon the boundary conditions imposed. If a zero value of ζ is imposed at outlet then boundary layers are avoided. This is consistent with the exponentially decreasing complementary function form for ζ when treated as ODE along the streamlines in the upstream direction.

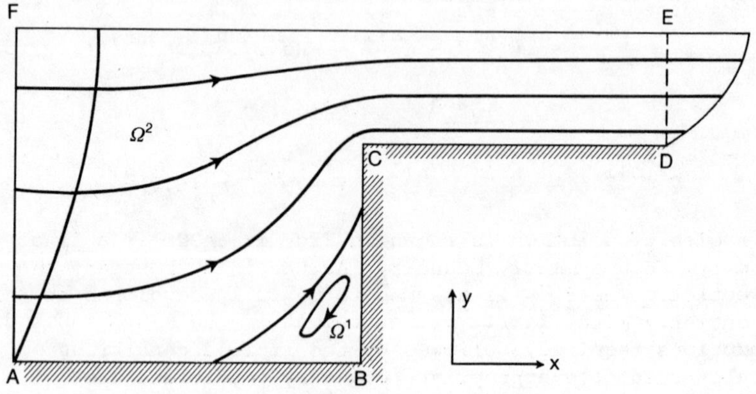

Fig. 1 Schematic flow diagram

4. NUMERICAL APPROXIMATION

4.1 Finite Difference Equations

The region Ω of Fig. 1 is covered by a regular square grid of side h and nodal values of the variables ψ, ω, ζ are denoted by $\psi_{i,j}$ etc. at nodes $x = ih$, $y = jh$. Values of h of $\frac{1}{8}$ and $\frac{1}{16}$ are used, with some 1400 mesh points and 16 mesh lengths at inflow for the former choice. Use is made of standard difference operator notation as follows:

$$\delta_x \psi_{i+\frac{1}{2},j} = \psi_{i+1,j} - \psi_{i,j} \; ; \quad \mu_x \psi_{i+\frac{1}{2},j} = \tfrac{1}{2}(\psi_{i+1,j} + \psi_{i,j}) \qquad (4.1)$$

with similar meaning for δ_y and μ_y. The difference approximations for the Poisson equations for ω and ψ (3.1b,c) are then

$$(\delta_x^{\,2} + \delta_y^{\,2})\, \omega_{i,j}^{n+1} = h^2 \zeta_{i,j}^{n+1} \; ; \quad (\delta_x^2 + \delta_y^2)\, \psi_{i,j}^{n+1} = - h^2 \omega_{i,j}^{n+1} \, , \qquad (4.2)$$

the standard five-point schemes: here superscript (n+1) denotes the stage of the outer iteration process, $\zeta^{n+1} \longrightarrow \omega^{n+1} \longrightarrow \psi^{n+1}$.

Two different schemes are used for the stress equation (3.1a): the box scheme in the recirculating flow region Ω_h^1 and the Crank-Nicolson type scheme in Ω_h^2 where the flow is predominantly in the x-direction. For the former, velocities are required at the centre of each cell and are given by the four-point formulae

$$u^n_{i+\frac{1}{2},j+\frac{1}{2}} = \frac{\mu_x}{h} \delta_y \psi^n_{i+\frac{1}{2},j+\frac{1}{2}} \, , \quad v^n_{i+\frac{1}{2},j+\frac{1}{2}} = - \frac{\mu_y}{h} \delta_x \psi^n_{i+\frac{1}{2},j+\frac{1}{2}} \, . \qquad (4.3a)$$

The box scheme approximation to (3.1a) is

$$W\left[(u^n \mu_y \delta_x + v^n \mu_x \delta_y)\, \zeta^{n+1}\right]_{i+\frac{1}{2},j+\frac{1}{2}} - h\left[\theta \mu_x \mu_y \zeta_{i+\frac{1}{2},j+\frac{1}{2}} + (1-\theta)\zeta_{i,j+1}\right]^{n+1}$$

$$= - R\left[(u^n \mu_y \delta_x + v^n \mu_x \delta_y)\omega^n\right]_{i+\frac{1}{2},j+\frac{1}{2}} \qquad (4.3b)$$

where the parameter θ permits weighting within the scheme on a pointwise basis and ensures stability for all possible velocity fields. The Crank-Nicolson scheme covers two neighbouring cells with the same x-coordinates and is therefore centred at $(i+\frac{1}{2},j)$: thus the velocities are given by the unsymmetric formulae

$$u^n_{i+\frac{1}{2},j} = \frac{\mu_x}{h}\mu_y\delta_y\psi^n_{i+\frac{1}{2},j}\ ,\ v^n_{i+\frac{1}{2},j} = -\frac{\delta_x}{h}\psi^n_{i+\frac{1}{2},j}\ ; \tag{4.4a}$$

and the approximation to (3.1a) is

$$\left[W(u^n\delta_x + v^n\mu_x\mu_y\delta_y)\zeta^{n+1} - h\hat{\zeta}^{n+1}\right]_{i+\frac{1}{2},j} =$$

$$- R\left[u^n\delta_x + v^n\mu_x\mu_y\delta_y)\omega^n\right]_{i+\frac{1}{2},j} \tag{4.4b}$$

where $$\hat{\zeta}_{i+\frac{1}{2},j} = \beta\zeta_{i+1,j} + (1-\beta)\ \zeta_{i,j}\ . \tag{4.4c}$$

The weighting parameter β is introduced to again ensure stability of the scheme for all velocity fields.

4.2 Boundary Condition Implementation

For the stream function ψ^n in (4.2), Dirichlet boundary conditions apply over the whole boundary except DE, from (3.2a,b,d). From (3.2c) Neumann conditions apply on DE (same for ω^n) and are incorporated in the following fashion:

$$\psi^n_{I,j} = \psi^n_{I-1,j}\ ,\ \omega^n_{I,j} = \omega^n_{I-1,j}. \tag{4.5}$$

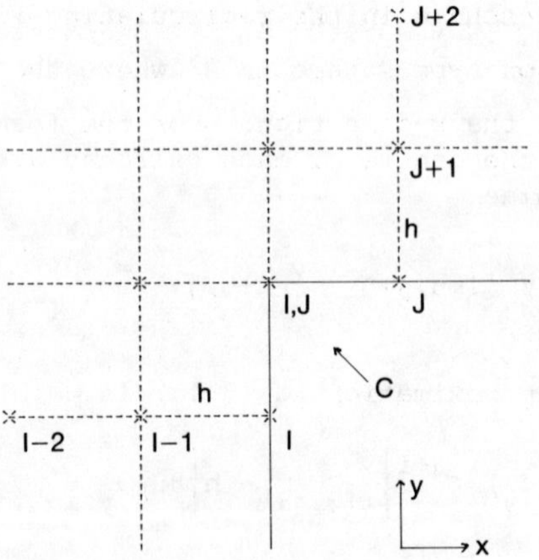

Fig. 2 Mesh near boundary

Likewise for ω^n in (4.2) Dirichlet conditions apply on AF and EF from (3.2b,d); only the specification of ω^n on the wall ABCD remains. This approximation is achieved by Taylor series expansions based on local wall velocity conditions. With reference to Fig. 2, use is made of the following derived formulae:

$$\omega^n_{i,J} = -\ 2h^{-2}(\psi_{i,J+1} - \psi_{i,J})^{n-1}, \tag{4.6a}$$

$$\omega^n_{i,J} = -\ 3h^{-2}(\psi_{i,J+1} - \psi_{i,J})^{n-1} + .5\ \omega^r_{i,J+1}, \tag{4.6b}$$

$$\omega^n_{I,J} = -\ 2h^{-2}(\psi_{I-1,J} + \psi_{I,J+1} - 2\ \psi_{I,J})^{n-1}, \tag{4.6c}$$

where I and J are wall coordinates, (I,J) is the re-entrant corner point C, and r in (4.6b) may be chosen at level n-l or n. Formula (4.6a) is formally first-order accurate, see (Thom 1933); formula (4.6b) is second-order accurate, see (Woods 1954), and is used with r = n at all wall points bar C; formula (4.6c) is attributed to Kawaguti (1965) and provides a finite estimate of ω^n_C based on the first-order form. At C, ω is unbounded; however the simple formula (4.6c) generates an estimate closely approximating that provided by an asymptotic expansion matching scheme (see Holstein and Paddon 1982).

For stability reasons the equations for ζ^{n+1} in (4.4b) are integrated from outflow to inflow, and from (3.2a,c,d) ζ^{n+1} is set to zero on all boundaries, except on the inflow boundary AF. Fully-developed conditions at inflow provide a compatability check on the solution generated and for consistency ζ^{n+1} should decay and vanish there. Providing the exit length CD is sufficiently large no boundary layers are encountered, and in fact with $R \equiv 0$, then $\zeta \equiv 0\ \forall W \geqslant 0$ as it should be. If the Crank-Nicolson scheme (4.4) were to be used over the whole region Ω_h <u>sufficient</u> data would now be available with $\{\omega,\psi\}^n_{i,j}$ specified at all interior and boundary points: the same is also true if the box scheme (4.3) is incorporated in a <u>limited</u> zone Ω^1_h based on ABC but not extending across to the symmetry boundary EF.

5. SOLUTION PROCEDURE

The nonlinear equation system is decoupled into a system of three linear equations. A three-step outer Picard-type iteration is introduced where each step constitutes the solution of a linear equation at the inner level. The process commences with the iterate $(\zeta^n,\omega^n,\psi^n)$ and the order of computation is to first calculate ζ^{n+1}, from which ω^{n+1} is generated, and finally ψ^{n+1} is obtained from ω^{n+1}. The cycle may then be repeated after recomputation of the new velocity field iterate $\underset{\sim}{u}^{n+1}$.

The difference equations (4.2) for ω^{n+1} and ψ^{n+1} are solved using a secondary inner SOR iteration with empirically estimated optimal relaxation factors of $1.6 \leq \rho_\psi \leq 1.8$ and $\rho_\omega = 1.0$ (as in Davies et al., 1979).

The stress ζ^{n+1} is calculated in the following manner. First, the box scheme (4.3) is introduced for the recirculation zone Ω_h^1. Proceeding in a point-by-point fashion starting near B, $\zeta_{i,j+1}^{n+1}$ is given by

$$c_1\zeta_{i,j+1}^{n+1} + c_2\zeta_{i,j}^{n+1} + c_3\zeta_{i+1,j+1}^{n+1} + c_4\zeta_{i+1,j}^{n+1} = g_{i+\frac{1}{2},j+\frac{1}{2}}^{n} \tag{5.1}$$

where $c_1 = (3\theta-4)\frac{h}{2} - W(u-v)_{i+\frac{1}{2},j+\frac{1}{2}}^{n}$; $c_m = -\theta\frac{h}{2} \pm W(u \mp v)_{i+\frac{1}{2},j+\frac{1}{2}}^{n}$, $m = 2, 3, 4$; and $0 \leq \theta \leq 1$. It is vital that Ω_h^1 is restricted to the recirculation zone defined by the known separation line value of ψ^n. Otherwise, large numerical oscillations in ζ^{n+1} occur cross-stream and are subsequently swept upstream. This is due to the sign switch of coefficient c_1, when v^n begins to dominate u^n. The factor θ may be altered from the average value of unity to the totally weighted zero value dependent upon a pointwise selection criterion; its choice guarantees the stability of the scheme for all variations in $(u-v)_{i+\frac{1}{2},j+\frac{1}{2}}^{n}$.

Second, the Crank-Nicolson scheme (4.4) is used in Ω_h^2 involving a direct line-by-line <u>marching</u> procedure from outflow to inflow. A tridiagonal matrix of equations results for each I-line and $\underset{\sim}{\zeta}_I^{n+1}$ is related to $\underset{\sim}{\zeta}_{I+1}^{n+1}$ by

$$\left[V\,\zeta_{j+1} - (U+h(1-\beta))\,\zeta_j - V\zeta_{j-1}\right]_I^{n+1} + \left[V\,\zeta_{j+1} + (U-\beta h)\,\zeta_j - V\zeta_{j-1}\right]_{I+1}^{n+1} = f_{I+\frac{1}{2},j}^{n} \tag{5.2}$$

where $V = \frac{W}{4} v^n_{I+\frac{1}{2},j}$ and $U = Wu^n_{I+\frac{1}{2},j}$. A fast tridiagonal solver is used to solve the system of equations (5.2). Numerical stability of this solver is guaranteed since the associated matrix is diagonally dominant almost everywhere in Ω^2_h, i.e. provided $\left[u^n_{I+\frac{1}{2},j}\right] \geq \frac{1}{2} \left[v^n_{I+\frac{1}{2},j}\right]$ for $\beta = 0$. From a basic 1-D analysis of simple channel flow, the condition $\beta \leqq \frac{U}{h}$ must be satisfied to avoid streamwise oscillations in the implementation of the Crank-Nicolson scheme. Furthermore, from a discrete Fourier analysis of the line-by-line marching sweep, the stability criterion $\beta \leq \frac{U}{h} + \frac{1}{2}$ emerges. Both conditions are satisfied and stability affirmed if $\beta = 0$ and $u^n_{I+\frac{1}{2},j} \geq 0$, (see Morton et al., 1985). The Crank-Nicolson scheme is only applicable for a zone predominantly in the x-direction, where streamlines can effectively be tracked. If it is extended to Ω^1_h a large cross-stream oscillation in ζ^{n+1} is observed due to small velocity field values, and leads to a failure to converge in the Picard iteration.

Relative tolerances in $[\zeta, \omega, \psi]$ of $[10^{-2}, 10^{-2}, 10^{-3}]$ are used to monitor iterative convergence: smoothing of one outer iterate with the next is also used. It is generally observed that as R is increased, convergence criteria become increasingly more difficult to satisfy. The convergence of the Picard iteration is directly related to R: for small R convergence may be obtained for any W, whilst for larger R (greater than 10) convergence of the Picard iteration deteriorates for larger W. These conclusions are in agreement with the arguments of section 2.

A selection of results is presented in Figs. 3 and 4. For creeping flow (R = 0) ζ vanishes $\forall\, W \geq 0$. For $R = 10^{-4}$ solutions are found for $10^{-4} \leq W \leq 10^{2}$: ζ behaves like $R\underset{\sim}{u}.\nabla\underset{\sim}{\omega}$ for $W \leqq \frac{1}{U}$ and like $\frac{R}{W}\omega$ for $W > \frac{1}{U}$, where U is a characteristic velocity value. The same relative behaviour for ζ is observed for $R \lesssim 1$. The solutions for ζ for $0.1 \leq W \leq 10$ at R = 1 and R = 10 are shown for comparison in Fig. 3. For completeness the solutions for ψ and ω are also given at W = 1 for R = 1 and R = 10 in Fig. 4. Solutions for ψ and ω change negligibly with W for $R \leq 1$.

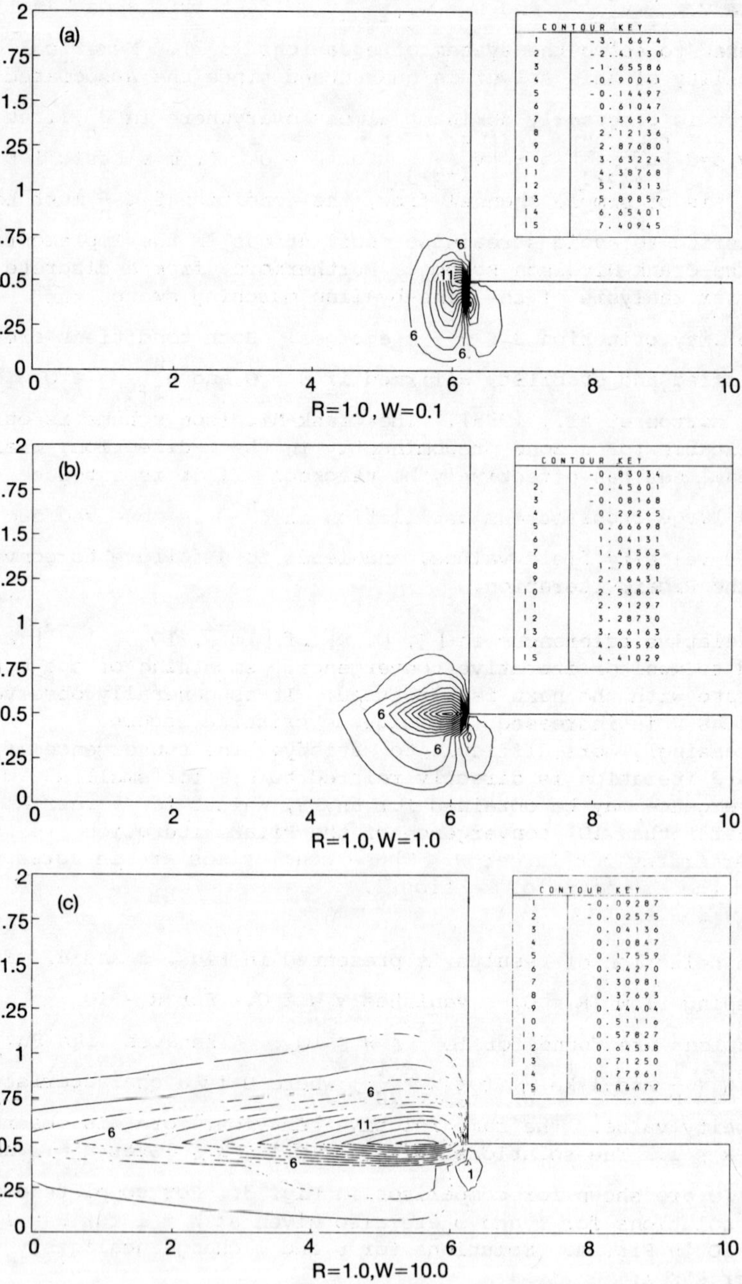

Fig. 3 a-c Stress contours

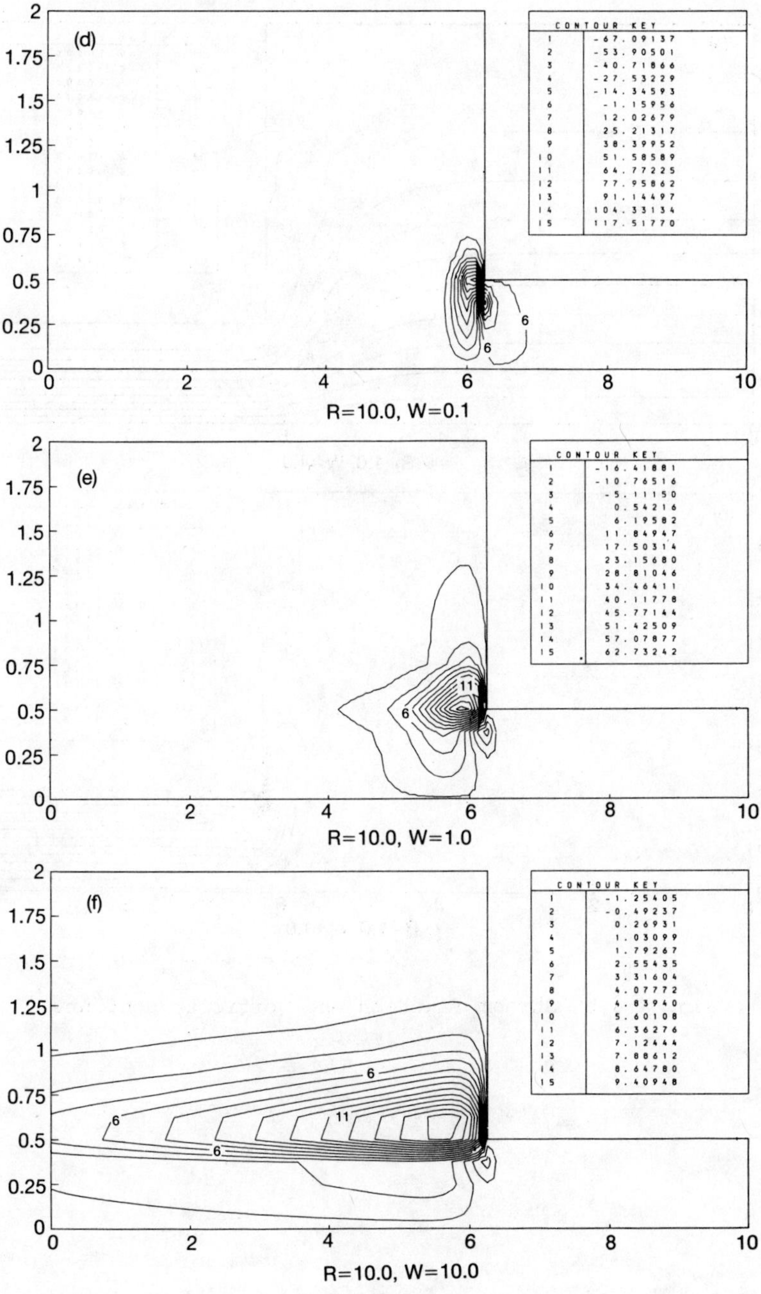

Fig. 3 d-f Stress contours

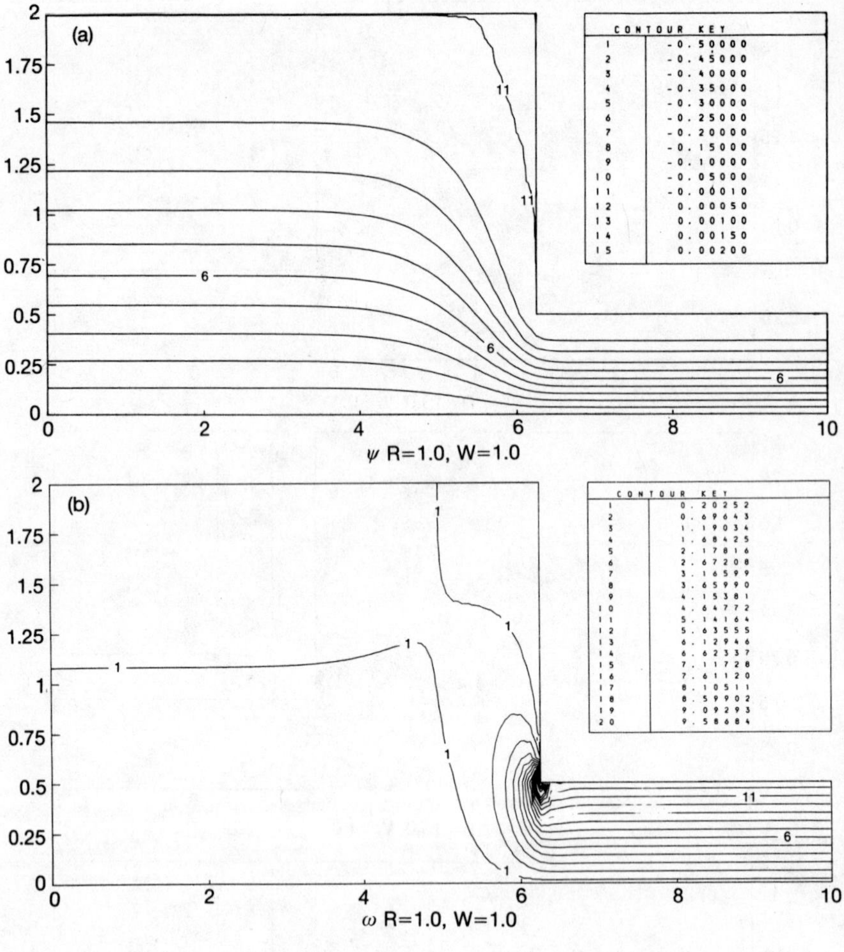

Fig. 4 a,b Stream function and vorticity contours

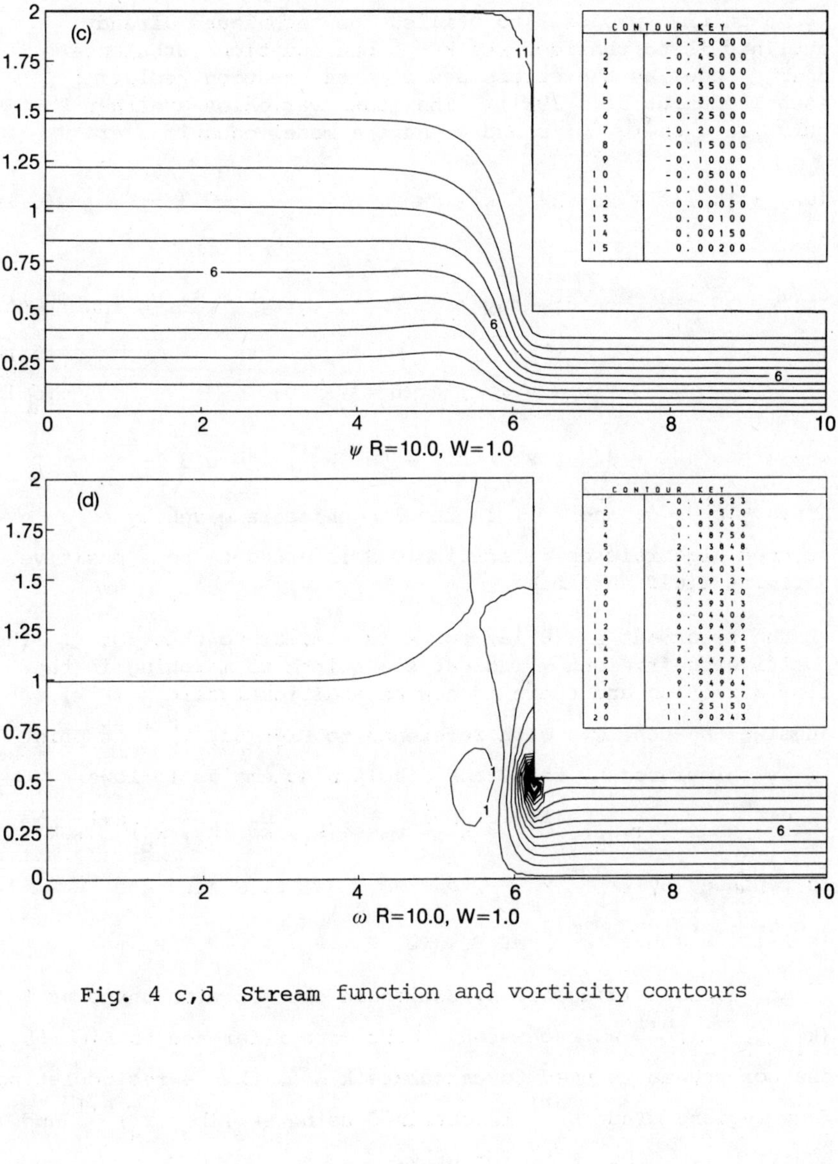

Fig. 4 c,d Stream function and vorticity contours

6. EXTENSION FOR A TURBULENCE MODEL

An initial proposal to utilise the techniques already outlined is to consider the $k \sim \ell$ one equation turbulence energy model as the stress description equation replacing ζ (see Thomas et al., 1981). The model variables are then turbulence energy k, ω and ψ and the model equations are

$$(\underset{\sim}{u}.\underset{\sim}{\nabla} + \gamma k^{\frac{1}{2}})\, k = \underset{\sim}{\nabla} B.\underset{\sim}{\nabla} k + B\nabla^2 k + \beta k^{\frac{1}{2}} S, \quad S = \frac{\partial u_i}{\partial x_j}\left(\frac{\partial u_i}{\partial x_j} + \frac{\partial u_j}{\partial x_i}\right), \tag{6.1a}$$

$$\underset{\sim}{u}.\underset{\sim}{\nabla}\omega - D\nabla^2\omega = \underset{\sim}{\nabla}.(D_x(\psi_{xx}-\psi_{yy}) + 2D_y\psi_{xy},\ 2D_x\psi_{xy} + D_y(\psi_{yy}-\psi_{xx}))\ , \tag{6.1b}$$

$$\nabla^2\psi = -\,\omega \quad , \qquad \underset{\sim}{u} = \underset{\sim}{\nabla} \times \psi, \tag{6.1c}$$

where $D = R^{-1} + \beta k^{\frac{1}{2}}$, $B = R^{-1} + \beta\sigma^{-1}k^{\frac{1}{2}}$, $\gamma = c_\mu \beta^{-1}$;

$R, \beta, \sigma, \gamma, c_\mu > 0$; and $\beta(\underset{\sim}{x})$ is the length scale given by an assumed algebraic specification. S is noted to be a positive definite field variable.

The treatment of (6.1a) for k is similar to that for ζ. The positive coefficient γ indicates a switch to marching in the flow direction and there is now an additional term $\underset{\sim}{\nabla}.(B\underset{\sim}{\nabla} k)$ to augment the scheme. With reference to Fig. 5a, k_2^{n+1} at point P_2 is calculated by the Crank-Nicolson scheme as follows; $\partial_x k^{n+1}$ uses $(k_2, k_5)^{n+1}$; $\partial_y k^{n+1}$ uses $(k_1, k_3, k_4, k_6)^{n+1}$; $k^{\frac{1}{2}}k$ is replaced by $(k_5^n)^{\frac{1}{2}} k_2^{n+1}$, and $\underset{\sim}{\nabla} B$, B and $\beta k^{\frac{1}{2}} S$ also use $(k^n)^{\frac{1}{2}}$; $B\nabla^2 k$ uses k^{n+1} at P_1, P_2, P_3, P_4, P_5, P_6, P_8 and k^n at P_7. A tridiagonal system again emerges with unknowns $(k_1, k_2, k_3)^{n+1}$ at each step. Next with reference to Fig. 5b, the box scheme is used to calculate k_1^{n+1} within a recirculating flow region. Thus k_1^{n+1} is obtained using $(k_2, k_3, k_4)^{n+1}$ and, additionally, $(k_5, k_6)^n$ for $B\nabla^2 k$.

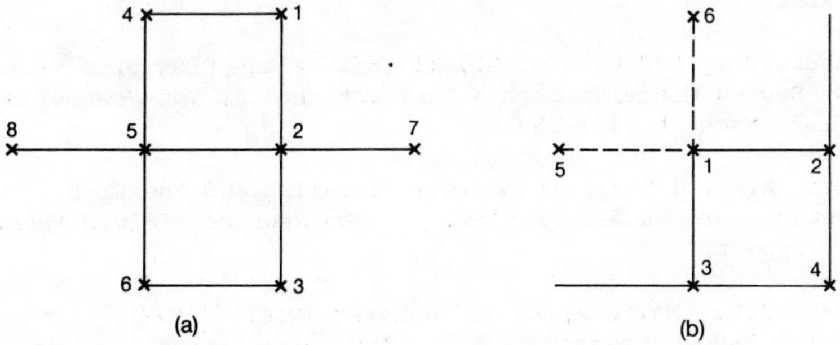

Fig. 5 Finite difference scheme stencil

Boundary conditions follow as before except for k and ω at the wall. Here the velocity field near the wall is used to derive such boundary conditions based on the universal logarithmic law of the wall and matching with a viscous sublayer, (see Turner, 1973). This necessitates the calculation of the wall shear stress τ_{WALL} from the known velocity field $\underset{\sim}{u}^n$ using the relationships

$$\omega_{WALL} = - R\tau_{WALL} \quad , \quad k_{WALL} = c_\mu^{-\frac{1}{2}}|\tau_{WALL}| . \qquad (6.2)$$

Also at attachment and separation points the normal derivative of k at the wall must vanish.

The approach adopted shall be to solve laminar flow for $R \leqq 2000$, using for the vorticity equation the centred difference scheme of Dennis and Smith (1980) with exponential coefficients. Beyond R = 2000 the switch to modelling turbulence with the proposed scheme can then be attempted.

7. ACKNOWLEDGEMENTS

The work reported here forms part of the research programme of the Oxford/Reading Institute for Computational Fluid Dynamics. Particular thanks must go to Professor K.W. Morton, Oxford University, and Dr. E.E. Süli, University of Belgrade, for their collaboration in this research.

REFERENCES

Crochet, M.J. and Pilate, G., (1976). Plane Flow of a Fluid of Second Grade through a Contraction. *J. Non-Newtonian Fluid Mech.* **1**, 247-258.

Davies, A.R., (1983). Numerical Filtering and the High Weissenberg Number Problem. *J. Non-Newtonian Fluid Mech.*, to appear.

Davies, A.R., Walters, K. and Webster, M.F., (1979). Long Range Memory Effects in Flows Involving Abrupt Changes in Geometry Part 3: Moving Boundaries, *J. Non-Newtonian Fluid Mech.* **4**, 325-344.

Dennis, S.C.R. and Smith, F.T., (1980). Steady Flow through a Channel with a Symmetrical Constriction in the Form of a Step. *Proc. Roy. Soc. Lond.* **A372**, 393-414.

Holstein, H. and Paddon, D.J., (1982). A Finite Difference Strategy for Re-entrant Corner Flow. In "Numerical Methods for Fluid Dynamics". Academic Press (K.W. Morton and M.J. Baines, eds.), pp 341-358, London.

Kawaguti, M., (1965). Mathematical Research Centre Report, No. 574, University of Wisconsin, Madison.

Morton, K.W., Süli, E.E. and Webster, M.F., (1985). In preparation, to appear in Int. J. Num. Meth. Enging.

Tanner, R.I., (1982). The Stability of some Numerical Schemes for Model Viscoelastic Fluids. *J. Non-Newtonian Fluid Mech.* **10**, 169-174.

Tanner, R.I., (1966). Plane Creeping Flows of Incompressible Second-Order Fluids. Phys. Fluids **9** (I), 1246-1247.

Thom, A., (1933). The Flow Past Circular Cylinder at Low Speeds. *Proc. Roy. Soc. Lond.* **A141**, 651-666.

Thomas, C.E., Morgan, K. and Taylor, C., (1981). A Finite Element Analysis of Flow over a Backward Facing Step. *Comput. Fluids* **9**, 265-278.

Turner, J.S., (1973). "Buoyancy Effect in Fluids". Cambridge University Press, New York and London.

Woods, L.C., (1954). A Note on the Numerical Solution of Fourth Order Differential Equations. *Aeronautical Quarterly* 5 (III), 176.

FINITE ELEMENT SOLUTION OF STEADY EULER TRANSONIC FLOW BY STREAM VECTOR CORRECTION

F. El Dabaghi
(INRIA, 78153 Le Chesnay, Cedex, France)

J. Periaux and G. Poirier
(Avion Marcel DASSAULT, 92210 St. Cloud, France)

O. Pironneau
(Université Paris XIII, France)

1. INTRODUCTION

1.1 Review

Starting from the Helmholtz decomposition of a given vector field u into

$$u = \nabla\phi + \nabla \times \psi$$

where ϕ is a scalar potential and ψ a stream vector, we propose to study the efficiency of such a decomposition for solving the steady compressible EULER equations; in that approach ψ is introduced as a correction to the transonic potential flows.

Let u be any solenoidal vector field ($\nabla . u = 0$), it is well known "see (De Rham, 1961)" that u is the curl of a stream vector ψ:

$$u = \nabla \times \psi. \tag{1.1}$$

In the 2-D case with suitable boundary conditions, ψ is uniquely defined from u and it has only one non zero component; this property has been introduced, among other authors, in (Amara, 1983) to calculate a correction for isentropic potential flows.

The ideas of Amara (1983), Ecer & Akay (1982), Grossman (1983), Lacor & Hirsch (1982), Sokhey (1980) and Papaillou et al. (1985) have been extended in (2-D) and (3-D) for general inviscid flows by El Dabaghi & Pironneau (1985) and good results have been obtained in subsonic rotational case by El Dabaghi (1984).

1.2 Notation

∇ : Gradient Δ: Laplacian $\nabla\cdot$: Divergence
∇x : Curl $(u \otimes v)_{ij} = u_i v_j$ u x v = Cross vector product between u and v
Ω : Bounded domain $\Gamma = \Gamma_{in} \cup \Gamma_{out} \cup \Gamma_B$: Boundary of Ω,
n its outward unit normal

u : Velocity ψ : Stream vector ϕ : scalar potential
ρ : Density p : Pressure E : Total energy
H : Enthalpy s : Specific entropy S : Modified entropy
ω : Vorticity γ : C_p/C_v : Specific Heat ratio (1.4 in air)

2. THEORETICAL RESULTS

2.1 Notations

Ω is a bounded open set of $R^m (m = 2,3)$, simply connected with boundary $\Gamma \in C^2 (\Gamma = \cup \Gamma_i, \Gamma_i$ connected component of Γ); on Γ we define n its outward unit normal.

(a,b) denotes the inner product in $L^2(\Omega)$, and $||.||_{o,\Omega}$ the associated norm.

$H^1(\Omega)_{/R} = H^1(\Omega)$ quotiented by the constants.

$H^1_o(\Omega) = \{v \in H^1(\Omega): v|_\Gamma = 0\}$.

2.2 Decomposition theorem

We will state the first result on Helmholtz decomposition which is a straightforward generalization of Bernadi (1979).

Theorem: Let u be a given vector field of $(L^2(\Omega))^2$. Let ϕ be the unique solution in $H^1(\Omega)_{/R}$ of

$$(P_1) \quad (\nabla\phi, \nabla w) = (u, \nabla w) \quad \forall\, w \in H^1(\Omega)_{/R} .$$

Let ψ be the unique solution in $H^1_o(\Omega)$ of

$$(P_2) \quad (\nabla\psi, \nabla w) = (u, \nabla x w) \quad \forall\, w \in H^1_o(\Omega) .$$

Then

$$u = \nabla\phi + \nabla x\psi \quad . \tag{2.1}$$

Proof "See (El Dabaghi, 1984)".

3. NUMERICAL INDUSTRIAL ASPECT

3.1 Motivation

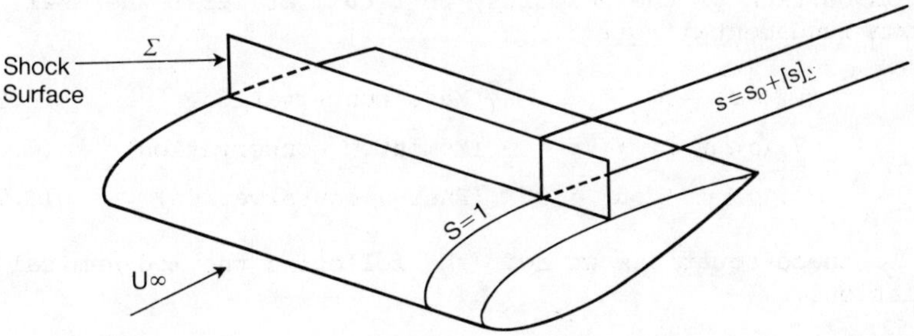

Fig. 3.1

(For simplicity, only a portion of wing between two walls is considered on Fig. 1).

Around non lifting bodies, for example, flows described by

$$\nabla.(\rho u) = 0 \quad (3.1)$$

are well approximated in transonic regimes by the isentropic gas law assumption:

$$\nabla\cdot[(H - (|u|^2/2)^{1/\gamma-1}\nabla\phi] = 0 , \quad (3.2)$$

where H is given and we assume

$$u = \nabla\phi \text{ and } \omega = \nabla x u = 0 \quad (3.3)$$

Equation (3.3) is valid only if we suppose that the entropy S is constant or if the jump is close to zero across the shock.

However when strong shocks are developed in the flow, S is no longer constant and the jump of S is no longer small and therefore the assumption $u = \nabla\phi$ fails; also in order to generalize the potential approach of Glowinski et al (1984) at high transonic regimes, we introduce a Helmholtz decomposition of u:

$$u = \nabla\phi + \nabla x\psi \quad (3.4)$$

in which the second term of the decomposition can be interpreted as a correction.

3.2 Steady Euler Equations for Inviscid Flows

In order to solve the Euler equations via a (ϕ , ψ) - decomposition of the velocity, we recall at first the well known fundamental equations:

$$\nabla.(\rho u) = 0 \qquad \text{(Mass conservation)} \qquad (3.5)$$

$$\nabla.(\rho u \otimes u) + \nabla p = 0 \quad \text{(Momentum conservation)} \qquad (3.6)$$

$$\nabla.(\rho E\, u + pu) = 0 \qquad \text{(Energy conservation)} \qquad (3.7)$$

To these equations we add the following thermodynamical relations:

$$p = \rho^{\gamma}\, S \qquad \text{(State gas law)} \qquad (3.8)$$

$$\nabla.(\rho s u) = 0 \qquad \text{(Second thermodynamic principle)} \qquad (3.9)$$

where

$$S = e^{s/c_v} \qquad (3.10)$$

$$E = H - p/\rho \qquad (3.11)$$

$$H = \frac{1}{2}\,|u|^2 + (\gamma/\gamma-1)(p/\rho)\ . \qquad (3.12)$$

Now in order to derive a convergent $(\phi-\psi)$- algorithm, the main idea is described as follows: assuming the velocity known, from the evaluation of ω, we calculate ψ and then we add $\nabla \times \psi$ to $\nabla\phi$ by a fixed point process to obtain a new value of u; more precisely (3.4) yields:

$$\nabla x \nabla x \psi = \omega\ ; \qquad (3.13)$$

assuming ψ solenoidal and adding suitable conditions on ψ, then ψ can be computed by (3.13) if ω is known.

<u>Equation for ω</u>

Developing (3.6) and combining (3.5), (3.8) and (3.12), we obtain the following relation

$$\rho\, \nabla H - \rho u \times \omega - C(\rho^{\gamma}/\gamma-1)\nabla S = 0\ ; \qquad (3.14)$$

then resolving the cross product of u with (3.14) and using the unknown scalar function λ (very close to helicity) defined by:

$$\lambda = \omega.u/\rho|u|^2 \qquad (3.15)$$

we obtain finally the following equation for ω:

$$\omega = \lambda \rho u + (u/|u|^2) \times (\rho^{\gamma-1}/_{\gamma-1}\ \nabla S - \nabla H) \qquad (3.16)$$

where λ, H, ρ and S are to be determined.

Equations for λ, H, ρ and S

It must be remembered by the definition of ω that

$$\nabla.\omega = 0 \ . \qquad (3.17)$$

Therefore combining equations (3.17) and (3.5) with (3.16) leads to the transport equation for λ:

$$\rho u.\nabla\lambda = -\nabla[(u/|u|^2) \times (p^{\gamma-1}/\gamma-1)\nabla S - \nabla H], \qquad (3.18)$$

On the other hand replacing (3.11) in (3.7) and using (3.5) we obtain also a transport equation for H:

$$u.\nabla H = 0 \qquad (3.19)$$

and taking the inner product of (3.14) with u, we find also a transport equation similar to (3.19) for S:

$$u.\nabla S = 0 \qquad (3.20)$$

where H and S can be computed respectively from their value at upstream infinity:

$$H(x) = H_\infty(Z(x)) \qquad (3.21)$$

$$S(x) = S_\infty(Z(x)) \qquad (3.22)$$

where H_∞ and S_∞ are respectively the given enthalpy and entropy on Γ_{in} and Z(x) is the upstream intersection of Γ_{in} with the stream line that passes at x.

Finally an equation for ρ can be derived from (3.12):

$$\rho = \left[\frac{\gamma-1}{\gamma_S}\left(H - \frac{1}{2}|u|^2\right)\right]^{\frac{1}{\gamma-1}} . \qquad (3.23)$$

At this point as u, H, λ, S and $\hat{\rho}$ are known ω can be evaluated from (3.16), then the rotational correction can be computed by solving (3.13) and finally u can be updated by (3.4) by solving:

$$\nabla(\rho\nabla\phi) = -\nabla\rho . \nabla x \psi \tag{3.24}$$

Remark 3.1

From the above considerations, it appears that the master piece of the associated algorithm, in order to solve the Euler equations via a (ϕ,ψ)-formulation, is the modified transonic equation (3.24).

Boundary Conditions

In order to solve (3.24), we prescribe on Γ Neumann boundary conditions corresponding physically to slip conditions:

$$\frac{\partial\phi}{\partial n}\Big|_\Gamma = u.n\,\Big|_\Gamma \tag{3.25}$$

which imply a non standard boundary condition on ψ added to (3.13):

$$(\nabla x\psi).n\big|_\Gamma = 0: \tag{3.26}$$

in fact the boundary condition (3.26) will be replaced by a natural one:

$$\psi\big|_\Gamma = 0 \quad \text{in the 2-D case} \tag{3.27a}$$

$$\psi x n\big|_\Gamma = 0 \text{ in the 3-D case.} \tag{3.27b}$$

Considering the above boundary conditions and the nature of the flow (shocks can occur in transonic regimes), the following remarks can be made.

Remark 3.2

If we deal with a homogeneous boundary conditions on ϕ of Neumann type:

$$\frac{\partial\phi}{\partial n}\Big|_\Gamma = 0 \tag{3.28}$$

(3.28) involves a non homogeneous boundary condition on ψ:

$$(\nabla x\psi).n\big|_\Gamma = u.n\big|_\Gamma \tag{3.29}$$

which implies on Γ the solution of a Laplace-Beltrami problem in order to find $\psi_{|\Gamma}$ "See (El Dabaghi & Pironneau, 1985)".

Remark 3.3

Equations derived for ω, λ, S and ρ are not valid across shocks: we have to add locally, where discontinuities occur, the Rankine-Hugoniot conditions detailed in (El Dabaghi, 1984) in order to compute the jump of the above variables across shocks.

For the following section we denote by $[v]_\Sigma$ the jump of a variable v across shocks.

3.3 Fixed point algorithm

In order to compute the rotational correction of a potential flow (described in section 3.2) we consider the following algorithm:

Step 0 Initialization

$\psi^o = 0,\ S^o = S_\infty,\ H^o = H_\infty,\ \lambda^o = \lambda_\infty$ (∞: at upstream infinity)

Compute a first guess for $u^o = \nabla\phi^o$ and ρ^o by solving the non linear transonic potential equation (3.2). Then for $m \geq 0$, assuming u^m known, compute S^{m+1}, H^{m+1}, λ^{m+1}, ω^{m+1}, ψ^{m+1}, ϕ^{m+1}, u^{m+1} and ρ^{m+1} by the following:

Step 1 Solve $u^m.\nabla S^{m+1} = 0$ in Ω, except at shocks.

Step 2 Solve $u^m.\nabla H^{m+1} = 0$ in Ω.

Step 3 Solve

$$u^m.\nabla\lambda^{m+1} = -\frac{1}{\rho m}\nabla\cdot[(u^m/|u^m|^2)\ x\ (((\rho^m)^{\gamma-1}/\gamma-1)\nabla S^{m+1} - \nabla H^{m+1})]$$

in Ω, except at shocks.

Step 4 Compute ω^{m+1} by (3.16).

Step 5 Solve $\nabla x \nabla x \psi^{m+1} = \omega^{m+1}$ in Ω
with the boundary conditions (3.27).

Step 6 Solve $\nabla\cdot(\rho^m\nabla\phi^{m+1}) = -\nabla\rho^m.\nabla x\psi^{m+1}$ in Ω
with the boundary conditions (3.25).

Step 7 Update u^{m+1} by (3.4) and ρ^{m+1} by (3.23), m = m+1, go to Step 1, until convergence of the entropy correction.

Remark 3.4

i) The above algorithm (Step 0-7) operates for 2-D and 3-D general situations with finite elements.

ii) In Step 0 and 6 solutions ϕ^{0} and ϕ^{m+1} are computed by least squares formulations using preconditioned conjugate gradient algorithm; physical shocks are obtained by upwinding of the density in the flow direction "See (Glowinski et al. 1984)".

iii) Transport equations of Steps 1, 2, 3 are resolved by characteristic methods using finite elements detailed in (Pironneau, 1982).

iv) Concerning the boundary conditions involved in step 5, the 2-D case is solved by a classical P^{1} finite element method but in the 3-D case, the situation is more complicated and a general recent study concerning the determination of ψ with numerical results can be found in (El Dabaghi & Pironneau, 1985).

v) In the 2-D case, $\lambda = 0$.

vi) For the numerical experiments we consider an isenthalpic fluid (H = Constant).

4. NUMERICAL EXPERIMENTS

In this section we present the results of a first numerical test using the method discussed in the above section. It appears from the experiments that the method is well suited to obtain in a efficient way, a transonic Euler solution as an entropy correction of a potential flow.

4.1 Transonic simulation of flow around a NACA 0012 airfoil

We have considered a transonic flow around a NACA 0012 airfoil at angle of attack $\alpha = 0^{\circ}$ and Mach number at infinity $M_{\infty} = 0.85$.

Fig. 4.1 shows a typical triangulation with enlargement near the body, made of 2354 nodes and 4568 elements. As an initial guess, a transonic potential flow with shock is computed, then the localization of the shock allows one to start the entropy correction process and the Euler solution is reached in a few iterations (Iter < 5).

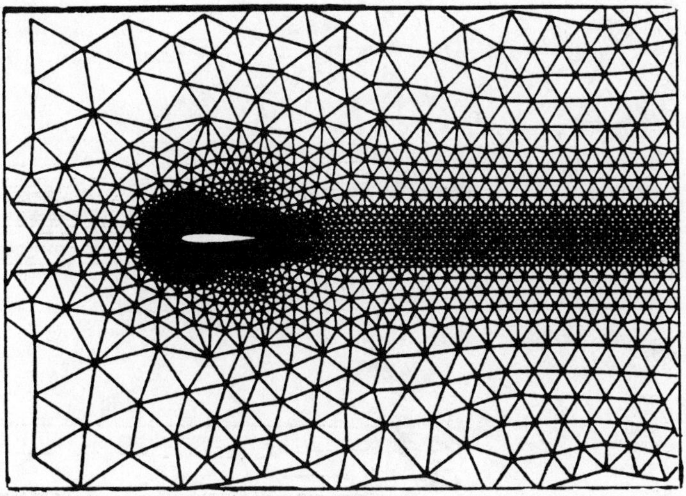

Fig. 4.1

Fig. 4.2 shows the Mach lines around the airfoil of the flow. We have presented on Fig. 4.3 the entropy lines created across the shock via the Rankine-Hugoniot condition and then transported downstream by the characteristics method. The vorticity in the flow created behind the shock is also plotted on Fig. 4.4 and we can observe a good symmetry of the correction stream vector on Fig. 4.5.

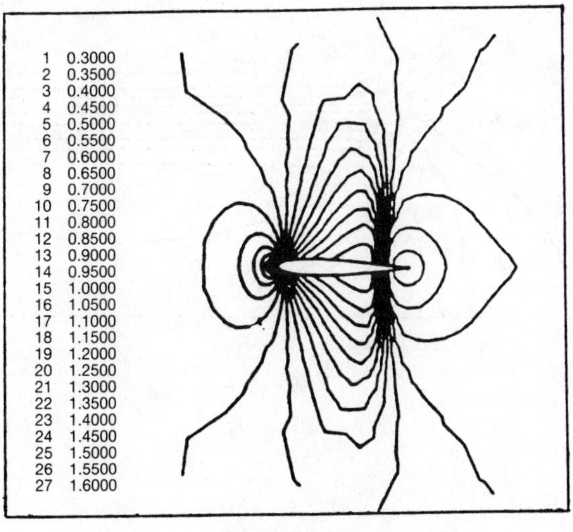

Fig. 4.2

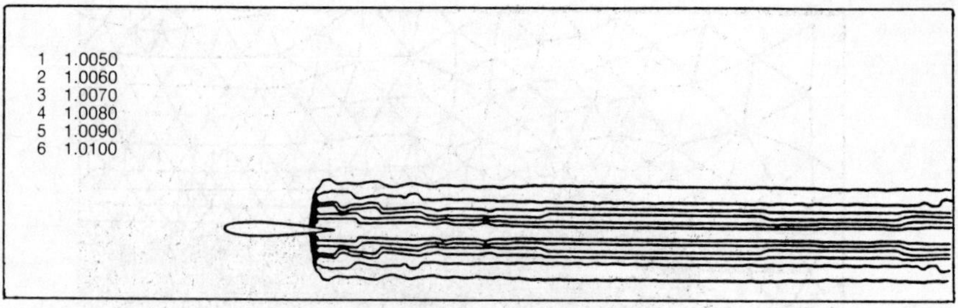

1 1.0050
2 1.0060
3 1.0070
4 1.0080
5 1.0090
6 1.0100

Fig. 4.3

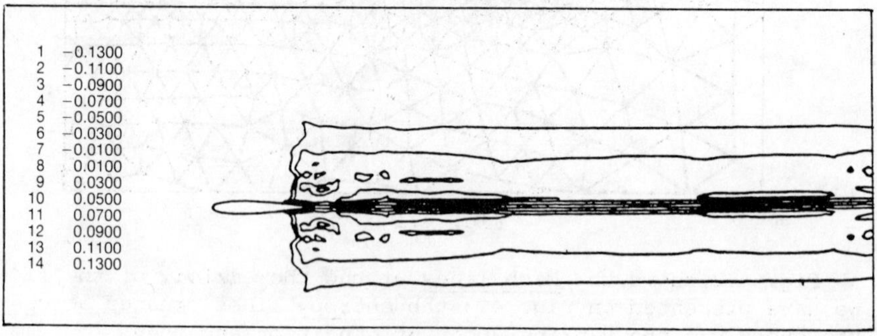

1 −0.1300
2 −0.1100
3 −0.0900
4 −0.0700
5 −0.0500
6 −0.0300
7 −0.0100
8 0.0100
9 0.0300
10 0.0500
11 0.0700
12 0.0900
13 0.1100
14 0.1300

Fig. 4.4

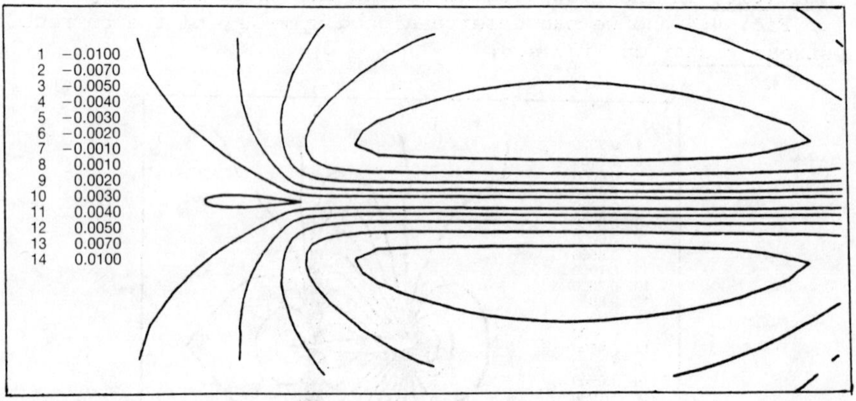

1 −0.0100
2 −0.0070
3 −0.0050
4 −0.0040
5 −0.0030
6 −0.0020
7 −0.0010
8 0.0010
9 0.0020
10 0.0030
11 0.0040
12 0.0050
13 0.0070
14 0.0100

Fig. 4.5

The shock location and the entropy level have been compared with an Euler solution entropy value on the skin of the airfoil, it appears to be in good agreement on Fig. 4.6 with the results obtained by Lerat (1981).

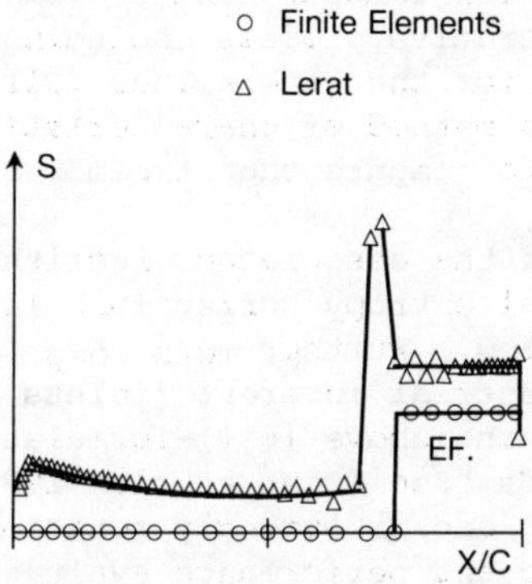

Fig. 4.6

Finally a convergence history of the solution is shown in Fig. 4.7 and we can observe that only 3 to 5 iterations are necessary to reach convergence which gives a CPU time ratio (3 to 5 units of time to get the potential solution) compared to the CPU time of a full Euler solution. This fact makes the correction method attractive for further 3-D applications.

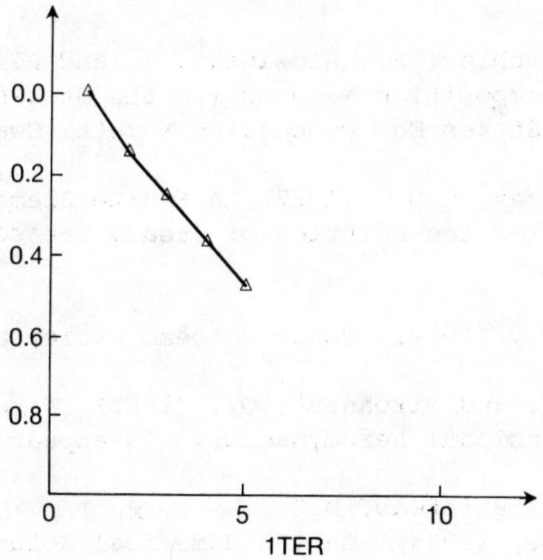

Fig. 4.7

4.2 Further comments and conclusion

We have presented in this paper an entropy correction method based on the velocity decomposition for the numerical solution of the Euler equations modelling transonic flows. With this method we can compute finite element Euler solutions via an efficient iterative process including a nonlinear least squares formulation for the generalized full potential equations and also a method of characteristics for the transport equations in order to compute the stream vector correction.

The modularity of the associated algorithm allows the introduction of local entropy corrections in selected subdomains of the flow. Further more complicated 3-D applications of industrial interest (inlets and complete aircraft) combining the above (ϕ,ψ)-formulation and domain decomposition methods "See (Dinh et al., 1985)" with finite elements for lifting and/or strongly rotational transonic flows are considered and performance evaluations will appear in a forthcoming paper.

REFERENCES

Amara, M., (1983) Thèse d'Etat, Univ. Paris Vl.

Bernardi, C., (1979) Thèse de 3ème Cycle, Univ. Paris Vl.

De Rham G., (1960) Variétés Différentiables, Ed. Hermann, Paris.

Dinh Q.V., Fischler, A., Glowinski, R. and Périaux, J., (1985) Domain Decomposition Methods for the Stokes Pb., Applications to Navier-Stokes Eq. Proceeding Numeta, Swansea, U.K.

Ecer A. and Akay, H.U., (1982) A Finite Element Formulation of Euler Eq. for the Solution of Steady Transonic Flows, AIAA, USA.

El Dabaghi, F., (1984) Thèse de 3ème Cycle, Univ. Paris Xlll.

El Dabaghi, F. and Pironneau, O., (1985) Stream Vectors in Three Dimensional Aerodynamics. To appear in Numer. Mat.

Glowinski, R., Bristeau, M.O., Periaux, J., Pironneau, O., and Poirier, G., (1984) On the Numerical Solution of Nonlinear Pb in Fluid Dynamics by Least Squares and Finite Element Methods (ll). Application to Transonic Flow Simulation, Int. Proceeding of Fenomech 84, Stuttgart, RFA.

Grossman, B., (1983) The Computation of Inviscid Rotational Gas Dynamic Flows Using an Alternate Velocity Decomposition, 6th AIAA CFD Conference, Danvers, USA.

Lacor, C. and Hirsch, Ch., (1982) Rotational Flow Calculations in Three Dimensional Blade Passage, ASME Report, 82-GT-316.

Lerat, A., (1981) Thèse d'Etat, Univ. Paris Vl.

Papaillou et al., (1984) DRET Contracts, Paris.

Pironneau, O., (1982) On the Transport-Diffusion Algorithm and its Applications to the Navier-Stokes Equations. *Numer. Mat.* 8, pp 309-332.

Sokhey, J.S., (1980) Transonic Flow Around Axisymmetric Inlets including Rotational Flow Effects, 18th AIAA, Pasadena, USA.

SEMI-INVERSE MODE BOUNDARY LAYER COUPLING

S.P. Newman and P. Stow
(Rolls-Royce Limited, Derby)

1. INTRODUCTION

The effect on the surface boundary layer on an inviscid blade-to-blade flow is in general important in terms of blockage and flow deviation and consequently the boundary layer and inviscid calculations must be coupled together iteratively. There are two main procedures for modelling the effect of the boundary layer on the inviscid flow, a displacement model or a transpiration model. In the former the blade geometry is altered by the boundary layer displacement thickness and the inviscid flow is calculated around the modified geometry. With the transpiration model transpiration of mass, momentum and energy through the blade surface is used to model the effects of the boundary layer. Matching conditions at the edge of the boundary layer determine the transpiration parameters to be used. The choice of model is often determined by properties of the inviscid method being used; for example the displacement model being adopted with methods where the use of a boundary streamline is essential eg. a stream-function approach or the streamline curvature method. With a finite volume or finite element inviscid method the transpiration model is easily incorporated.

With either a finite difference or an integral boundary later method the iterative coupling with the inviscid mainstream calculation needs to be given special consideration irrespective of whether a displacement or transpiration model is adopted. In cases where the boundary layer remains attached and a second order effect then direct mode coupling can be adopted; in other cases inverse or semi-inverse coupling needs to be used. These are discussed below in the context of a full potential blade-to-blade method that adopts a finite element solution procedure.

2. BOUNDARY LAYER INCLUSION

The inviscid calculation adopted in this paper is a finite element full potential flow method presented for two-dimensional flow by Whitehead and Newton (1985) extended to include quasi-three dimensional effects by Cedar and Stow (1985a).

A transpiration boundary layer model is adopted using an integral method based on the method of Luxton and Young (1958) for laminar flow and the lag-entrainment method of Green, Weeks and Brooman (1973) for turbulent flow. Also included are correlations to account for laminar separation and reattachement and to handle the start and end of transition.

For the inviscid flow the continuity equation is

$$\nabla.\rho \underset{\sim}{W} h = 0 \tag{2.1}$$

where, ρ is the density, $\underset{\sim}{W}$ the velocity vector relative to the rotating blade and h is the stream-tube height.

Equation (2.1) is solved using a Galerkin weighted residual approach leading to a system of equations of the form,

$$\sum_{e(I)} \iint \rho \underset{\sim}{W} h.\nabla N_I dA = \sum_{e(I)} \int \rho \underset{\sim}{W} h.\underset{\sim}{n} N_I \; dS \tag{2.2}$$

for each node I, where N_I represents the shape function, S represents the perimeter of the domain of integration and where the summation is over all elements containing I as a node. The right-hand side of equation (2.2) is non-zero only for nodes on the blade surface and at inlet and outlet to the domain. With the transpiration boundary layer model it can be shown that,

$$\rho W_n = \frac{1}{rh} \frac{d}{ds} [rh\rho W_s \delta^*] \tag{2.3}$$

where W_s and W_n are the streamwise and normal components of the velocity.

A velocity potential is introduced and three node straight sided triangular elements adopted so that the integrals involved in equation (2.2) and the left-hand side of equation (2.3) are easily evaluated, see (Whitehead and Newton, 1985). The final system of equations may be written as

$$\underset{\sim}{A}(\underset{\sim}{\phi}) = \underset{\sim}{C} - B \underset{\sim}{\delta}^* \tag{2.4}$$

where ϕ is the vector of unknown potentials, $\underset{\sim}{A}$ is an N*1 column vector, N being the total number of nodes, $\underset{\sim}{C}$ is formed from the boundary conditions (being non-zero) only for nodes on the inlet and exit boundaries), $\underset{\sim}{\delta}^*$ is the vector of displacement thicknesses and B is an $N*N_s$ matrix where N_s is the number of surface nodes.

3. DIRECT MODE COUPLING

Both B and $\underset{\sim}{\delta}^*$ depend on the surface velocity and with direct mode coupling are evaluated using the last cycle values. The scheme may be written as

$$\underset{\sim}{A}\,(\underset{\sim}{\phi}^{(n+1)}) = \underset{\sim}{C} - (B\underset{\sim}{\delta}^*)^{(n)} \qquad (3.1)$$

for the inviscid flow and

$$\underset{\sim}{\delta}^{*(n+1)} = \underset{\sim}{\delta}^*(\underset{\sim}{W}_s^{(n+1)}) \qquad (3.2)$$

for the boundary layer, where the superscript denotes the iteration number; Fig. 1. illustrates the linking adopted. The system of equations (3.1) are solved using a Newton-Raphson linearization technique which putting

$$\underset{\sim}{\phi}^{(n+1)} = \bar{\underset{\sim}{\phi}} + \underset{\sim}{\phi}' \qquad (3.3)$$

may be written as

$$J.\underset{\sim}{\phi}' = [\underset{\sim}{C} - \underset{\sim}{A}(\underset{\sim}{\phi}) - (B\underset{\sim}{\delta}^*)^{(n)}] \qquad (3.4)$$

where J is the Jacobian matrix

$$J_{ij} = \frac{\partial A_i}{\partial \phi_j} \qquad (3.5)$$

and can be obtained analytically. The iterative procedure is repeated until convergence is achieved. Once a new approximation to the blade surface velocities is obtained $\underset{\sim}{\delta}^*$ can be calculated using the boundary layer method and the right-hand side of equation (3.4) up-dated. In practice under relaxation of the procedure is used.

This system is stable and rapid convergence is found for attached boundary layers but in regions where the boundary layer is separated or near separation then it either fails to converge or requires such heavy under relaxation that it is impracticable to use. To avoid these convergence difficulties it is possible to employ 'fixes', such as imposing an upper limit on the form factor, but this puts the accuracy of the results in doubt.

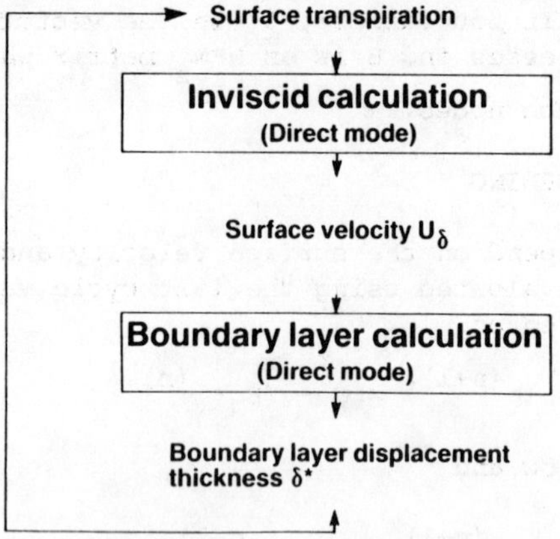

Fig. 1 Direct Mode Boundary Layer Coupling

4. SEMI-INVERSE MODE COUPLING

To overcome the problems associated with separation, it is necessary to adopt a different procedure in the region where the boundary layer is separated or near separation, while maintaining the direct mode coupling in the rest of the flow field. There are two obvious alternatives which could be adopted, these are fully inverse mode coupling as shown in Fig. 2, and semi-inverse mode coupling (see Fig. 3). The major difference between these two procedures is the treatment of the inviscid calculation.

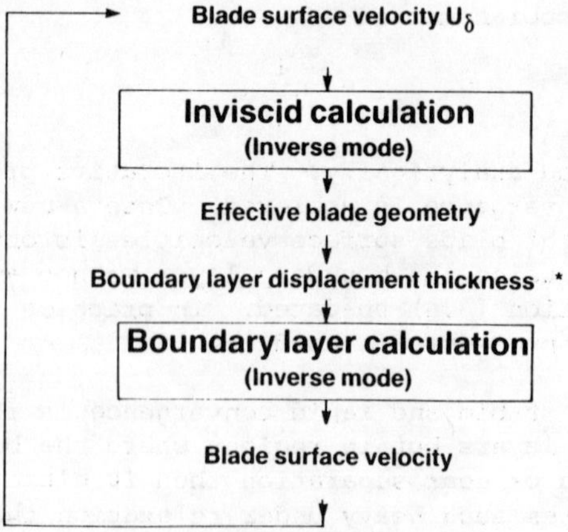

Fig. 2 Inverse Mode Boundary Layer Coupling

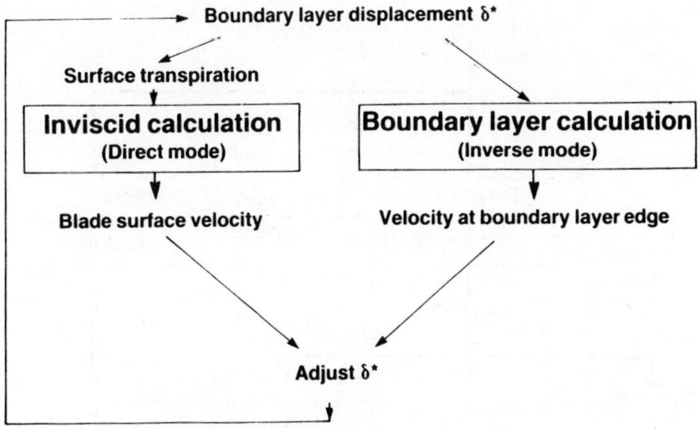

Fig. 3 Semi-Inverse Boundary Layer Coupling

The fully inverse mode coupling requires an inverse inviscid calculation. This has already been developed for the finite element method, by Cedar and Stow (1985b), in the form of an influence matrix, which relates changes in velocity to changes in displacement thickness. While, it would be possible to implement a fully inverse mode coupling procedure, it would be complicated to link it to the direct mode coupling.

The semi-inverse mode coupling is inherently much simpler to link with the direct mode because both procedures use the direct inviscid calculation. Due to the relative ease of implementation, the semi-inverse mode coupling was developed. The form of the linking between the direct mode coupling and semi-inverse mode coupling is shown in Fig. 4. The Semi-Inverse Mode Coupling requires an inverse boundary layer calculation, the one used is based on the method presented by Lock and Firmin (1981), which basically recasts the equations used in the direct boundary layer calculation. Besides the inverse boundary layer calculation, the semi-inverse mode coupling also requires an updating procedure to complete each iterative cycle, as can be seen in Fig. 3. Several updating procedures were tried, including those of Le Balleur (1981) and Carter (1979), but these were found to be slow to converge. The analysis, from which Le Balleur produces his procedure, is based upon single aerofoil theory rather than blade-to-blade analysis required here. A new procedure was therefore sought which would meet the requirements of blade-to-blade schemes.

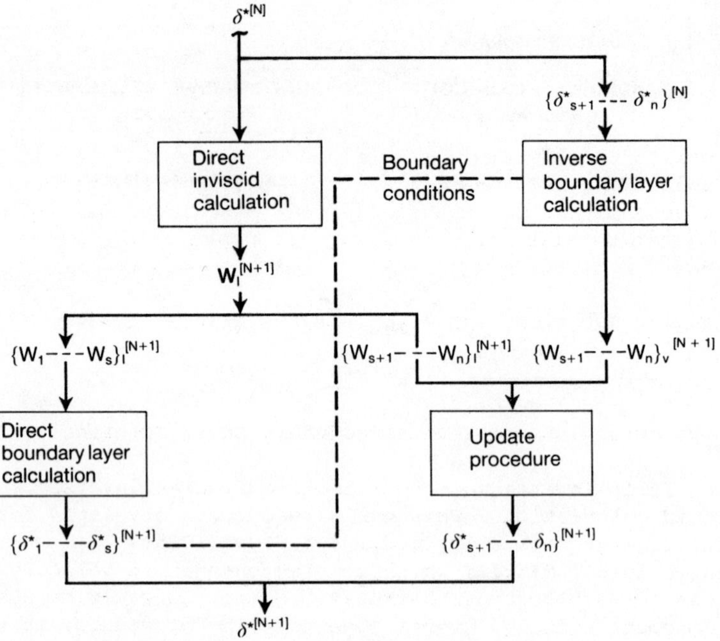

Fig. 4

The updating procedure is required to estimate a correction $\underset{\sim}{\delta}^{*\prime}$, to the displacement thickness distribution, $\underset{\sim}{\delta}^{*}$ such that both the inviscid calculation and the inverse boundary layer calculation will generate the same velocity distribution in the next iteration.

First consider the direct inviscid calculation. Suppose there is a solution $\overline{\underset{\sim}{\phi}}$ to equation (2.4), then a change to the displacement thickness $\underset{\sim}{\delta}^{*\prime}$ can be related to a change in velocity potential $\underset{\sim}{\phi}'$ by equation (3.4), giving:-

$$J(\overline{\underset{\sim}{\phi}})\ \underset{\sim}{\phi}' = B(\overline{\underset{\sim}{\phi}})\ \underset{\sim}{\delta}^{*\prime} \qquad (4.1)$$

The change in surface velocity, $\underset{\sim}{W}'$, can be related to the change in velocity potential by:-

$$\underset{\sim}{W}' = D\underset{\sim}{\phi}' \qquad (4.2)$$

Hence

$$\begin{aligned} \underset{\sim}{W}' &= DJ^{-1}B\underset{\sim}{\delta}^{*\prime} \\ &= K\,\underset{\sim}{\delta}^{*\prime} \end{aligned} \tag{4.3}$$

K is called the influence matrix because it gives the influence of displacement thickness on the surface velocity distribution. The above analysis is essentially that given by Cedar & Stow (1985b) and details are given in their paper on how to calculate K.

Changing notation, equation (4.3) is written as

$$\underset{\sim}{W}_I' = K\,\underset{\sim}{\delta}^{*\prime} \tag{4.4}$$

where $\underset{\sim}{W}_I$ denotes the velocity distribution calculated from the inviscid analysis.

The inverse boundary layer equations are expressed as

$$\underset{\sim}{W}_v = f(\underset{\sim}{\delta}^{*}) \tag{4.5}$$

where $\underset{\sim}{W}_v$ denotes the velocity distribution calculated from the boundary layer equations. From equation (4.5) an approximate expression for $\underset{\sim}{W}_v'$ can easily be obtained.

$$\underset{\sim}{W}_v' = \frac{d\underset{\sim}{W}_v}{d\underset{\sim}{\delta}^{*}} \tag{4.6}$$

where $\dfrac{d\underset{\sim}{W}_v}{d\underset{\sim}{\delta}^{*}}$ is called the boundary layer influence matrix because it gives the influence of a change in velocity on the displacement thickness. Given that there is a 'true' velocity distribution, $\underset{\sim}{W}_T$ then

$$\underset{\sim}{W}_T = \underset{\sim}{W}_v + \underset{\sim}{W}_v' = \underset{\sim}{W}_I + \underset{\sim}{W}_I' \tag{4.7}$$

An expression for $\underset{\sim}{\delta}^{*\prime}$ can be obtained by substituting equations (4.4) and (4.6) into equation (4.7) to give

$$\underset{\sim}{\delta}^{*\prime} = \left[K - \frac{d\underset{\sim}{W}_v}{d\underset{\sim}{\delta}^{*}}\right]^{-1} (\underset{\sim}{W}_v - \underset{\sim}{W}_I) \tag{4.8}$$

Equation (4.8) represents a global correction for the displacement thickness.

It remains to generate the boundary layer influence matrix. Examining equation (4.6), it can be seen that if $\underset{\sim}{\delta}^{*\prime}$ is replaced by

$$\underset{\sim}{\delta}^{*\prime} = \begin{pmatrix} 0 \\ \vdots \\ 0 \\ 1 \\ 0 \\ \vdots \\ 0 \end{pmatrix} - j^{th} \text{ element} \tag{4.9}$$

then the elements of the j th column of the boundary layer influence matrix can be written as

$$\frac{dW_{v_i}}{d\delta^*_j} = W'_{v_i} \qquad \text{for all i.} \tag{4.10}$$

So, by perturbing a member of δ^* and using the resulting distribution as input for an inverse boundary layer calculation, a column of the boundary layer influence matrix can be generated. By repeated application of this procedure the full matrix can be obtained, the resulting matrix being lower triangular in form.

5. RESULTS

The semi-inverse mode coupling scheme is being tested against experimental results. To illustrate the capability of the method, two sets of predictions are given here with the corresponding experimental data. The test cases are from a compressor cascade from the Compressor Dept., Rolls Royce Ltd., the blade section being shown in Fig.5 with a finite element mesh. The first case shows a small turbulent separation on the suction surface of the blade. For this case the predicted boundary layer parameters compare well with the experimental results (see Fig. 6a). It should be noted that for this case convergence can be obtained using the direct mode coupling, with heavy under relaxation in 30 iterations while the semi-inverse mode coupling only requires 10 iterations.

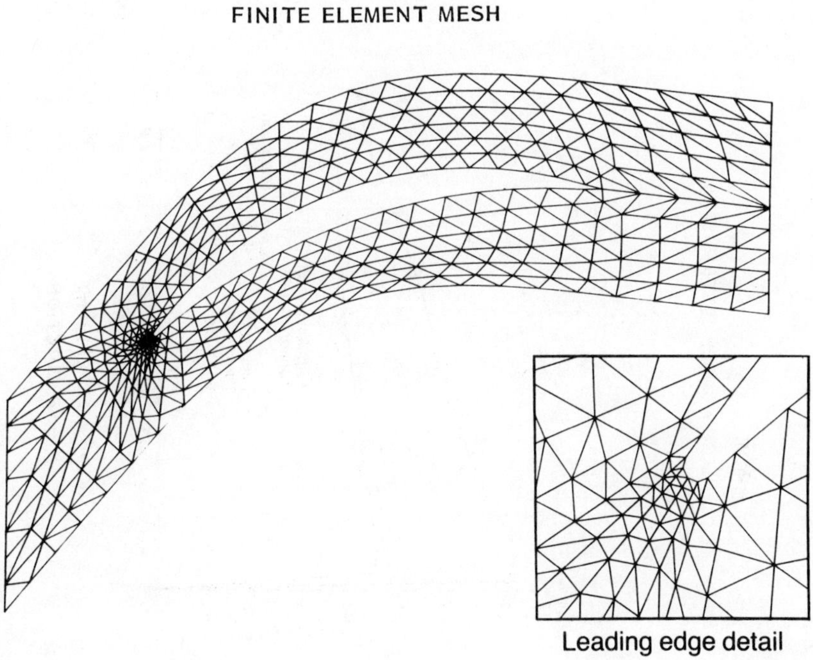

Fig. 5

The second case has a pronounced suction surface separation. This case will not converge with the direct mode coupling but converges in 13 iterations with the semi-inverse mode coupling. It can be seen that although boundary layer growth is underestimated (see Fig. 7a), the results are encouraging.

The contraction ratios measured on the cascade were imposed as linear variations in streamtube height. This simple approximation will account for some of the discrepancies seen in the Mach number distributions, Figs. 6b and 7b.

6. CONCLUSIONS

The semi-inverse mode coupling has been successfully implemented in a finite element blade-to-blade program. The global update procedure provides fast convergence, allowing the program to be run interactively. The comparisons with experimental data show that predictions can now be easily achieved for separated turbulent boundary layers.

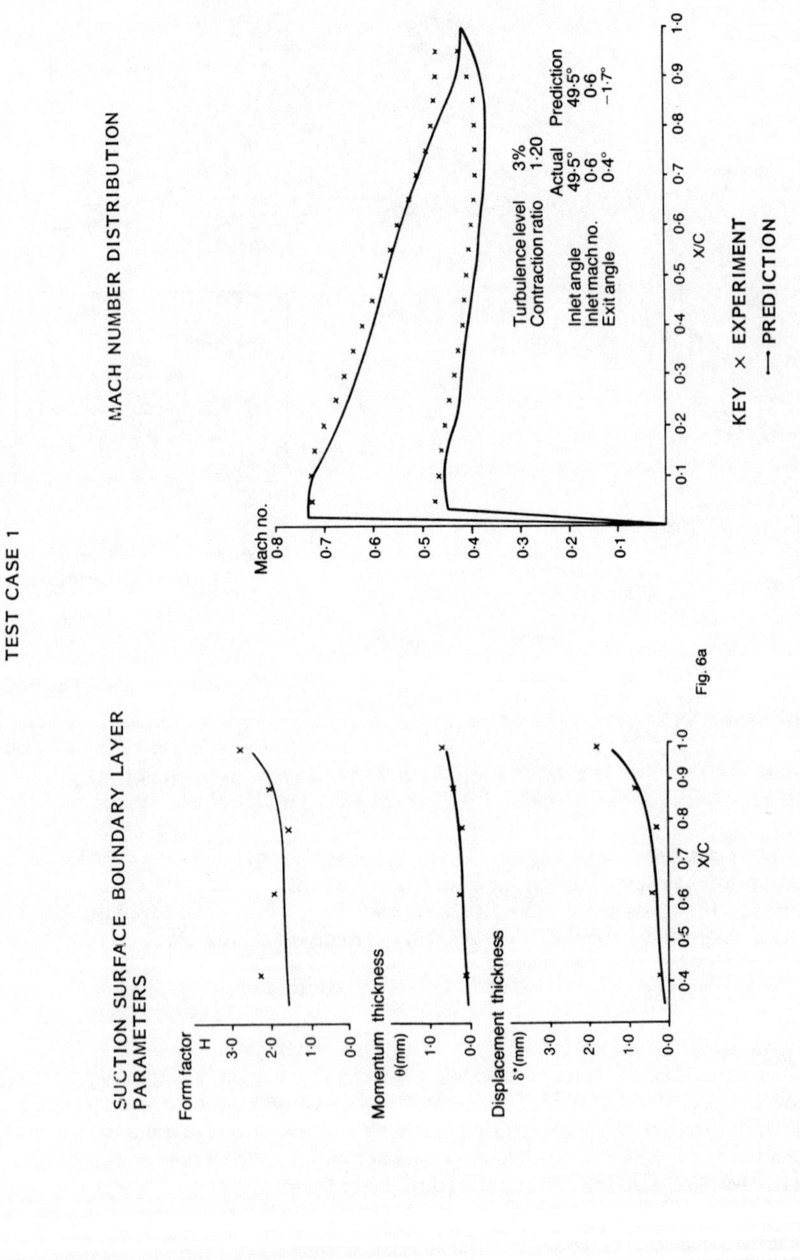
TEST CASE 1
SUCTION SURFACE BOUNDARY LAYER PARAMETERS
Form factor H
Momentum thickness θ(mm)
Displacement thickness δ*(mm)
X/C
Fig. 6a
MACH NUMBER DISTRIBUTION
Mach no.
Turbulence level 3%
Contraction ratio 1·20
Actual Prediction
Inlet angle 49·5° 49·5°
Inlet mach no. 0·6 0·6
Exit angle 0·4° −1·7°
X/C
Fig. 6b
KEY × EXPERIMENT
— PREDICTION

Fig. 6

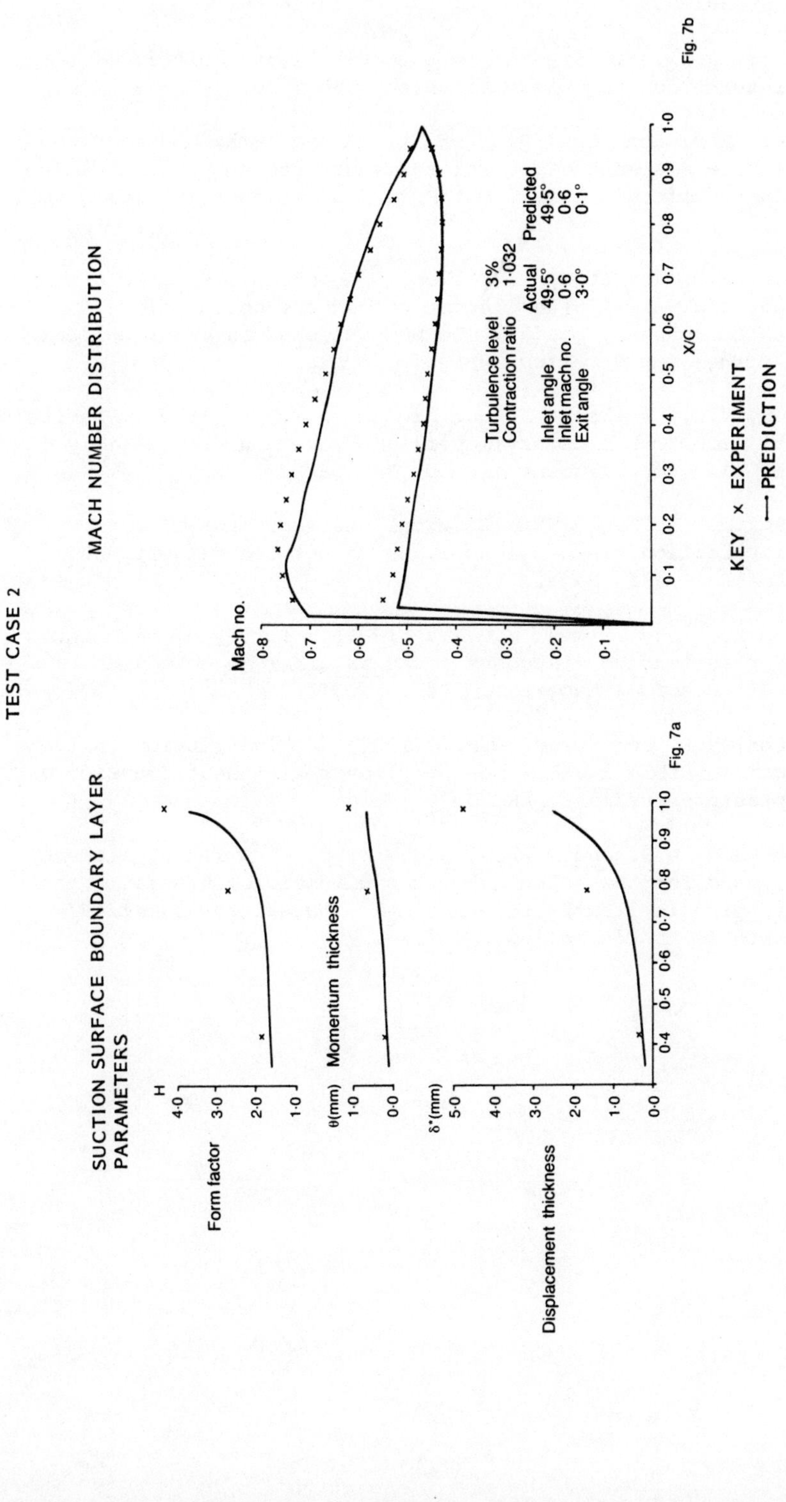
TEST CASE 2
SUCTION SURFACE BOUNDARY LAYER PARAMETERS
Form factor
H
4·0
3·0
2·0
1·0
Momentum thickness
θ(mm)
1·0
0·0
Displacement thickness
δ*(mm)
5·0
4·0
3·0
2·0
1·0
0·0
0·4 0·5 0·6 0·7 0·8 0·9 1·0
Fig. 7a
MACH NUMBER DISTRIBUTION
Mach no.
0·8
0·7
0·6
0·5
0·4
0·3
0·2
0·1
0·1 0·2 0·3 0·4 0·5 0·6 0·7 0·8 0·9 1·0
X/C
Turbulence level 3%
Contraction ratio 1·032
Actual Predicted
Inlet angle 49·5° 49·5°
Inlet mach no. 0·6 0·6
Exit angle 3·0° 0·1°
Fig. 7b
KEY × EXPERIMENT
— PREDICTION

Fig. 7

7. REFERENCES

Carter, J.E., (1979). A new boundary layer inviscid interaction technique for separated flow, *AIAA* 79-1450.

Cedar, R.D. and Stow, P. (1985a). A quasi-three dimensional finite element method for transonic and supersonic flows. *International Journal for Numerical Methods in Fluids* Vol. 5. 101-114

Cedar, R.D. and Stow, P. (1985b). A compatible mixed design and analysis finite element method for design of turbomachinery blades. To be published in *International Journal for Numerical Methods in Fluids*.

Green, J.E., Weeks, D.J. and Brooman J.W.F., (1973) Prediction of turbulent boundary layers and wakes in compressible flow by a lag-entrainment method. RAE TR72231.

Le Balleur, J.C., (1981). Calcul des ecoulment a forte interaction visqueuse an moyen de methode de couplage, *AGARD CP-291*, Paper 1.

Lock, R.C. and Firmin, M.C.P., (1981). Survey of techniques for estimating viscous effects in external aerodynamics". RAE Technical Memorandum Aero 1900.

Luxton, R.E. and Young, A.D., (1958). Skin friction in the compressible laminar boundary layer with heat transfer and pressure gradient, ARC20336.

Whitehead, D.S. and Newton, S.G., (1985). A finite element method for the solution of two dimensional transonic and supersonic flows. *International Journal for Numerical Methods in Fluids*, Vol. 5 115-132.

ADAPTIVE CONSTRUCTION OF GRID-SYSTEMS FOR FLOW SIMULATIONS

L. Fuchs

(Department of Gasdynamics, The Royal Institute of Technology, Stockholm, Sweden)

1. INTRODUCTION

The accuracy of numerical solutions to differential equations modelling flows depends on physical parameters (e.g. Mach number, Reynolds number), the geometrical shape and numerical factors (local mesh spacing, type of discretization and order of approximation). These factors are inter-related and only in very simple cases can one determine the effects of each factor a priori. A basic parameter in all cases is the ratio of the numerical length scale (i.e. the local mesh spacing) and the local physical length scale. The last parameter may vary between just a few mean-free-paths (i.e. the shock thickness) to as much as ten orders of magnitude larger scales (near free-stream conditions). Generally, a priori grid generation techniques will lead to either inaccurate numerical solutions due to the coarseness of the grid, or to a waste of computational resources without any gain in accuracy. Thus, in most cases one has to use some adaptive technique for solving numerically, non-trivial problems.

Sophisticated methods for 1-D (steady and time-dependent) problems have been proposed by Pereyra et al. (1975), Chong (1978) and Davis et al. (1982). An adaptive grid generation procedure for multidimensional problems has been developed by Dwyer et al. (1980). This scheme is based upon the placement of grid points in proportion to the gradients in the dependent variables. Brandt (1977, 1979 and 1984) describes a rather general way of using Multi-Grid (MG) methods in an adaptive manner. The basic principles are simple but the implementation of Brandt's adaptive criteria have been limited to 1-D problems. These criteria can adapt the local mesh and the local order (of finite-difference approximations) so that the numerical resultsare optimal in some sense (e.g. minimal amount of computational effort for a given level of accuracy). The generalization of the scheme to higher dimensions makes it much more complicated and the total efficiency would depend on the proper choice of the residual weighting function.

The present scheme is based also on the MG algorithm. Locally refined grids are introduced whenever the local truncation errors are 'large'. These local grids may be defined completely independently from the grid on which the solution is given (approximately), by using some proper local coordinate system. In this way the amount of the additional memory required to store grid data is minimum. The criteria for determining the location of the local grids and the stage when the process should be stopped are kept as simple as possible. Higher order approximations can be incorporated into the scheme, in a simple manner, without any need to modify the controlling criteria. The relative error of the numerical results can also be estimated. In the following we describe the MG scheme with controlled local mesh refinements. The application of the scheme for some flow problems is discussed.

2. MULTI-GRID METHODS WITH LOCAL MESH REFINEMENTS

Basic MG methods use a sequence of m grids G_k, $(1 \leq k \leq m)$ where each coarse grid is derived from a finer one by doubling the mesh spacing. The differential problem, $Lu=R$ is approximated by the set of problems $L_k u_k = R_k$. These discrete problems (e.g. defined by some finite differences) are interconnected through the following relations:

$$R_k = I_k^{k+1}(R_{k+1} - L_{k+1}u_{k+1}) + L_k J_k^{k+1} u_{k+1} \qquad (2.1)$$

Converged solutions, on different grids, are related through $u_k = J_k^{k+1} u_{k+1}$. The operator J_k^{k+1} denotes a restriction operator for the transfer of the dependent variable(s). The operator I_k^{k+1} is a corresponding operator for the residual transfer. The use of the dependent variables on coarse grids together with (2.1) results in the FAS scheme of Brandt (1977 and 1984). As pointed out by Brandt, expression (2.1) contains also information about the local truncation errors since

$$\tau_k^{k+1} = L_k J_k^{k+1} u_{k+1} - I_k^{k+1} L_{k+1} u_{k+1} \qquad (2.2)$$

where τ_k^{k+1} is proportional to the difference between the truncation errors of L_{k+1} and L_k. The accuracy of the converged solution on coarse grids, u_k, is the same as the accuracy of the fine grid solution u_m. In this way the function of the fine grids may be viewed as providing truncation error corrections for the coarse grid solvers. In the more conventional sense, coarse grids are regarded as means for accelerating the iterative process. We use this "dual view" of MG methods both for constructing locally refined grids to improve the accuracy of relatively (uniform) coarse grid solutions, and for solving efficiently the discrete problem.

Recently (Fuchs, 1985a), we have introduced a local mesh refinement technique in the MG solution process. The basic scheme has also been extended to include general overlapping local grids (Fuchs, 1985b). The scheme requires (for uniform subgrids) the storage of the coordinates of one node point, the relative angle of rotation of the local grid, and the number of node points in that grid. In all, the scheme requires the storage of seven scalars for each sub-grid. To describe accurately certain geometries, some of the sub-grids may use general, curve-linear coordinates. In such cases, the parameters of the node points must be stored only for those particular sub-grids.

3. ADAPTIVE CONSTRUCTION OF GRID-SYSTEMS

For accurate numerical solutions, with small amount of computational effort, we use certain grid and relaxation controlling criteria. These criteria are kept as simple as possible so that the controlling part of the algorithm requires negligible amount of computational work.

The solution procedure starts on a uniform and relatively coarse grid. The grid is refined (by halving the mesh globally) and the problem is solved again on the new grid. The initial approximation is better compared to an arbitrary guess, since the solution on the coarser grid is used to generate the new approximation (by interpolation). After repeating this global refinement process several times, the grid in larger parts of the domain is fine enough, and refinements are done only at certain subdomains. Solutions on the composed (uniform and locally refined subgrids) grids are computed by the same basic MG solver that is used to solve the problem on uniform grids. The sequence of solutions u_i that is computed on successively refined grids gives direct information about the relative accuracy of the numerical results. The relative error, ε_i, in the solution u_i, is given by:

$$\varepsilon_i = \|(u^* - u_i)\|/\|u_i\|; \quad (1 \leq i \leq m) \tag{3.1}$$

where u* is the (extrapolated) "zero-mesh" solution. This relative error estimate is used to terminate the construction of more refined grid-systems. That is, when ε_m, (i.e. the error on the finest grid-system) is less than some prescribed value (say 1%) the solution is accepted and the adaptive solution procedure is terminated.

The mean accuracy of numerical solutions can be improved by reducing the local truncation errors. It must be observed that successive construction of the numerical solution gives information where the errors in the solutions are large. However, this information cannot be used to determine where the mesh should be refined. This is so since the largest errors in the solution and the largest truncation errors do not necessarily coincide. The only real measure of the error of the approximation is the relative size of the truncation errors. On the other hand, the truncation errors cannot be used to determine the accuracy levels of numerical solutions. Thus, both these measures must be used for controlling the adaptive procedure.

Optimally, the level of the truncation errors on the final grid-system is uniform. This goal may be used to define regions with 'large' local truncation errors. During the MG solution procedure one can estimate [eq. (2.2)] the local truncation errors during the transfer of the problem to coarser grids. The sequence of local truncation errors, at some subregions, reveal when the discrete approximation converges with the theoretical (asymptotical) rate. This asymptotical convergence rate is achieved only when the local scales of the flow field, are being resolved. Therefore, it is important to ensure that this is the case indeed (or otherwise simulate the subgrid scales by some other model). The regions where the grid should be refined is determined by the local truncation error, τ_1. That is, if

$$\tau_1 \geq C\,\bar{\tau} \tag{3.2}$$

the grid in that region should be refined. C is a constant of order one, and $\bar{\tau}$ is the (global) mean of the truncation error.

Controlling the local mesh refinements by relation (3.2) works well in smooth cases. However, many flow problems are associated with 'singularities' of different types. Inner, incompressible viscous flows, with errors in the outflow boundary conditions, are associated with mesh independent, non-vanishing truncation errors, near the outflow boundary. In the case of transonic flows with shocks, the largest 'truncation errors' in the vicinity of the shocks behave as $O(h^{-1})$ (where h is the typical local mesh spacing). In the case of driven cavity flows (with viscous incompressible fluids), the local truncation errors near the corners close to the moving wall behave as $O(h^{-2})$. In these examples (discussed in Sec. 4), and especially in the last case, relation (3.2) is not proper for controlling the local mesh. In those cases where the truncation error estimates do not decrease as the mesh is refined, we use another mesh controlling parameter. In these cases we identify the location of the boundaries of the region that should be refined. (Usually, this is a region around the singularity.) We compute the mean truncation errors near the boundaries of a subdomain. This truncation error on the grid G_k is denoted by τ_b^k. The region (in which the mesh is to be refined) should be enlarged when

$$\tau_b^k \geqq \beta \, \tau_k^{k-1} \tag{3.3}$$

where β is a constant. Optimally, the ratio of the estimated truncation errors would be 2^{-p} (where p is the order of the discrete approximation on both grids). Thus, $\beta=2^{-p}$ is a sufficient condition that the local subregion is large enough. By numerical experiments we have found that β may be considerably larger. For the transonic cases below, it was found that $\beta \leqq 1$ is adequate for solutions that are not affected by variations in the extent of the locally refined region.

In all, the simple criteria (3.1)-(3.3) enable us to control the process of constructing grids during the solution procedure. When the accuracy of the solution (3.1) is not used for terminating the refinement process, the intermediate grid solutions can be computed by about a single MG cycle. Thus, the total amount of computational work to obtain the final solution is determined largely by the number of unknowns in the finest grid-system.

4. COMPUTED EXAMPLES

The basic adaptive scheme for constructing proper systems of grids have been applied both for the computation of incompressible viscous flows (both in terms of the streamfunction and the vorticity, and in terms of the primitive variables) and the simulation of potential transonic flows past airfoils. Some examples representing both these different flow regimes are given in the following.

The basic MG schemes for the solution of incompressible viscous flows are described by Thunell et al (1981) and Fuchs (1984). Three-dimensional extensions are given in (Fuchs, 1985c). The management of local grid refinement is described in (Fuchs, 1985a) and the application to multi-zonal grids in (Fuchs, 1985b). Here, we study the application of the scheme to computation of the flow in a channel with a backward facing step, since the problems have been used in many computations (see e.g. Morgan et al (1984)). A typical dilemma in these computations is the choice of a proper grid-system. A closer study of the different grids that have been used in (Morgan et al, 1984) reveal that intuitively one would place a very fine grid in the recirculation region, and the grid elsewhere is determined, more or less, arbitrarily (depending on the method for generating the grid).

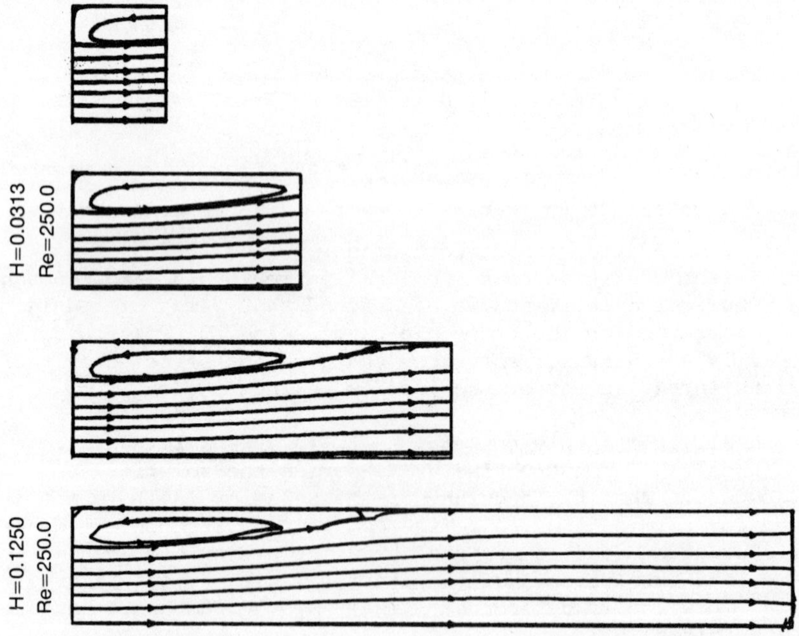

Fig. 1 The streamline pattern in the channel with a backward facing step (local mesh refinements). Re=250

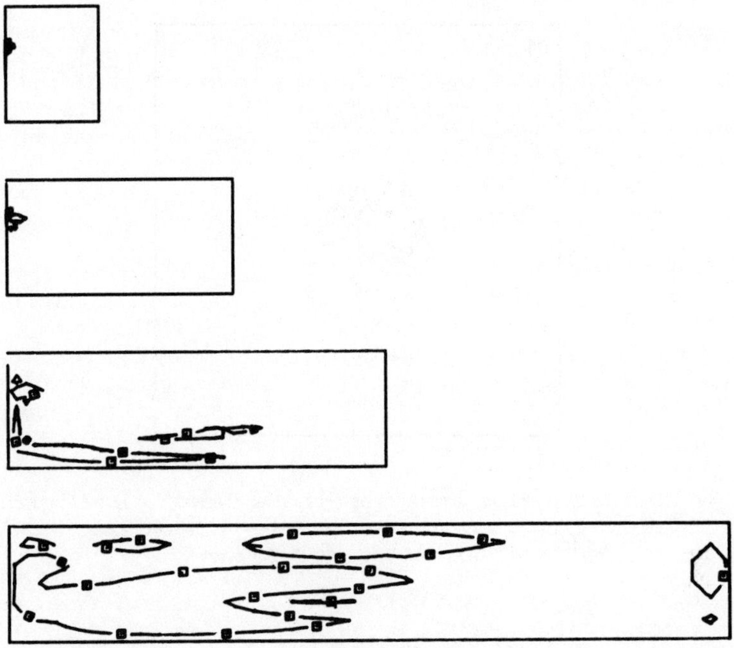

Fig. 2 The truncation error map of the cases corresponding to Fig. 1

Our computed streamline pattern (Re=250) on a sequence of grids (a global grid and three levels of local refinements) is shown in Fig. 1. The truncation error maps, on these grids, are shown in Fig. 2. On the global (uniform) grid one can notice that the largest truncation errors are located close to the entrance and near the outflow boundary (and not in the recirculating region). The errors located in the outflow region cannot be reduced by local mesh refinements. These errors are due to the application of free-stream conditions at a too short distance. Using other types of boundary conditions (see Fuchs et al, 1984) will produce small "truncation errors". By local mesh refinements one can reduce the genuine truncation errors in a rational and efficient manner (Fig. 2).

Next, consider the computation of potential transonic flows past airfoils. The full details of the MG scheme are given in (Gu, 1985). For a lifting supercritical case (flow about a NACA-0012 airfoil at free-stream Mach number, $M_\infty = 0.80$, and angle of attack, $\alpha = 1^\circ$) solutions have been computed on a sequence of grids (with local refinements around the airfoil). The truncation error levels on a uniform grid and after 2 and 4 steps of local mesh refinements, respectively, are shown in Fig. 3. The extent of the truncation errors is reduced, with

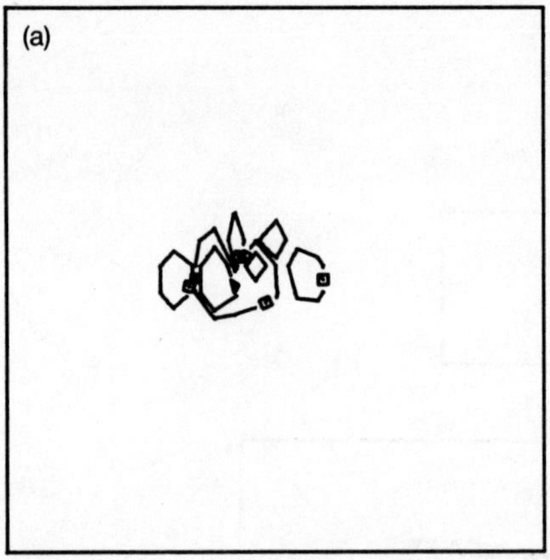

Fig. 3a The truncation error map on the finest global grid (with l/h=8, the number of intervals on each side of the NACA-0012 airfoil). M=0.80, $\alpha=1^{\circ}$

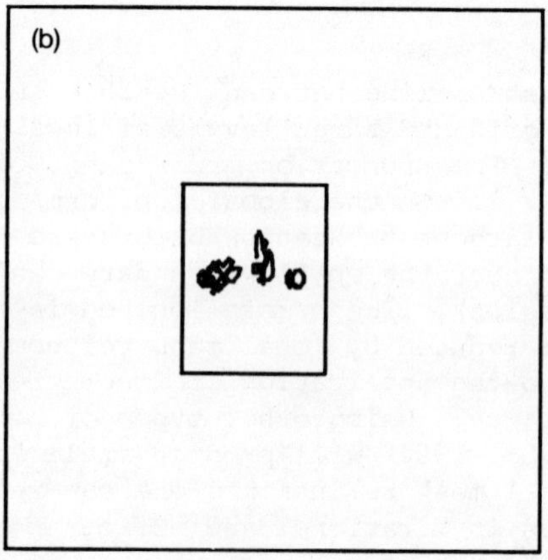

Fig. 3b The same case as in Fig. 3a, after two steps of local mesh refinements around the airfoil (l/h=32)

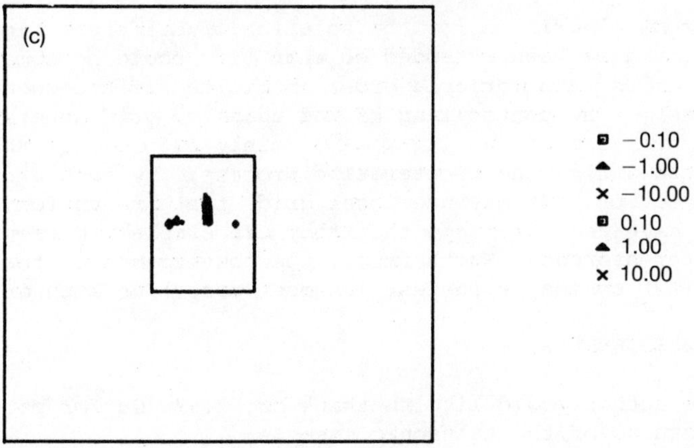

Fig. 3c The same case as in Fig. 3b, after two additional steps of local mesh refinements around the airfoil (1/h=128)

increased refinement. Three clear zones emerge: leading- and trailing edges, and the shock zones. The largest values of the truncation errors cannot be reduced by further refinements (because of the singularities near stagnation points and shocks). In this case relation (3.3) has to be used to produce correct criteria for local grid refinements. When β in (3.3) is too large, high frequency oscillations appear in the solution. These oscillations are detected easily as (large amplitude) oscillations in the truncation errors on the local grid (see Fig. 4). In such cases, one may increase the extent of the local fine grid until (3.3) is satisfied.

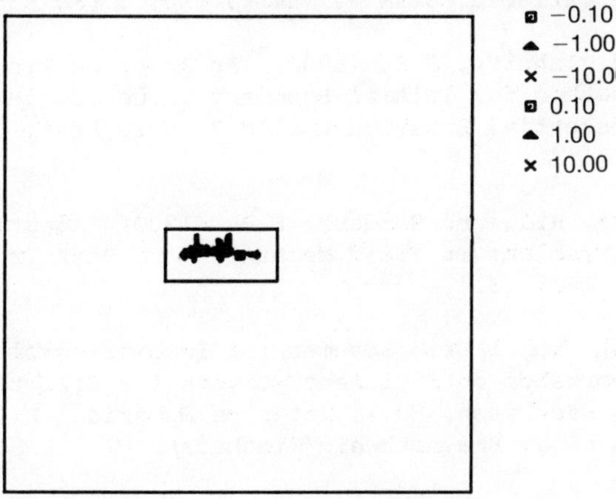

Fig. 4 The same case as in Fig. 3c, using locally refined grids in too small regions

6. CONCLUDING REMARKS

Our MG algorithms for the solution of different flow problems have been extended so that they could accommodate local grids (and variable order of finite difference approximations). The controlling of the adaptive grid generation process is simple and it requires mainly information that is produced during the MG iterative process. By such a grid construction, one may use local grids that are uniform. These grids have the advantage that they are simple and require least computer storage. Furthermore, the convergence of the iterative (MG) scheme is fastest (in most cases) on such grids.

ACKNOWLEDGEMENTS

The author would like to thank Mr. C.-Y. Gu for producing the results of the transonic cases.

REFERENCES

Brandt, A. (1977) Multi-Level Adaptive Solutions to Boundary Value Problems. *Mathematics of Computations,* **31**, 333-390.

Brandt, A. (1979) Multi-Level Adaptive Computations in Fluid Dynamics. *AIAA J.,* 79-1455.

Brandt, A. (1984) Multigrid Techniques; 1984 Guide. Computational Fluid Dynamics Lecture Series at the von-Karman Institute.

Chong, T.H. (1978) A Variable Mesh Finite-Difference Method for Solving a Class of Parabolic Differential Equations in One Space Variable. *SIAM J. Numer. Anal.,* **15**, 835.

Davis, S.F. and Flaherty, J.E. (1982) An Adaptive Finite-Difference Method for Initial-Boundary Value Problems for Partial Differential Equations. *SIAM J. Sci. Stat. Comput.,* **3**, 6-27.

Dwyer, H.A., Kee, R.J. and Sanders, B.R. (1980) Adaptive Grid Methods for Problems in Fluid Mechanics and Heat Transfer. *AIAA J.,* **18**, 1205-1212.

Fuchs, L. (1984) Multi-Grid Schemes for Incompressible Flows, Proc. GAMM-workshop on Efficient Solvers for Elliptic Systems. (W. Hackbusch, Ed.), Notes on Numerical Fluid Mechanics, Vieweg, Braunschweig/Wiesbaden, **10**, 38-51.

Fuchs, L. and Zhao, H.-S. (1984) Solution of Three-Dimensional Incompressible Flows by a Multi-Grid Method. *Int. J. Numerical Methods in Fluids,* **4**, 539-555.

Fuchs, L. (1985a) A Local Mesh-Refinement Technique for Incompressible Flows. *Computers and Fluids,* To appear.

Fuchs, L. (1985b) Numerical Flow Simulations Using Zonal-Grids. *AIAA,* 85-1518.

Fuchs, L. (1985c) Computation of Viscous Laminar Flows in Cavities. Proc. Conf. Numerical Methods in Laminar and Turbulent Flow, Swansea 1985, To appear.

Gu, C.-Y. (1985) Numerical Methods for Transonic Potential Flows Past Airfoils. The Royal Inst. Tech. Report. TRITA-GAD-6.

Morgan, K., Periaux, J. and Thomasset, F. (1984) Analysis of Laminar Flow Over a Backward Facing Step. Notes on Numerical Fluid Mechanics, Vol. **10**, Vieweg, Braunschweig.

Pereyra, V. and Sewell, E.G. (1975) Mesh Selection for Discrete Solution of Boundary Problems in Ordinary Differential Equations. *Numer. Math.,* **23**, 261.

Thunell, T. and Fuchs, L. (1981) Numerical Solution of the Navier-Stokes Equations by Multi-Grid Techniques. Numerical Methods in Laminar and Turbulent Flow-II, (C. Taylor and B.A. Schrefler, Eds.), Pineridge Press, Swansea, 141-152.

EPIC - BEYOND THE ULTIMATE DIFFERENCE SCHEME

J.W. Eastwood and W. Arter
(Culham Laboratory (Euratom/UKAEA Fusion Association), Abingdon, Oxford)

1. INTRODUCTION

EPIC - Extended or Ephemeral Particle-in-Cell schemes combine ideas of particle-mesh methods and finite element methods to give numerical schemes which outperform state-of-the-art finite difference schemes in both cost and quality.

Finite element and particle-mesh methods were initially combined in an elegant Hamiltonian formulation by Lewis (1970; Hockney and Eastwood 1981). In multidimensional phase-fluids, integration is performed most economically using monte-carlo techniques, in which sampling points are commonly interpreted as particles. In (magneto-) fluid dynamics, the need to use particles as the principal carriers of information becomes less compelling, in which case the particles can become ephemeral, as in EPIC (Eastwood 1982). In 1-D, or with fractional timestepping, integrals become easy to do exactly, in which case particles become merely conceptual devices and EPIC schemes appear similar to the characteristic finite element methods (Labadie et al. 1981: Morton 1982: Pironneau 1982).

In present-day fusion research, devices such as tokamaks operate with small diffusivities, yet resistive instabilities can lead to the convective redistribution of heat on short timescales. Strong magnetic fields cause large anisotropy of heat and particle flows. Changing field topology and, in 3-D, ergodic flux-lines make field-aligned meshes impractical. We believe EPIC offers an effective solution to these problems: anisotropies are well handled by finite elements, and the almost-hyperbolic limit favours our schemes which treat advection accurately without instability.

2. EPIC

2.1 Lagrangian description

For low dissipation systems it is natural to think in terms of motions of parcels of fluid. We therefore follow the ideas of particle-mesh methods and transform the governing (almost-) hyperbolic equations into co-ordinate mappings.

We recall that if a parcel of fluid initially at $\underset{\sim}{x}_o$ moves to $\underset{\sim}{x}$ at later time t, so that

$$\underset{\sim}{x} = \underset{\sim}{x}(\underset{\sim}{x}_o, t), \tag{2.1.1}$$

then the infinitesimal volume $d\tau_o$ at $\underset{\sim}{x}_o$ is mapped to

$$d\tau = |D| d\tau_o, \tag{2.1.2}$$

where

$$D_{ij} = \frac{\partial x_i}{\partial x_{oj}}, \quad D_{ij}\Big|_{t=o} = \delta_{ij}. \tag{2.1.3}$$

$\underset{\approx}{D}$ is the displacement gradient matrix, and $|D|$ is its determinant. Similarly, the transformation of a vector element of surface area $d\underset{\sim}{S}_o \rightarrow d\underset{\sim}{S}$ is described by

$$d\underset{\sim}{S} = d\underset{\sim}{S}_o . \underset{\approx}{D}^{-1} |D|. \tag{2.1.4}$$

It follows that the density ρ satisfies

$$\rho = |D|^{-1} \rho_o \tag{2.1.5}$$

in the absence of source terms, and the magnetic field $\underset{\sim}{B}$ obeys

$$\underset{\sim}{B} = |D|^{-1} \underset{\approx}{D} . \underset{\sim}{B}(\underset{\sim}{x}_o, 0) \tag{2.1.6}$$

in the absence of resistivity. Sources s add to e.g. (2.1.5) the term

$$|D|^{-1} \int_0^t |D| s \, dt', \tag{2.1.7}$$

which we neglect in order to keep the exposition simple, but which can be calculated using standard numerical techniques.

2.2 Discretisation

We project (2.1.5) onto the set of basis functions $\{W_k(\underset{\sim}{x},t)\}$ to give the equations for the amplitudes $\{\rho\}$ in form:-

$$\int W_k[\tilde{\rho} - \rho_0|D|^{-1}]d\tau = -\int W_k \varepsilon d\tau = 0, \tag{2.2.1}$$

where $\tilde{\rho}$ is the trial function approximation to ρ, i.e. $\rho = \tilde{\rho} + \varepsilon$, $\tilde{\rho} = \rho_\ell \phi_\ell$, $\{\phi_\ell\}$ are trial functions and the summation convention is used. Depending on the choice of time-dependence of $\{W_k\}$, (2.2.1) describes (i) a Lagrangian finite element method, (ii) an Eulerian EPIC scheme or (iii) a moving mesh EPIC scheme.

For (i) $\{W_k\}$ satisfies $W_k(\underset{\sim}{x}) = W_k(\underset{\sim}{x}_0)$ and (2.2.1) becomes

$$\left(\int W_k(\underset{\sim}{x})\phi_\ell(\underset{\sim}{x})d\tau\right)\rho_\ell = \int W_k(\underset{\sim}{x}_0)\rho(\underset{\sim}{x}_0)d\tau_0, \tag{2.2.2}$$

while for (ii) $\partial W_k/\partial t = 0$ so

$$\left(\int W_k(\underset{\sim}{x})\phi_\ell(\underset{\sim}{x})dt\right)\rho_\ell = \int W_k(\underset{\sim}{x})\rho(\underset{\sim}{x}_0)d\tau. \tag{2.2.3}$$

(2.2.2) and (2.2.3) may each be written as

$$A_{k\ell}\rho_\ell = m_k \tag{2.2.4}$$

which is to be solved for unknowns $\{\rho_\ell\}$. m_k may be interpreted as the mass associated with node k. For (i) node masses are fixed and the mass matrix $A_{k\ell}$ changes at each timestep (because the geometry and connectivity of the elements vary), vice versa for (ii), while both $A_{k\ell}$ and m_k change in (iii): (iii) corresponds to adaptive elements which e.g. align themselves according to the direction of the magnet field locally (Eastwood and Lee-Hsaio 1983). In each case, summing over k shows the scheme is conservative.

For the remainder of the paper we shall assume the Galerkin approximation i.e. $\{W_k\}$ and $\{\phi_k\}$ are the same functions. (2.2.1) then minimises the mean-square error in ρ, and in addition nonlinear instability is suppressed because positive definite quadratic quantities can be shown to be non-increasing. We specialise to case (ii) since studying arbitrarily moving elements introduces complications but no fundamental differences.

The lowest order elements which conform for both advection and diffusion terms are piecewise linear. In 1-D, elements of unit width are given by:

$$W_k(x) = W(x-x_k), \quad W(x) = \begin{cases} 1-|x| & ; \quad |x| \leqslant 1 \\ 0 & ; \quad \text{otherwise.} \end{cases} \qquad (2.2.5)$$

The right-hand-side of (2.2.3) becomes

$$m_k = \int W_k(x+\xi) W_n(x)\,dx\, \rho_{on}, \qquad (2.2.6)$$

where ξ is the displacement of point x_o in the time interval $(t, t+\Delta t)$. Representing ξ by piecewise linear functions reduces the integrand in (2.2.6) to a piecewise cubic; (2.2.6) is given by the overlap of two "triangle" functions (Fig. 1). The mass matrix on the left-hand-side of (2.2.3) is tridiagonal with stencil (1/6, 2/3, 1/6), so (2.2.3) is easily inverted to give ρ at time $t+\Delta t$. This gives the timestep loop:

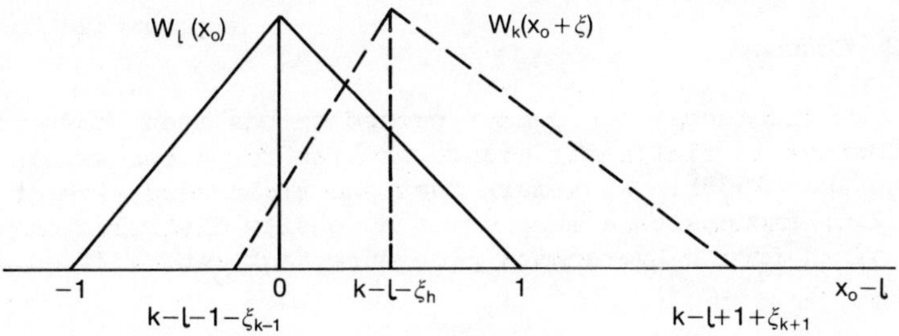

Fig. 1 Contributions to $A_{kl}(\xi)$ are given by the overlap integral of $W_\ell(x_o)$ with the displaced and deformed triangle function $W_k(x_o+\xi)$.

i) $t := t+\Delta t$

ii) find displacements ξ

iii) compute node masses $\{m_k\}$

iv) solve matrix equation $\rho = A^{-1}m$ to get $\{\rho_k(t+\Delta t)\}$

v) go to (i)

3. LINKS WITH OTHER SCHEMES

3.1 FCT

Steps (iii) and (iv) in the above timestep loop are like flux-corrected-transport (Boris and Book 1976 and refs. therein) in that m_k may be interpreted as a low order approximation to new densities $\rho_k(t+\Delta t)$ and inverting the mass matrix as anti-diffusion. Explicitly, (2.2.4) with linear Eulerian elements becomes

$$(\rho_{k-1}+4\rho_k+\rho_{k+1})/6 = m_k,$$

$$\text{i.e. } \rho_k - m_k = -\frac{1}{6}(\rho_{k+1}-2\rho_k+\rho_{k-1}), \tag{3.1.1}$$

which has the form of a diffusion equation with coefficient -1/6.

3.2 Godunov

The "Lagrange then remap" procedure has been shown to be effective in finite difference schemes for compressible flows (van Leer 1979). We remark that the right-hand-side of (2.2.6) amounts to a mapping of a density distribution obtained from a Lagrangian procedure back onto a fixed mesh (cf. Fig. 1).

3.3 PIC

The right-hand-side of (2.2.6) is the "charge-assignment" operation in the jargon of particle-mesh methods. Such methods are concerned with multidimensional calculations where the integral is hard to evaluate analytically, so quadrature is used instead. Letting $\delta m_i = \rho(x_{Oi})\delta\tau_{Oi}$ be the mass associated with quadrature point i, the troublesome integral is approximated by

$$m_k = \sum_i W_k(\underset{\sim}{x}_{Oi} + \underset{\sim}{\xi})\delta m_i . \qquad (3.3.1)$$

This may be interpreted as: take particle i at x_{Oi}, move it to $x_i = x_{Oi}+\xi$ and assign mass to node k according to the assignment function W_k.

The diffusive nature of conventional fluid PIC follows because in effect it lumps the mass matrix. However, when terms of the form (2.1.7) are treated by one- or two-point quadrature at each timestep, it is natural to use the advection/diffusion splitting of PIC, and if nodal values are regarded as principal variables, particles can be interpreted simply as quadrature points for awkward integrals.

4. EPIC COMPARED TO OTHER SCHEMES

4.1 Linear analysis

Linear analysis gives the variation with wavenumber of numerical dispersion and dissipation. The best schemes generally have small damping for wavenumbers with small phase errors, and large dissipation where phase errors are large.

Applying the usual procedure to EPIC for a problem with uniform advection (ξ = constant) and node spacing, we derive first the linearised equation, a convolution,

$$I(p-q)\rho_q(t+\Delta t)=I(p-q-\xi)\rho_q(t)\,, \qquad (4.1.1)$$

where for triangle functions

$$I(z) = \begin{cases} (4-2z^2[2-|z|])/6 & ; \ |z|\leqslant 1 \\ (2-|z|)^3/6 & ; \ 1\leqslant |z|\leqslant 2 \\ 0 & ; \ \text{otherwise} \end{cases} \qquad (4.1.2)$$

Fourier transforming (4.1.1) gives the dispersion relation

$$\lambda = e^{i\omega\Delta t} = \tilde{I}(k,\xi)/\tilde{I}(k,0)\,, \qquad (4.1.3)$$

which is to be compared with the exact result

$$\omega\Delta t = k\xi. \qquad (4.1.4)$$

Linear stability is assured since λ lies in the unit circle for all ξ.

The dispersion of EPIC with linear basis functions is compared with that of Lax-Wendroff and upwinding in Fig. 2(a) and with minimum phase error /FCT and the high order Godunov scheme PPM (Colella and Woodward 1984) in Fig. 2(b). All curves are drawn for a Courant number C = 0.25: larger timesteps are more flattering to EPIC, but we have to choose a value for which the other schemes are stable.

Lax-Wendroff has large phase errors and small damping so we expect (and find cf. Section 4.2) spurious wiggles and incorrect propagation speeds. Upwind has smaller phase errors but much larger dissipation. FCT has about the same phase errors as EPIC, but greater damping at shorter wavelengths and is weakly unstable at long wavelengths: PPM has slightly larger phase errors and dissipation intermediate between EPIC

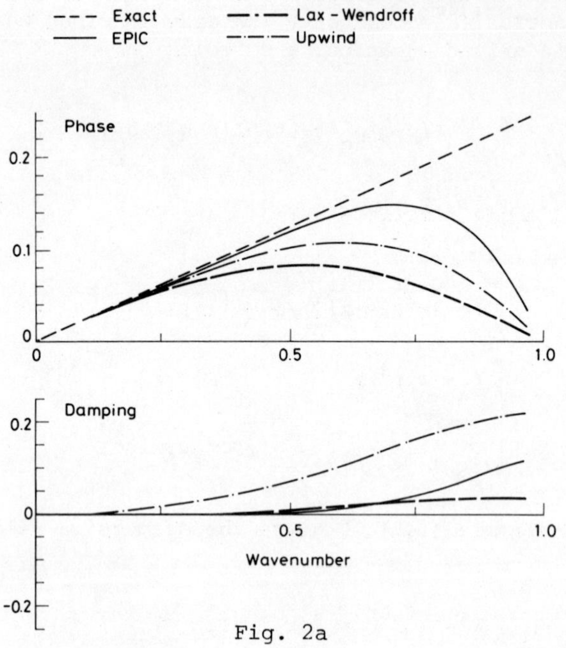

Fig. 2a

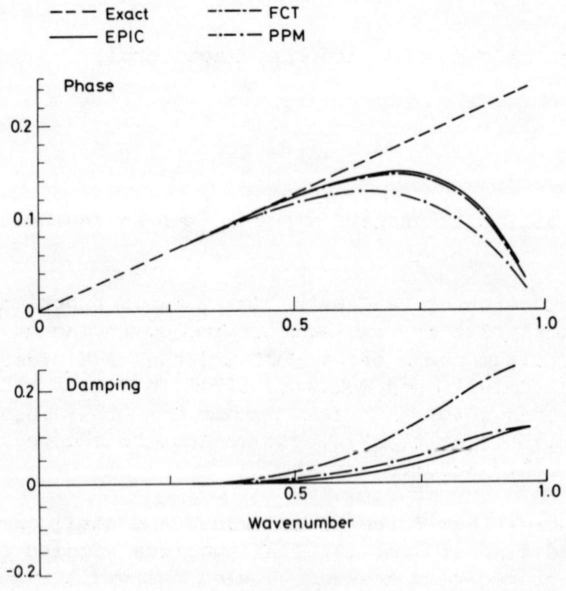

Fig. 2b

(a) and (b) Real (phase) and imaginary (damping) parts of the frequency plotted against wavenumber for uniform advection. The curves are identified in the legends.

and FCT. For wavelengths greater than 4 node spacings (wavenumbers < 0.5), Fig. 2(b) shows there is little to choose between EPIC, FCT and PPM on the basis of linear analysis.

4.2 Kinematic test problem

We have developed a test problem based on the 1-D continuity equation to compare schemes. A gaussian hump is transported through a unit length periodic box by a velocity field $u(x)$ that provides both compression and expansion. We take $u = 2-\sin 2\pi x$, $N = 100$ nodes in a spatial period, $C = 0.25$, and a gaussian density hump placed initially at the velocity maximum ($\rho(t=0) = \exp(-[x-1/4]^2/\sigma^2)$, $\sigma = 0.1$).

The compression reduces the effective σ of the gaussian to approximately 3 node spacings at smallest (proportionately increasing its height to conserve total mass). The compressed ρ Fourier transforms to a gaussian, the amplitude of which is $< 2\times 10^{-10}$ of peak value for wavelengths less than two mesh-spacings, i.e. ρ can always be resolved to within round-off error (of 4-byte arithmetic) by the chosen mesh. Specifying $u(x)$ avoids the special case of uniform advection while eliminating the velocity field as a source of error. Thus our test problem should provide a sensitive measure of the performance of a scheme.

Fig. 3 shows the outcome for the five schemes of Section 4.1. The broken curves are the initial conditions and the solid ones the result after transport through a distance of 3 periods (about 2000 steps); broken and solid should be identical in the absence of error. Fig. 3(a) and 3(b) show that Lax-Wendroff and (conservative) upwinding perform badly, as predicted in Section 4.1. Fig. 3(c) illustrates the profile steepening that is characteristic of FCT. From Fig. 3(d) PPM has more accurate slopes than FCT, but Fig. 3(e) reveals that EPIC clearly gives the best approximation. The cpu times taken were 0.21 for (a), 0.26 for (b), 0.44 for (c) and 1.39 for (e), all measured relative to EPIC. Thus EPIC is of comparable speed to the schemes it outperforms.

4.3 Nonlinear test problem

We apply the linear EPIC scheme to Burgers' equation

$$\frac{\partial v}{\partial t} + v\,\frac{\partial v}{\partial x} = R^{-1}\,\frac{\partial^2 v}{\partial x^2}\,, \qquad (4.3.1)$$

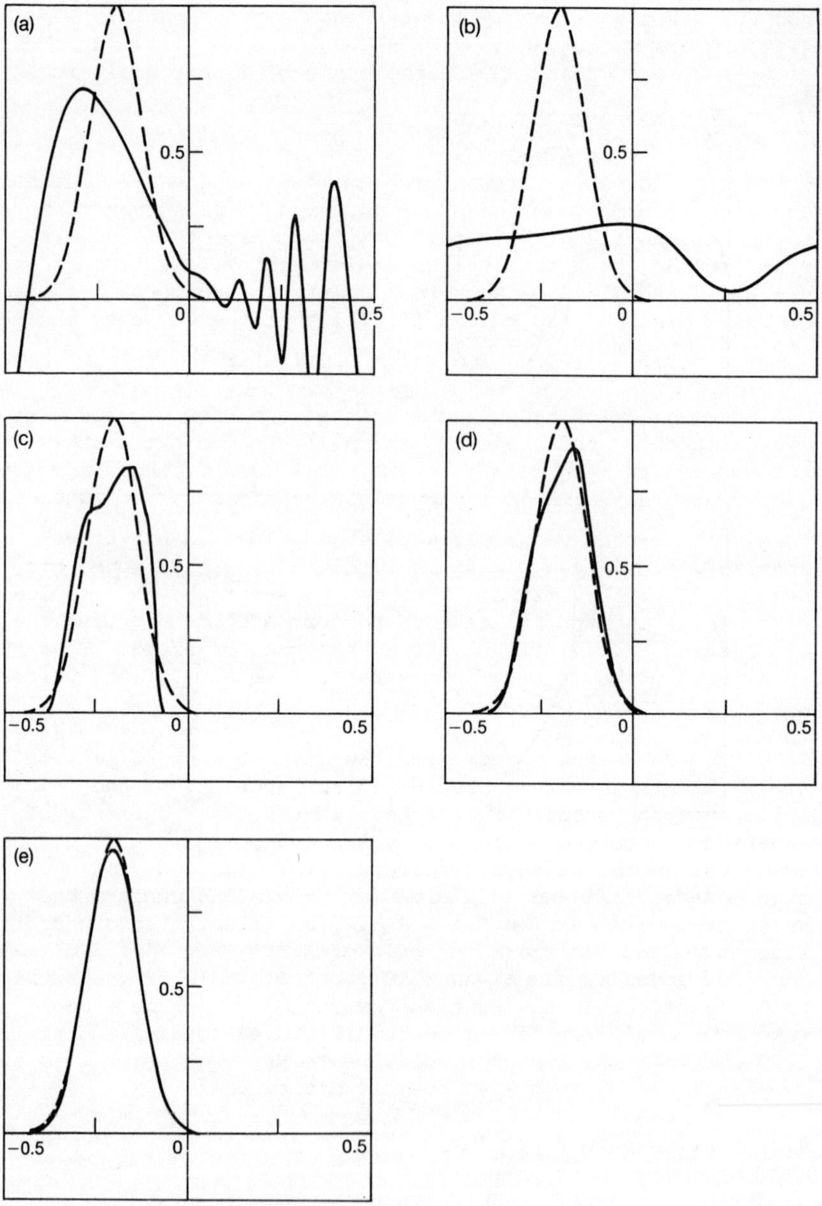

Fig. 3 Initial (dashed curve) and final (solid curve) profiles of a gaussian convected by a sinusoidally varying velocity field as described in Section 4.2, for (a) Lax-Wendroff, (b) upwind, (c) FCT, (d) PPM and (e) linear basis function EPIC schemes.

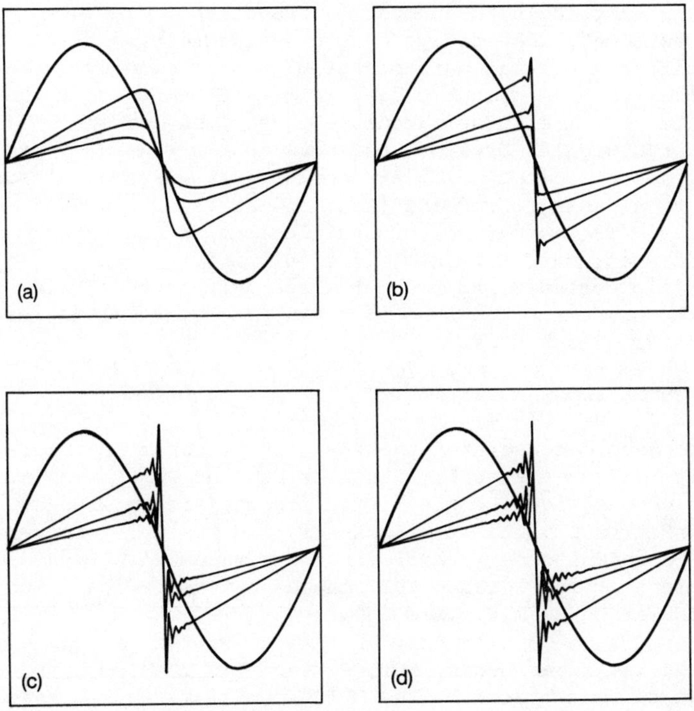

Fig. 4 Sequences showing overall reduction in amplitude as time increases of the solutions to Burgers' equation described in Section 4.3. Linear basis function EPIC is used with predictor $\xi = v\Delta t$; $C = 0.5$, $N = 100$, and the curves are drawn at intervals of $100\Delta t$.
$R =$ (a) 10^2, (b) 10^3, (c) 10^4 and (d) 10^5.

on the unit periodic interval with initial conditions $v = -\sin 2\pi x$. ξ is evaluated as $v(t)\Delta t$, otherwise the procedure of Section 4.2 is copied.

For large Reynolds number R, the sinusoid rapidly becomes a "sawtooth", then decays on the resistive timescale. The significant spatial harmonic of highest frequency in the problem has wavenumber 0.75R. Fig. 4(a) shows that when R=100, solutions v are indeed accurately resolved and smooth. At higher R wiggles develop, but note that unlike many finite difference schemes which are nonlinearly unstable at such R, due to aliasing error (Eastwood and Arter 1985), EPIC remains stable. The appearance of oscillations is the Gibbs phenomenon, caused by neglect of significant sub-grid-scale harmonic content. Appropriate, sub-grid-scale physics models can be used with EPIC (as with any other scheme) to remove these wiggles.

5. CONCLUSIONS

We have demonstrated that EPIC outperforms state-of-the-art finite difference schemes such as FCT and PPM for about the same computational costs. EPIC schemes are consistent, give unconditionally stable kinematics, and the Galerkin variants can be shown to be free from nonlinear instabilities. Optimal accuracy arises because EPIC causes errors in the principal variables to be minimised.

The key ideas behind FCT, PPM and PIC methods have been shown to be embraced by the EPIC formalism, whilst knowledge of particle-mesh techniques provides insight into the best ways of tackling tricky overlap integrals that appear in EPIC. The finite element formulation used also allows a variety of element connectivities and geometries to be employed (although the flexibility has to be paid for in increased computational costs associated with indirect addressing, element geometry, etc.). We expect EPIC to be useful over the whole gamut of fluid and MHD problems.

REFERENCES

Boris, J.P. and Book, D.L. (1976) Flux-corrected transport III: Minimal error FCT algorithms. *J. Comput. Phys.*, **20**, 397-431.

Collela, P. and Woodward, P.R. (1984) The piecewise parabolic method (PPM) for gas-dynamical simulations. *J. Comput. Phys.*, **54**, 174-201.

Eastwood, J.W. (1982) In "Proceedings of Astrophysical Radiation Hydrodynamics Workshop, Munich". (K.H. Winkler, ed.), Springer-Verlag, Berlin.

Eastwood, J.W. and Arter, W. (1985) Spurious singularities in numerically computed fluid flows. Submitted to *IMA J. Numer. Anal.*

Eastwood, J.W. and Lee-Hsiao (1983) In "Proceedings of 10th Conference on the Numerical Simulation of Plasmas, San Diego". (J. Helton, ed.), paper IC13.

Hockney, R.W. and Eastwood, J.W. (1981) Computer Simulation using Particles. McGraw-Hill, New York.

Labadie, G., Benque, J.P. and Latteux, B. (1981) In "Proceedings, 2nd International Conference on Numerical Methods in Laminar and Turbulent Flow". (C. Taylor and B.A. Schrefler, eds.), Pineridge Press, Swansea, 681-692.

Lewis, H.R. (1970) Application of Hamilton's principle to the numerical analysis of Vlasov plasmas". *Meth. Computational Phys.*, **9**, 307-338.

Morton, K.W. (1982) In "Numerical Methods for Fluid Dynamics". (K.W. Morton and M.J. Baines, eds.), Academic Press, London, 1-32.

Pironneau, O. (1982) On the transport-diffusion algorithm and its applications to the Navier-Stokes equations". *Numer. Math.*, **38**, 309-332.

van Leer, B. (1979) Towards the ultimate conservative difference scheme. V. A second-order sequel to Godunov's method. *J. Comput. Phys.*, **32**, 101-136.

GRID GENERATION AND FLOW CALCULATIONS FOR COMPLEX AERODYNAMIC SHAPES

N.P. Weatherill, J.A. Shaw and C.R. Forsey
(Aircraft Research Association Ltd, Bedford)

1. INTRODUCTION

The generation of coordinate systems about complex configurations is generally recognised as a pacing item for computational aerodynamics. Although the literature on grid generation is now extensive, few methods have been shown to be applicable to general three-dimensional configurations. In addition, grid generation has too frequently been viewed as an isolated topic with little attention being given to the hosted flow algorithm. In this paper we describe a multi-block method of generating body conforming grids for complex three-dimensional configurations and give flow calculation results for realistic aircraft using a finite volume flow algorithm for the Euler equations.

2. MULTI-BLOCK - GENERAL PRINCIPLES

The philosophy we have adopted for grid generation has been detailed previously (Weatherill and Forsey, 1984). The whole flow field, between the surfaces of the configuration and some outer farfield boundary, is broken down into a set of blocks. The union of these blocks fills the entire flow field without either holes or overlaps. Each block is chosen to be topologically equivalent to a cuboid, in that it has six faces and eight vertices, and can be mapped into a unit cube in computational space. The mapping between computational and physical spaces is performed using the elliptic grid generation approach of Thompson et al (1974). A user constructed topology file specifies the arrangment of blocks and thereby defines the topological structure of the grid. To simplify the internal program logic, a single boundary condition is imposed on each face of each block.

Block boundaries which are part of the configuration or outer-boundary have the grid specified on them a priori by a surface grid generator. They are referred to as Dirichlet block boundaries. Block boundaries, which are not part of the configuration or outer boundary, are purely notional, in that they have no physical significance, their position in space evolving in a similar manner to any point interior to a block. They are referred to as continuity block boundaries. On such boundaries, internal program logic, driven by the topology file which contains the relevant information about adjacent block number, face number and relative orientations of adjacent faces, is used to access the coordinates of points in neighbouring blocks required by the solution procedure. Across these block boundaries, the grid is as smooth as at any point inside a block.

3. SURFACE GRIDS

An intrinsic part of our approach to surface grid generation is a geometry package based on bi-cubic surface patches. The geometry of each component is input in terms of cross-sections and converted internally to a continuous function of two parametric coordinates S = S(s,t) using a Coons patch formulation (Coons, 1967). This transformation from a discrete to a continuous description of the surface enables component intersections to be calculated using a Newton-Raphson routine. The resulting parametric coordinates of the intersection lines are then used as fixed Dirichlet data for the surface grid generation.

The surface grids, whose topological structures are derived from the field grid topology information, are generated as solutions of the two-dimensional elliptic equations of Thompson et al (1974), expressed in parametric coordinates S = S(s,t)

$$g^{ij} S_{\xi^i \xi^j} = -p^i S_{\xi^i} \tag{3.1}$$

where g^{ij} are the metric terms, p^i the control functions and ξ^i the computational coordinates with i,j summed over 1 and 2 in the usual tensor fashion. The non-linear algebraic equations resulting from the discretisation of equations (3.1) are solved using a successive point over-relaxation iterative scheme, with all points in a given block being updated before advancing to the next block.

The solution values are mapped back to physical space via the Coons patches, which ensures mathematical consistency with the geometry definition. They are then used as fixed Dirichlet data for the field grid generation.

4. FIELD GRIDS

The positions of points in the field are determined as the solutions of the three-dimensional elliptic grid generation equations which are equivalent to (3.1) with $S = S(s,t)$ replaced by $X = X(x,y,z)$ and the tensor summation for i,j taken over the values, 1, 2 and 3. In a similar manner to the surface grid technique, the equations are solved iteratively with a point over-relaxation scheme using block-by-block update.

Inherent in our multi-block approach is the acceptance of singular points and singular lines where the Jacobian of the transformation is zero. In physical space such singularities manifest themselves as points in two dimensions at which other than 4 lines meet or in three dimensions where other than 6 lines intersect. These points are identified automatically by analysing the topology file and their position determined as the average of the points surrounding them.

For the field grid generator, as for the surface grid generator and the Euler flow code, indirect addressing, which we call a pointer scheme, is used to access variables from adjacent blocks required by the solution scheme on continuity boundaries. For a given topology, this offers a considerable saving in computer storage compared with a halo type scheme which requires all variables from adjacent blocks needed by the solution scheme to be stored in a halo around each current block. In this way, the halo scheme effectively increases all block dimensions by two for the grid generators and four for the Euler flow code. The drawback with a pointer scheme is that it is more difficult to take advantage of the vectorisation capabilities of modern computers. Nevertheless, for the field grid generator we have been successful in vectorising such a scheme and have reduced run times by a factor of five compared with an unvectorised halo scheme.

The field grid generation has highlighted three particular problems: (i) the difficulty in viewing and presenting three-dimensional grids, which indicates the need for very sophisticated interactive graphics software, (ii) the grid cross-over property, exhibited by the discrete form of Thompson's equations, which occurs particularly in 3D around sharp edges such as wing tips, and (iii) the satisfaction of a particular convergence criterion.

To overcome problems of grid cross-over, which are particularly apparent in the more complex topologies we have examined, we have fixed additional block boundaries in the field. These boundaries are chosen to produce surfaces of approximately constant x, y or z. The grids on these surfaces are generated using the surface grid generator for two of the variables, x and y, say, and the approximately constant z values found by solving a two-dimensional Laplacian equation in z. Fixing such boundaries has the additional benefit of improving convergence and provides an effective technique for grid control. A drawback of this approach would be if the flow algorithm was sensitive to slight discontinuities in the slope of the grid and illustrates the importance of considering the hosted algorithm in the grid generation process.

5. FLOW ALGORITHM

The flow algorithms now available for solving the Euler equations combine satisfactory accuracy with acceptable convergence rates, making the prediction of transonic flow over a complete aircraft a realistic proposition. A finite volume explicit Runge-Kutta time stepping scheme based on the work of Jameson et al (1981) has been coded to work within a multi-block framework. As with the grid generator, all variables are stored in a one-dimensional array and a pointer scheme is used to balance fluxes across continuity block boundaries. Again, the topology file gives information about the boundary conditions for the flow code on each block face.

An important aspect of our Euler multi-block work is the block-by-block update of all flow variables. In a single block grid, all variables in the field are updated to an intermediate time level before advancing to the next stage in the Runge-Kutta scheme. However, in the multi-block scheme, the flow variables are updated to the new time level block-by-block to save on computer storage. Consequently, on a continuity block boundary, the variables required from an adjacent block could either be frozen at the old time level or have already been advanced to the new time level. This mixing of new and old values could impair convergence rate as the number of blocks is increased. However, our experience suggests that convergence rate is only slightly degraded.

No special formulation of this cell centred Euler algorithm is required at singular lines and points and limited investigations have indicated that the code is insensitive to such features.

6. RESULTS

We have used the three main types of mapping, O-grids, C-grids and H-grids, as the basis for constructing various grid topologies around civil and combat aircraft. Each mapping has its attractions and limitations. O-grids are computationally efficient, requiring fewer grid points to fill the same volume, but topologically very difficult to construct around all components. C-grids are less efficient than O-grids but offer an improved grid structure around areas such as wing trailing edges and wakes. H-grids are computationally inefficient, their appeal lying in the simplicity of adding extra components to a configuration by opening up a slit in a cartesian-like grid.

At present, both the field grid generator and the Euler code run in core on a 1 megaword Cray 1S and hence are restricted to rather coarse grids. Thus, the following examples are more illustrative of the types of grid topologies obtainable rather than actual, practically usable, grids. In Figs 1, 2 and 3 we show three different grid topologies which have been constructed for the RAE wing A body B2 test case (Treadgold et al, 1979). In Fig. 1 we have a C-grid in the chordwise direction and an H-grid in the spanwise direction. The topology is made up of 24 blocks and there are 34,000 points in the field.

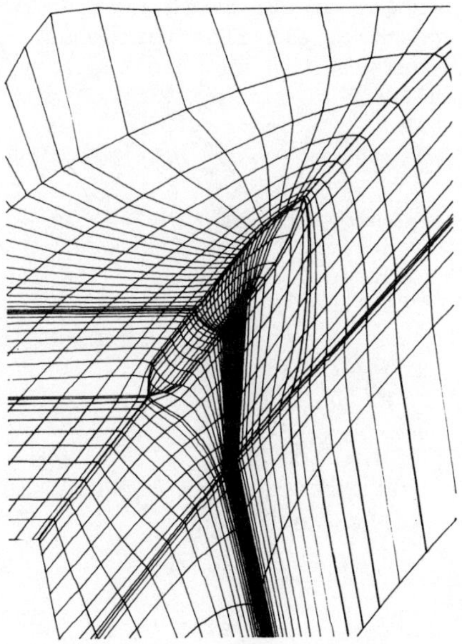

Fig. 1. Wing A/B2 C-H Grid

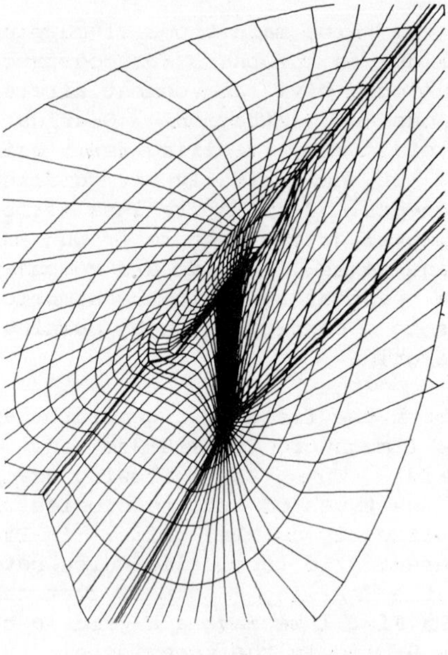

Fig. 2. Wing A/B2 O-O Grid

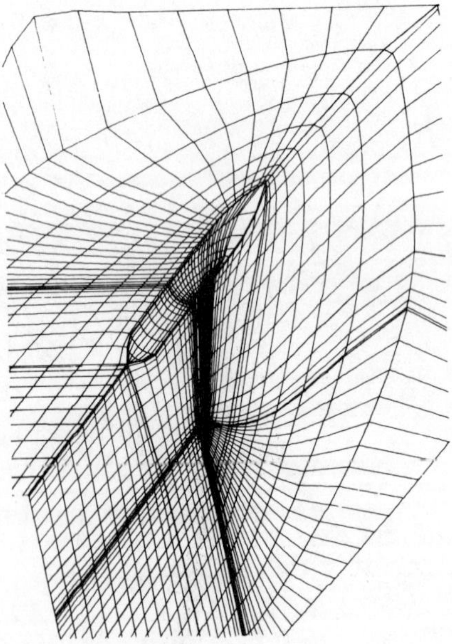

Fig. 3. Wing A/B2 C-O Grid

Fig. 2 shows an eight block grid with 17,000 field points; the chordwise and spanwise grids are both O-grid structures. In Fig. 3 we have chosen an O-grid in the spanwise direction and a C-grid in the chordwise direction. This topology is comprised of 24 blocks with 32,000 field grid points.

The Euler flow code was run for 1000 time steps on all these grids for a subcritical case, $M = 0.4$, $\alpha = 0^\circ$. A comparison of the results obtained for this case, along with experimental results and a single block full potential result using the ARA FP wing/body code with an equivalent number of grid points, are shown in Fig. 4. Given the coarseness of the grids, the results are quite encouraging. A transonic flow result, $M = 0.9$, $\alpha = 0^\circ$, is shown in Fig. 5.

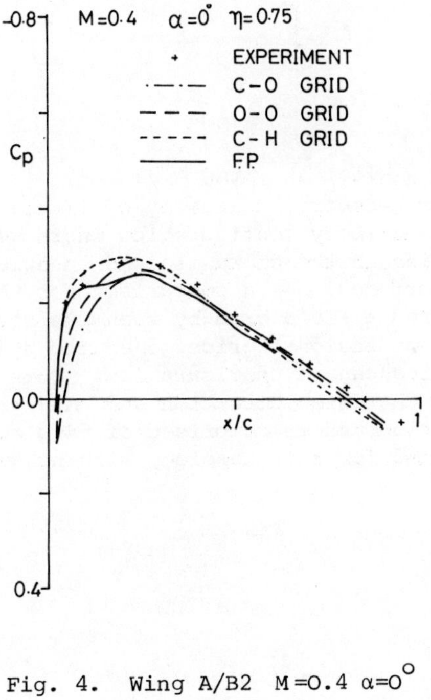

Fig. 4. Wing A/B2 $M=0.4$ $\alpha=0^{\circ}$

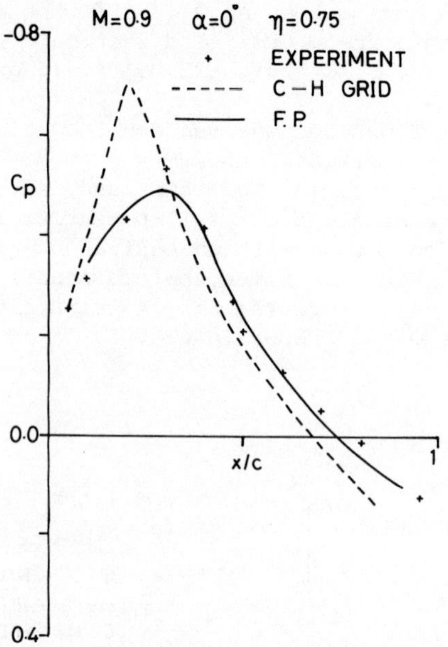

Fig. 5. Wing A/B2

A particular geometry, currently of interest, is a canard/ forward swept wing/body configuration where we have chosen to have an O-mapping in the chordwise and spanwise direction of both lifting surfaces. As a comparison for the flow results, we have generated a field grid by spanwise stacking of strea streamwise two-dimensional grids generated by the surface grid program on a wing/canard configuration. This topology has C-grids in the chordwise direction and an H-grid in the spanwise direction and is comprised of 54 blocks. Fig 7 shows the grid obtained for this topology with no grid control.

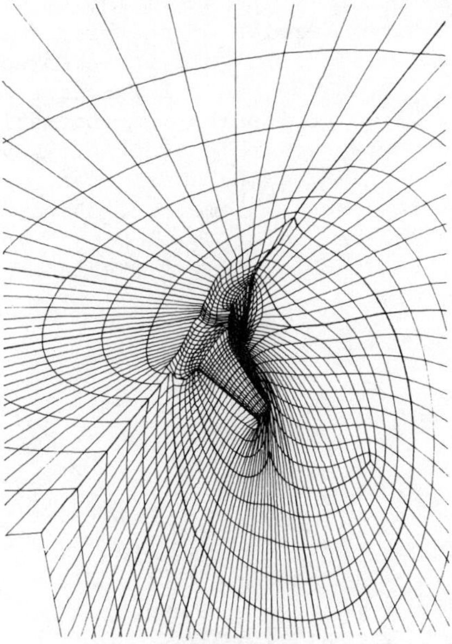

Fig. 6. Wing/Body/Canard O-O-O Grid

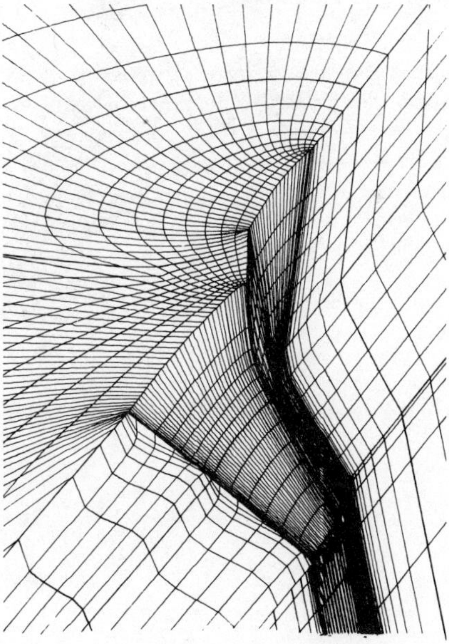

C-C-H Grid

Fig. 7. Wing/Body/Canard (No Control)

To provide a mechanism for controlling grid spacing we have developed a multi-block version of the grid control technique of Thomas and Middlecoff (1980) and implemented it in the surface grid generator. The control functions are made continuous across block boundaries and automatically modified in regions of high curvature to prevent crossover. Fig. 8 shows the grid generated for the configuration shown in Fig. 7 when grid control has been implemented. A comparison of the flow results obtained on the grids of Figs 6 and 8 for M = 0.7 and $\alpha = 4^{\circ}$ are shown in Fig 9 for a station threequarters of the way along the wing.

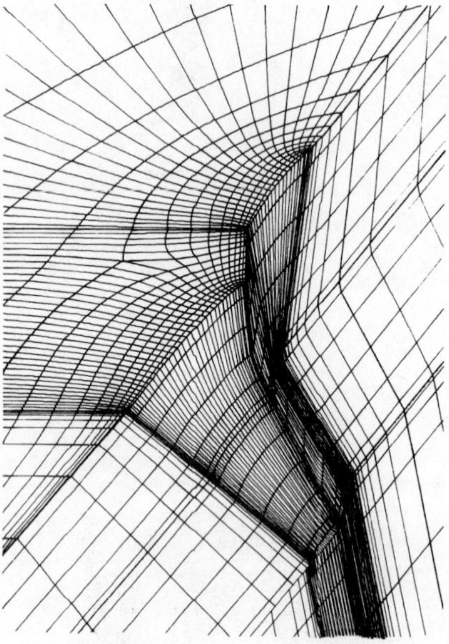

C-C-H Grid

Fig. 8. Wing/Body/Canard (With Control)

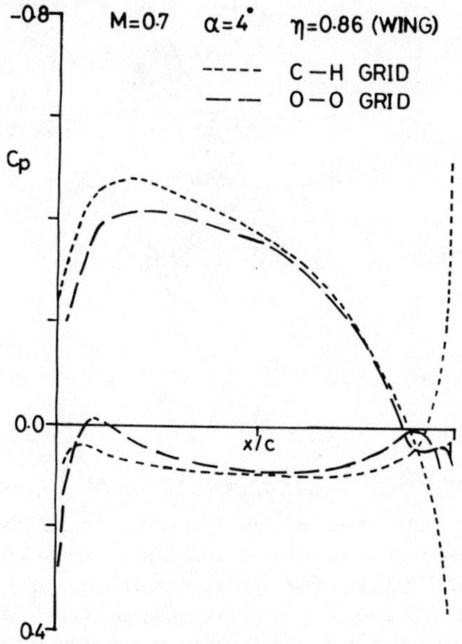

Fig. 9. Wing/Body/Canard

For those not familiar with multi-block grids, the concept of visualising 30 blocks mapped in O-O structures is not easy. In Figs. 10 and 11 we show schematics of the grids shown respectively in Figs. 6 and 8.

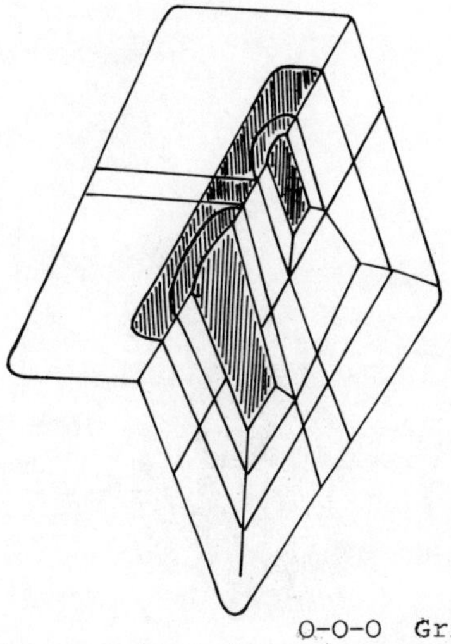

O-O-O Grid
Fig. 10. Wing/Body/Canard (Schematic)

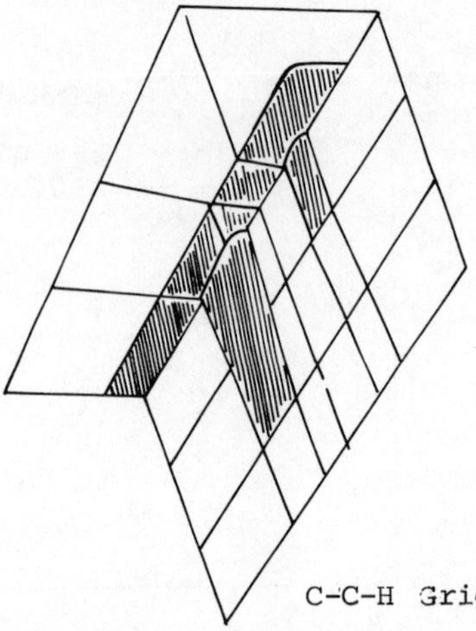

Fig. 11. Wing/Body/Canard (Schematic)

It is clear from the above results that the work is still very much at an initial stage but that our multi-block approach provides a sound basis for grid and flow calculations for complex configurations. Finally, for completeness, we show two civil aircraft configurations which are currently under study using the multi-block approach. Fig. 12 shows a 30 block wing-body-tailplane grid and Fig. 13 a 150 block wing-body-nacelle-pylon grid. The topology file for this final case, made up of C-grids and H-grids, was generated automatically, offering a considerable saving over manual specification of the topology file.

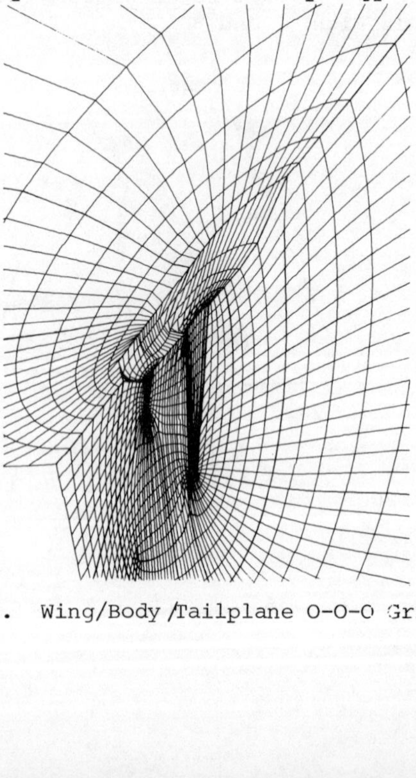

Fig. 12. Wing/Body/Tailplane O-O-O Grid

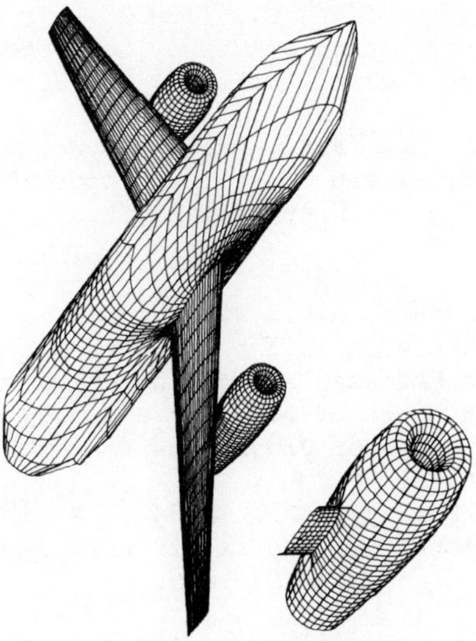

Fig. 13. Wing/Body/Pylon/Nacelle Surface Grid

7. SUMMARY

A general multi-block grid generation method has been described for computing coordinate systems (grids) which are aligned with each of the components making up an aircraft. An established formulation for solving the Euler equations has been structured to accept a multi-block grid and preliminary results on various topologies indicate that the methods described can form a system which can successfully predict the flow around a wide variety of complex aerodynamic shapes.

8. ACKNOWLEDGEMENTS

This work has been carried out with the support of the Procurement Executive, Ministry of Defence. The authors are indebted to Dr. K.E. Rose for his development work on the graphics software which forms an intrinsic part of our work.

9. REFERENCES

Coons, S.A., (1967), Surfaces for Computer-Aided Design of Space Forms", MIT MAC-TR-41.

Jameson, A., Schmidt, W., Turkel, E., (1981), Numerical Solutions of the Euler Equations by Finite Volume Methods using Runge-Kutta Time Stepping Schemes, *AIAA* Paper 81-1259.

Thomas, P.D., Middlecoff, J.F., (1980), Direct Control of the Grid Point Distribution of Meshes Generated by Elliptic Equations, *AIAA Journal,* Vol. 18, No. 6.

Thompson, J.F., Thames, F.C., Mastin, C.W., (1974), Automatic Control of Body Fitted Curvilinear Coordinate Systems for Field Containing any Number of Arbitrary 2D Bodies, *J. Comput. Phys.,* Vol. 15.

Treadgold, D.A., Jones, A.F., Wilson, K.H., (1979). Pressure Distribution measured in the RAE 8 ft x 6 ft Transonic Wind Tunnel on RAE Wing A in Combination with an Axisymmetic Body at Mach Numbers of 0.4, 0.8 and 0.9, in Experimental Data Base for Computer Program Assessment, AGARD-AR-138.

Weatherill, N.P., Forsey, C.R., (1984), Grid Generation and Flow Calculations for Complex Aircraft Geometries using a Multi-block Scheme, *AIAA* 84-1665.

FCT APPLIED TO THE 2-D FINITE ELEMENT SOLUTION OF TRACER TRANSPORT BY SINGLE PHASE FLOW IN A POROUS MEDIUM

A.K. Parrott and M.A. Christie
(British Petroleum Exploration, London)

1. INTRODUCTION

This paper describes a new adaptation of the Flux Corrected Transport technique of Boris and Book (1973, 1975), Zalesak (1979) to the finite element solution of a 2-D advection - diffusion equation. Results for the problem of propagating step discontinuities show the method to be monotone and to give sharp resolution of fronts.

The examples derive from the problem of modelling the inter-well flow of tracer in water injected into a petroleum reservoir. These problems are very similar to those of modelling pollutant transport in groundwater, an area in which finite element methods have been used for some time (Pinder, 1975). It is well established that for small dispersion coefficients numerical oscillations occur in the standard Galerkin formulation and resort is usually made to upstream weighting of some kind: see e.g. (Sun and Yeh, 1983). To avoid excessive numerical dispersion these methods effectively interpolate between full upstream weighting and standard Galerkin using a parameter which will in general depend both on the mesh variation and the problem type. Specifying this parameter in advance can constitute a drawback for such techniques.

The flux corrected transport technique described below is capable of resolving fronts over 3 nodes without any numerical oscillation. The FCT algorithm used is that of Zalesak (1979) with some minor modifications to extend it to triangular meshes and finite element flux terms. It effectively constitutes a parameter-free non-linear interpolation between full upstream weighting and standard Galerkin.

2. FORMULATION

2.1 Concentration Equation

The following non-reaction chemical transport equation provides a simple model of the tracer problem we are interested in solving. We have assumed a constant thickness single phase reservoir with a number of injection and production wells represented as delta functions. i.e.

$$\frac{\partial c}{\partial t} + \underline{\nabla}.t(\underline{u}c) = \alpha_{\ell}\underline{\nabla}.(\underline{\underline{D}}\ (\underline{u})\ \underline{\nabla}c) + Q \qquad (2.1.1)$$

for all $(x,y)\ \varepsilon\Omega$, with boundary condition

$$u_n C + D_n \frac{\partial c}{\partial n} = 0 \qquad \text{for all } (x,\ y)\ \varepsilon\Gamma \qquad (2.1.2)$$

where Γ is the boundary of Ω. The well term Q is a sum of source and sink terms as follows:

$$Q = \sum_{r=1}^{n} q_r C_r^{inj}\ \delta(\underline{x} - \underline{x}_r) - \sum_{s=1}^{m} q_s C_s \partial\ (\underline{x} - \underline{x}_s) \qquad (2.1.3)$$

where C_r^{inj} is the injection concentration at each of the n injection wells and $C_s = C\ (\underline{x}_s,\ t)$ is the <u>in-situ</u> concentration at each of the m production wells. The dispersivity tensor has the standard form

$$\underline{\underline{D}}\ (\underline{u}) = \begin{bmatrix} u_x^2 & u_x u_y \\ u_x u_y & u_y^2 \end{bmatrix} \qquad (2.1.4)$$

where α_{ℓ} is the longitudinal dispersion coefficient.

2.2 Velocity Field

The velocity field $\underline{u}$ is obtained via Darcy's law from the pressure field. This is calculated initially and for an incompressible single phase system remains constant throughout the tracer calculations, providing the rates do not change. The pressure equation is solved conventionally using linear finite elements on the same mesh as the concentration equation and using local singularity subtraction at the wells as described in (Spivak et al., 1977). The singularity subtraction is facilitated by the use of a mesh generator which automatically matches node positions to well positions. The approximate pressure field has the form

$$P(x,y) = \sum_{j=1}^{N} P_j \phi_j(x,y) + \sum_{s=1}^{m+n} \frac{q_s}{2\pi K_s} \log \frac{[\underline{x} - \underline{x}_s]}{R_s} \phi_{j(s)}(x,y) \tag{2.2.1}$$

where $\phi_{j(s)}$ is the linear basis function for the nodes coincident with the position of well s, R_s is the maximum extent of $\phi_{j(s)}$, and K_s is the permeability at well s. The velocity field is then given simply by

$$\underline{u} = K \underline{\nabla} p = K \sum_{j=1}^{N} P_j \underline{\nabla} \phi_j + K \sum_{s=1}^{m+n} \frac{q_s}{2\pi K_s} \underline{\nabla} (\log([\underline{x} - \underline{x}_s]/R_s) \phi_{j(s)}) \tag{2.2.2}$$

$$= \underline{u}_p + \underline{u}_\ell \tag{2.2.3}$$

where $\underline{u}_\ell$ is non-zero only for those elements having a well at a vertex.

2.3 Upstream Weighted Finite Elements

In this section a "low-order" finite element approximation to the concentration equation is derived which uses mass lumping and a convenient form of upstream weighting. Since high Peclet number flows are of most interest the physical dispersion coefficient is set to zero from hereon and the limiting hyperbolic case considered.

The mass lumping factors Ω_j are obtained by summing 1/3 of the area of each triangle with node j as a vertex. Equation (2.1.1) is then approximated by

$$\Omega_j(C_j^{n+1} - C_j^n) = \langle \underline{u}_p C^n, \underline{\nabla}\phi_j\rangle + \langle \underline{u}_\ell C^{n+1}, \underline{\nabla}\phi_j\rangle + \langle Q^{n+1}, \underline{\nabla}\phi_j\rangle$$

$$\text{for } j=1,\ldots,N \qquad (2.3.1)$$

where C_j^n is the nodal value of the concentration at time $t = n\Delta t$ and the L_2 inner products $\langle \underline{u}\, C, \underline{\nabla}\phi_j\rangle$ are integrated in the following upstream sense. Consider the element flow integral,

$$f_j^e = \int_{T_e} \underline{u}.(\underline{\nabla}\phi_j)\; C\, dx\, dy \qquad (2.3.2)$$

on an element with node j as a vertex, where T_e is the element area. An upstream approximation to f_j^e is then

$$(f_j^e)^{ups} = (C_j^e)^{ups} \int_{T_e} \underline{u}.(\nabla\phi_j)\; dx\, dy \qquad (2.3.3)$$

$$= (C_j^e)^{ups}\, \nu_j^e \qquad (2.3.4)$$

where the upstream concentration $(C_j^e)^{ups}$ with respect to node j is given by

$$(C_j^e)^{ups} = \begin{cases} C_j & \text{for } \nu_j^e \geq 0 \\ \dfrac{w_k\, C_k + w_i\, C_i}{w_k + w_i} & \text{for } \nu_j^e < 0 \end{cases} \qquad (2.3.5)$$

$w_j^e = \max(0, \nu_j^e)$, and the nodes of the element are denoted by i,j,k. The quantity ν_j^e is simply the flow vector $\underline{u}$ resolved perpendicular to the side opposite node j.

These upstream element flow integrals are assembled in the standard manner to give an upstream evaluation of the flow terms in equation (2.3.1), i.e. if

$$F_j^L(\underline{u},C) = \sum_e (f_j^e)^{ups} \tag{2.3.6}$$

for all the elements having node j as a vertex, then (2.3.1) can be written as

$$\Omega_j(C_j^{n+1} - C_j^n) = F_j^L(\underline{u}_p, C^n) + F_j^L(\underline{u}_\ell, C^{n+1}) + \langle Q^{n+1}, \underline{\nabla}\phi_j\rangle \quad \text{for } j=1,\ldots,N \tag{2.3.7}$$

The solution of this equation is monotone provided that the timestep is kept within the appropriate CFL limit. It is however very dispersive as can be seen in Fig. 1. The implicit terms in (2.3.7) are easily dealt with since $\underline{u}_\ell$ is radial around wells; no matrix inversion is required.

2.4 A Finite Element FCT Algorithm

Equation (2.3.7) contains three sources of inaccuracy, namely the upstream weighted flow terms, the first-order time integration used, and the mass lumping. These inaccuracies can be corrected in a monotonicity preserving fashion using Zalesak's FCT algorithm providing they can be expressed as explicit conservative fluxes. We have only corrected the first two inaccuracies: however Donea (1984) describes an explicit correction to mass lumping which is conservative and could easily be incorporated in what follows. The high-order fluxes used to correct (2.3.7) are taken as

$$F_j^H = \int_\Omega (\underline{u}_p\cdot\underline{\nabla}\phi_j)\, C^n(n,y)\, dx\, dy + \tfrac{1}{2}\Delta t \int_\Omega \underline{\underline{D}}(\underline{u}_p)\underline{\nabla}C^n\cdot\underline{\nabla}\phi_j\, dx\, dy \quad j=1,\ldots,N \tag{2.4.1}$$

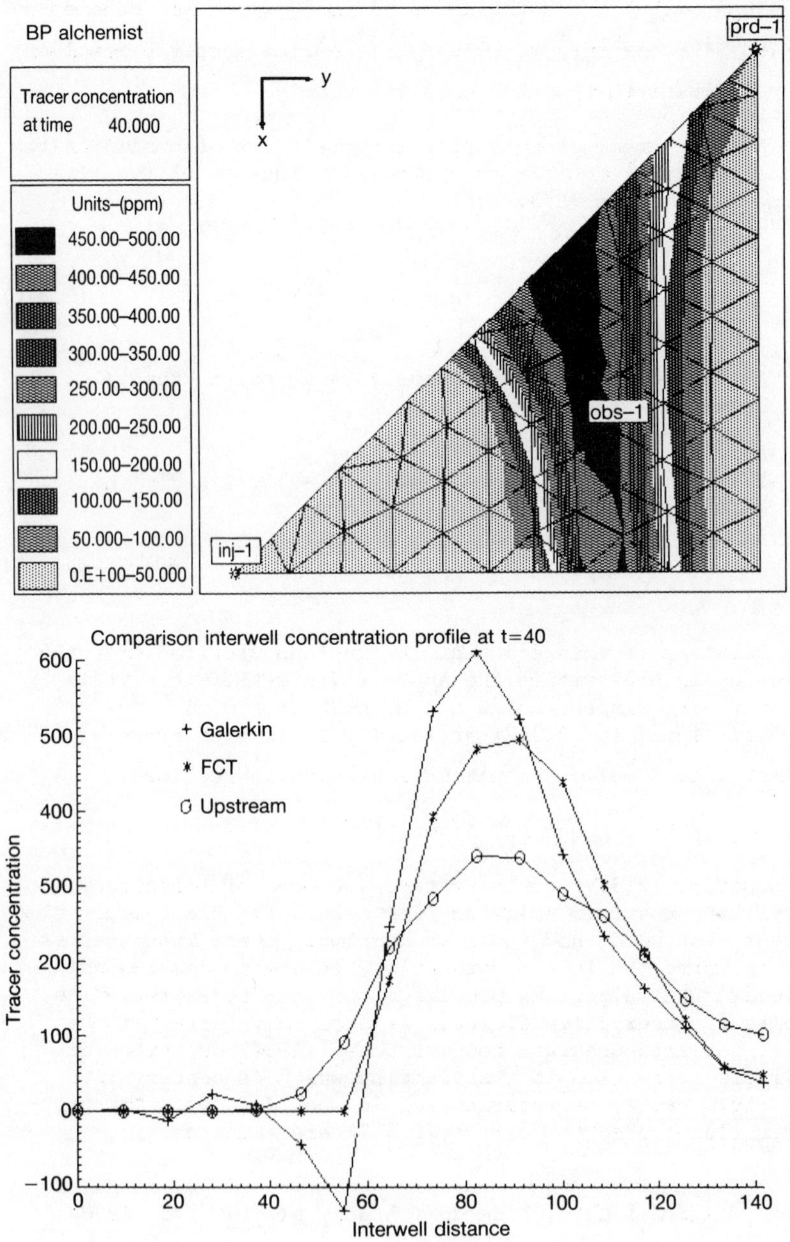
BP alchemist
Tracer concentration
at time 40.000
Units–(ppm)
450.00–500.00
400.00–450.00
350.00–400.00
300.00–350.00
250.00–300.00
200.00–250.00
150.00–200.00
100.00–150.00
50.000–100.00
0.E+00–50.000
y
x
prd–1
obs–1
inj–1
Comparison interwell concentration profile at t=40
+ Galerkin
* FCT
O Upstream
Tracer concentration
Interwell distance
600
500
400
300
200
100
0
−100
0
20
40
60
80
100
120
140

Fig. 1

where only the $\underline{u}_p$ component of the flow field is used in the corrections as the $\underline{u}_\ell$ component would involve implicit terms. The second term is a Lax-Wendroff correction term which conveniently depends on the physical dispersion tensor defined in (2.1.4).

The element flux limiting algorithm is a straightforward adaptation of the FCT algorithm, as follows

(i) compute the upstream weighted $C^{up} = C^{n+1}$ from equation (2.3.7)

(ii) compute the high-order fluxes from equation (2.4.1) and then calculate the anti-diffusive flux corrections

$$a_j = F_j^L (\underline{u}_p, C^n) - F_j^H (\underline{u}_p, C^n) = \sum_e a_j^e \tag{2.4.2}$$

(iii) limit the individual element flux corrections

$$a_j^{e\prime} = \beta^e a_j^e, \qquad 0 < \beta^e < 1,$$

and hence limit the total flux correction for node j,

$$a_j' = \sum_e a_j^{e\prime}$$

(iv) correct the upstream concentrations at $t=(n+1)\Delta t$ using

$$\Omega_j (C_j^{n+1} - C_j^{up}) = a_j' \tag{2.4.3}$$

Clearly for $\beta = 1$ in step (iii) the usual oscillatory solution would be obtained and so the selection of limiters is crucial to the properties of FCT. The choice of limiter for each element is based on the multidimensional FCT limiter of Zalesak which is restated here for convenience with minor adaptations for triangular elements.

For each node j define the following

$P_j^+ = \Sigma \max(0, a_j^e)$ for each element having node j as a vertex,

$$Q_j^+ = \Omega_j\,(C_j^{max} - C_j^{up})/\Delta t, \qquad (2.4.4)$$

$$R_j^+ = \begin{cases} \min(1, Q_j^+/P_j^+) & P_j^+ > 0 \\ 0 & P_j^+ = 0 \end{cases}$$

and similarly

$$P_j^- = -\Sigma \min(0, a_j^e)\ ,$$

$$Q_j^- = \Omega_j\,(C_j^{up} - C_j^{min})/\Delta t\ , \qquad (2.4.5)$$

$$R_j^- = \begin{cases} \min(1, Q_j^-/P_j^-) & P_j^- > 0 \\ 0 & P_j^- = 0 \end{cases}$$

The R_j^+ and R_j^- terms are then sorted for each node of a given element into element values R_j^e where

$$R_j^e = \begin{cases} R_j^+ \text{ if } a_j^e > 0 \\ R_j^- \text{ if } a_j^e < 0\ . \end{cases} \qquad (2.4.6)$$

One choice of element limiter β^e is then

$$\beta^e = \min\,[R_i^e, R_j^e, R_k^e] \qquad (2.4.7)$$

where i, j, k are the element nodes.

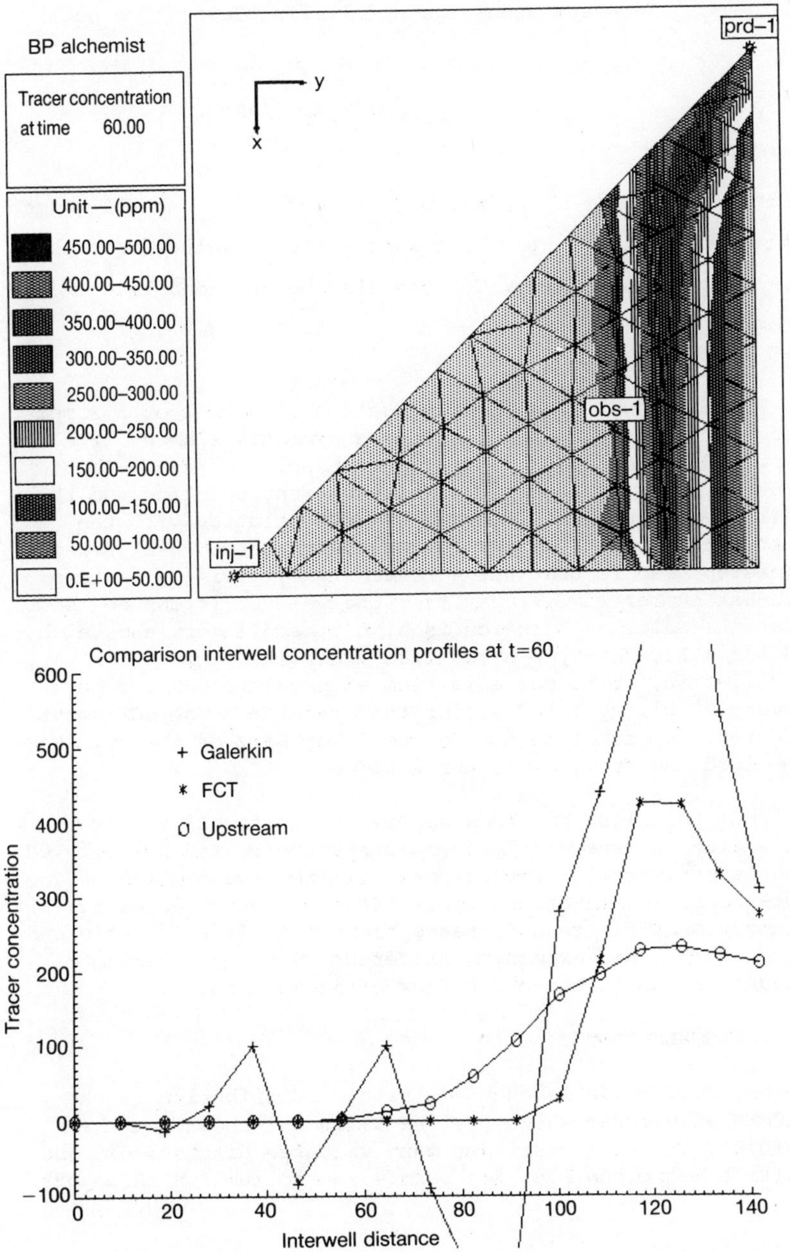
BP alchemist
Tracer concentration
at time 60.00
Unit — (ppm)
450.00–500.00
400.00–450.00
350.00–400.00
300.00–350.00
250.00–300.00
200.00–250.00
150.00–200.00
100.00–150.00
50.000–100.00
0.E+00–50.000
y
x
prd–1
obs–1
inj–1
Comparison interwell concentration profiles at t=60
+ Galerkin
* FCT
O Upstream
Tracer concentration
Interwell distance
600
500
400
300
200
100
0
−100
0
20
40
60
80
100
120
140

Fig. 2

It remains to specify the maximum and minimum values C_j^{max}, C_j^{min} for each node: again following Zalesak we put

$$C_j^{max} = \max \{C_j^{up}, C_{j'}^{up}, C_{j''}^{up}, \ldots\} \qquad (2.4.8)$$

and similarly for C_j^{min}, where j', j'' etc. run over all nodes connected to node j in the triangulation. Values from the previous timestep (i.e. C_j^n) can also be included into (2.4.7).

3. DISCUSSION

Figs. 1 and 2 show the propagation of a bank of 500 ppm tracer from left to right across a symmetry element of a repeated 5-spot injection/production well pattern. The contour plots show the FCT finite element solution and the well-to-well profiles compare the FCT solution with the upstream scheme (β=0) and the Galerkin Scheme (β=1). The timestep used is the same for each case, giving a maximum Courant number of 0.7 away from the wells. It can be seen that the Galerkin solution is highly oscillatory and, although adding a Lapidus-type artificial viscosity damps this oscillation, it is not effective at breakthrough and gave severe stability problems for this problem. The FCT solution is precisely monotone and in the lower part of the grid has resolved the fronts over three nodes.

This technique has been applied to multiwell tracer injection in irregularly bounded reservoirs and has behaved consistently well. Future work will involve testing on the two-phase flow equations where FCT has proved successful with conventional finite difference techniques (see Christie and Bond, 1985), and examining different types of upstream weighting and higher-order flux correction terms.

4. ACKNOWLEDGEMENTS

Dr. I.R. White is thanked for his contribution to the ALCHEMIST program which was the basis for obtaining these results, Dr. A. Settari for many valuable discussions, and British Petroleum Plc for permission to publish this work.

5. REFERENCES

Book, D.L., Boris, J.P. and Hain, K.H. (1975) Flux Corrected Transport II: Generalisation of the method. *J. Comput. Phys.*, **18**, 248-283.

Boris, J.P. and Book, D.L. (1973) Flux Corrected Transport I: Shasta, a fluid transport algorithm that works. *J. Comput. Phys.*, **11**, 38-69.

Christie, M.A. and Bond, D.J. (1985) Multi-dimensional Flux Corrected Transport for reservoir simulation. SPE13505 Eighth SPE Symposium on Reservoir Simulation, Dallas.

Donea, J. (1984) A Taylor-Galerkin Method for Convective Transport Problems. *Int. J. Num. Meth. Eng.*, **20**, 104-119.

Gray, W.G. and Pinder, G.F. (1976) An analysis of the numerical solution of the transport equation. *Water Res. Research*, **12**, 547-555.

Spivak, A., Price, H.S. and Settari, A. (1976) Solution of the Equations for Multidimensional Two-Phase Flow by Variational Methods. SPE5123 4th SPE Symposium on Reservoir Simulation, Los Angeles.

Sun, N.-Z. and Yeh, W.W.G. (1983) A proposed upstream weight numerical method for simulating pollutant transport in groundwater". *Wat. Res. Research*, **19**, 1489-1500.

Zalesak, S.T. (1979) Fully multidimensional Flux-Corrected Transport algorithm for fluids. *J. Comput. Phys.*, **31**, 335-362.

NUMERICAL MODELLING OF DISCONTINUOUS ATMOSPHERIC FLOWS

M.J.P. Cullen
(Meteorological Office, Bracknell, U.K.)

1. INTRODUCTION

Numerical modelling of large scale atmospheric flow is very successful. Useful forecasts of the flow pattern are obtained out to 3 days in nearly every case and to 5 days in about half the cases. The problems occur in converting these forecasts into useful statements about actual weather. In order to do this the detailed vertical structure of the atmosphere must be accurately predicted so that different air masses can be clearly identified.

A large scale weather map, such as those published in the newspapers, appears smooth. The dynamical theory of vertically averaged atmospheric motion shows that the behaviour is essentially the same as two dimensional incompressible flow. That system of equations, given smooth initial data, has smooth solutions indefinitely: see (Kato, 1965). A correct numerical solution can be obtained by any stable consistent finite difference or Galerkin scheme. The most effective way of obtaining nonlinear stability in this case is to use the conservative schemes of Arakawa (1966), or a Galerkin method. The concept of smooth solutions extends to three dimensional flow provided that the scale remains large and the stratification strong. However, detailed observations of the vertical structure of the atmosphere indicate that the solutions are far from smooth. The structure is often like a series of layers of air with markedly different properties and sharp interfaces between them: see (Danielson, 1959). These interfaces support waves and turbulence. The amount of moisture and dust in the different layers is usually very different, leading to marked changes in weather when different layers reach the surface.

A system of equations whose scaling allows for the presence of discontinuities has been developed by Hoskins (1982) and the existence of solutions for piecewise constant initial data shown by Cullen and Purser (1984). There is a need to solve the complete equations of atmospheric motion in a way which will generate these piecewise constant solutions and model the waves and turbulence on the interfaces. As in other branches of fluid dynamics when discontinuities develop, it is essential to choose the correct conservation law form of the equations if numerical solutions are not to converge to unphysical solutions. It is also necessary to refine the solution procedure to help it to capture the discontinuities.

In this paper this is illustrated for three situations. In forecasting atmospheric fronts, a convergence to the correct solution can be obtained by absorbing certain acceleration terms into the turbulence model. In the case of flow over large scale mountains, the difficulty is to confine the influence of the mountains to a shallow layer above its top when the stratification is strong. In hemispheric scale forecasting, absorption of extra terms into the turbulence model can lend to a substantial reduction in error through eliminating spurious solutions. However, some useful detail is lost at the same time.

2. GOVERNING EQUATIONS

2.1 Primitive Equations

The normal form of equations used in atmospheric modelling are the equations of compressible gas flow with the hydrostatic approximation. For present purposes we also make the Boussinesq approximation and treat the effect of the Earth's rotation as constant in space. The resulting system of equations, as used by Gent and McWilliams (1983), is also appropriate for the ocean:

$$\frac{Du}{Dt} + \frac{\partial\phi}{\partial x} - fv = 0 \qquad (2.1.1)$$

$$\frac{Dv}{Dt} + \frac{\partial\phi}{\partial y} + fu = 0 \qquad (2.1.2)$$

$$\frac{D\theta}{Dt} = 0 \qquad (2.1.3)$$

$$\frac{\partial\phi}{\partial z} + g\theta/_{\theta_0} = 0 \qquad (2.1.4)$$

$$\frac{\partial u}{\partial x} + \frac{\partial v}{\partial y} + \frac{\partial w}{\partial z} = 0 \qquad (2.1.5)$$

$$w = 0 \quad \text{at} \quad z = 0, H \qquad (2.1.6)$$

The vertical coordinate z is a function of pressure, and the lower boundary condition is simplified so that it is applied at a constant pressure surface near the ground, rather than the Earth's surface. ϕ represents the height of a constant pressure surface, θ is the potential temperature and f the Coriolis parameter. The rest of the notation is standard.

2.2 The geostrophic momentum approximation

A scale analysis appropriate to large-scale atmospheric flow suggests that the velocities in (2.1.1) to (2.1.6) can be approximated by their geostrophic values.

$$u_g = -\partial\phi/\partial y \qquad (2.2.1)$$

$$v_g = \partial\phi/\partial x \qquad (2.2.2)$$

In accordance with the prescription of Lighthill (1961), this approximation is only made in the advected quantities, not in the trajectories themselves. This allows uniform validity of the approximation for large times. Hoskins (1982) shows that the resulting system is still valid in the presence of discontinuities provided the scale parallel to the discontinuity is large. Therefore we replace $(\frac{Du}{Dt}, \frac{Dv}{Dt})$ in (2.2.2) and (2.1.2) by $(\frac{Du_g}{Dt}, \frac{Dv_g}{Dt})$.

2.3 Lagrangian conservation form

Equations (2.2.2) to (2.2.6) subject to the geostrophic momentum approximation can be written in the following Lagrangian form:

$$\frac{DM}{Dt} = - fu_g \tag{2.3.1}$$

$$\frac{DN}{Dt} = - fv_g \tag{2.3.2}$$

$$\frac{D\theta}{Dt} = 0 \tag{2.3.3}$$

$$\frac{D\tau}{Dt} = 0 \tag{2.3.4}$$

where $$(M,N,\theta) = \nabla(\phi + \frac{1}{2} f^2(x^2 + y^2)) \tag{2.3.5}$$

and τ is the specific volume. The natural boundary condition is that no fluid crosses the boundary of the integration domain. The work of Hoskins shows that this system of equations can generate a discontinuity in a finite time from smooth initial data. After this time we assume that the Lagrangian form remains valid while the Eulerian form does not. The correctness of this can only be tested by actual experiment.

2.4 Implications for finite difference solutions

If it is assumed that (2.3.1) to (2.3.5) hold for almost all the fluid volume, then a finite difference solution of the original equations (2.1.1) to (2.1.6) must be made consistent with these conservation laws. The requirement of conservation of quantities along trajectories cannot only be met in an Eulerian scheme by upwinding, however this violates the requirement that the volume of fluid with a particular value of a conserved quantity is also conserved. The latter requirement cannot be met exactly in a conventional scheme. The best practical procedure is to use a quadratic conservative scheme (Arakawa, 1966) and to use an artificial viscosity which acts as little as possible on the vertical component of the vorticity, essentially the geostrophic part of the flow.

3. RESULTS

3.1 Fronts

Fig. 1 shows a solution of (2.3.1) to (2.3.5) obtained by representing the initial data as piecewise constant, in the manner of Glimm (1965) for hyperbolic conservation laws, integrating the ordinary differential equations for each segment, and fitting the results together in the way described by Cullen and Purser (1984). This method is like a moving finite element method except that the volumes rather than the shapes of elements are specified.

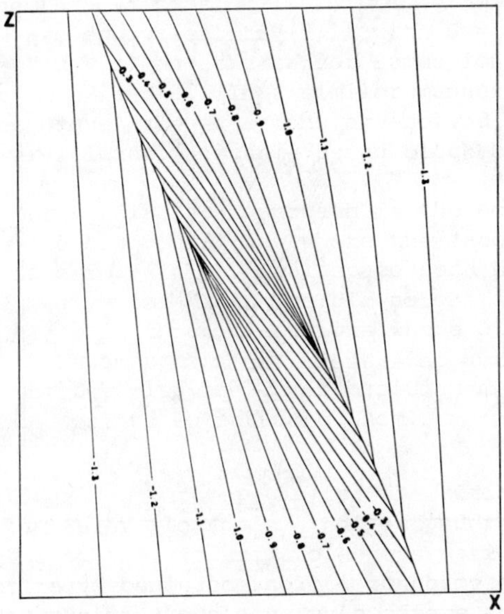

Fig. 1 Cross section of a front obtained by a Lagrangian method, contours of potential temperature.

Fig. 2 shows a solution of the Eulerian equations (2.1.1) to (2.1.6) obtained by adding artificial viscosity of the form

$$\frac{D(u-u_z)}{Dt} + K\,\nabla^2 u \qquad (3.1.1)$$

to equation (2.1.1), an equivalent term to equation (2.1.2) and $K\nabla^2\theta$ to equation (2.1.3). This has the effect of enforcing the geostrophic momentum approximation on the scale of the discontinuity and allows a value of K to be used which is 100 times smaller than that required if the extra acceleration term is not included in the viscosity. The solution does not capture the change in slope with height of the front very well. As the resolution is increased to 200 x 40, there is very slow convergence towards the solution shown in Fig. 1. The adaptive method has a clear advantage here.

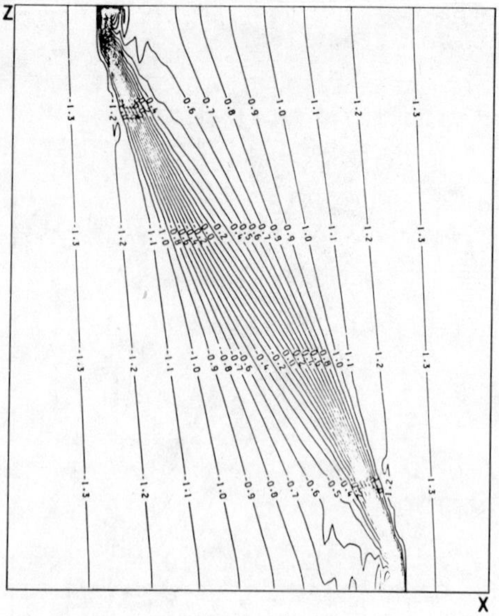

Fig. 2 Cross section of a front obtained by a finite difference method using a 200 x 20 grid, contours of potential temperature.

3.2 Mountain flow

Fig. 3 shows a cross section of the flow over the Alps predicted by a limited area forecast model which uses equations (2.1.1) to (2.1.6) and also includes the effect of moisture. The wind component across the Alps is about 15 ms^{-1}. A large standing wave is generated, also transient waves.

This structure is only observed in the atmosphere on much smaller horizontal scales. On the scale of the Alps, the constraint of the Earth's rotation makes the flow tend to go round, rather than over the mountain. Even when flow crosses the mountain, observations of clouds suggest that the rapid flow is confined to a thin layer near the ridge crest.

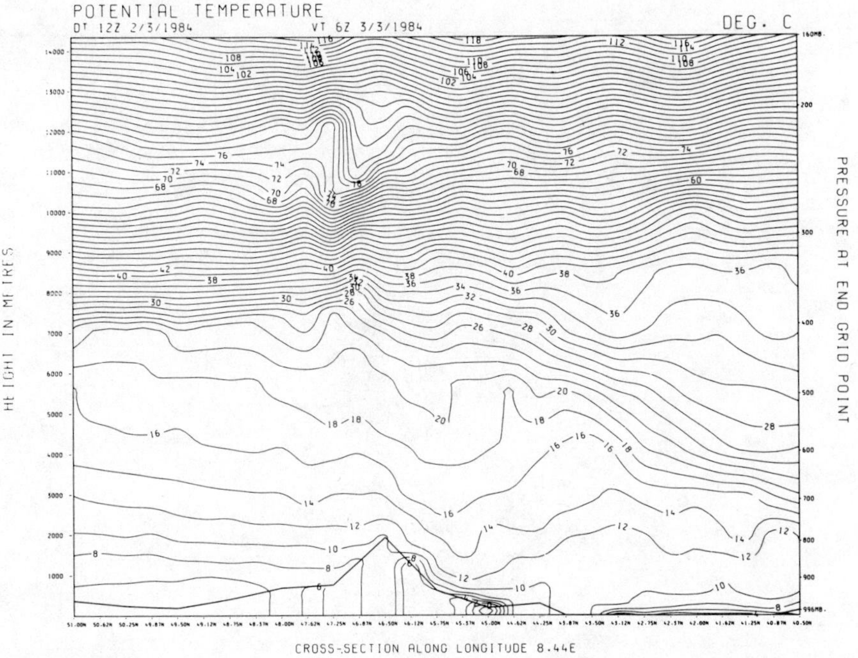

Fig. 3 18 hour forecast potential temperature cross section along $8^{\circ}E$.

Fig. 4 shows a two dimensional finite difference solution of the equations using a viscosity of the form (3.1.1). This suppresses the waves, but the disturbance due to the mountain extends up to the top of the atmosphere.

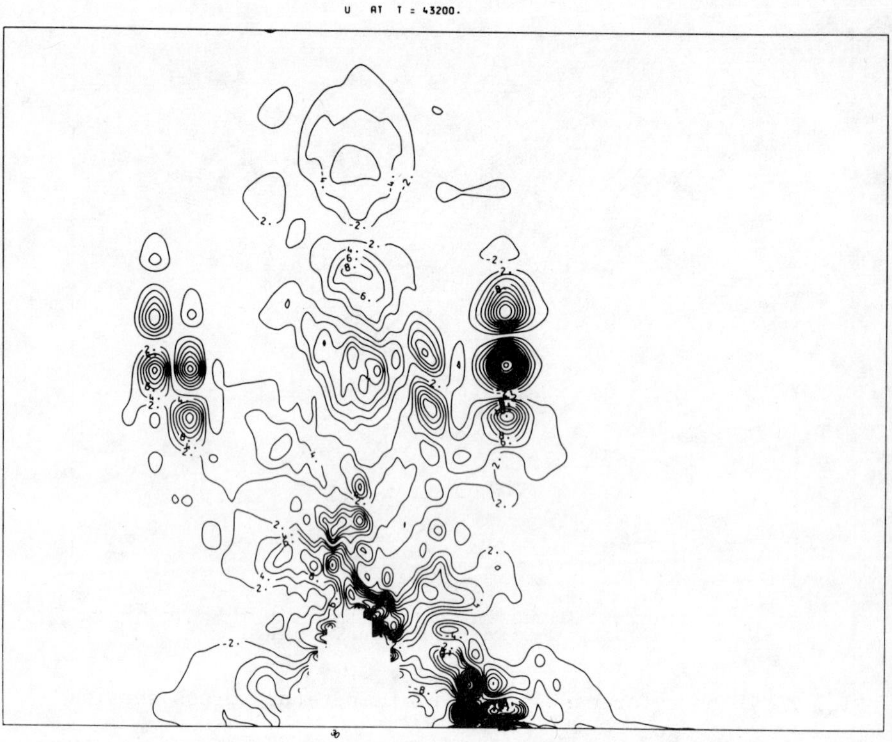

Fig. 4 Cross-mountain wind using finite difference model on 200 x 20 grid.

A solution using an adaptive Lagrangian method by S. Chynoweth (1985) prevents this extensive disturbance which may play a part in generating the spurious waves in the forecast model.

3.3 Short range forecasts

Figs. 5 and 6 show 24 hour surface pressure forecasts made using the U.K. operational hemispheric model, in the first case using (2.1.1) to (2.1.6) with artificial viscous terms added, and in the second case with acceleration terms absorbed into the viscosity as in equation (3.1.1). The verifying analysis is shown in Fig. 7.

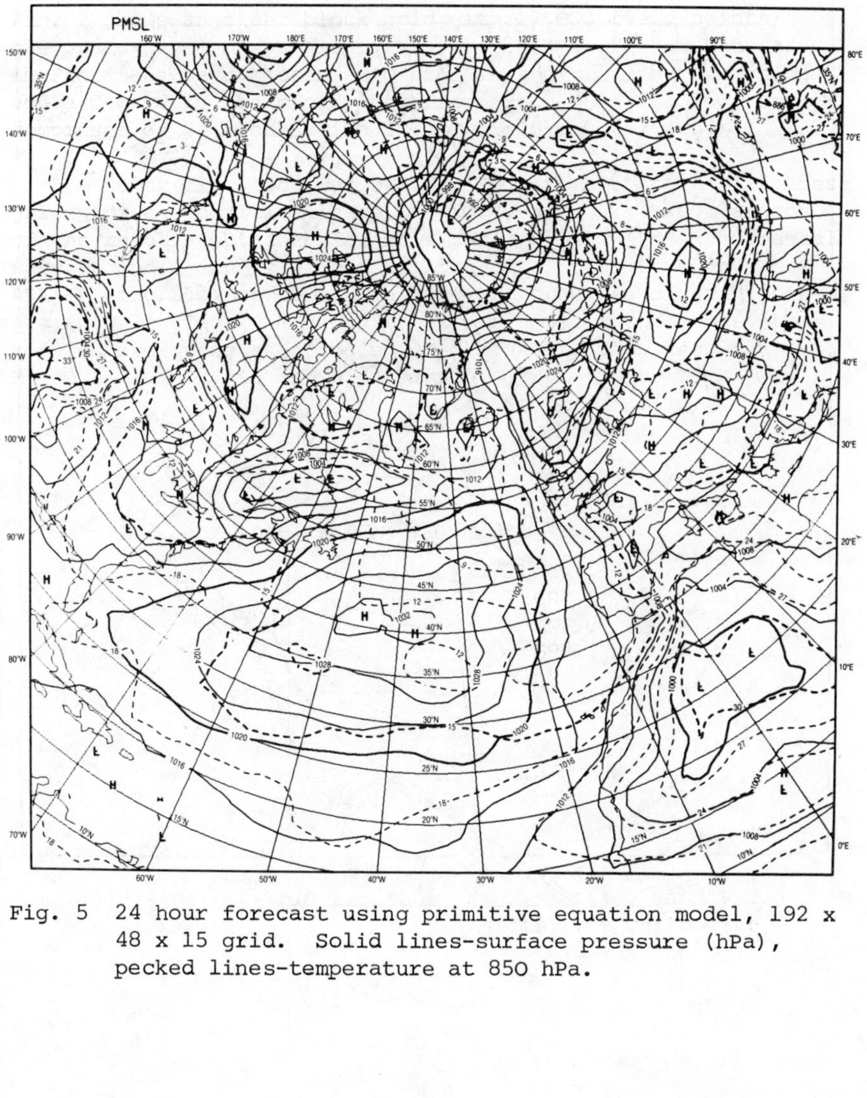

Fig. 5 24 hour forecast using primitive equation model, 192 x 48 x 15 grid. Solid lines-surface pressure (hPa), pecked lines-temperature at 850 hPa.

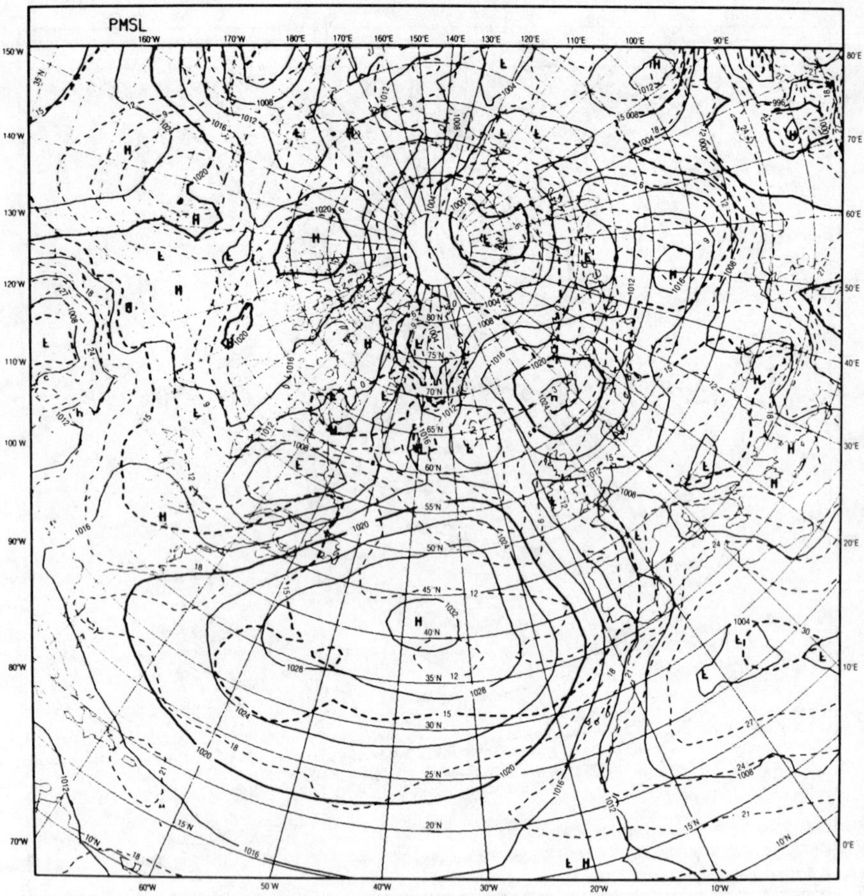

Fig. 6 24 hour forecast using model as Fig. 5 with artificial viscosity (3.1.1).

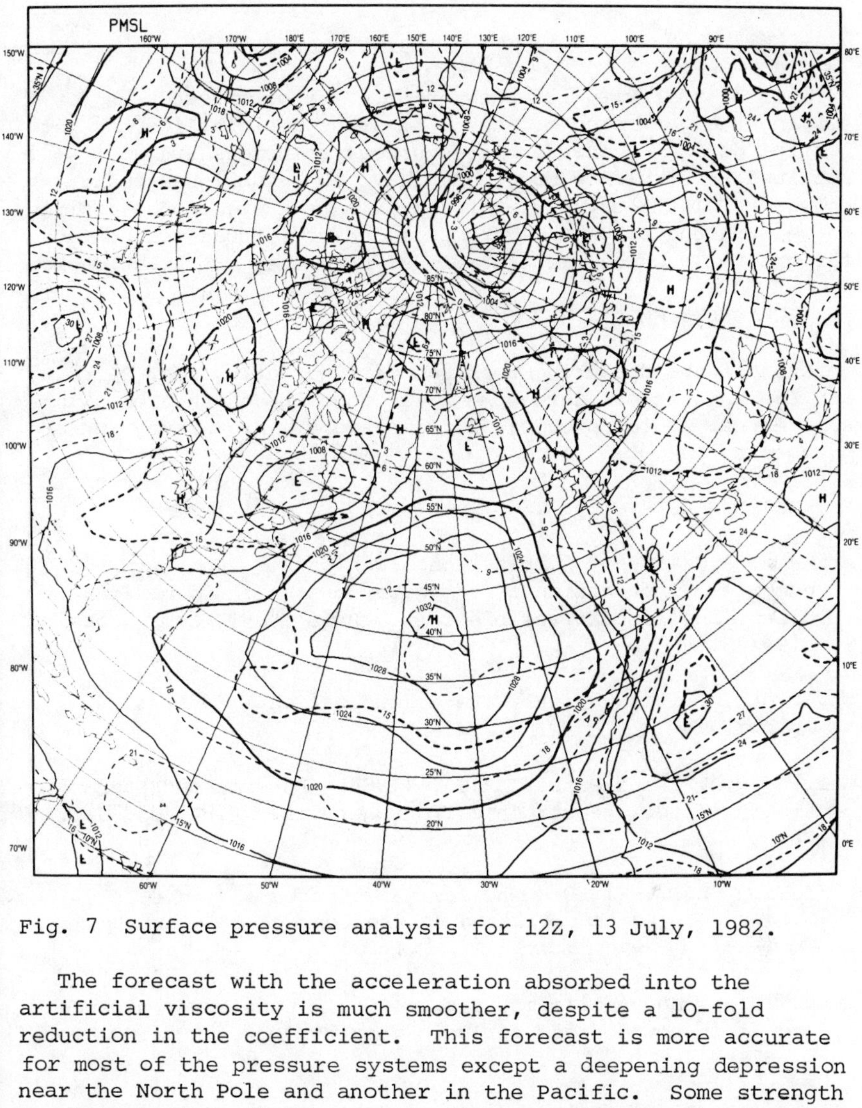

Fig. 7 Surface pressure analysis for 12Z, 13 July, 1982.

The forecast with the acceleration absorbed into the artificial viscosity is much smoother, despite a 10-fold reduction in the coefficient. This forecast is more accurate for most of the pressure systems except a deepening depression near the North Pole and another in the Pacific. Some strength has been lost in these. The improvement reflects the removal of much spurious noise, though it has been achieved at the cost of losing some correct detail from the forecast shown in Fig. 5. The L_2 error over the region 70°W to 20°E, 30° to 70°N has been reduced from 3.2 to 2.4 hPa.

4. CONCLUSIONS

This brief review has shown that numerical procedures for solving the governing equations of atmospheric motion may need careful design to ensure that spurious solutions are not obtained. Turbulence models may have to be designed using knowledge of the structure of approximate systems of equations as well as the complete system. Adaptive methods can achieve far greater accuracy in situations where they are practicable.

5. ACKNOWLEDGEMENTS

Much of the computing work was carried out by C.A. Parrett, and the theoretical development was done in conjunction with R.J. Purser (Meteorological Office) and Dr. J. Norbury (Oxford University).

6. REFERENCES

Arakawa, A., (1966) Computational design for long-term numerical integration of the equations of fluid motion: two dimensional incompressible flow. Part 1". *J. Comp. Phys.* **1**, 119-143.

Chynoweth, S., (1985) Private Communication. Address as author.

Cullen, M.J.P. and Purser, R.J., (1984) An extended Lagrangian theory of semi-geostrophic frontogenesis. *J. Atmos. Sci.*, **41**, 1477-1497.

Danielson, E.F., (1959) The laminar structure of the atmosphere and its relation to the concept of a tropopause. *Arch. Met. Geoph. Biokl. A.*, **11**, 293-332.

Gent., P.R. and McWilliams, J.C., (1983) Consistent balanced models in bounded and periodic domains, *Dyn. Atmos. Oceans*, **7**, 67-93.

Glimm, J., (1965) Solutions in the large for non-linear hyperbolic systems of equations, *Comm. Pure Appl. Math*, **18**, 697-715.

Hoskins, B.J., (1982) The mathematical theory of frontogenesis, *Ann. Rev. Fluid Mech.* **14**, 131-151.

Kato, T., (1965) On classical solutions of the two-dimensional non-stationary Euler equations, *Arch. Rat. Mech. Anal.* **25**, 188-200.

Lighthill, M.J., (1961) A technique for rendering approximate solutions to physical problems uniformly valid. *Z. Flugwiss*, **9**, 267-275.

THE IMPLEMENTATION OF MOVING POINT METHODS FOR CONVECTION-DIFFUSION EQUATIONS

C.L. Farmer and R.A. Norman
(Oil Recovery Projects Division, AEE Winfrith, Dorset)

1. INTRODUCTION

Moving point methods are able to solve high Peclet number, multi-dimensional, convection-diffusion equations with very small levels of numerical diffusion, overshoot or phase error and without excessive mesh refinement. The method uses a fixed mesh for the approximation of diffusion processes and a set of moving points for the approximation of convection.

We classify moving point methods into either pure or hybrid methods. Pure methods make minimal use of the fixed mesh and include the original moving point method of Garder, Peaceman and Pozzi (1964) which fails to converge under mesh refinement (Price, Cavendish and Varga, 1968). Many pure methods are inconsistent but can be modified by appropriate treatment of the diffusion terms (Farmer, 1985a). Unfortunately these convergent pure methods are too complicated for practical use.

Hybrid methods make maximal use of the fixed mesh. Convergence of these methods is trivial because at sufficiently low mesh Peclet numbers the methods are designed to become identical to an underlying, convergent, fixed mesh method. Relative to pure methods, the hybrid methods are straightforward to implement and possess the advantage that local, adaptive management of the set of moving points may be accomplished. This is done by the insertion or deletion of moving points according to the local gradient of the solution.

A detailed description of the hybrid moving point method may be found in (Farmer, 1985b).

2. EQUATIONS TO BE SOLVED

Throughout the following we shall consider various functions of a spatial point x in a two-dimensional Euclidean space, E_2. The components of x will be denoted by x_i $(i = 1,2)$. Let Ω be a finite, singly connected, subset of E_2 with boundary $\partial\Omega = \partial\Omega_1 U \partial\Omega_2$. The symbol t denotes time. Let $u = u(x,t)$ be a continuous time-dependent solenoidal vector field in E_2, called the velocity.

We wish to find a scalar field $c = c(x,t)$, called the concentration, satisfying the conditions;

$$\frac{\partial c}{\partial t} + u.\nabla c = \nabla.D.\nabla c \qquad x\varepsilon\Omega, t > 0 \tag{2.1}$$

$$c(x,0) = c_o(x) \qquad x\varepsilon\Omega \tag{2.2}$$

$$c(x,t) = 1 \qquad x\varepsilon\partial\Omega_1, t \geqslant 0 \tag{2.3}$$

$$\frac{\partial c}{\partial n} = 0 \qquad x\varepsilon\partial\Omega_2, t \geqslant 0 \tag{2.4}$$

where D is a positive second rank tensor called the diffusion tensor. c_o is a function defining the initial condition and the symbol $\partial/\partial n$ denotes the outward pointing normal spatial derivative. We assume that on $\partial\Omega_1$ $(\partial\Omega_2)$ characteristics enter (do not enter) the region Ω.

3. THE MOVING POINT METHOD

3.1 Definitions

Let Ω be a rectangular region which is partitioned into a union of M identical rectangular <u>mesh cells</u>, Ω_m $(m = 1, \ldots, M)$. Let x^m denote the coordinates of the centre of Ω_m. Introduce the discrete time $t_n = n\tau$, $\tau \geqslant 0$, $n = 0,1,2,\ldots$ where τ is the <u>time step</u>. Introduce a mesh function with values C_m^n which is interpreted as an approximation to the concentration at the point x^m at a time t_n.

Introduce a set of moving points $\{P_k : k = 1,2,\ldots,N\}$ with distinct coordinates X_k^n in E_2 at time t_n. Let A_k^n be the concentration on moving point P_k at time t_n, which is interpreted as an approximation to $c(X_k^n, t_n)$. Let ν_m^n be the total number of moving points with co-ordinates in Ω_m at t_n. If the point P_k is nearer to x^m than any other moving point then we write $x_k^n \varepsilon\, \Omega_m$.

3.2 Fixed Mesh Interpolation

Let I_F^s denote the fixed mesh interpolator which from the fixed mesh values F_ℓ^n constructs a function G^n according to

$$G^n(X) = I_F^s\,[F_\ell^n, X] \qquad X\varepsilon\Omega \tag{3.1}$$

where the index ℓ ranges over the indices of the fixed mesh points. s is the order of the interpolator. Examples of such interpolators are piecewise constant interpolation (s = 1) and piecewise bi-linear interpolation (s = 2).

3.3 Fixed Mesh Algorithm

If X_k^n, A_k^n and C_j^n are known at t_n then the hybrid moving point algorithm requires the construction of a first approximation $\tilde{C}_j^{n+1}$ to C_j^{n+1}. We will use the finite difference version of the modified method of characteristics (Huffenus and Khaletzky, 1981; Douglas and Russell, 1982; Hartree, 1953). Convergence of the algorithm has been proved by Douglas and Russell (1982).

We will denote the construction of $\tilde{C}_j^{n+1}$ by the fixed mesh algorithm using the notation

$$\tilde{C}_j^{n+1} = \Phi_j\,[C_\ell^n] \tag{3.2}$$

3.4 Moving Point Interpolation

Let I_M^s denote the moving point interpolator which from moving point function values G_k^n and fixed mesh values F_i^n constructs a function Q^n according to

$$Q^n(X) = I_M^s\,[G_k^n, F_i^n, X] \qquad X\varepsilon\Omega \tag{3.3}$$

where the indices k and i range over the indices of the moving and fixed mesh points respectively and s is the order of the interpolator. A variety of interpolation methods are possible but in problems in which the concentration possesses some smoothness the following algorithm is appropriate. This is based upon a Taylor expansion about the nearest moving point to a cell centre using spatial derivatives evaluated on the fixed mesh,

$$Q^n(X) = \begin{cases} G_k^n - (X_k^n - X).\nabla I_F^s[F_\ell^n, X] & X\varepsilon\Omega_m,\ k\bar{\varepsilon}\Omega_m,\ \nu_m^n > 0 \\ I_F^2[F_\ell^n, X] & X\varepsilon\Omega_m,\ \nu_m^n = 0 \end{cases} \quad (3.4)$$

where $s \geqslant 3$. In most applications we use $s = 3$ but in very smooth problems we use $s = 4$.

In problems in which the concentration is not smooth relative to the fixed mesh then the use of the nearest grid point interpolator (NGP interpolation) is appropriate;

$$Q^n(X) = \begin{cases} G_k^n & X\varepsilon\Omega_m,\ k\varepsilon\Omega_m,\ \nu_m^n > 0 \\ I_F^2[F_\ell^n, X] & \nu_m^n = 0 \end{cases} \quad (3.5)$$

In our codes we combine these interpolators according to the magnitude of the local concentration gradient such that if it exceeds a user defined value we use NGP interpolation and otherwise the Taylor method.

3.5 Point Moving Algorithm

To obtain X_k^{n+1} from X_k^n we solve the equation

$$\frac{dz}{dt} = u(z, t_n) \quad (3.6)$$

with the initial conditions $z(0) = X_k^n$ $(k = 1, \ldots, N)$ using T steps of the classical 4-th order explicit Runge-Kutta method with time step τ/T. T is an integer which may vary during the course of a calculation.

We denote the operation of moving the points by the symbol E and we write

$$X_k^{n+1} = E\,[X_k^n,\ u(X,t_n),\ \tau/T] \tag{3.7}$$

3.6 *Local Insertion-Deletion of Moving Points*

In many problems the concentration is almost constant, with values of nearly zero or unity, in a large fraction of Ω and possesses sharp fronts separating the regions of different values. One only requires moving points in the regions occupied by, and near to, sharp fronts. It is thus useful to implement some strategy for the insertion or deletion of moving points in Ω.

The operation of insertion and deletion will be denoted by the symbol J and we write

$$X_k^n = J\,[X_\ell^n] \tag{3.8}$$

A description of an insertion-deletion algorithm may be found in (Farmer, 1985b).

3.7 *The Hybrid Moving Point Algorithm*

Using the initial condition we assign C_m^o for each fixed mesh cell, X_k^o and A_k^o for each moving point. The X_k^o are chosen so that there are, say, 10 points per cell near to regions of large spatial gradient and, say, 1 point per cell away from such fronts. The points are most conveniently placed in the cells using a uniform random number generator.

At time t_n suppose C_m^n, X_k^n and A_k^n to be known. The C_m^{n+1}, X_k^{n+1} and A_k^{n+1} are constructed as follows.

<u>Step 1</u> - Update the moving point positions to obtain X_k^{n+1} by

$$X_k^{n+1} = E\,[X_k^n,\ u(X,t_n),\ \tau/T] \tag{3.9}$$

choosing T so that over any time interval, τ/T, no moving point travels a distance of, say, more than 1/3rd (or other user specified value) of a mesh cell in any direction.

<u>Step 2</u> - Perform the insertion-deletion procedure

$$X_k^{n+1} = J\,[X_\ell^{n+1}] \tag{3.10}$$

Step 3 - Update the fixed mesh concentrations using the fixed mesh algorithm to obtain the first approximation

$$\tilde{C}_m^{n+1} = \Phi_m[C_\ell^n] \tag{3.11}$$

Step 4 - Calculate the changes, δC_m^n, due to diffusion by

$$\delta C_m^n = \tau\Delta.(D\Delta)C_m^n \tag{3.12}$$

where $\Delta.(D\Delta)$ is the central difference approximation to $\nabla.(D\nabla)$. We must test that τ is small enough to satisfy the usual stability condition for an explicit central difference method for a diffusion equation. δC_m^n can be obtained as part of Step 3.

Step 5 - Update the moving point concentrations by setting, for each k,

$$A_k^{n+1} = A_k^n + I_F^2[\delta C_\ell^n, X_k^{n+1}] \tag{3.13}$$

Step 6 - Construct the fixed mesh concentration C_m^{n+1} by the interpolation

$$C_m^{n+1} = \lambda I_M^s[A_k^{n+1}, C_\ell^{*n+1}, x^m] + (1-\lambda)\, C_m^{n+1}$$

where

$$\begin{aligned} C_\ell^{*n+1} &= \tilde{C}_\ell^{n+1} & \ell \geqslant m \\ &= C_\ell^{n+1} & \ell < m \end{aligned} \tag{3.14}$$

and $\geqslant$ and $<$ in (3.14) denotes an ordering relation on the mesh cell indices. λ is a weighting factor, dependent upon the lowest upper bound, Pe_h, of the mesh Peclet number. The function λ is chosen via numerical experiments in such a way that the hybrid moving point method is trivially convergent. A suitable λ is

$$\begin{aligned} \lambda &= 1 & Pe_h \geqslant 10^{-3} \\ &= 0 & Pe_h < 10^{-3} \end{aligned} \tag{3.15}$$

4. REMARKS CONCERNING IMPLEMENTATION

In a two-dimensional problem we require three arrays, X, Y and A, storing the x- and y- coordinates and concentrations on the moving points. It is useful to scale X and Y so that the integer parts of the elements of X and Y give the indices of the mesh cells in some suitable ordering.

In our codes we utilise 'buffer' cells all around the physical cells with zero velocities on their outward facing boundaries. This is so that when moving points leave the physical region they may still be referenced in the inner loop executing the Runge-Kutta method for moving the points.

In the insertion-deletion step we test each point and delete it from the data structure if

(i) It is outside the physical region.

(ii) There are more than the maximum number of moving points in the fixed mesh cell.

A convenient method of deletion is to decrement by one the index of all the remaining moving points.

Finally, we would assign a moving point to the centre of each fixed mesh cell which contains fewer than a required minimum number of moving points. The concentration assigned to the moving point is that of the cell centre.

As work space we require two arrays of the same size as the fixed mesh concentration array to store the index of the moving point nearest to the centre of each mesh cell and the total number of moving points in each cell.

5. NUMERICAL TESTS

In this section the components of x will be denoted by (x,y) and the components of u by (u,v). The diffusion tensor is assumed diagonal and constant with common element, D.

5.1 One-Dimensional Test

A solution of the constant coefficient convection-diffusion equation in an infinite interval is

$$c_s(x,t) = \tfrac{1}{2} - \tfrac{1}{2}\,\mathrm{erf}\left[\frac{x - ut}{2\sqrt{Dt}}\right] \qquad (5.1)$$

If we choose constants u, D and t_o appropriately then the function

$$c(x,t) = c_s(x,t+t_o) \tag{5.2}$$

is, for small enough t, a good approximation to an exact solution on a finite interval. The solution (5.1) has the form of a smoothed step function with values between 0 and 1.

In Table 1 we give numerical results which investigate the effect of mesh refinement. At the initial time we have placed four moving points in each mesh cell using a uniform random number generator. We have used a second order Taylor moving point interpolator. These results were obtained using $t_o = 0.25$, $u = 1.0$ and $D = 5.10^{-4}$. The <u>width</u>, w, is the number of mesh cells in which the mesh cell values are in the interval [.1, .9] at t = 0. The width is a measure of how smooth the front is.

A more detailed investigation of convergence behaviour may be found in (Farmer, (1985b).

Pe_h	Width (cells)	Distance Convected (cells)	Maximum absolute error as percentage of unity
100	1	10	9.174
50	2	20	4.100
25	4	40	1.004
12.5	6	80	0.1947

Table 1:

<u>Maximum Absolute Error as a Function of Mesh Peclet Number</u>

5.2 Two-Dimensional Test

To test the performance of the hybrid moving point algorithm on an example with curved characteristics we study the standard rotating cone problem with (u,v) = (-y,x). We used a 43 x 43 mesh (41 x 41 physical mesh) of square cells of unit side. Reflection boundary conditions were used and the origin was placed at the centre of the mesh. At the initial time a single moving point was placed at the centre of each mesh cell. The insertion-deletion procedure ensured that at the end of each time step there was at least one moving point in each mesh cell. 100 time steps were used for a complete revolution. The

Runge-Kutta method computed the position of the moving points with an absolute error in each component of less than 10^{-5} of a mesh length. As the initial condition we used the function

$$c_o = e^{-\alpha[(x-9)^2 + y^2]} \tag{5.3}$$

where α is a parameter controlling the width of the front.

In Table 2 we show the maximum absolute error during the rotation and the maximum absolute error at the end of a complete revolution. A second order Taylor moving point interpolator was used except for the first result where the NGP method was used. In Table 2 the width parameter, w, is the number of cells across a diameter of the cone parallel to the x-axis on which the mesh cell centre concentration is greater than 0.1 at t = 0. The increase in error for the smoothest cone is a result of the increasing influence of the approximate reflection boundary conditions.

α	Width (cells)	Maximum Absolute Error as percentage of unity: During revolution	At completion
1/15	1	97.34	0.0 (NGP interpolation)
1	3	41.30	3.391 10^{-4}
5	7	14.02	2.318 10^{-4}
15	11	6.492	1.934 10^{-2}

Table 2:

Maximum Absolute Errors in the Rotating Cone Problem

6. CONCLUDING REMARKS

A hybrid moving point method has been described and some features of computer implementation discussed. The method has been shown to be very accurate in both one- and two-dimensional problems.

The structure of the method is such that the convection-diffusion problem is essentially reduced to a problem of interpolation on scattered data sets. This interpolation is the main source of error and so in any practical problem the expected numerical error can be estimated from estimates of the solution gradient and the order of the moving point interpolator.

The hybrid algorithm is suitable for application to practical problems (Farmer, 1985b) and we are currently using it in the two-dimensional simulation of flow in oil reservoirs.

7. ACKNOWLEDGEMENTS

This work was funded by the U.K. Department of Energy.

8. REFERENCES

Douglas, Jr., J., and Russell, T.F., (1982). Numerical methods for convection-dominated diffusion problems based on combining the method of characteristics with finite element or finite difference procedures. *SIAM J. Numer. Anal.* **19**(5), 871-885.

Farmer, C.L., (1985a). A moving point method for arbitrary Peclet number multi-dimensional convection-diffusion equations. *IMA J. Numer. Anal.* To be published.

Farmer, C.L., (1985b). A moving point method for the numerical calculation of miscible displacement. *UKAEA Winfrith* report number AEEW - R 1895.

Garder, Jr., A.O., Peaceman, D.W. and Pozzi, Jr., A.L., (1964). Numerical calculation of multi-dimensional miscible displacement by the method of characteristics. SPEJ. March 1964, 26-36.

Hartree, D.R., (1953). Some practical methods of using characteristics in the calculation of non-steady compressible flow. Report LA-HU-1, Harvard University.

Huffenus, J.P. and Khaletzky, D., (1981). The Lagrangian approach to advective term treatment and its application to the solution of the Navier-Stokes equations. *Int. J. Num. Meth. Fluids.* **1**, 365-387.

Price, H.S., Cavendish, J.C. and Varga, R.S., (1968). Numerical methods of higher-order accuracy for diffusion-convection equations. *SPEJ. Sept.* 1968, 293-303.

THREE DIMENSIONAL FREE LAGRANGIAN HYDRODYNAMICS

H.E. Trease
Computational Physics Group (X-7),
Los Alamos National Laboratory, New Mexico)

1. INTRODUCTION

The purpose of the following discussion is to describe the development of a 3-D Lagrangian hydrodynamics algorithm. The 3-D algorithm is an outgrowth of the 2-D free Lagrange model that is fully described in Trease (1981). Only the more pertinent issues of the free Lagrange algorithm will be presented; the details of the rest of the code development project are interesting but not appropriate in the context of a free Lagrange discussion. Let it suffice to say that a complete production code is being developed to support the free Lagrange algorithm to be described. A graphic description that outlines this code development project is presented in Figure 1.

The main objective of this project is to develop a computer model that can be used to simulate fluid flow in three dimensions. The inspiration for using free Lagrange as a basis for a hydrodynamics code was gained from the work of Crowley (1971), Fritts and Boris (1979), and Kirkpatrick (1976). The 2-D code, described in Trease (1981), was based on this previous work. This two dimensional model showed several attractive features about free Lagrange. First, it showed that free Lagrange can be used to handle fluid flow problems that exhibit strong shearing forces which classically could only be handled by Eulerian type algorithms.

Second, the accuracy of a free Lagrange algorithm was shown to be sufficient, with the irregular mesh, to produce credible solutions. Third, due to arbitrary connectivity of the free Lagrangian logic, meshes can be variably zoned. This allows the user to put the resolution where it is needed.

All three of these features of free Lagrange have been exploited to extend the 2-D model to include the third dimension. In doing so many aspects of large scale code development have been investigated. Several modern software tools have been used to make the manipulation of the arbitrary connectivity matrix, that is associated with free Lagrange, easier. For reference these tools include; a dynamic (heap) memory manager, a storage block manager and also a relational data base manager.

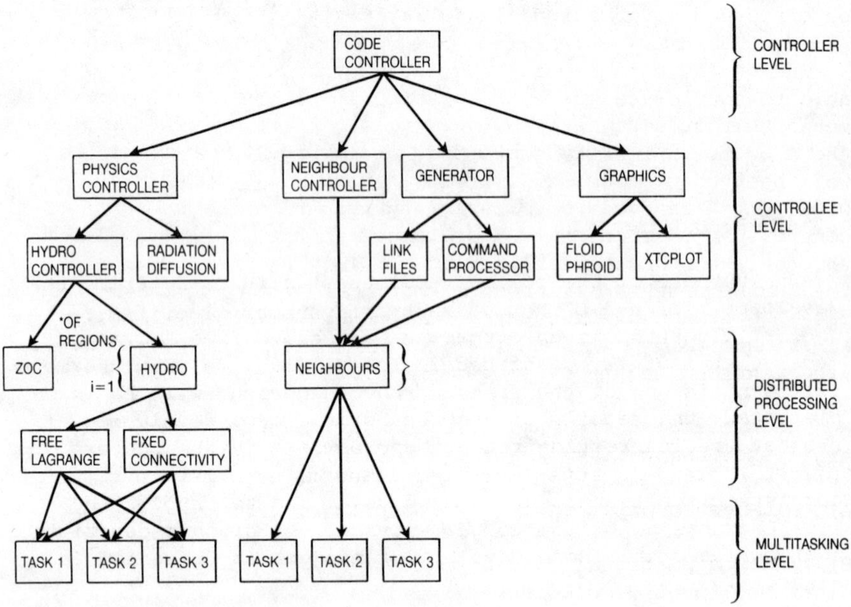

Fig. 1 Block structure of the 3-D code, illustrating the four levels that make up the code.

2. DESCRIPTION OF THE FREE LAGRANGE ALGORITHM

The main feature of the free Lagrange algorithm that identifies it from a standard Lagrange algorithm is the connectivity matrix that both defines the nearest neighbours for each point and the shape of the computational cells over which the fluid equations are integrated. A construction technique, that creates a VORONOI mesh, is used to identify nearest neighbours and define the mesh cells. The Voronoi mesh that is constructed has several properties that make it an excellent choice for maintaining the connectivity matrix for the 3-D code. These are:

A) The set of resulting polyhedra map the space defined by the mass points and bounding surfaces uniquely, i.e., none of the polyhedra overlap and nearest neighbours are guaranteed to be reciprocal.

B) Each polyhedron remains convex. This is accomplished by changing the area and the number of faces, i.e., neighbour swapping.

C) The volume and surface area of each polyhedron changes continuously. These and other aspects of the Voronoi mesh will be discussed more thoroughly in later sections.

One of the more important goals of the 3-D code is to be able to couple various hydrodynamic algorithms together. This means that free Lagrangian hydro will be used in regions where the flow field is most distorted. Then, in the (more) well behaved regions we will use an adaptive rezoning technique, with a mesh composed of mass points that have a fixed connectivity. These two algorithms will then be coupled through a third algorithm called a ZOC. The free Lagrange and the ZOC algorithms will now be described in detail. The detailed hydrodynamic equations, that are solved by these algorithms, along with their finite difference representations are discussed in detail in Appendix A.

The basic features that describe the free Lagrange algorithm are listed below:

A) All mesh quantities are cell centred.

B) The computational domain is described by an arbitrary distribution of mass points.

C) The code automatically constructs its connectivity matrix.

D) Mass points can be merged and/or added to the mesh.

The code determines its connectivity matrix by constructing a unique polyhedron about each mass point. The resulting polyhedral mesh is known as a Voronoi mesh. The faces of the polyhedron determine the set of "nearest" neighbours with which a mass point interacts. The faces of the polyhedron are represented by intersecting perpendicular bisecting planes between a given mass point and each of its neighbours. The details of this construction process are described more fully in Appendix B. Figure 2 shows several examples of Voronoi cells. The set of polyhedra that describe the mesh completely and uniquely span the space over which the mass points are distributed. Figure 3 shows a 2-dimensional projection of a 3-dimensional mesh, where the arbitrary polyhedra reduce to arbitrary polygons.

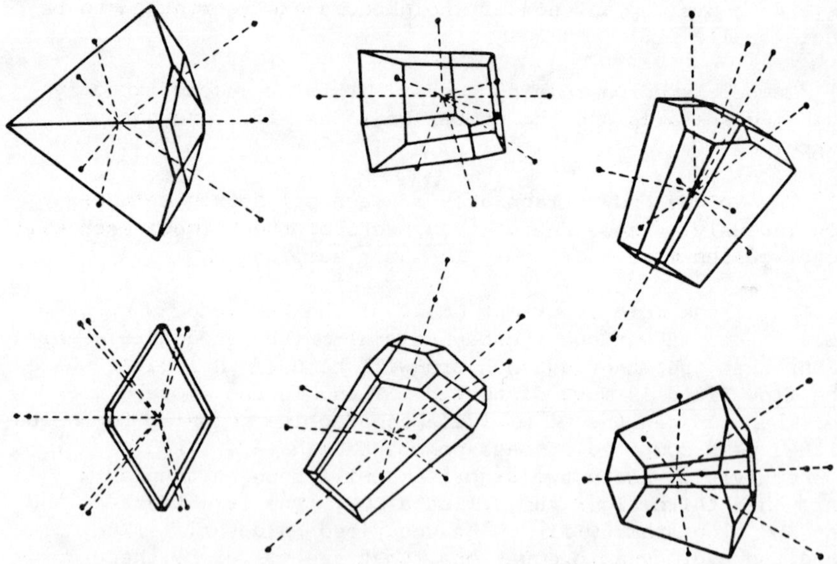

Fig. 2 Examples of several Voronoi cells. The Voronoi cells are the polyhedral shaped objects. The straight lines that end at a point represent the connections between the central mass point and its "nearest" neighbours.

Due to the fact that all physical quantities are carried at cell centred mass points, each point can change the set of "nearest" neighbours that it associates with by changing the shape of the polyhedron surrounding it while still retaining its Lagrangian definition. The neighbour changing process is smooth and continuous because of the integral nature of the algorithm. Two points become neighbours when a face with "epsilon" surface area appears between them. These points will drop each other as "nearest" neighbours when (and if) this face area shrinks below "epsilon".

There are several advantages and disadvantages associated with free Lagrange hydro. The advantages are obvious to anyone doing hydrodynamic calculations. Due to the arbitrary connectivity of the mesh and the ability to change this connectivity, highly distorted flows can be modelled by using a Lagrangian algorithm. Also, since the mesh maintains itself, no manual rezoning is needed (this is extremely important in a 3-dimensional code). Complex geometries that require variable zoning can be set up relatively easily since the code calculates the connectivity matrix from an arbitrary distribution of mass points. The main disadvantage of this method is the overhead associated with maintaining and processing the connectivity lists, but since the neighbour lists are unique, they are very amenable to a calculation using a multi-tasking algorithm (i.e., the neighbour searches can be done in any order, but the resulting global connectivity matrix is the same). Also, future machines that support hardware gather-scatter operations will improve the efficiency of this algorithm.

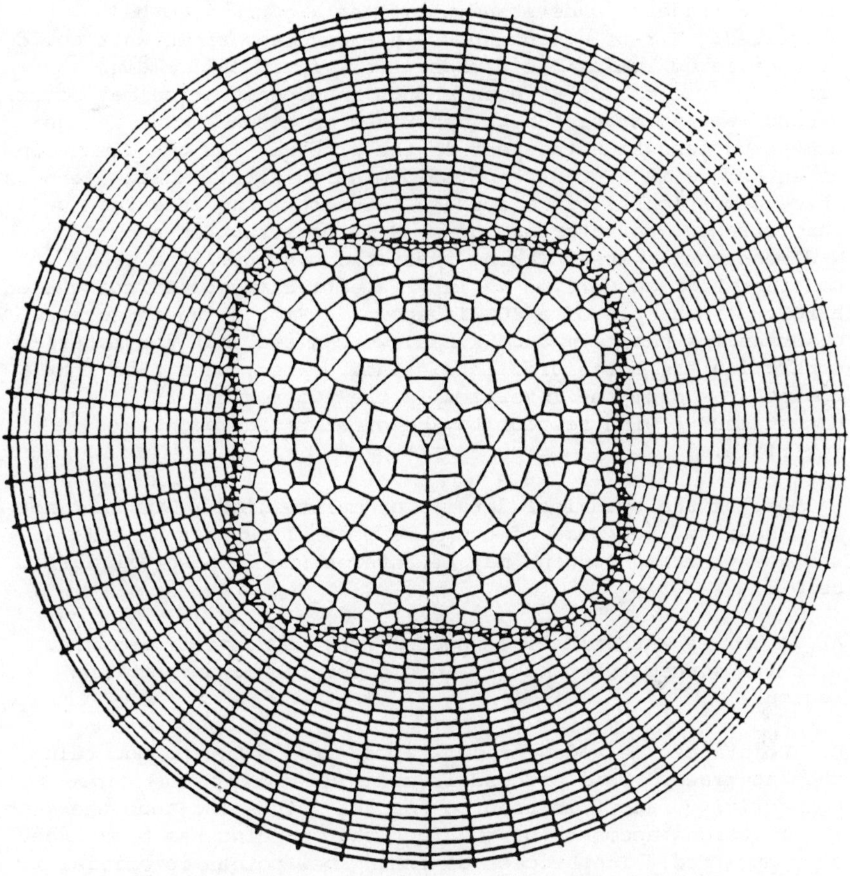

Fig. 3 2-dimensional projection of a 3-dimensional Voronoi mesh.

One of the weakest aspects of the free Lagrange method described is the treatment of continuous interfaces. This results from the fact that the edges of the computational volumes are arbitrarily defined to be midway between two "nearest" neighbours. This definition, while consistent with the Lagrangian equations, leads to a poor definition of a continuous interface. The realization of this fact suggested that an interface tracking algorithm was needed to follow the motion of interfaces. We will now describe the algorithm that is being used and how we intend to develop it into an all encompassing interface tracking algorithm, along with some of the positive side effects, in relation to distributed processing and slip-line treatment.

A little reflection on two key properties of interfaces will help the reader's understanding of the algorithm to be described. First, a continuous interface separates what could be considered immiscible fluids, i.e., fluid "A" remains distinct from fluid "B" even though interpenetration may occur. Second, an interface can be described in a space that is one dimension less than the rest of the problem. In one dimension an interface is a point, in two dimensions it is a line, and in three dimensions it is a surface. Generalizing this idea we can represent an interface as a (N-1) construct in a N-dimensional space. Taking these two concepts into consideration we derived an interface tracking construct called a ZOC. An example of a ZOC is shown in Fig. 4, where several observations can be made. First, we can see that the interface separating the two fluids is distinct. Also, we notice that zoning away from the interface in the two regions is discontinuous with respect to the other region and the interface.

Most of the technical details of maintaining a ZOC will not be discussed, but some of the more general aspects of this surface tracking concept may be interesting. These are listed below:

A) A ZOC is essentially a special free Lagrange region that uses its connectivity matrix to connect to the surrounding regions.

B) Points can be added or subtracted from a ZOC to maintain the interface. This process is made especially easy since a connectivity matrix is used to connect points and thus the mesh reconnections account for the fact that a point has been added or subtracted. The process of adding new points is trivial because the Voronoi mesh indicates when a new point should be added and where the new point should be located. The rest of the detail of adding a point involves the redistribution of mass, momentum, and energy in a local region of space.

C) A ZOC will work in two dimensions, to maintain a line interface, just as well as in three dimensions.

D) The treatment of slip-lines should be automatic with a ZOC since there is no restriction on the tangential velocity of the fluid on either side of the interface.

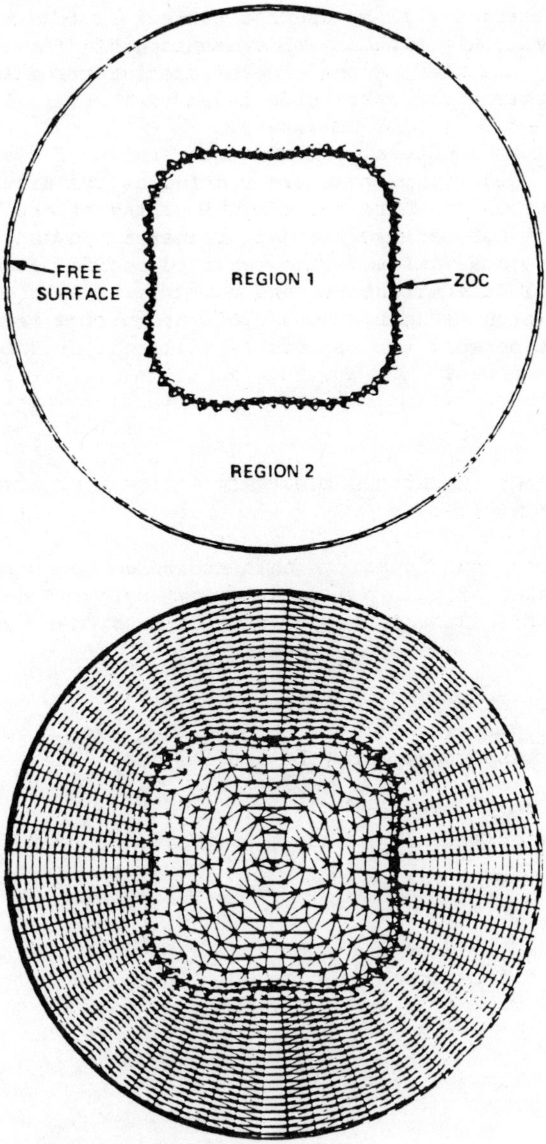

Fig. 4 Example of ZOC. The upper figure shows the ZOC in relation to the free surface. The lower figure shows the corresponding grid.

E) This interface will be used to connect regions that use different hydro algorithms. This means that a free Lagrange algorithm can be used on one side of the interface and a fixed mesh algorithm on the other side.

F) A ZOC makes a natural communication buffer for connecting two separate algorithms that are running as distributed processes. This is where the dimensionality of the ZOC becomes important because the data transfer between processes must be kept to a minimum. The separate region processes are N-dimensional data structures and a ZOC is a (N-1) data structure, which means the amount of information being communicated between the regions is small compared to the regions themselves.

APPENDIX A

THE HYDRODYNAMIC EQUATIONS AND THEIR FINITE DIFFERENCE REPRESENTATION

The hydrodynamic equations that are solved on this free Lagrange system of mass points are given below. These equations represent the conservation of mass, momentum, and specific internal energy, respectively;

Continuity Equation;

$$\frac{1}{\rho}\frac{D\rho}{Dt} = -\overline{\nabla}.\overline{u} \ , \qquad (A.1)$$

Conservation of Momentum;

$$\rho \frac{D\overline{u}}{Dt} = -\overline{\nabla}p - [\overline{\nabla}.\overline{\overline{\gamma}}] \ , \qquad (A.2)$$

Conservation of Internal Energy;

$$\rho \frac{D\varepsilon}{Dt} = -p(\overline{\nabla}.\overline{u}) - (\overline{\overline{\gamma}}:\nabla u) \ , \qquad (A.3)$$

where

ρ = fluid density,
p = fluid pressure,
ε = internal energy/unit mass,

$\overline{u}$ = fluid velocity vector,
$\overline{\overline{\gamma}}$ = total stress tensor, and
$= \overline{\overline{\tau}} + \overline{\overline{Q}}$,

where

$\bar{\bar{\tau}}$ = stress tensor and

$\bar{\bar{Q}}$ = artificial viscosity tensor

$$\bar{\bar{Q}} = \begin{cases} \ell^2 \rho \, \mathrm{div}(\bar{u}) \{ (\overline{\nabla u}) - \frac{1}{3} \mathrm{div}(\bar{u}) \bar{\bar{e}} \} & \text{, if div } (\bar{u}) < 0 \\ 0 & \text{, if div } (\bar{u}) \geq 0, \end{cases} \tag{A.4}$$

and where

$\mathrm{div}(\bar{u})$ = divergence of velocity vector $\bar{u}$,

$\overline{\nabla u}$ = dyadic product of the differential operator $\bar{\nabla}$ and the velocity vector $\bar{u}$,

$\bar{\bar{e}}$ = unit tensor , and

ℓ = constant * local grid spacing.

The algorithm used to solve these equations can be outlined as follows. First, Equations 1, 2 and 3 are integrated over an arbitrary volume (in reference to the code we integrate over a computational cell). The volume integrals are transformed to surface integrals by using the divergence theorem. The mean-value theorem from calculus is used to obtain average quantities. These resulting equations are then cast into finite difference form as shown below. The following notation will be used in writing the finite difference representation of the fluid flow equations;

i = mass point i, spatial position at (XI, YI, ZI),

j = j^{th} nearest neighbour of mass point i, spatial position at (XJ, YJ, ZJ),

J = total number of nearest neighbours associated with mass point i,

n = present time step (time = t),

$n+1$ = next time step (time = t + Δt),

$R_{i,j}$ = distance from mass point i to nearest neighbour j,

$A_{i,j}$ = area of polygon face separating mass point i from nearest neighbour j,

$V_{i,j}$ = volume of the polyhedron associated with mass point i and nearest neighbour j,

M_i = mass of fluid associated with mass point i,

$P_{i,j}$ = fluid pressure at the face associated with mass point i and nearest neighbour j,

ρ_i = fluid density in cell i,

$\overline{U}_{i,j}$ = fluid velocity at the face associated with mass point i and nearest neighbour j, and

$\hat{n}_{i,j}$ = normal vector to the face associated with mass point i and nearest neighbour j.

$$V_i^{n+1} \left(\frac{D\rho}{Dt}\right)_i = -\rho_i^n \sum_{j=1}^{J} n_{i,j} \cdot \overline{U}_{i,j}^{n+1} A_{i,j} \,, \qquad (A.5)$$

$$M_i \left(\frac{D\overline{U}}{Dt}\right)_i = -\sum_{j=1}^{J} P_{i,j} \hat{n}_{i,j} A_{i,j} - \sum_{j=1}^{J} \hat{n}_{i,j} \cdot \overline{\overline{\gamma}}_{i,j} A_{i,j} \,, \qquad (A.6)$$

$$M_i \left(\frac{D\varepsilon}{Dt}\right)_i = -P_i \sum_{j=1}^{J} \hat{n}_{i,j} \cdot \overline{U}_{i,j}^{n+1} A_{i,j} - \sum_{j=1}^{J} \hat{n}_{i,j} \cdot [\overline{\overline{\gamma}}.U]_{i,j} A_{i,j} +$$

$$+ \overline{U}_{i,j} \cdot \sum_{j=1}^{J} \hat{n}_{i,j} \cdot \overline{\overline{\gamma}}_{i,j} A_{i,j} \,. \qquad (A.7)$$

APPENDIX B

MESH CONNECTIVITY (NEAREST NEIGHBOUR CALCULATIONS)

The purpose of this Appendix is to describe the manner in which the connectivity matrix for the free Lagrange algorithm is calculated. The connectivity matrix contains the "nearest" neighbours for all the mass points. These connections are used for calculating surface areas and volumes of the Voronoi cells that make up the computation mesh. In addition to describing the geometry of the cells, the connectivity matrix indicates which cells will interact hydrodynamically.

Each Voronoi cell is made of an arbitrary number of intersecting planes. These planes construct a convex polyhedron with an arbitrary number of "faces" about each mass point. Each face forms a polygon with an arbitrary number of "edges". The trick is to devise an algorithm that can calculate the connectivity matrix from an arbitrary distribution of points. As we proceed through this discussion the two following definitions should be kept in mind.

A) "face" neighbour: Any two points that are "nearest" neighbours are separated, in 3 dimensions, by a polygon shaped "face". A "face" neighbour therefore refers to the "nearest" neighbour that is across a given "face" from a given central point "I". There is a one-to-one correspondence between the number of "face" neighbours that are associated with a point and the number of "faces" that make up the polyhedral cell surrounding that point.

B) "edge" neighbour: Each of the "faces" of a Voronoi cell is a polygon. Each vertex of a given polygon "face" is found to be the centre of a sphere that passes through four points. The four points are: the central point "I", its "face" neighbour "j", and two other points that are called "edge" neighbours (they are referred to as "k-1" and "k"). Also, since each polyhedron is a closed figure each "edge" neighbour must also be a "face" neighbour. It should be noted that there is a one-to-one correspondence between the number of "edges" on a given polygonal "face" and the number of "edge" neighbour associated with that "face".

The main idea in discovering the "nearest" neighbours of a point is to identify the list of "edge" neighbours for each "face" of the polyhedron surrounding that point. As each set of "edge" neighbours is discovered they are put on a stack of "face" neighbours, then the next "face" neighbour to be looked at is pulled from this stack. When the stack of "face" neighbours is empty the Voronoi polyhedron is complete and the connectivity for this given point has been found.

The process of constructing the connectivity matrix for a given point is described more completely in the following steps:

A) Assemble a list of "possible" neighbours. This process depends on whether or not the point has been calculated before.

a) If the point hasn't been calculated before:

1) use any "logical" neighbour information to select as many potential neighbours as possible.

2) use a proximity rule to select a set of closest points based on filters such as distance between and material type.

b) If the point has been calculated before:

1) recall the old connectivity matrix for a given point.

2) form the list of possible "nearest" neighbours from the old "nearest" neighbours plus the "nearest" neighbours of the old "nearest" neighbours.

B) From this list of "possible" neighbours we identify the 1st "face" neighbour by the procedure outlined below.

a) First all points are translated so the coordinate axes are centred at the central mass point, point "I", by setting

$$\vec{X}_\ell = \vec{X}_\ell - \vec{X}_I \quad (\ell=1,\ldots,nkin) \tag{A.8}$$

where

$\vec{X}_\ell$ = coordinates of a possible neighbour "ℓ",

$\vec{X}_I$ = coordinates of point "I", and

nkin= number of points on the list of "possible" neighbours for point "I".

b) Select the first "face" neighbour by defining following equation,

$$N_1 = \text{Index of } [\min_\ell |\vec{X}_\ell|] \quad (\ell=1,\ldots,nkin) \tag{A.9}$$

where

$|\vec{X}_\ell|$ = the distance to point ℓ and

N_1 = index of the 1st "nearest" neighbour (this is the index into the global mesh arrays).

c) Select the first "edge" neighbour for the first "face" by defining

$$E_{1,1} = \text{Index of } [\min_\ell \{W_1 |(\vec{X}_V)_1 \times (\vec{X}_V)_\ell| + \tag{A.10}$$

$$+ W_2 \cos((\vec{X}_N)_1, \vec{X}_\ell) + W_3 |\vec{X}_\ell|\}] \quad (\ell=1,\ldots,nkin)$$

where

$E_{1,1} \neq N_1$,

$E_{k,j}$ = the kth "edge" neighbour of the jth "face" neighbour,

$(\vec{X}_N)_j$ = the coordinates of the "face" neighbour "N_j",

W_i = weighting parameters $\left(W_1 > W_2 > W_3\right)$,

$(\vec{X}_V)_\ell$ = the coordinates of the ℓth Voronoi point. This point is found as the centre of a circle which passes through three points. These points are: the central point "I", the "face" neighbour "N_j" and the next possible neighbour "ℓ" ($N_\ell \neq N_j$). The equations, in matrix form, that determine the (X,Y,Z) - coordinates of the ℓth Voronoi point are:

$$\begin{Bmatrix} (\vec{X}_N)_j \\ \\ \vec{X}_\ell \\ \\ (\vec{X}_N)_j \times \vec{X}_\ell \end{Bmatrix} \cdot \begin{pmatrix} \\ \\ (\vec{X}_V)_\ell \\ \\ \\ \end{pmatrix} = -\tfrac{1}{2} \begin{pmatrix} |(\vec{X}_N)_j|^2 \\ \\ |\vec{X}_\ell|^2 \\ \\ 0 \end{pmatrix} .$$

NOTE: This step describes the "nearest" neighbour algorithm in 2 dimensions (Z=0). Here we just use it to "start" the 3-dimensional algorithm described below.

d) The process of selecting the rest of the "nearest" neighbours for point "I" serves a dual purpose. First, we finish calculating the rest of the "face" neighbours and we also discover the list of "edge" neighbours that make up the polygonal "face" that separates points "I" and "j". An important point to notice is that the list of "nearest" neighbours for point "I" are contained within the sets of "edge" neighbours, i.e., the "nearest" neighbours are a subset of the "edge" neighbours. Therefore, by sifting the "edge" neighbours we obtain a list of unique points that represent the "face" neighbours for point "I". The "edge" and "face" neighbour lists will bind each other to completely describe the polyhedral cell surrounding point "I". The 3-D "nearest" neighbour selection algorithm is described below.

1) We already know the first "face" neighbour, N_1, and the first "edge" neighbour, $E_{1,1}$, for face 1 for point "I". These were found in Steps (B.b) and (B.c).

2) Now we calculate the "edge" neighbours for "face" j.

$$E_{k,j} = \text{Index of } [\min_{\ell} \{W_1 \ (\vec{X}_N)_j \cdot [[(\vec{X}_V)_{k-1} - \tfrac{1}{2}(\vec{X}_N)_j]$$

$$\times [(\vec{X}_V)_\ell - \tfrac{1}{2}(\vec{X}_N)_j]] + W_2 \cos((\vec{X}_N)_j, \vec{X}_\ell) + W_3 \cos(\vec{X}_{k-1}, \vec{X}_\ell)$$

$$+ W_4 |\vec{X}_\ell|\}] \ (\ell=1,\ldots,nkin) \qquad \text{(A.11)}$$

where $E_{k,j} \neq N_j$, $E_{k,j} \neq N_{1,j}$,

j = index of the current face,

k = 2 to K (until $E_{k+1,j} = E_{1,j}$, where K=number of "edge" neighbours),

W_i = weighting parameters $\{W_1 > W_2 > W_3 > W_4\}$

$(\vec{X}_N)_j$ = coordinates of nearest j,

$(\vec{X}_V)_\ell$ = coordinates of the ℓth Voronoi point and

$$\begin{Bmatrix} (\vec{X}_N)_j \\ \vec{X}_{k-1} \\ \vec{X}_\ell \end{Bmatrix} \cdot \begin{pmatrix} \\ (\vec{X}_V)_\ell \\ \\ \end{pmatrix} = -\tfrac{1}{2} \begin{pmatrix} |(\vec{X}_N)_j|^2 \\ |\vec{X}_{k-1}|^2 \\ |\vec{X}_\ell|^2 \end{pmatrix} .$$

3) From the list of "edge" neighbours, (E_k, k=1, K), we add the unique indices of the list of "face" neighbours,(N_j, j=1, J).

4) Increment j and continue with Step (d.2). This continues until all the "face" neighbours have been calculated, i.e., J > j. J is the number of "face" neighbours that are associated with point "I".

REFERENCES

Crowley, W.P., (1971) FLAG: A Free Lagrange Method for Numerically Simulating Hydrodynamic Flow in Two Dimensions, Proceedings of the Second International Conference on Numerical Methods in Fluid Dynamics, Lecture Notes in Physics, Vol. **8**, pp. 37-43, Springer-Verlag, New York.

Fritts, M.J. and Boris, J.P., (1979) The Lagrangian Solution of Transient Problems in Hydrodynamics using a Triangular Mesh, *Journal of Computational Physics*, Vol. 3, pp. 319-343, Academic Press, New York.

Kirkpatrick, R.C., (1976) FREE FLOW HYDRO, Internal Report (TD-2), Los Alamos National Laboratory, Los Alamos, New Mexico.

Trease, H.E., (1981) A Two-Dimensional Free Lagrangian Hydrodynamics Model, *Ph.D. Thesis*, University of Illinois, Urbana-Champaign.

PARAMETRIZATION OF VISCOSITY IN THREE DIMENSIONAL VORTEX METHODS AND FINITE - DIFFERENCE MODELS

S.P. Ballard
(Meteorological Office, Bracknell, U.K.)

1. INTRODUCTION

Three dimensional vortex methods, such as described by Leonard (1980) and Chorin (1982), provide a description of inviscid, incompressible fluid flow in terms of the evolution of vortex filaments or line vortices with small but finite cores.

This paper will discuss the results of inviscid simulations that identify the close alignment of opposite signed vorticity as important in allowing rapid stretching of vorticity whilst conserving energy. This suggests a method of parametrizing the effects of viscosity on the small scales of motion cancelling these sections of the filaments and relinking the remainder.

Results of simulations of the interactions of vortex rings, with the above parametrization of viscosity included, are compared with the observations of Fohl and Turner (1975).

The simulations are being used in order to evaluate turbulence models used in finite difference integrations of three dimensional flow. The implications are discussed.

2. CHORIN'S THREE DIMENSIONAL VORTEX METHOD

The Euler equations for three dimensional, incompressible, inviscid flow in vorticity formulation are

$$\frac{D\vec{\omega}}{Dt} = \frac{\partial\vec{\omega}}{\partial t} + (\vec{u}.\nabla)\vec{\omega} = (\vec{\omega}.\nabla)\vec{u} \qquad (2.1)$$

where

$$\vec{\omega} = \nabla \times \vec{u}, \quad \nabla.\vec{u} = 0 \qquad (2.2)$$

with $\vec{\omega}$ the vorticity vector, $\vec{u}$ the velocity and ∇ the differentiation vector.

The solution of equations 2.2 is the Biot-Savart Law for the velocity field,

$$\vec{u}(\vec{x},t) = -\frac{1}{4\pi}\int \frac{(\vec{x}-\vec{x}') \times \vec{\omega}(\vec{x}',t)\, d\vec{x}'}{|\vec{x} - \vec{x}'|^3} \tag{2.3}$$

The vorticity can change with time, from equation 2.1, due to the advection and stretching of vorticity.

The vortex method used here is based on that described by Chorin (1982). It is assumed that the vorticity field can be described in terms of a sum of M vortex filaments of small but finite cross-section, σ, with equal and constant circulations, Γ. The vortex filaments are further approximated by N_i straight line segments of length Δs_j^i, i=1,M, j=1,Ni. The motions of the end points of the segments, nodes, are followed so that vortex stretching is represented by the increase in length of the line segments. The maximum length of the segments is limited to some value, S_o. Once this value is exceeded the segment is split and an extra node is followed at its midpoint. Therefore the number of nodes followed increases as the length of the filaments, and hence vorticity, increases.

The velocities are calculated using a smoothed and discretized form of the Biot-Savart Law.

$$\vec{u}(\vec{x},t) = -\frac{1}{4\pi}\sum_{i=1}^{M}\Gamma_i \sum_{j=1}^{N_i} \frac{(\vec{x}-\vec{x}_j^i) \times \psi(\vec{x}-\vec{x}_j^i)\,\Delta s_j^i\, d\vec{s}_j^i}{|\vec{x}-\vec{x}_j^i|^3} \tag{2.4}$$

where $d\vec{s}_j^i$ is the unit vector directed along the tangent to the jth segment of the ith filament and $\vec{x}_j^i$ is the midpoint of the segment. $\psi(\vec{x}-\vec{x}_j^i)$ is a smoothing function of width σ required to remove the singularity of the integral whilst distorting close vortex interactions.

3. INVISCID SIMULATION TO STUDY VORTEX STRETCHING

3.1 Initial data

A simple situation in which to study vortex stretching is a periodic unit cube containing two perpendicular rectilinear filaments separated by distance 0.1 in the mutually perpendicular direction. In this situation there is initial, mutually induced, strain along both filaments in the direction of their vorticity vectors producing stretching.

A constant value of $\sigma = 0.05$ was used and

$$\begin{aligned} \psi(\vec{a}) &= 0 & |\vec{a}|>0.495 \\ &= 1 & \sigma<|\vec{a}|<0.495 \\ &= \frac{|\vec{a}|^3}{\sigma^3} & |\vec{a}|<\sigma \end{aligned} \tag{3.1.1}$$

similar to the form used by Chorin (1982). Image contributions within distance 0.495 were included in the velocity integral and a fourth order Runge-Kutta time integration scheme was used.

3.2 Results

As the flow evolved the filaments stretched and became increasingly chaotic and tangled. By the end of the integration some segments had stretched by a factor of 10^5. However, despite the rapid increase in vorticity, energy, as calculated on an 80^3 grid, had been conserved in the flow.

Chorin (1982), simulating the evolution of a turbulent vortex, observed that the constraint of energy conservation forced the filament into folds or tight tangles. In this simulation we observe principally pairing of opposite signed vorticity, where sections are aligned and very close, between either the same or both filaments, see Fig. 1.

In these dipole pairs the contributions of their vorticity to the velocity fields away from their location almost cancel and so they only have a small net effect. They are also partly responsible for the observed rapid stretching of the filaments as they move through the fluid, with their own induced velocities, away from the unpaired sections.

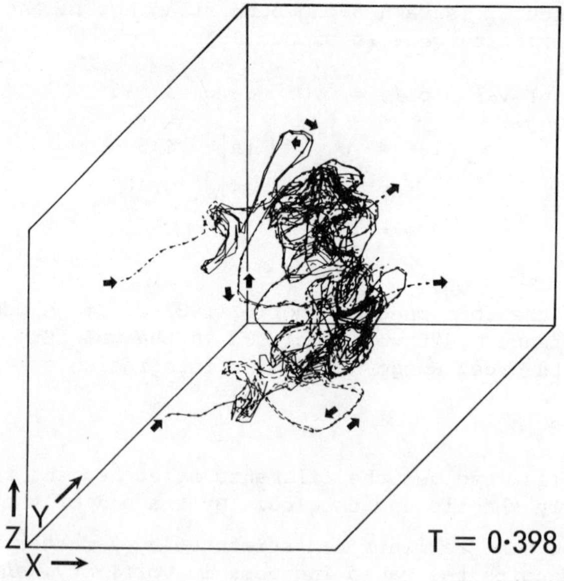

Fig. 1 Three dimensional representation of configuration of two initially perpendicular vortex filaments at t = 0.398 for σ constant.
——— Filament 1 initially at (0,y,-0.05)
_ _ _ _ Filament 2 initially at (x,0,0.05)
➡ Direction of vorticity vector

3.3 Accuracy of solution

Vortex filament methods do have limitations (Leonard, 1985) and are unlikely to be accurate once the radius of curvature or separations of filaments are less than σ. In Ballard (1985) it was found that, when different smoothing functions were used in the above simulation, the solutions diverged once separations became less than σ. However, all solutions showed evidence of dipole pairing. Runs where σ was allowed to reduce as the segments stretched indicated that the use of a constant value of σ underestimated the rate of increase of vorticity. The process of dipole pairing occurred earlier and more markedly as shown in Fig. 2. Obviously further numerical, theoretical or analytical calculations are required to produce a rigorous justification of dipole pairing but logical arguments and observations can be put forward in favour of the feature and are discussed in the next section.

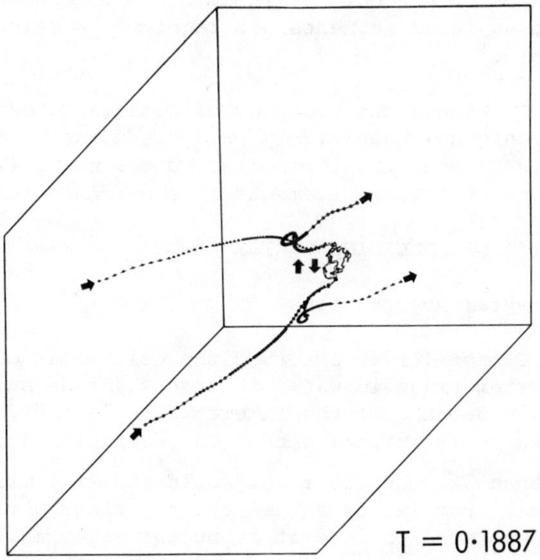

Fig. 2 As Fig. 1 but at t = 0.1887 for σ varying with local segment stretching.

——— Filament 1, — — — Filament 2

➡ Direction of vorticity vector

4. SUGGESTED PARAMETRIZATION OF EFFECTS OF VISCOSITY

The effects of viscosity in producing diffusion of vorticity can be treated by including viscous fattening of the core in the calculations, (Leonard, 1980), (Shirayama and Kuwahara, 1984) or by applying a random walk to markers on the filament at each time step, (Chorin, 1980), (Anderson and Greengard, 1985). However, the above methods cannot deal with the observed merging and relinking of vortex rings, see (Fohl and Turner, 1975), (Oshima and Asaka, 1977), or trailing vortices, (Sarpkaya, 1983) if whole filaments are tracked.

Assuming that dipole pairing is a real feature of the flow this suggests a plausible method of parametrizing the effects of viscosity in vortex filament calculations by removing any dipole paired sections of filaments and relinking the remainder. This is justified because in these regions the diffusive effects of viscosity will be most effective since the vorticity density and hence velocity gradients are very high. The two cores of opposite signed vorticity can be expected to diffuse into each other and result in zero net vorticity. They are also associated with highly stretched vorticity and hence the smallest scales of motion. It is also computationally advantageous as fewer segments are required to calculate the velocity field.

The need to change the topology of closely interacting vortex filaments has been widely recognised and Leonard (1975) implemented a reconnection method as did Schwarz (1982) in the collisions of vortex filaments in superfluid helium.

5. SIMULATION OF COLLIDING VORTEX RINGS

5.1 Experimental Observations

Fohl and Turner (1975) observed the collisions of two identical vortex rings in water at high Reynolds numbers and found that the details of the interactions depended on the angle between their initial paths, the collision angle ϕ. If ϕ was less than 32° the two rings would normally merge and remain as one. For larger angles the two rings merged and then eventually separated to move at right angles to the original paths.

5.2 Numerical Method

The same vortex method was implemented but using a smoother function $\psi(\vec{a})$ and with the rings in free space so that

$$\psi(\vec{a}) = \tanh(|\vec{a}|^3/\sigma^3) \quad \text{for all } a \tag{5.2.1}$$

The rings had the same proportion of radii and separation as in the experiments, with 60 segments per ring. A value of Γ was used to produce approximately the observed velocity scaled to the radii with $\sigma = 1.5\ \Delta s^i_j$ kept constant. A simple segment removal algorithm was implemented removing segments if the separation of their mid points was less than α and the angles between the vorticity vectors greater than θ_0. This involves detailed book-keeping.

5.2 Results

Choosing $\alpha = 0.02$ and $\theta_0 = 175^\circ$, the parametrization worked in simple situations and, for example, correctly predicted the behaviour when $\phi = 40^\circ$ and 20° as shown in Figs. 3 and 4. However, problems can occur due to the generation of short wavelength structure along the filaments and a more sophisticated parametrization and vortex method should be used.

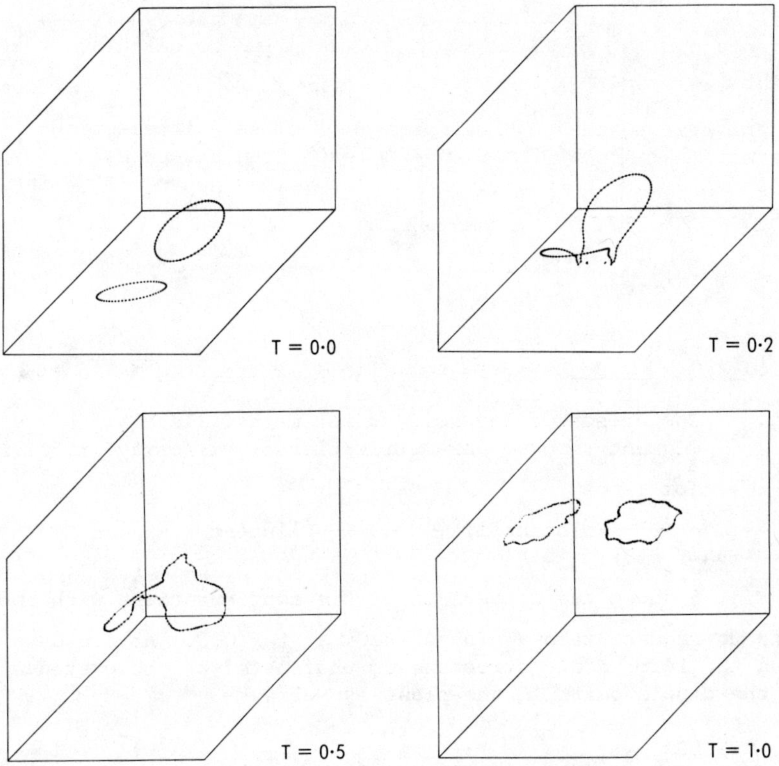

Fig. 3 Interaction of 2 vortex rings, $\phi = 40^\circ$

——— Filament 1, — — — Filament 2

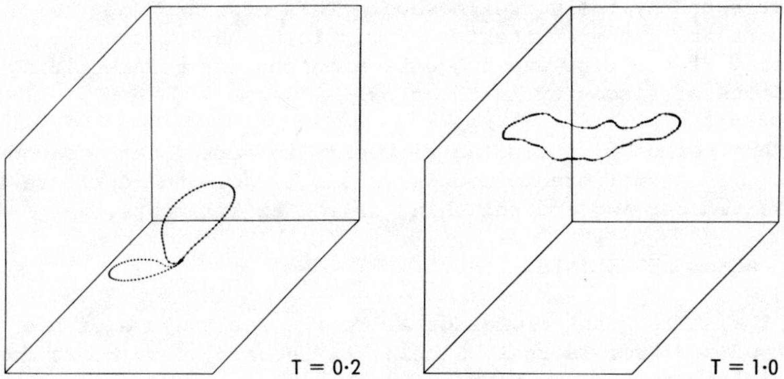

Fig. 4 As Fig. 3 but with $\phi = 20^\circ$

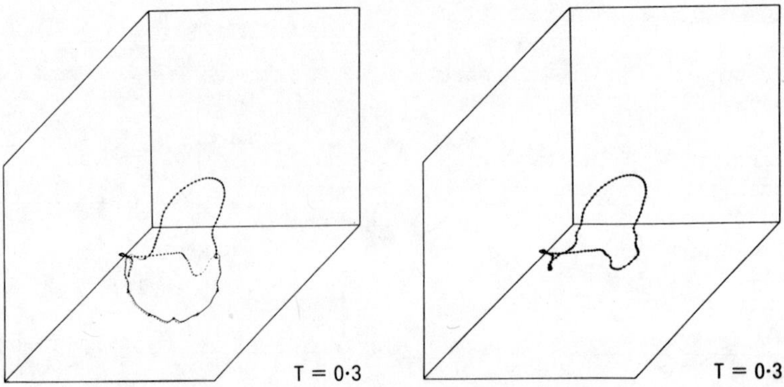

Fig. 5 Comparison of inviscid vortex method (left) with segment removal parametrization of viscosity (right) for $\phi = 40^{\circ}$

——— Filament 1, — — — Filament 2

Fig. 5 shows the comparison of the configurations with and without segment removal for $\phi = 40^{\circ}$ at $t = 0.3$. As can be seen the large scale structure is unaffected by the presence of the dipole pairs in the plane symmetry.

6. CONCLUSIONS

The observation of dipole pairing in inviscid vortex filament calculations is consistent with observations of filament relinking, although the close interactions are mistreated by the vortex method. This mechanism may be important in the stretching of vorticity whilst conserving the energy of the flow and suggests a method for parametrizing the effects of viscosity in the flow.

Any method of including viscosity in vortex methods and also finite difference models should be able to reproduce the observed merging and relinking of vortex filaments.

7. ACKNOWLEDGEMENTS

I wish to thank Professor A. Chorin for the use of his method and Professors A. Chorin, A. Leonard, D.W. Moore for their comments on my work.

REFERENCES

Anderson, C. and Greengard, C., (1985). On vortex methods. *SIAM. J. Num. Anal.* **22**, 413-440.

Ballard, S.P., (1985). Three dimensional vortex methods and their application to the direct simulation of turbulence. In preparation.

Chorin, A.J., (1980). Vortex models and boundary layer instability. *SIAM J. Sci. Stat. Comput.* **1**, 1-21.

Chorin, A.J., (1982). The evolution of a turbulent vortex. *Commum. Math. Phys.* **83**, 517-535.

Fohl, T. and Turner, J.S., (1975). Colliding vortex rings *Physics of Fluids.* **18** , 433-436.

Leonard, A., (1975). Numerical simulation of interacting, three dimensional vortex filaments. Lecture Notes in Physics Part 35, 245-250.

Leonard, A., (1980). Vortex methods for flow simulation. *J. Comp. Phys.* **37**, 289-335.

Leonard, A., (1985). Computing three-dimensional incompressible flows with vortex elements. *Ann. Rev. Fluid Mech.* **17**, *523-559*.

Oshima, Y. and Asaka, S., (1977) Interaction of multi-vortex rings. *J. Phys. Soc.* Japan **42**, 1391-1395.

Sarpkaya, T., (1983). Trailing vortices in homogeneous and density-stratified media. *J. Fluid. Mech.* **136**, 85-109.

Shirayama, S. and Kuwahara, K., (1984). Proceedings Ninth International Conference on numerical methods in fluid dynamics, June 1984, Suclay, France.

Schwarz, K.W., (1982). Generation of superfluid turbulence deduced from simple dynamical rules. *Phys. Rev. Lett.* **49**, 283-285.

AN EFFICIENT BOUNDARY-INTEGRAL METHOD FOR STEEP UNSTEADY WATER WAVES

J.W. Dold and D.H. Peregrine
(School of Mathematics, University of Bristol)

1. INTRODUCTION

A new method for computing the unsteady motion of a water surface, including the overturning of water waves as they break, has been developed. It is based on a Cauchy theorem boundary integral for the evaluation of multiple time-derivatives of the surface motion. The integrals are cast into the form of full matrices which can be solved iteratively for the instantaneous motion of discrete surface particles. The size of each time-step is determined to maintain a specified order of accuracy by using a truncated Taylor-series to perform explicit time-stepping. The numerical implementation of the method is efficient and accurate. For sufficiently large time-steps a growing "sawtooth" instability appears in the region of greatest point density, but the method is found to be stable for small enough time steps. A deliberate reduction in time step size to restore stability without smoothing normally only becomes necessary with the formation of the jet of a breaking wave where surface particles tend to become strongly concentrated.

Boundary-integral methods can be used to reduce the calculation of the inviscid incompressible irrotational motion of a body of fluid to the evaluation of the motion of its surface alone. In a numerical scheme it is possible therefore to solve for the fluid motion using only a point discretisation of the surface, thereby substantially reducing the number of unknown variables in the problem. This approach has been implemented by Longuet-Higgins and Cokelet (1976), Vinje and Brevig (1981), Baker, Meiron and Orszag (1981) and Roberts (1983), using a number of different integral formulations to model the motion of two-dimensional gravity waves on water.

In order to obtain accurate solutions for complicated surface motions, ranging from breaking waves to instabilities in a train of travelling waves it is, nevertheless, necessary to use fairly large numbers of surface points. (For example, a steepening wave in shallow water, part of which is shown in Fig. 1 was computed using up to 200 points). This involves running times per time step which vary as the cube of the number of points for algorithms which solve the integral equation using matrix inversion or factorisation methods, or as the square, using iterative techniques. It is therefore important to ensure computational efficiency particularly for more complicated wave phenomena.

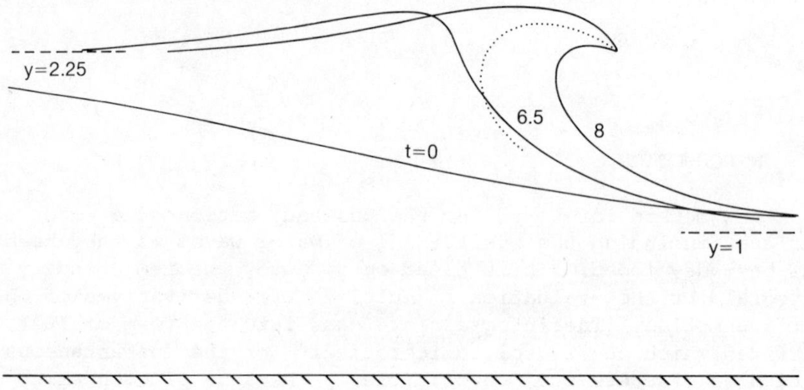

Fig. 1 Nonlinear steepening of an initially gently sloping surface between uniform levels, y = 1 and y = 2.25. The dotted line marks the path of a surface particle. Only about 20% of the computed region is shown.

2. MODEL AND ALGORITHM:

With the fluid velocity determined by the gradient of a velocity potential ϕ, Laplace's equation is satisfied in the body of the fluid,

$$\nabla^2\phi = 0. \tag{2.1}$$

With the values of ϕ known on the surface, $\underline{r} = \underline{R}(\xi)$, equation (2.1) can be solved to give the gradient of ϕ on the surface. Hence the kinematic boundary condition,

$$\frac{D\underline{R}}{Dt} = \underline{u} = \nabla\phi \tag{2.2}$$

and the dynamic boundary condition, given by Bernoulli's equation,

$$\frac{D\phi}{Dt} = \frac{\underline{u}^2}{2} - \left(\frac{P}{\rho} + gy\right) \tag{2.3}$$

(in which the pressure P may be taken to be constant on the surface) provide the basic information for time-stepping the surface motion.

Equation (2.1) is not only satisfied by ϕ, but also by derivatives of ϕ, e.g. ϕ_t, ϕ_{tt}, etc. Since Bernoulli's equation in the form,

$$\phi_t = -\left(\frac{\underline{u}^2}{2} + \frac{P}{\rho} + gy\right) \tag{2.4}$$

gives ϕ_t on the surface once $\nabla\phi$ is known, solving Laplace's equation for $\nabla\phi_t$ in the same way can be used to find the second time derivatives of $\underline{R}$ and ϕ. In principle the procedure can be carried further by using appropriate derivatives of equations (2.1) to (2.4) to calculate any order of time derivative of $\underline{R}$ and ϕ. Using truncated Taylor series to perform the time-stepping this additional information improves efficiency by permitting relatively large time-steps to be made while maintaining a desired degree of accuracy.

If the fluid is considered to be two dimensional with an impermeable flat bottom at $y = -h$, then the fluid may be considered to extend downwards to a reflection of the surface in the bottom. In complex notation, with $x+iy = R(\xi)$ on the surface (R being complex-valued), the condition

$$q(R^*-2ih) = q^*(R) \quad : \quad q = \phi_x - i\phi_y \tag{2.5}$$

must be satisfied, where * denotes the complex conjugate. With this definition the complex velocity q is an analytic function of x+iy in the region between the fluid surface and its relection.

3. BOUNDARY INTEGRALS

The arclength gradient ϕ_s of ϕ is easily determined since the values of $\phi(\xi)$ are known and their rate of change along the known surface, $\underline{r} = \underline{R}(\xi)$, can be found. There are three forms of boundary integral method available for determining the normal gradient ϕ_n of ϕ.

1. The use of Green's indentity,

$$\phi = \oint [\,\phi G_n - G\phi_n]\, ds \tag{3.1}$$

together with the Green's function, $\ln|\underline{r}-\underline{r}'|$, provides one approach. This method, however, is not easily solved by iterative means so that matrix inversion or L-U factorisation has been used by Longuet-Higgins and Cokelet (1976) and by New, McIver and Peregrine (1985). The resulting penalty of running times which increase as the cube of the number of points is severe for large numbers of points.

2. In a slightly different formulation of the problem, the surface can be considered to form a vortex sheet with potential flow on either side of the sheet as in (Baker et al., 1982) and (Roberts, 1983). Although the integral equation formed by this method can be solved iteratively, solving for higher time-derivatives is not as straightforward.

3. Since the complex velocity q(x+iy) is analytic it is possible to make use of Cauchy's integral theorem, as in (Vinje and Brevig, 1981). This can be cast into the following integral equation involving principal values,

$$\pi\phi_n = ⨍_C \mathrm{Im}\left[\frac{Z_s}{Z'-Z}\right] \phi'_n ds' + ⨍_C \mathrm{Re}\left[\frac{Z_s}{Z'-Z}\right] \phi'_s ds' \tag{3.2}$$

where the arclength s is measured in a clockwise sense, the closed contour C surrounds the fluid, and ϕ_n is the inward normal gradient of ϕ. This equation can be solved iteratively and can be used to calculate the gradients of higher derivatives of $\dot{\phi}$.

Because of these advantages the latter method was chosen as the basis of a numerical scheme for following the evolution of the water surface.

4. APPLICATION OF CAUCHY'S THEOREM:

Assuming for simplicity that the surface is periodic, and that time and space dimensions are suitably scaled to make this period exactly 2π in x then the infinite fluid surface can be transformed into a finite closed contour by a conformal transformation, giving:

$$\Omega(\xi) = e^{-iR(\xi)}. \tag{4.1}$$

Applying the formula (3.2) to this contour and the transformation of the reflected surface leads to the following integral equation:

$$\pi\phi_\nu = \fint \mathrm{Im}\left[\frac{\Omega_\xi}{\Omega-\Omega'} + \frac{\Omega_\xi}{\Omega-e^{-2h}/\Omega'^*}\right]\phi'_\nu d\xi'$$
$$+ \int \mathrm{Re}\left[\frac{\Omega_\xi}{\Omega-\Omega'} - \frac{\Omega_\xi}{\Omega-e^{-2h}/\Omega'^*}\right]\phi'_\xi d\xi' \qquad (4.2)$$

where ϕ_ξ is simply the derivative of ϕ with respect to ξ along the surface, while ϕ_ν is the outward normal gradient of ϕ scaled by $|\Omega_\xi|$. Having solved for ϕ_ν the complex potential gradient of ϕ in the physical plane is given by

$$\phi_x + i\phi_y = i\,\frac{\Omega^*}{\Omega^*_\xi}\,(\phi_\xi + i\phi_\nu). \qquad (4.3)$$

It can be noted that the term arising from the bottom condition in the integral kernels of equation (4.2) does not vanish as $h \to \infty$. The limit, Ω_ξ/Ω, however makes no overall contribution to the integrals. It can therefore be subtracted out so that the term arising from the bottom condition can be altered to the following:

$$\frac{e^{-2h}\,\Omega_\xi/\Omega'^*}{\Omega(\Omega-e^{-2h}/\Omega'^*)} \quad . \qquad (4.4)$$

This makes a considerable difference to the diagonal dominance of the matrices in equation (5.3) below. The iterative behaviour improves from being divergent, when this alteration is not made, to being convergent.

5. NUMERICAL SOLUTION

In order to solve equation (4.2) numerically it is now convenient to identify ξ as a point-label parameter taking integer values for points $\underline{R}(\xi)$ on the surface. By considering a Taylor expansion for Ω' it can be seen that

$$\frac{\Omega_\xi}{\Omega-\Omega'} = \frac{1}{\xi-\xi'} + \frac{\Omega_{\xi\xi}}{2\Omega_\xi} + O(\xi-\xi'). \qquad (5.1)$$

Thus only the real part of $\Omega_\xi/(\Omega-\Omega')$ is genuinely singular. Moreover, since

$$\lim_{\xi'\to\xi} \frac{\phi'_\xi-\phi_\xi}{\xi-\xi'} = -\phi_{\xi\xi} \tag{5.2}$$

the effect of this singularity is easily taken into account. Because of the periodic nature of problem, the best available quadrature formula representing the integrals of equation (4.2) can be shown to give

$$\pi\phi_\nu(\xi) = \Sigma\, A(\xi,\xi')\phi_\xi(\xi') - \phi_{\xi\xi}(\xi) + \Sigma\, B(\xi,\xi')\phi_\nu(\xi') \tag{5.3}$$

where

$$A + iB = -\left[\frac{e^{-2h}\ \Omega_\xi/\Omega'^*}{\Omega(\Omega-e^{-2h}/\Omega'^*)}\right]^* + \begin{cases} \dfrac{\Omega_\xi}{\Omega-\Omega'} & : \quad \xi' \neq \xi \\[2ex] \dfrac{\Omega_{\xi\xi}}{2\Omega_\xi} & : \quad \xi' = \xi\ . \end{cases} \tag{5.4}$$

A program was written using this formula to solve for the potential gradients. Given a smooth periodic discretisation of the surface $\underline{R}(\xi)$ the transformation (4.1) is made and the matrices of equation (5.3) are calculated. With $\phi(\xi)$ defined on the surface the same matrices are used to directly calculate up to the third Lagrangian time derivatives of $\underline{R}$ and ϕ by the method outlined between equations (2.1) and (2.5). After two time-steps quadratic backward differencing is used to estimate up to the fifth time derivative and Taylor series are used to perform explicit time-stepping . The size of time-step for this is calculated as the geometric mean of the time-steps which would make the maximum contributions to the Taylor series of the third and fourth derivatives less than a given small precision parameter ε.

6. STABILITY AND ACCURACY

A particular problem which has been reported with other methods is the generation of a "sawtooth" numerical instability which has to be kept under control using smoothing techniques. Even for very steep waves this instability appears to be absent from this method when time step sizes are sufficiently small. For step sizes greater than some critical value (which depends on point density) sawtooth instability does appear. It is first noticeable as a localised "roughness" in the highest derivative calculated while the corresponding roughness in the profile is still extremely small. A correspondingly small degree of smoothing of this profile can be used to keep the instability in check although the critical time step is not normally found to be restrictive and smoothing is not normally used.

As the jet of a breaker develops the points near the tip tend to move together and become more densely distributed. Under such circumstances the sawtooth instability is found to become important. It can be completely controlled without smoothing by reducing the time step size as necessary. This, however, leads to an increase in running time so that it is generally more efficient to use smoothing techniques even though some information is lost in doing so.

It is found that the ability to accurately calculate higher and higher derivatives is eventually limited by the adequacy of the discretisation of the profile. In regions of large surface extension, such as beneath the jet of a breaking wave where surface particles are drawn apart, a particle-following time-stepping technique leads to poor resolution of the motion and the highest derivatives are the first to show signs of inaccuracy.

7. SPEED

The running time is empirically found to vary in the proportion,

$$\text{C.P.U. time} = O(\tau N^2 \varepsilon^{-0.3}) \tag{7.1}$$

with an accuracy of about $\tau\varepsilon^{.85}$ after τ wave periods, provided the number of points N is adequate to describe the surface. For more complicated surfaces the running time thus increases as the square of the number of points required to describe the surface. For wave propagation covering many periods proportionally smaller values of ε are required to maintain a given final accuracy. Since the running times are found to vary more-or-less as $\varepsilon^{-0.3}$ the penalty for this is not great. A reduction in ε by a factor of 10 roughly doubles the running time.

An example is given by the wave in Fig. 1, showing part of a computed wave profile. This was calculated on a computer operating at about 80,000 floating point operations per second. Using 120 points the program calculated up to the time, $t = 6.5$, in 7 minutes of C.P.U. time. It took only 21 time steps to reach this stage with an overall accuracy of about 0.2% showing the effect of calculating higher derivatives. For comparison, Baker, Meiron and Orszag (1981) quote a time of about 30 seconds on a Cray-1 (equivalent to about 10 hours on our computer) for a similar calculation using 128 points and 400 time-steps. The remainder of the wave motion shown in Fig. 1. was calculated with 200 points using both smoothing and progressive reduction in step-size to control the instability. A further 47 time steps were made, taking 50 C.P.U. minutes.

8. CONCLUSIONS

Cauchy's integral theorem is used to set up a boundary integral equation relating the normal potential gradient with the tangential potential gradient at a periodic wave surface. Discretisation of the integral kernels leads to quadrature formulae involving full matrices in a form suitable for iterative solution. The same matrices are used to calculate the velocity of surface particles as well as acceleration and the rate of change of acceleration. The technique can be successively continued to calculate any higher time derivatives of the surface motion. Taylor series are used to explicitly step the wave surface forward in time using fairly large time-step sizes for a given accuracy.

The combination of relatively large time steps and iterative solution of the boundary-integral equation leads to a numerical scheme which is fast and accurate. The scheme is efficient enough to be able to compute lengthy and complicated surface motions with only moderate computer resources.

9. REFERENCES

Baker, G.R., Meiron, D.I. and Orszag, S.A., (1981). Applications of a generalised vortex method to nonlinear free-surface flows. *3rd Intnl. Conf. on Numerical Ship Hydrodynamics, Paris.*

Baker, G.R., Meiron, D.I. and Orszag, S.A., (1982). Generalised vortex methods for free-surface flow problems. *J. Fluid Mech.* **123**, 477-501.

Dold, J.W. and Peregrine, D.H., (1984). Steep unsteady water waves: An efficient computational scheme. *19th Intnl. Conf. on Coastal Engineering, Houston.*

Longuet-Higgins, M.S. and Cokelet, E.D., (1976). The deformation of steep surface waves on water; 1. A numerical method of computation. *Proc. Roy. Soc. Lond. A* **350**, 1 - 26.

New, A.L., McIver, P. and Peregrine, D.H., (1985). Computations of overturning waves. *J. Fluid Mech.* **150**, 233-252.

Roberts, A.J., (1983). A stable and accurate numerical method to calculate the motion of a sharp interface between fluids. *IMA Jnl. Appl. Math.* **31**, 13-36.

Vinje, T. and Brevig, P., (1981). Numerical simulation of breaking waves. *Adv Water Resources* **4** 77-82.